SELF-SIMILAR NETWORK TRAFFIC AND PERFORMANCE EVALUATION

SELF-SIMILAR NETWORK TRAFFIC AND PERFORMANCE EVALUATION

Edited by

KIHONG PARK
Purdue University

WALTER WILLINGER
AT&T Labs-Research

A Wiley-Interscience Publication
JOHN WILEY & SONS, INC.
New York • Chichester • Weinheim • Brisbane • Singapore • Toronto

This book is printed on acid-free paper. ∞

Published simultaneously in Canada.

Library of Congress Cataloging-in-Publication Data:

Self-similar network traffic and performance evaluation / [edited by]
Kihong Park, Walter Willinger.
p. cm.
"A Wiley-Interscience publication."
Includes bibliographical references and index.
ISBN 0-471-31974-0 (alk. paper)
1. Computer networks- -Evaluation. 2. Telecommunication- Traffic- -Evaluation. 3. Queuing theory. 4. Stochastic analysis. I. Park, Kihong, 1964– . II. Willinger, Walter, 1956– .
TK5105.5.S422 2000 99-15962
004.6- dc21

Printed in the United States of America.

10 9 8 7 6 5 4 3 2 1

CONTRIBUTORS

Abdelnaser Adas, Conexant, Inc., Newport Beach, California, USA

P. Abry, CNRS UMR 5672, École Normale Supérieure de Lyon, Laboratoire de Physique, Lyon, France

O. J. Boxma, Eindhoven University of Technology, Eindhoven, The Netherlands and CWI, Amsterdam, The Netherlands

F. Brichet, France Télécom, CNET, Issy-Moulineaux, France

J. W. Cohen, CWI, Amsterdam, The Netherlands

Mark E. Crovella, Boston University, Boston, Massachusetts, USA

N. G. Duffield, AT&T Labs–Research, Florham Park, New Jersey, USA

Anja Feldmann, University of Saarbrücken, Saarbrücken, Germany

P. Flandrin, CNRS UMR 5672, École Normale Supérieure de Lyon, Laboratoire de Physique, Lyon, France

Daniel P. Heyman, AT&T Labs, Middleton, New Jersey, USA

Philippe Jacquet, INRIA, Le Chesnay, France

P. R. Jelenković, Columbia University, New York, New York, USA

Gitae Kim, Boston University, Boston, Massachusetts, USA

T. V. Lakshman, Bell Laboratories, Lucent Technologies, Holmdel, New Jersey, USA

Guang-Liang Li, The University of Hong Kong, Hong Kong, China

Victor O.K. Li, The University of Hong Kong, Hong Kong, China

N. Likhanov, Institute for Problems of Information Transmission, Russian Academy of Science, Moscow, Russia

Lester Lipsky, University of Connecticut, Storrs, Connecticut, USA

Armand M. Makowski, University of Maryland, College Park, Maryland, USA

L. Massoulié, Microscoft Research Ltd., Cambridge, United Kingdom

Amarnath Mukherjee, Knoltex Corporation, San Jose, California, USA

Ilkka Norros, VTT Information Technology, Espoo, Finland

Kihong Park, Purdue University, West Lafayette, Indiana, USA

Minothi Parulekar, University of Maryland, College Park, Maryland, USA

R. H. Riedi, Rice University, Houston, Texas, USA

Sidney Resnick, Cornell University, Ithaca, New York, USA

J. W. Roberts, France Télécom, CNET, Issy-Moulineaux, France

Gennady Samorodnitsky, Cornell University, Ithaca, New York, USA

A. Simonian, France Télécom, CNET, Issy-Moulineaux, France

M. S. Taqqu, Boston University, Boston, Massachusetts, USA

Tsunyi Tuan, Purdue University, West Lafayette, Indiana, USA

D. Veitch, Software Engineering Research Centre, Carlton, Victoria, Australia

W. Whitt, AT&T Labs–Research, Florham Park, New Jersey, USA

Walter Willinger, AT&T Labs–Research, Florham Park, New Jersey, USA

CONTENTS

PREFACE

The recent discovery of scaling phenomena in modern communication networks involving self-similarity or fractals and power-law or heavy-tailed distributions is yet another realization of Benoit Mandelbrot's vision of order in physical, social, and engineered systems characterized by scaling laws. Since the seminal paper by Leland, Taqqu, Willinger and Wilson in 1993 which set the groundwork for considering self-similarity an ubiquitous feature of empirically observed network traffic and an important notion in the understanding of the traffic's dynamic nature for modeling analysis and control of network performance, an explosion of work has ensued investigating the multifaceted nature of this phenomenon.

Despite the fact that data networks such as the Internet are drastically different from legacy public switched telephone networks, the long held paradigm in the communication and networking research community has been that data traffic—analogous to voice traffic—is adequately described by certain Markovian models which are amenable to accurate analysis and efficient control. This supposition has been instrumental in shaping the optimism permeating the late 1980s and early 1990s regarding the ability of achieving efficient traffic control for quality of service provisioning in modern high-speed communication networks. The discovery and, more importantly, succinct formulation and recognition that actual data traffic may, in fact, be fundamentally different in nature from the hereto accustomed telephony traffic has significantly influenced the networking research landscape, necessitating a reexamination and revamping of some of its basic premises.

This book is a collection of chapter contributions which brings together relevant past works spanning a cross-section of topics covering traffic measurement, modeling, performance analysis, and traffic control for self similar network traffic. The primary objective of the book is to present a comprehensive yet cohesive account of some of the principal developments and results concerning self-similar network traffic across its various facets, with the aim of serving as a reflective milestone that captures the state-of-the-art in the field. The book is organized around three main subtopics—traffic modeling, queueing-based performance analysis, and traffic control. By and large, the chapters reflect how research in these areas has reacted when faced with the new scientific discoveries involving self similarity and ubiquitous presence of heavy-tailed phenomena in networked systems.

The spectrum of reactions ranges from evolutionary—holding on to traditional frameworks and tested concepts, and trying to extend, generalize them in the presence of unfamiliar characteristics that, in many ways, contradict conventional wisdom—all the way to revolutionary, which embrace the novel and, at times, surprising features giving rise to new questions, research problems, and challenges both on theoretical and practical fronts of relevance to the future Internet. Overall, the reader may find the majority of book chapters to be of an evolutionary rather than revolutionary nature: Many of the problems that have been considered in the past and have been assumed to fit into the powerful, but also mathematically convenient, framework of Markovian analysis are being reformulated and analyzed to incorporate the slowly improving understanding of data traffic. More fundamental issues such as whether or not these problems are still relevant in light of the stark contrast between hereto assumed properties of network traffic and observed reality has attracted less attention to date. In this sense, the book chapters give a sense of how science, in many instances, works when faced with new discoveries and realities, and they also illustrate how a "give and take" between traditional approaches, on the one side, and unconventional thinking on the other side can lead to progress, thus advancing our overall understanding in the various subtopics covered in this book. It will be interesting to observe if, and when, future developments in these areas will require more concentrated focus on revolutionary ideas and approaches to networking research and practice, especially as far as network performance analysis and traffic control are concerned.

The chapter contributions have been organized into three parts: (i) estimation and simulation, (ii) queueing with self-similar input, and (iii) traffic control and resource provisioning. The threefold categorization is not strict in the sense that some chapters encompass subject matters that cross the set boundaries. Chapter 1, in addition to serving as an introductory chapter which provides the necessary background and technical know-how to understanding self-similar traffic that is common to many of the chapters, also gives a bird's eye view of each chapter, how they fit into the overall picture, and comments on the role and potential relevance for future advances. The remaining two chapters in Part I deal with traffic characterization, estimation, and modeling issues. Wavelet analysis is introduced as a powerful technique for both modeling and estimation in self-similar traffic. Augmenting the theme of traffic modeling are issues surrounding simulations such as those arising in the generation of self-similar traffic and workloads which entails, in many instances, sampling from heavy-tailed distributions requiring special considerations.

The second part of the book consists of ten chapters and focuses on traditional performance evaluation issues, in particular, queueing behavior of finite and infinite buffer systems when fed with long-range dependent input. Due to the breakdown of

Markovian assumptions which are key to achieving tractable analysis in traditional queueing analysis, the technical challenges encountered with self-similar input are great, and this part of the book exposes what is known about queueing with self-similar input, above and beyond the phenomenon that queue length distribution decays polynomially and not exponentially. The traffic models employed, to a large extent, can be viewed as variants of on/off renewal reward processes where session arrivals are allowed to be Poisson, however, on- or off-periods which correspond to busy and idle transmission times, respectively, are heavy-tailed. Starting with Chapter 4, many of the chapters employ asymptotic techniques to investigate tail behavior in queueing systems which, in turn, are related to buffer overflow or packet drop probabilities. Chapters 8, 9, and 10 provide asymptotic bounds on the tail probability. Chapter 12 discusses a traditional, Markovian view of modeling and analyzing variable bit rate video traces which represents a form of extreme adherence to conventional techniques and world view which has its roots in telephony traffic. Chapter 13 provides a form of transient analysis which, in spite of its elementary nature, is a useful exercise and points toward the need for nonequilibrium analysis.

A total of six chapters make up the third part of the book which is mainly concerned with traffic control and dynamic resource provisioning issues that arise under self-similar traffic conditions. There are two aspects to the question, one centered on the problem of resource provisioning/dimensioning and ensuing trade-off relations, and the other based on the traditional traffic control framework of feedback control and its implementation in network protocols. With respect to resource provisioning, due to the amplified queueing delay incurred when employing buffer dimensioning, an alternative resource provisioning strategy based on bandwidth dimensioning as the central control variable has been advanced. A high-level discussion is provided in Chapter 16. Chapter 17 provides analysis of bufferless systems and long-range dependent processes whose future behavior is conditioned on past behavior which are relevant to on-line resource provisioning and traffic control. Chapter 19 describes a concrete resource provisioning architecture based on framing. Feedback traffic control presents a more subtle challenge to traffic management where the central idea revolves around exploiting correlation structure at multiple time scales, as afforded by long-range dependence and self-similarity, to affect traffic control decisions executed at smaller time scales. Chapter 14 discusses the influence of the protocol stack and network traffic, and Chapter 15 gives a detailed characterization of TCP based connection arrivals and network traffic which constitutes the bulk of current Internet traffic. Chapter 18 introduces the multiple time scale congestion control framework and its use in self-similar traffic for throughput maximization.

We conclude the book with two overview chapters which seek to take stock of known results, and point toward research avenues and open problems that may benefit from concerted efforts by the research community. Chapter 20 gives a broad overview of traffic characterization and modeling issues, with focus on achieving a comprehensive and refined understanding of network traffic spanning both long and short time scales. Chapter 21 describes a set of research problems and themes categorized into workload characterization, performance analysis, and traffic control.

Some problems are more aptly described as research programs whereas other are more focused in their scope and nature.

As co-editors, we greatly appreciate the generous efforts of all the contributors to this volume. Because of their cooperation, flexibility, and willingness in helping us achieve a measure of coherence and balanced representation, this project has been a productive and timely occasion, and a delightful experience for us. We are confident that despite the rapidly changing conditions that have become a trademark of modern communication networks, this book contains insights and lessons that are less transient and will withstand the test of time. We hope the book will be of service as a comprehensive, in-depth, and up-to-date reference on self-similar network traffic for the larger networking and communication research communities. Our work would have been much more difficult and time consuming without the help of Wiley and its professional staff, especially, Andrew Smith who participated in the initial idea of the book and Rosalyn Farkas who provided critical editing support. We would like to extend our appreciation and thanks.

KIHONG PARK
WALTER WILLINGER

Purdue University
AT&T Labs
May 2000

SELF-SIMILAR NETWORK TRAFFIC AND PERFORMANCE EVALUATION

1

SELF-SIMILAR NETWORK TRAFFIC: AN OVERVIEW

KIHONG PARK
Network Systems Lab, Department of Computer Sciences, Purdue University, West Lafayette, IN 47907

WALTER WILLINGER
Information Sciences Research Center, AT&T Labs—Research, Florham Park, NJ 07932

1.1 INTRODUCTION

1.1.1 Background

Since the seminal study of Leland, Taqqu, Willinger, and Wilson [41], which set the groundwork for considering self-similarity an important notion in the understanding of network traffic including the modeling and analysis of network performance, an explosion of work has ensued investigating the multifaceted nature of this phenomenon.[1] The long-held paradigm in the communication and performance communities has been that voice traffic and, by extension, data traffic are adequately described by certain Markovian models (e.g., Poisson), which are amenable to accurate analysis and efficient control. The first property stems from the well-developed field of Markovian analysis, which allows tight equilibrium bounds on performance variables such as the waiting time in various queueing systems to be found. This also forms a pillar of performance analysis from the queueing theory side [38]. The

[1]For a nontechnical account of the discovery of the self-similar nature of network traffic, including parallel efforts and important follow-up work, we refer the reader to Willinger [71]. An extended list of references that includes works related to self-similar network traffic and performance modeling up to about 1995 can be found in the bibliographical guide [75].

Self-Similar Network Traffic and Performance Evaluation, Edited by Kihong Park and Walter Willinger
ISBN 0-471-31974-0

second feature is, in part, due to the simple correlation structure generated by Markovian sources whose performance impact—for example, as affected by the likelihood of prolonged occurrence of "bad events" such as concentrated packet arrivals—is fundamentally well-behaved. Specifically, if such processes are appropriately rescaled in time, the resulting coarsified processes rapidly lose dependence, taking on the properties of an independent and identically distributed (i.i.d.) sequence of random variables with its associated niceties. Principal among them is the exponential smallness of rare events, a key observation at the center of large deviations theory [70].

The behavior of a process under rescaling is an important consideration in performance analysis and control since buffering and, to some extent, bandwidth provisioning can be viewed as operating on the rescaled process. The fact that Markovian systems admit to this avenue of taming variability has helped shape the optimism permeating the late 1980s and early 1990s regarding the feasibility of achieving efficient traffic control for quality of service (QoS) provisioning. The discovery and, more importantly, succinct formulation and recognition that data traffic may not exhibit the hereto accustomed scaling properties [41] has significantly influenced the networking landscape, necessitating a reexamination of some of its fundamental premises.

1.1.2 What Is Self-Similarity?

Self-similarity and fractals are notions pioneered by Benoit B. Mandelbrot [47]. They describe the phenomenon where a certain property of an object—for example, a natural image, the convergent subdomain of certain dynamical systems, a time series (the mathematical object of our interest)—is preserved with respect to scaling in space and/or time. If an object is self-similar or fractal, its parts, when magnified, resemble—in a suitable sense—the shape of the whole. For example, the two-dimensional (2D) Cantor set living on $A = [0, 1] \times [0, 1]$ is obtained by starting with a solid or black unit square, scaling its size by $1/3$, then placing four copies of the scaled solid square at the four corners of A. If the same process of scaling followed by translation is applied recursively to the resulting objects ad infinitum, the limit set thus reached defines the 2D Cantor set. This constructive process is illustrated in Fig. 1.1. The limiting object—defined as the infinite intersection of the iterates—has the property that if any of its corners are "blown up" suitably, then the shape of the zoomed-in part is similar to the shape of the whole, that is, it is *self-similar*. Of

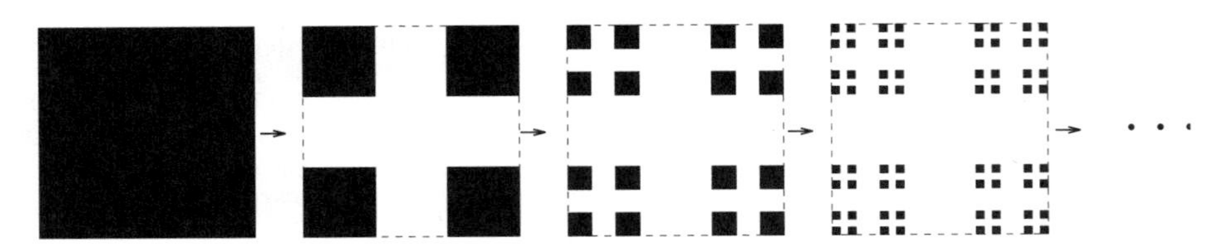

Fig. 1.1 Two-dimensional Cantor set.

course, this is not too surprising since the constructive process—by its recursive action—endows the limiting object with the scale-invariance property.

The one-dimensional (1D) Cantor set, for example, as obtained by projecting the 2D Cantor set onto the line, can be given an interpretation as a traffic series $X(t) \in \{0, 1\}$—call it "Cantor traffic"—where $X(t) = 1$ means that there is a packet transmission at time t. This is depicted in Fig. 1.2 (left). If the constructive process is terminated at iteration $n \geq 0$, then the contiguous line segments of length $1/3^n$ may be interpreted as *on periods* or packet trains of duration $1/3^n$, and the segments between successive on periods as *off periods* or absence of traffic activity. Nonuniform traffic intensities may be imparted by generalizing the constructive framework via the use of probability measures. For example, for the 1D Cantor set, instead of letting the left and right components after scaling have identical "mass," they may be assigned different masses, subject to the constraint that the total mass be preserved at each stage of the iterative construction. This modification corresponds to defining a probability measure μ on the Borel subsets of [0, 1] and distributing the measure at each iteration nonuniformly left and right. Note that the classical Cantor set construction—viewed as a map—is not measure-preserving. Figure 1.2 (middle) shows such a construction with weights $\alpha_L = \frac{2}{3}$, $\alpha_R = \frac{1}{3}$ for the left and right

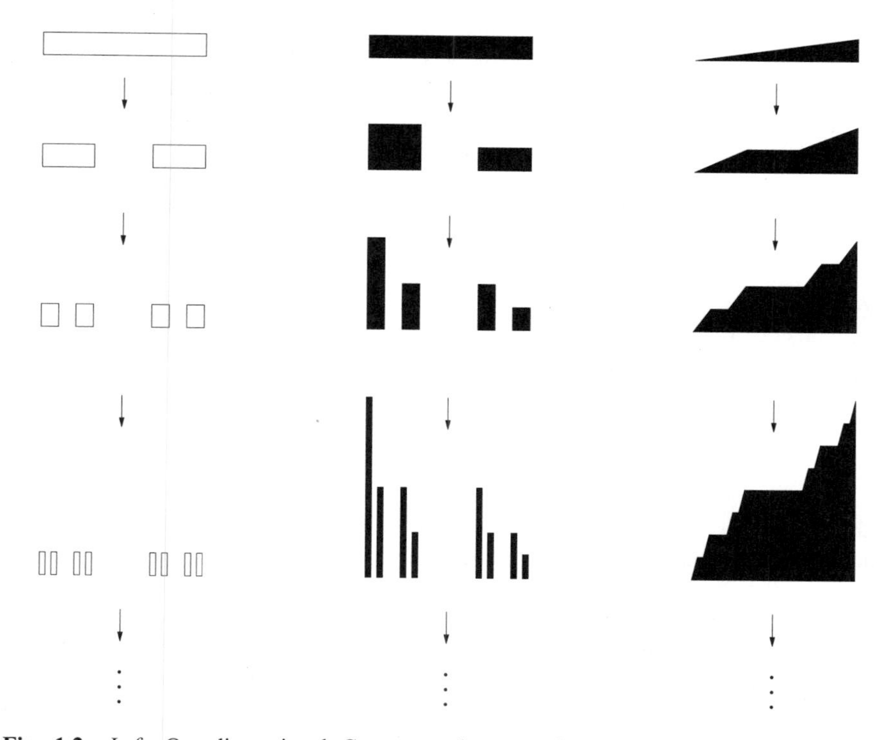

Fig. 1.2 *Left*: One-dimensional Cantor set interpreted as on/off traffic. *Middle*: One-dimensional nonuniform Cantor set with weights $\alpha_L = \frac{2}{3}$, $\alpha_R = \frac{1}{3}$. *Right:* Cumulative process corresponding to 1D on/off Cantor traffic.

components, respectively. The probability measure is represented by "height"; we observe that scale invariance is exactly preserved. In general, the traffic patterns producible with fixed weights α_L, α_R are limited, but one can extend the framework by allowing possibly different weights associated with every edge in the weighted binary tree induced by the 1D Cantor set construction. Such constructions arise in a more refined characterization of network traffic—called multiplicative processes or cascades—and are discussed in Chapter 20. Further generalizations can be obtained by defining different affine transformations with variable scale factors and translations at every level in the "traffic tree." The corresponding traffic pattern is self-similar if, and only if, the infinite tree can be compactly represented as a finite directed cyclic graph [8].

Whereas the previous constructions are given interpretations as traffic activity *per unit time*, we will find it useful to consider their corresponding *cumulative* processes, which are nondecreasing processes whose differences—also called increment process—constitute the original process. For example, for the on/off Cantor traffic construction (cf. Fig. 1.2 (left)), let us assign the interpretation that time is discrete such that at step $n \geq 0$, it ranges over the values $t = 0, 1/3^n, 2/3^n, \ldots, (3^n - 1)/3^n, 1$. Thus we can equivalently index the discrete time steps by $i = 0, 1, 2, \ldots, 3^n$. With a slight abuse of notation, let us redefine $X(\cdot)$ as $X(i) = 1$ if, and only if, in the original process $X(i/3^n) = 1$ and $X(i/3^n - \varepsilon) = 1$ for all $0 < \varepsilon < 1/3^n$. That is, for i values for which an on period in the original process $X(t)$ begins at $t = i/3^n$, $X(i)$ is defined to be zero. Thus, in the case of $n = 2$, we have

$$\begin{array}{lllll} X(0) = 0, & X(1) = 1, & X(2) = 0, & X(3) = 1, & X(4) = 0, \\ X(5) = 0, & X(6) = 0, & X(7) = 1, & X(8) = 0, & X(9) = 1. \end{array}$$

Now consider the continuous time process $Y(t)$ shown in Fig. 1.2 (right) defined over $[0, 3^n]$ for iteration n. $Y(t)$ is nondecreasing and continuous, and it can be checked by visual inspection that

$$X(i) = Y(i) - Y(i-1), \qquad i = 1, 2, \ldots, 3^n,$$

and $X(0) = Y(0) = 0$. Thus $Y(t)$ represents the total traffic volume *up to* time t, whereas $X(i)$ represents the traffic intensity during the ith interval. Most importantly, we observe that exact self-similarity is preserved even in the cumulative process. This points toward the fact that self-similarity may be defined with respect to a cumulative process with its increment process—which is of more relevance for traffic modeling—"inheriting" some of its properties including self-similarity.

An important drawback of our constructions thus far is that they admit only a strong form of recursive regularity—that of *deterministic* self-similarity—and needs to be further generalized for traffic modeling purposes where stochastic variability is an essential component.

1.1.3 Stochastic Self-Similarity and Network Traffic

Stochastic self-similarity admits the infusion of nondeterminism as necessitated by measured traffic traces but, nonetheless, is a property that can be illustrated visually. Figure 1.3 (top left) shows a traffic trace, where we plot throughput, in bytes, against time where time granularity is 100 s. That is, a single data point is the aggregated traffic volume over a 100 second interval. Figure 1.3 (top right) is the same traffic series whose first 1000 second interval is "blown up" by a factor of ten. Thus the truncated time series has a time granularity of 10 s. The remaining two plots zoom in further on the initial segment by rescaling successively by factors of 10.

Unlike deterministic fractals, the objects corresponding to Fig. 1.3 do not possess exact resemblance of their parts with the whole at finer details. Here, we assume that the measure of "resemblance" is the shape of a graph with the magnitude suitably normalized. Indeed, for measured traffic traces, it would be too much to expect to observe exact, deterministic self-similarity given the stochastic nature of many network events (e.g., source arrival behavior) that collectively influence actual network traffic. If we adopt the view that traffic series are sample paths of stochastic processes and relax the measure of resemblance, say, by focusing on certain statistics of the rescaled time series, then it may be possible to expect exact similarity of the mathematical objects and approximate similarity of their specific realizations with respect to these relaxed measures. Second-order statistics are statistical properties

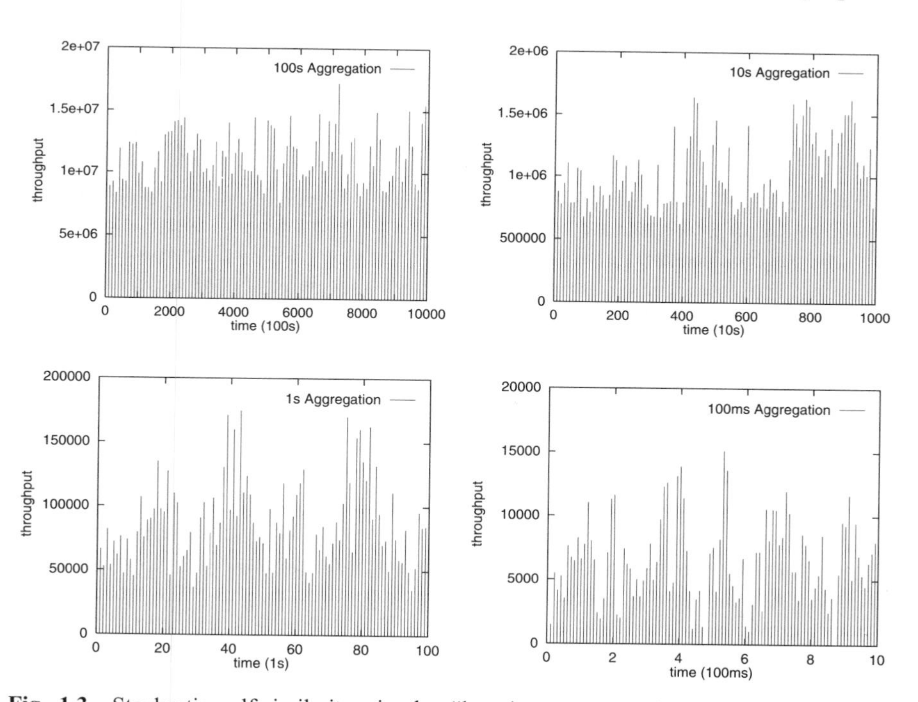

Fig. 1.3 Stochastic self-similarity—in the "burstiness preservation sense"—across time scales 100 s, 10 s, 1 s, 100 ms (top left, top right, bottom left, bottom right).

that capture burstiness or variability, and the autocorrelation function is a yardstick with respect to which scale invariance can be fruitfully defined. The shape of the autocorrelation function—above and beyond its preservation across rescaled time series—will play an important role. In particular, correlation, as a function of time lag, is assumed to decrease polynomially as opposed to exponentially. The existence of nontrivial correlation "at a distance" is referred to as *long-range dependence*. A formal definition is given in Section 1.4.1.

1.2 PREVIOUS RESEARCH

1.2.1 Measurement-Based Traffic Modeling

The research avenues relating to traffic self-similarity may broadly be classified into four categories. In the first category are works pertaining to *measurement-based traffic modeling* [13, 26, 34, 42, 56, 74], where traffic traces from physical networks are collected and analyzed to detect, identify, and quantify pertinent characteristics. They have shown that scale-invariant burstiness or self-similarity is an ubiquitous phenomenon found in diverse contexts, from local-area and wide-area networks to IP and ATM protocol stacks to copper and fiber optic transmission media. In particular, Leland et al. [41] demonstrated self-similarity in a LAN environment (Ethernet), Paxson and Floyd [56] showed self-similar burstiness manifesting itself in pre-World Wide Web WAN IP traffic, and Crovella and Bestavros [13] showed self-similarity for WWW traffic. Collectively, these measurement works constituted strong evidence that scale-invariant burstiness was not an isolated, spurious phenomenon but rather a persistent trait existing across a range of network environments.

Accompanying the traffic characterization efforts has been work in the area of statistical and scientific inference that has been essential to the detection and quantification of self-similarity or long-range dependence.[2] This work has specifically been geared toward network traffic self-similarity [28, 64] and has focused on exploiting the immense volume, high quality, and diversity of available traffic measurements; for a detailed discussion of these and related issues, see Willinger and Paxson [72, 73]. At a formal level, the validity of an inference or estimation technique is tied to an underlying process that presumably generated the data in the first place. Put differently, correctness of system identification only holds when the data or sample paths are known to originate from specific models. Thus, in general, a sample path of unknown origin cannot be uniquely attributed to a specific model, and the main (and only) purpose of statistical or scientific inference is to deal with this intrinsically ill-posed problem by concluding whether or not the given data or sample paths are consistent with an assumed model structure. Clearly, being consistent with an assumed model does not rule out the existence of other models that may conform to the data equally well. In this sense, the aforementioned works on measurement-based traffic modeling have demonstrated that self-similarity is

[2]The relationship between self-similarity and long-range dependence—they need not be one and the same—is explained in Section 1.4.1.

consistent with measured network traffic and have resulted in adding yet another class of models—that is, self-similar processes—to an already long list of models for network traffic. At a practical level, many of the commonly used inference techniques for quantifying the degree of self-similarity or long-range dependence (e.g., Hurst parameter estimation) have been known to exhibit different idiosyncrasies and robustness properties. Due to their predominantly heuristic nature, these techniques have been generally easy to use and apply, but the ensuing results have often been difficult to interpret [64]. The recent introduction of wavelet-based techniques to the analysis of traffic traces [1, 23] represented a significant step toward the development of more accurate inference techniques that have been shown to possess increased sensitivity to different types of scaling phenomena with the ability to discriminate against certain alternative modeling assumptions, in particular, nonstationary effects [1]. Due to their ability to localize a given signal in scale and time, wavelets have made it possible to detect, identify, and describe *multifractal* scaling behavior in measured network traffic over fine time scales [23]: a nonuniform (in time) scaling behavior that emerges when studying measured TCP traffic over fine time scales, one that allows for more general scaling phenomena than the ubiquitous self-similar scaling property, which holds for a range of sufficiently large time scales.

1.2.2 Physical Modeling

In the second category are works on *physical modeling* that try to explicate the physical causes of self-similarity in network traffic based on network mechanisms and empirically established properties of distributed systems that, collectively, collude to induce self-similar burstiness at multiplexing points in the network layer. In view of traditional time series analysis, physical modeling affects model selection by picking among competing and—in a statistical sense—equally well-fitting models that are most congruent to the physical networking environment where the data arose in the first place. Put differently, physical modeling aims for models of network traffic that relate to the physics of how traffic is generated in an actual network, is capable of explaining empirically observed phenomena such as self-similarity in more elementary terms, and provides new insights into the dynamic nature of the traffic. The first type of causality—also the most mundane—is attributable to the arrival pattern of a single data source as exemplified by variable bit rate (VBR) video [10, 26]. MPEG video, for example, exhibits variability at multiple time scales, which, in turn, is hypothesized to be related to the variability found in the time duration between successive scene changes [25]. This "single-source causality," however, is peripheral to our discussions for two reasons: one, self-similarity observed in the original Bellcore data stems from traffic measurements collected during 1989–1991, a period during which VBR video payload was minimal—if not nonexistent—to be considered an influencing factor[3]; and two, it is

[3]The same holds true for the LBL WAN data considered by Paxson and Floyd [56] and the BU WWW data analyzed by Crovella and Bestavros [13].

well-known that VBR video can be approximated by short-range dependent traffic models, which, in turn, makes it possible to investigate certain aspects of the impact on performance of long-range correlation structure within the confines of traditional Markovian analysis [32, 37].

The second type of causality—also called *structural causality* [50]—is more subtle in nature, and its roots can be attributed to an empirical property of distributed systems: the heavy-tailed distribution of file or object sizes. For the moment, a random variable obeying a *heavy-tailed* distribution can be viewed as giving rise to a very wide range of different values, including—as its trademark—"very large" values with nonnegligible probability. This intuition is made more precise in Section 1.4.1. Returning to the causality description, in a nutshell, if end hosts exchange files whose sizes are heavy tailed, then the resulting network traffic at multiplexing points in the network layer is self-similar [50]. This causal phenomenon was shown to be robust in the sense of holding for a variety of transport layer protocols such as TCP—for example, Tahoe, Reno, and Vegas—and flow-controlled UDP, which make up the bulk of deployed transport protocols, and a range of network configurations. Park et al. [50] also showed that research in UNIX file systems carried out during the 1980s give strong empirical evidence based on file system measurements that UNIX file systems are heavy-tailed. This is, perhaps, the most simple, distilled, yet high-level physical explanation of network traffic self-similarity. Corresponding evidence for Web objects, which are of more recent relevance due to the explosion of WWW and its impact on Internet traffic, can be found in Crovella and Bestavros [13].

Of course, structural causality would be meaningless unless there were explanations that showed why heavy-tailed objects transported via TCP- and UDP-based protocols would induce self-similar burstiness at multiplexing points. As hinted at in the original Leland et al. paper [41] and formally introduced in Willinger et al. [74], the *on/off model* of Willinger et al. [74] establishes that the superposition of a large number of independent on/off sources with heavy-tailed on and/or off periods leads to self-similarity in the aggregated process—a fractional Gaussian noise process—whose long-range dependence is determined by the heavy tailedness of on or off periods. Space aggregation is inessential to inducing long-range dependence—it is responsible for the Gaussian property of aggregated traffic by an application of the central limit theorem—however, it is relevant to describing multiplexed network traffic. The on/off model has its roots in a certain renewal reward process introduced by Mandelbrot [46] (and further studied by Taqqu and Levy [63]) and provides the theoretical underpinning for much of the recent work on physical modeling of network traffic. This theoretical foundation together with the empirical evidence of heavy-tailed on/off durations (as, e.g., given for IP flow measurements [74]) represents a more low-level, direct explanation of physical causality of self-similarity and forms the principal factors that distinguish the on/off model from other mathematical models of self-similar traffic. The linkage between high-level and low-level descriptions of causality is further facilitated by Park et al. [50], where it is shown that the application layer property of heavy-tailed file sizes is preserved by the protocol stack and mapped to approximate heavy-tailed busy periods at the network layer. The interpacket spacing within a single session (or equivalently transfer/connection/flow), however, has been observed to exhibit its own distinguishing variability. This refined short time scale structure and its possible causal attribution to the feedback control mechanisms of TCP are investigated in Feldmann et al. [22, 23] and are the topics of ongoing work.

1.2.3 Queueing Analysis

In the third category are works that provide mathematical models of long-range dependent traffic with a view toward facilitating performance analysis in the queueing theory sense [2, 3, 17, 43, 49, 53, 66]. These works are important in that they establish basic performance boundaries by investigating queueing behavior with long-range dependent input, which exhibit performance characteristics fundamentally different from corresponding systems with Markovian input. In particular, the queue length distribution in infinite buffer systems has a *slower-than-exponentially* (or *subexponentially*) decreasing tail, in stark contrast with short-range dependent input for which the decay is exponential. In fact, depending on the queueing model under consideration, long-range dependent input can give rise to *Weibullian* [49] or *polynomial* [66] tail behavior of the underlying queue length distributions. The analysis of such non-Markovian queueing systems is highly nontrivial and provides fundamental insight into the performance impact question. Of course, these works, in addition to providing valuable information into network performance issues, advance the state of the art in performance analysis and are of independent interest. The queue length distribution result implies that buffering—as a resource provisioning strategy—is rendered ineffective when input traffic is self-similar in the sense of incurring a disproportionate penalty in queueing delay vis-à-vis the gain in reduced packet loss rate. This has led to proposals advocating a *small buffer capacity/large bandwidth* resource provisioning strategy due to its simplistic, yet curtailing influence on queueing: if buffer capacity is small, then the ability to queue or remember is accordingly diminished. Moreover, the smaller the buffer capacity, the more relevant short-range correlations become in determining buffer occupancy. Indeed, with respect to first-order performance measures such as packet loss rate, they may become the dominant factor. The effect of small buffer sizes and finite time horizons in terms of their potential role in delimiting the scope of influence of long-range dependence on network performance has been studied [29, 58].

A major weakness of many of the queueing-based results [2, 3, 17, 43, 49, 53, 66] is that they are *asymptotic*, in one form or another. For example, in infinite buffer systems, upper and lower bounds are derived for the tail of the queue length distribution as the queue length variable approaches infinity. The same holds true for "finite buffer" results where bounds on buffer overflow probability are proved as buffer capacity becomes unbounded. There exist interesting results for zero buffer capacity systems [18, 19], which are discussed in Chapter 17. Empirically oriented studies [20, 33, 51] seek to bridge the gap between asymptotic results and observed behavior in finite buffer systems. A further drawback of current performance results is that they concentrate on first-order performance measures that relate to (long-term) packet loss rate but less so on second-order measures—for example, variance of packet loss or delay, generically referred to as *jitter*—which are of importance in multimedia communication. For example, two loss processes may have the same first-order statistic but if one has higher variance than the other in the form of concentrated periods of packet loss—as is the case in self-similar traffic—then this can adversely impact the efficacy of packet-level forward error correction used in the QoS-sensitive transport of real-time traffic [11, 52, 68]. Even less is known about transient performance measures, which are more relevant in practice when convergence to long-term steady-state behavior is too slow to be of much value for engineering purposes. Lastly, most queueing results obtained for long-range dependent input are for open-loop systems that ignore feedback control issues present in actual networking environments (e.g., TCP). Since feedback can shape and influence the very traffic arriving at a queue [22, 50], incorporating their effect in feedback-controlled closed queueing systems looms as an important challenge.

1.2.4 Traffic Control and Resource Provisioning

The fourth category deals with works relating to the control of self-similar network traffic, which, in turn, has two subcategories: resource provisioning and dimensioning, which can be viewed as a form of open-loop control, and closed-loop or feedback traffic control. Due to their feedback-free nature, the works on queueing analysis with self-similar input have direct bearing on the resource dimensioning problem. The question of quantitatively estimating the marginal utility of a unit of additional resource such as bandwidth or buffer capacity is answered, in part, with the help of these techniques. Of importance are also works on statistical multiplexing using the notion of effective bandwidth, which point toward how efficiently resources can be utilized when shared across multiple flows [27]. A principal lesson learned from the resource provisioning side is the ineffectiveness of allocating buffer space vis-à-vis bandwidth for self-similar traffic, and the consequent role of short-range correlations in affecting first-order performance characteristics when buffer capacity is indeed provisioned to be "small" [29, 58].

On the feedback control side is the work on *multiple time scale congestion control* [67, 68], which tries to exploit correlation structure that exists across multiple time scales in self-similar traffic for congestion control purposes. In spite of the negative performance impact of self-similarity, on the positive side, long-range dependence admits the possibility of utilizing correlation at large time scales, transforming the latter to harness predictability structure, which, in turn, can be affected to guide congestion control actions at smaller time scales to yield significant performance gains. The problem of designing control mechanisms that allow correlation structure at large time scales to be effectively engaged is a nontrivial technical challenge for two principal reasons: one, the correlation structure in question exists at time scales typically an order of magnitude or more above that of the feedback loop; and two, the information extracted is necessarily imprecise due to its probabilistic nature.[4] Tuan and Park [67, 68] show that large time scale correlation structure can be employed to yield significant performance gains both for throughput maximization—using TCP and rate-based control—and end-to-end QoS control within the framework of adaptive redundancy control [52, 68]. An important by-product of this work is that the *delay–bandwidth product problem* of broadband networks, which renders reactive or feedback traffic controls ineffective when subject to long round-trip times (RTT), is mitigated by exercising control across multiple time scales. Multiple time scale congestion control allows uncertainty stemming from outdated feedback information to be compensated or "bridged" by predictability structure present at time scales exceeding the RTT or feedback loop (i.e., seconds versus milliseconds). Thus even though traffic control in the 1990s has been occupied by the dual theme of large delay–bandwidth product and self-similar traffic burstiness, when combined, they lend themselves to a form of attack, which imparts proactivity transcending the limitation imposed by RTT, thereby facilitating the metaphor of "catching two birds with one stone."

A related, but more straightforward, traffic control dimension is *connection duration prediction*. The works from physical modeling tell us that connections or flows tend to obey a heavy-tailed distribution with respect to their time duration or lifetime, and this information may be exploitable for traffic control purposes. In particular, heavy tailedness implies that most connections are short-lived, but the bulk of traffic is contributed by a few long-lived flows [50]. By Amdahl's Law [4], it becomes relevant to carefully manage the impact exerted by the long-lived flows even if they are few in number.[5] The idea of employing "connection" duration was first advanced in the context of load balancing in distributed systems where UNIX processes have been observed to possess heavy-tailed lifetimes [30, 31, 40]. In contrast to the exponential distribution whose memoryless property renders prediction obsolete, heavy tailedness implies *predictability*—a connection whose measured time duration exceeds a certain threshold is more likely to persist into the future. This information can be used, for example, in the case of load balancing, to decide whether it is worthwhile to migrate a process given the fixed, high overhead cost of process migration [31]. The ensuing opportunities have numerous applications in traffic control, one recent example being the discrimination of long-lived flows from short-lived flows such that routing table updates can be biased toward long-lived flows, which, in turn, can enhance system stability by desensitizing against "transient" effects of short-lived flows [61]. In general, the connection duration information can also come from directly available information in the application layer—for example, a Web server, when servicing a HTTP request, can discern the size of the object in question—and if this information is made available to lower layers, decisions such as whether to engage in open-loop (for short-lived flows) or closed-loop control (for long-lived flows) can be made to enhance traffic control [67].

[4]We remark that understanding the correlation structure of network traffic at time scales below the feedback loop may be of relevance but remains, at this time, largely unexplored [22].

[5]A form of Amdahl's Law states that to improve a system's performance, its functioning with respect to its most frequently encountered states must be improved. Conversely, performance gain is delimited by the latter.

1.3 ISSUES AND REMARKS

1.3.1 Traffic Measurement and Estimation

The area of traffic measurement—since the collection and analysis of the original Bellcore data [41]—has been tremendously active, yielding a wealth of traffic measurements across a wide spectrum of different contexts supporting the view that network traffic exhibits self-similar scaling properties over a wide range of time scales. This finding is noteworthy given the fact that networks, over the past decades, have undergone significant changes in their constituent traffic flows, user base, transmission technologies, and scale with respect to system size. The observed robustness property or insensitivity to changing networking conditions justified calling self-similarity a *traffic invariant* and motivated focusing on underlying physical explanations that are mathematically rigorous as well as empirically verifiable. Robustness, in part, is explained by the fact that the majority of Internet traffic has been TCP traffic, and while in the pre-WWW days the bulk of TCP traffic stemmed from FTP traffic, in today's Internet, it is attributable to HTTP-based Web traffic. Both types of traffic have been shown to transport files whose size distribution is heavy-tailed [13, 56]. Physical modeling carried out by Park et al. [50] showed that the transport of heavy-tailed files mediated by TCP (as well as flow-controlled UDP) induces self-similarity at multiplexing points in the network layer; it also showed that this is a robust phenomenon insensitive to details in network configuration and control actions in the protocol stack.[6] Measurement work has culminated in refined workload characterization at the application layer, including the modeling of user behavior [6, 7, 24, 48]. At the network layer, measurement analyses of IP traffic over fine time scales have led to the multifractal characterization of wide-area network traffic, which, in turn, has bearing on physical modeling raising new questions about the relationship between feedback congestion control and short-range correlation structure of network traffic [22, 23]. The tracking of Internet workload and its characterization is expected to remain a practically important activity of interest in its own right. Demonstrating the relevance of ever refined workload models to networking research, however, will loom as a nontrivial challenge.

As with experimental physics, the measurement- or data-driven approach to networking research—rejuvenated by Leland et al. [41]—provides a balance to the more theoretical aspects of networking research, in the ideal situation, facilitating a constructive interplay of "give-and-take." A somewhat less productive consequence has been the discourse on short-range versus long-range dependent mathematical models to describe measured traffic traces starting with the original Bellcore Ethernet data. At one level, both short-range and long-range dependent traffic models are parameterized systems that are sufficiently powerful to give rise to

[6]Not surprisingly, extremities in control actions and resource configurations do affect the property of induced network traffic, in some instances, diminishing self-similar burstiness altogether [50]. Moreover, refined structure in the form of multiplicative scaling over sub-RTT time scales has only recently been discovered [23].

sample paths in the form of measured traffic time series. Mathematical system identification, under these circumstances, therefore, is an intrinsically ill-posed problem. Viewed in this light, the fact that different works can assign disparate modeling interpretations to the same measurement data, with differing conclusions, is not surprising [26, 33]. Put differently, it is well known that with a sufficiently parameterized model class, it is always possible to find a model that fits a given data set. Thus, the real challenge lies less in mathematical model fitting than in *physical modeling*, an approach that in addition to describing the given data provides insight into the causal and dynamic nature of the processes that generated the data in the first place. On the positive side, the discussions about short-range versus long-range dependence have brought out into the open concerns about nonstationary effects [16]—3 p.m. traffic cannot be expected to stem from the same source behavior conditions as 3 a.m. traffic—that can influence certain types of inference and estimation procedures for long-range dependent processes. These concerns have spurned the development and adoption of estimation techniques based on wavelets, which are sensitive to various types of nonstationary variations in the data [1]. What is not in dispute are computed sample statistics—for example, autocorrelation functions of measured traffic series—which exhibit nontrivial correlations at time lags on the order of seconds and above. Whether to call these time scales "long range" or "short range" is a matter of subjective choice and/or mathematical convenience and abstraction. What impact these correlations exert on queueing behavior is a function of how large the buffer capacity, the level of traffic intensity, and link capacity—among other factors—are [29, 58]. As soon as one deviates from empirical evaluation based on measurement data and adopts a model of the data, one is faced with the same ill-posed identification problem.

1.3.2 Traffic Modeling

There exist a wide range of mathematical models of self-similar or long-range dependent traffic each with its own idiosyncrasies [5, 21, 23, 35, 43, 49, 53, 59, 74]. Some facilitate queueing analysis [43, 49, 53], some are physically motivated [5, 23, 74], and yet others show that long-range dependence may be generated in diverse ways [21, 35]. The wealth of mathematical models—while, in general, an asset—can also distract from an important feature endowed on the networking domain: the physics and causal mechanisms underlying network phenomena including traffic characteristics. Since network architecture—either by implementation or simulation—is *configurable*, from a network engineering perspective physical traffic models that trace back the roots of self-similarity and long-range dependence to architectural properties such as network protocols and file size distribution at servers have a clear advantage with respect to predictability and verifiability over "black box" models associated with traditional time series analysis. Contrast this with, say, economic systems where human behavior cannot be reprogrammed at will to test the consequences of different assumptions and hypotheses on system behavior. Physical models, therefore, are in a unique position to exploit this "reconfigurability trait" afforded by the networking domain, and use it to facilitate an intimate, mechanistic understanding of the system.

The on/off model [74] is a mathematical abstraction that provides a foundation for physical traffic modeling by advancing an explicit causal chain of verifiable network properties or events that can be tested against empirical data. For example, the factual basis of heavy-tailed on periods in network traffic has been shown by Willinger et al. [74], a corresponding empirical basis for heavy-tailed file sizes in UNIX file systems of the past whose transport may be the cause of heavy-tailed on periods in packet trains has been shown by Park et al. [50], and a more modern interpretation for the World Wide Web has been demonstrated by Crovella and Bestavros [13]. One weakness of the on/off model is its assumption of *independence* of on/off sources. This has been empirically addressed [50] by studying the influence of dependence arising from multiple sources coupled at bottleneck routers sharing resources when the flows are governed by feedback congestion control protocols such as TCP in the transport layer. It was found that coupling did not significantly impact long-range dependence. A more recent study [22] shows that dependence due to feedback and interflow interaction may be the cause for multiplicative scaling phenomena observed in the short-range correlation structure, a refined physical characterization that may complement the previous findings, which focused on coarser structure at larger time scales. We remark that the on/off model is able to induce both fractional Gaussian noise—upon aggregation over multiple flows and normalization—and a form of self-similarity and long-range dependence called asymptotic second-order self-similarity—a single process with heavy-tailed on/off periods—which constitute two of the most commonly used self-similar traffic models in performance analysis.

Finally, physical models, because of their grounding in empirical facts, influence the general argument advanced in Section 1.3.1 on the ill-posed nature of the identification problem. They can be viewed as tilting the scale in favor of long-range dependent traffic models. That is, since file sizes in various network related contexts have been shown to be heavy-tailed and the physical modeling works show that resulting traffic is long-range dependent, other things being equal, empirical evidence afforded by physical models biases toward a more consistent and parsimonious interpretation of network traffic as being long-range dependent as opposed to the mathematically equally viable short-range dependence hypothesis. Thus physical models, by virtue of their casual attribution, can also influence the choice of mathematical modeling and performance analysis.

1.3.3 Performance Analysis and Traffic Control

The works on queueing analysis with self-similar input have yielded fundamental insights into the performance impact of long-range dependence, establishing the basic fact that queue length distribution decays slower-than-exponentially vis-à-vis the exponential decay associated with Markovian input [2, 3, 17, 43, 49, 53, 66]. In conjunction with observations advanced by Grossglauser and Bolot [29] and Ryu and Elwalid [58] on ways to curtail some of the effect of long-range dependence, a very practical impact of the queueing-based performance analysis work has been the growing adoption of the resource dimensioning paradigm, which states that buffer capacity at routers should be kept small while link bandwidth is to be increased. That is, the marginal utility of buffer capacity has diminished significantly vis-à-vis that of bandwidth. This is illustrated in Fig. 1.4, which shows mean queue length as a function of buffer capacity at a bottleneck router when fed with self-similar input with varying degrees of long-range dependence but equal traffic intensity (roughly, α values close to 1 imply "strong" long-range dependence whereas α values close to 2 correspond to "weak" long-range dependence). In other words, when long-range correlation structure is weak, a buffer capacity of about 60 kB suffices to contain the input's variability and, moreover, the average buffer occupancy remains below 5 kB. However, when the long-range correlation structure is strong, an increase in buffer capacity is accompanied by a corresponding increase in buffer occupancy with the buffer capacity horizon at which the mean queue length saturates pushed out significantly.

In spite of the fundamental contribution and insight afforded by queueing analysis, as a practical matter, all the known results suffer under the limitation that the analysis is asymptotic in the buffer capacity: either the queue is assumed to be infinite and asymptotic bounds on the tail of the queue length distribution are derived, or the queue is assumed to be finite but its overflow probability is computed as the buffer capacity is taken to infinity. There is, as yet, a chasm between these asymptotic results and their finitistic brethren that have alluded tractability. It is unclear whether the asymptotic formulas—beyond their qualitative relevance—are also practically useful as resource provisioning and traffic engineering tools. Further work is needed in this direction to narrow the gap. Another significant drawback of the performance analysis results—also related to the asymptotic nature of queueing

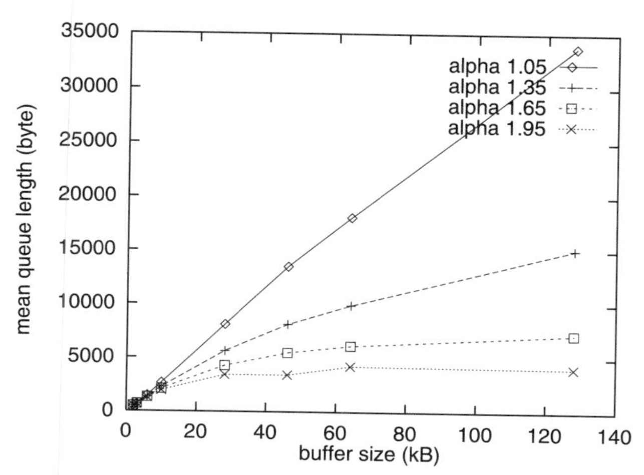

Fig. 1.4 Mean queue length as a function of buffer capacity for input traffic with varying long-range dependence ($\alpha = 1.05, 1.35, 1.65, 1.95$).

results—is the focus on first-order performance indicators such as packet loss rate and mean queue length, which is even true in experimental studies. Second-order performance measures such as packet loss variance or delay variance—generically denoted as jitter—play an important role in multimedia payload transport with real-time constraints. Even when a small buffer capacity resource provisioning policy is adopted to delimit the queueing aspect of self-similar traffic, if time-sensitive traffic flows are subject to concentrated periods of packet loss or severe interpacket delay variation (even though packet loss rate may be small), then performance—as reflected by QoS—has degraded. The effectiveness of real-time QoS control techniques such as packet-level forward error correction are directly impacted by burstiness structure [11, 52, 68] and explicit incorporation of second-order performance measures must be effected to yield a balanced account of the performance impact question.

On the traffic control front, self-similarity—in spite of its detrimental performance aspect—implies the existence of correlation structure at a distance, which may be exploitable for traffic control purposes. The framework of multiple time scale traffic control [67–69] exercises control actions across multiple time scales, using the information extracted at large time scales to modulate the output behavior of feedback congestion controls acting at the time scale of RTT. An important by-product of multiple time scale congestion control is the mitigation of the delay-bandwidth product problem, which has been a pariah of reactive controls due to the outdatedness of feedback information in WAN environments, which diminishes the effectiveness of reactive control actions. Fig. 1.5 shows the performance gain of imparting multiple time scale capabilities on top of TCP Reno, Vegas, and Rate (a rate-based version of TCP) as a function of RTT. We observe that as RTT increases, performance enhancement vis-à-vis ordinary TCP due to multiple scale congestion control is amplified accordingly.

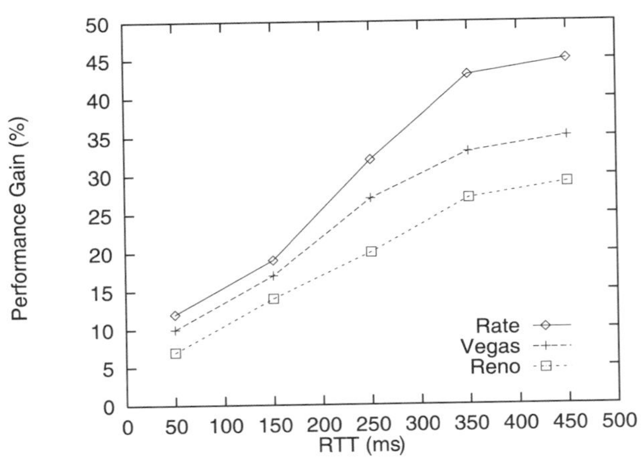

Fig. 1.5 Performance gain of TCP Reno, Vegas, and Rate, when endowed with multiple time scale capabilities as a function of RTT.

The area of self-similar traffic control faces a number of challenges. First, self-similar traffic control, in the past, has received less attention than measurement/estimation, traffic modeling, and queueing analysis, which is not too surprising since the problem of control is, in some sense, a natural continuation of research into "what is" type questions followed by "what if" questions. Research into utilizing predictability stemming from long-range dependence and heavy-tailed connection durations is far from exhaustive, and further work is needed to explore the wide array of traffic control possibilities. Second, whereas long-lived connections—although few in number but contributing the bulk of traffic—constitute the primary target of traffic control, the effective management of short-lived connections—due to their sheer number—looms as an important problem. Maintenance of a *persistent* state at end systems that is shared across multiple flows is a promising avenue that would allow open-loop traffic control to be sensitive to network state, thus imparting a measure of proactivity. Last but not least, analysis of feedback loop systems with respect to their stability and optimality including those arising in multiple time scale traffic control for self-similar traffic remains a challenge. New ideas and approaches are needed to succeed in our attempts to tractably analyze and understand large-scale, coupled, interacting complex systems such as the Internet.

1.4 TECHNICAL BACKGROUND

1.4.1 Self-Similar Processes and Long-Range Dependence

1.4.1.1 Second-Order Self-Similarity and Stationarity Consider a discrete time stochastic process or time series $X(t)$, $t \in \mathbb{Z}$, where $X(t)$ is interpreted as the traffic volume—measured in packets, bytes, or bits—*at* time instance t. Of interest is also the interpretation that $X(t)$ is the total traffic volume *up to* time t, say, from time 0. To minimize confusion, when a "cumulative" view is taken, we will denote the process by $Y(t)$. We will then reserve $X(t)$ to be the *increment process* corresponding to $Y(t)$, that is, $X(t) = Y(t) - Y(t-1)$.

For traffic modeling purposes, we would like $X(t)$ to be "stationary" in the sense that its behavior or structure is invariant with respect to shifts in time. In other words, t's responsibility as an *absolute* reference frame is relieved. Without some form of stationarity, "anything" is allowed and a model loses much of its usefulness as a compact description of (assumed) tractable phenomena. $X(t)$ is *strictly stationary* if $(X(t_1), X(t_2), \ldots, X(t_n))$ and $(X(t_1+k), X(t_2+k), \ldots, X(t_n+k))$ possess the same joint distribution for all $n \in \mathbb{Z}_+$, $t_1, \ldots, t_n, k \in \mathbb{Z}$. Denoting the k-shifted process or time series X_k, X and X_k are said to be equivalent in the sense of *finite-dimensional disributions*, $X =_d X_k$. Imposing strict stationarity, it turns out, is too restrictive and we will be interested in a weaker form of stationarity—*second-order stationarity*[7]—which requires that the autocovariance function $\gamma(r, s) = E[(X(r) - \mu)(X(s) - \mu)]$ satisfies translation invariance, that is, $\gamma(r, s) = \gamma(r+k, s+k)$ for all $r, s, k \in \mathbb{Z}$.

[7]Equivalent names are *weak*, *covariance*, and *wide sense* stationarity.

The first two moments are assumed to exist and be finite, and we set $\mu = E[X(t)]$, $\sigma^2 = E[(X(t) - \mu)^2]$ for all $t \in \mathbb{Z}$. We will also assume $\mu = 0$. Since, by stationarity, $\gamma(r, s) = \gamma(r - s, 0)$, we denote the autocovariance by $\gamma(k)$.

To formulate scale invariance, first define the *aggregated process* $X^{(m)}$ of X at aggregation level m,

$$X^{(m)}(i) = \frac{1}{m} \sum_{t=m(i-1)+1}^{mi} X(t).$$

That is, $X(t)$ is partitioned into nonoverlapping blocks of size m, their values are averaged, and i is used to index these blocks. Let $\gamma^{(m)}(k)$ denote the autocovariance function of $X^{(m)}$. Under the assumption of second-order stationarity we arrive at the following definitions of second-order self-similarity.

Definition 1.4.1 **(Second-Order Self-Similarity)** $X(t)$ is *exactly second-order self-similar* with Hurst parameter H $(1/2 < H < 1)$ if

$$\gamma(k) = \frac{\sigma^2}{2}((k+1)^{2H} - 2k^{2H} + (k-1)^{2H}) \tag{1.1}$$

for all $k \geq 1$. $X(t)$ is *asymptotically second-order self-similar* if

$$\lim_{m \to \infty} \gamma^{(m)}(k) = \frac{\sigma^2}{2}((k+1)^{2H} - 2k^{2H} + (k-1)^{2H}). \tag{1.2}$$

It can be checked that Eq. (1.1) implies $\gamma(k) = \gamma^{(m)}(k)$ for all $m \geq 1$. Thus, second-order self-similarity captures the property that the correlation structure is exactly—condition (1.1)—or asymptotically—the weaker condition (1.2)—preserved under time aggregation. The form of $\gamma(k) = ((k+1)^{2H} - 2k^{2H} + (k-1)^{2H})\sigma^2/2$ is not accidental and implies further structure—long-range dependence—to which we will return later. Second-order self-similarity (in the exact or asymptotic sense) has been a dominant framework for modeling network traffic and this is also reflected in the chapters of this book.

1.4.1.2 An Allegory into Distributional Self-Similarity To understand the particular form of $\gamma(k)$ in the definition of second-order self-similarity, we will make a short detour and discuss self-similar processes in slightly more generality. Further extensions and detailed treatments can be found in Beran [9] and Samorodnitsky and Taggu [60].

Consider the cumulative process $Y(t)$, albeit in continuous time $t \in \mathbb{R}$. Following is a definition of self-similarity for continuous-time processes in the sense of finite-dimensional distributions.

Definition 1.4.2 (H-**ss**) $Y(t)$ is *self-similar with self-similarity parameter, that is, Hurst parameter*, H $(0 < H < 1)$, denoted H-*ss*, if for all $a > 0$ and $t \geq 0$,

$$Y(t) =_d a^{-H} Y(at). \tag{1.3}$$

Thus $Y(t)$ and its time scaled version $Y(at)$—after normalizing by a^{-H}—must follow the same distribution. In the traffic modeling context, it is convenient to think of $Y(t)$ as the cumulative or total traffic up to time t. For $a > 1$—time is stretched or dilated—a contraction factor a^{-H} is applied to make the magnitude of $Y(at)$ comparable to that of $Y(t)$. For $a < 1$, the opposite holds true. As a varies, the scaling exponent H remains invariant. This is a most natural definition; however, it has an important drawback: unless $Y(t)$ is degenerate, that is $Y(t) = 0$ for all $t \in \mathbb{R}$, $Y(t)$ cannot be stationary due to the normalization factor a^{-H}. Its increment process $X(t) = Y(t) - Y(t-1)$, however, is another matter. In particular, consider the case where $Y(t)$ is H-ss and has *stationary increments*; in this case we say $Y(t)$ is H-*sssi*. Let us further assume that $Y(t)$ has finite variance. It can be checked that $E[Y(t)] = 0$, $E[Y^2(t)] = \sigma^2 |t|^{2H}$, and

$$\gamma(k) = \frac{\sigma^2}{2}(|t|^{2H} - |t-s|^{2H} + |s|^{2H}). \tag{1.4}$$

This is achieved by noting that[8]

$$Y(t) =_d t^H Y(1),$$

from which it follows $E[Y^2(t)] = \sigma^2 t^{2H}$. The latter, then, can be used in the derivation of the autocovariance function (1.4). The increment process $X(t)$ has mean 0 and autocovariance $\gamma(k)$ as given in Eq (1.1). The derivation is similar to that of $Y(t)$.

How does distributional self-similarity (of a continuous time process) tie in with second-order self-similarity (of a discrete time process), which requires exact or asymptotic invariance with respect to second-order statistical structure of the aggregated time series $X^{(m)}$? A key observation lies in noting that $X^{(m)}$ can be viewed as computing a sample mean

$$\begin{aligned} X^{(m)} &= \frac{1}{m}\sum_{t=1}^{m} X(t) = m^{-1}(Y(m) - Y(0)) \\ &=_d m^{-1} m^H (Y(1) - Y(0)) = m^{H-1} X. \end{aligned}$$

Thus, if $Y(t)$ is a H-sssi process then its increment process $X(t)$ satisfies

$$X =_d m^{1-H} X^{(m)}, \tag{1.5}$$

[8]From $a^H Y(t) =_d Y(at)$, substitute $t = 1$ and $a = t$.

which shows how $X^{(m)}$ is related to X via a simple scaling relationship involving H in the sense of finite-dimensional distributions. Equations (1.1) and (1.2), then, express the fact that X and $m^{1-H}X^{(m)}$ are required to have exactly or asymptotically the same second-order structure. As a result, depending on whether a discrete time process $X(t)$ satisfies Eq. (1.5) for all $m \geq 0$ or only in the limit as $m \to \infty$, $X(t)$ is said to be *exactly self-similar* or *asymptotically self-similar.* Note that in the Gaussian case, this definition coincides with second-order self-similarity.

As a lead-in to the role of the parameter H, recall that the variance of the sample mean $\bar{Z}$ of a random variable Z satisfies $\text{var}(\bar{Z}) = \sigma_Z^2/m$, where m is the sample size. From Eq. (1.5) it follows that $\text{var}(X^{(m)}) = \sigma^2 m^{2H-2}$. When viewed as a sample mean where the samples are drawn *independently*, $\text{var}(X^{(m)})$ reduces to $\sigma^2 m^{-1}$ if $H = \frac{1}{2}$. If $H \neq \frac{1}{2}$, in particular, $\frac{1}{2} < H < 1$, then

$$\text{var}(X^{(m)}) = \sigma^2 m^{-\beta}$$

with $0 < \beta < 1$ (and $H = 1 - \beta/2$), which hints at a certain—and not just any—*dependency structure* in the "samples" (i.e., time series in our case) that causes $\text{var}(X^{(m)})$ to converge to zero slower than the rate m^{-1}.

1.4.1.3 Long-Range Dependence Thus far we have focused on explicating the role of self-similarity in the second-order stationary and distributional senses with little regard to the role of H and its range of values. Let us return to the definition of second-order self-similarity and its autocovariance $\gamma(k)$. Let $r(k) = \gamma(k)/\sigma^2$ denote the *autocorrelation function*. For $0 < H < 1$, $H \neq \frac{1}{2}$, it holds

$$r(k) \sim H(2H-1)k^{2H-2}, \qquad k \to \infty. \tag{1.6}$$

In particular, if $\frac{1}{2} < H < 1$, $r(k)$ asymptotically behaves as $ck^{-\beta}$ for $0 < \beta < 1$, where $c > 0$ is a constant, $\beta = 2 - 2H$, and we have

$$\sum_{k=-\infty}^{\infty} r(k) = \infty. \tag{1.7}$$

That is, the autocorrelation function decays slowly—that is, hyperbolically—which is the essential property that causes it to be not summable. When $r(k)$ decays hyperbolically such that condition (1.7) holds, we call the corresponding stationary process $X(t)$ *long-range dependent*. $X(t)$ is *short-range dependent* if the autocorrelation function is summable[9]. An essentially equivalent definition can be given in the

[9]Technically, more subtle definitions of long-range dependence are possible, but in this book, we will mainly rely on our working definition involving condition (1.7).

frequency domain where the *spectral density* $\Gamma(\nu) = (2\pi)^{-1} \sum_{k=-\infty}^{\infty} r(k)e^{ik\nu}$ is required to satisfy the property

$$\Gamma(\nu) \sim c|\nu|^{-\alpha}, \qquad \nu \to 0.$$

Here $c > 0$ is a constant and $0 < \alpha = 2H - 1 < 1$. Thus $\Gamma(\nu)$ diverges around the origin, implying ever larger contributions by low-frequency components.

Following are some simple facts regarding the value of H and its impact on $r(k)$. First, if $H = \frac{1}{2}$, then $r(k) = 0$, and $X(t)$ is trivially short-range dependent by virtue of being completely uncorrelated. In the case where $0 < H < \frac{1}{2}$, we have $\sum_{k=-\infty}^{\infty} r(k) = 0$, an artificial condition rarely encountered in applications. $H = 1$ is uninteresting since it leads to the degenerate situation $r(k) = 1$ for all $k \geq 1$. Finally, H-values bigger than 1 are prohibited due to the stationarity condition on $X(t)$.

1.4.1.4 Self-Similarity Versus Long-Range Dependence The preceding discussion indicates that there are self-similar processes that are not long-range dependent, and vice versa. For example, Brownian motion is $\frac{1}{2}$-sssi with white Gaussian noise as its increment process, but the latter is not long-range dependent. Conversely, certain fractional ARIMA time series generate long-range dependence but they are not self-similar in the distributional sense. In the case of asymptotic second-order self-similarity, however, by the restriction $\frac{1}{2} < H < 1$ in the definition, self-similarity implies long-range dependence, and vice versa. It is for this reason and the fact that asymptotic second-order self-similar processes are employed as "canonical" traffic models, that we sometimes use *self-similarity* and *long-range dependence* interchangeably when the context does not lead to confusion.

1.4.2 Impact of Heavy Tails

1.4.2.1 Heavy-Tailed Distribution There is an intimate relationship between heavy-tailed distributions and long-range dependence, which we will discuss in the next sections. First, a few definitions and basic facts. A random variable Z has a *heavy-tailed distribution* if

$$\Pr\{Z > x\} \sim cx^{-\alpha}, \qquad x \to \infty, \tag{1.8}$$

where $0 < \alpha < 2$ is called the *tail index* or *shape parameter* and c is a positive constant[10]. That is, the tail of the distribution, asymptotically, decays hyperbolically. This is in contrast to *light-tailed distributions*—for example, exponential and Gaussian—which possess an exponentially decreasing tail. A distinguishing mark of heavy-tailed distributions is that they have infinite variance for $0 < \alpha < 2$, and if

[10]Technically, more subtle definitions involving slowly varying functions are possible and can be found in some chapters of this book. However, for practical purposes and to convey the main ideas, our working definition, centered around condition (1.8), will suffice.

$0 < \alpha \leq 1$, they also have an unbounded mean. In the networking context, we will be primarily interested in the case $1 < \alpha < 2$. A frequently used heavy-tailed distribution is the *Pareto distribution* whose distribution function is given by

$$\Pr\{Z \leq x\} = 1 - \left(\frac{b}{x}\right)^{\alpha}, \qquad b \leq x,$$

where $0 < \alpha < 2$ is the shape parameter and b is called the *location parameter*. The mean is given by $\alpha b/(\alpha - 1)$. We remark that there are distributions—for example Weibull and log normal—that have *subexponentially* decreasing tails but possess finite variance.

The main characteristic of a random variable obeying a heavy-tailed distribution is that it exhibits extreme variability. Practically speaking, a heavy-tailed distribution gives rise to very large values with nonnegligible probability so that sampling from such a distribution results in the bulk of values being "small" but a few samples having "very" large values. Not surprisingly, heavy-tailedness impacts sampling by slowing down the convergence rate of the sample mean to the population mean, dilating it as the tail index α approaches 1. For example, pending on the sample size m, the sample mean $\bar{Z}_m$ of a Pareto distributed random variable Z may significantly deviate from the population mean $\alpha k/(\alpha - 1)$, oftentimes underestimating it. In fact, the absolute estimation error $|\bar{Z}_m - E(Z)|$ asymptotically behaves as $m^{(1/\alpha)-1}$ (see, e.g., Crovella and Lipsky [15]), and thus for α values close to 1, care must be given when sampling from heavy-tailed distributions such that conclusions about network behavior and performance attributable to sampling error are not advanced. A more detailed discussion of sampling issues is given in Chapter 3.

1.4.2.2 Heavy Tails and Predictability Heavy-tailedness of certain network-related variables—for example, file sizes and connection durations—can be shown to underlie the root cause of long-range dependence and self-similarity in network traffic. First, let us examine a simple fact on the intrinsic predictability associated with heavy-tailed random variables. Let Z be a heavy-tailed random variable interpreted as the *duration* or *lifetime* of a network connection (e.g., TCP connection, IP-flow, or session). Since connection durations are physically measurable events, assume that we observe—in time—that a connection has been active for $\tau > 0$ seconds. To simplify the discussion, assume time is discrete ($t \in \mathbb{Z}_+$) and $A : \mathbb{Z}_+ \rightarrow \{0, 1\}$ is an indicator function such that $A(t) = 1$ iff $Z \geq t$. We are interested in the probability that the connection will persist into the future given that it has been active for τ seconds. That is, we would like to estimate the conditional probability

$$\mathcal{L}(\tau) = \Pr\{A(\tau + 1) = 1 | A(t) = 1, 1 \leq t \leq \tau\}. \tag{1.9}$$

$\mathcal{L}(\tau)$ can be expressed as

$$\mathcal{L}(\tau) = 1 - \frac{\Pr\{Z = \tau\}}{\Pr\{Z \geq \tau\}}. \tag{1.10}$$

Let us first compute $\mathcal{L}(\tau)$ for light tails, in particular, distributions with asymptotically exponential tails $\Pr\{Z > x\} \sim c_1 e^{-c_2 x}$, where $c_1, c_2 > 0$ are constants. The second term in Eq. (1.10) is computed by

$$\frac{\Pr\{Z = \tau\}}{\Pr\{Z \geq \tau\}} \sim \frac{c_1 e^{-c_2 \tau} - c_1 e^{-c_2(\tau+1)}}{c_1 e^{-c_2 \tau}} = 1 - e^{-c_2}$$

for large τ, and we get $\mathcal{L}(\tau) \sim e^{-c_2}$. Thus for exponentially light tails, prediction is not enhanced by conditioning on ever longer periods of observed activity. For heavy tails, the corresponding derivations are

$$\frac{\Pr\{Z = \tau\}}{\Pr\{Z \geq \tau\}} \sim \frac{c\tau^{-\alpha} - c(\tau+1)^{-\alpha}}{c\tau^{-\alpha}} = 1 - \left(\frac{\tau}{\tau+1}\right)^{\alpha},$$

which yields

$$\mathcal{L}(\tau) \nearrow 1, \quad \tau \to \infty. \tag{1.11}$$

Thus the longer the period of observed activity, the more certain that it will persist into the future. In fact, it is straightforward to generalize Eq. (1.9) so that we can measure the *persistence* of activity $\delta \geq 1$ time units into the future, that is

$$\mathcal{L}(\tau) = \Pr\{A(\tau + s) = 1, 1 \leq s \leq \delta | A(t) = 1, 1 \leq t \leq \tau\}.$$

This does not change the qualitative results: for the light-tailed case, $\mathcal{L}(\tau) \sim e^{-c_2\delta}$; for the heavy-tailed case, $\mathcal{L}(\tau)$'s asymptotic behavior follows $(1 + \delta/\tau)^{-\alpha} \nearrow 1$. Since $(1 + \delta/\tau)^{-\alpha} \leq e^{-\alpha\delta/\tau}$, we observe that in both cases predictability is exponentially sensitive to the prediction interval δ. However, in the heavy-tailed case, for any desired δ time unit "peek into the future," by conditioning the prediction on a sufficiently long past observation of activity, the prediction error can be reduced to an arbitrarily small level.

We remark that the mathematical implications of asymptotic analysis need not deter from the practical relevance of its conclusions, even considering the fact that tails are always finite in a physical network environment. First, if heavy tails are modeled using the Pareto distribution, then its shape is hyperbolic across its *entire* range—not just asymptotically—and accurate finitary computations can be carried out. Second, given an empirical distribution with finite support, the fact that it has a finite cut off point will not significantly influence the predictability computations carried out in practice as long as the tail is "sufficiently"—for example, several orders of magnitude beyond the mean—long. As with time series, the identification problem of whether an empirical distribution is best modeled by heavy-tailed or light-tailed distributions is intrinsically ill-posed and secondary to the fact that the predictability structure as computed by Eq. (1.10) from *empirical distributions* is significant.

1.4.2.3 Heavy Tails and Long-Range Dependence As we saw in the previous section, heavy tails lead to predictability, and for a related reason, they lead to long-range dependence in network traffic. First, we give a definition of fractional Brownian motion (FBM) and its increment process—fractional Gaussian noise (FGN)—which are Gaussian self-similar processes with, in general, long-range dependence, first introduced by Mandelbrot [45]. Their Gaussian structure renders them especially useful as *aggregate* traffic models where aggregation of independent traffic sources—by the central limit theorem—leads to the Gaussian property. In practice, of course, traffic flows need not be independent if they engage in feedback control and share common resources at bottleneck routers. The definitions of FBM and FGN are couched in the framework of distributional self-similarity given in Section 1.4.1.2.

Definition 1.4.3 **(FBM)** $Y(t), t \in \mathbb{R}$, is called *fractional Brownian motion* with parameter $H, 0 < H < 1$, if $Y(t)$ is Gaussian and H-sssi.

Definition 1.4.4 **(FGN)** $X(t), t \in \mathbb{Z}_+$, is called *fractional Gaussian noise* with parameter H if $X(t)$ is the increment process of FBM with parameter H.

By the definition of H-sssi, FBM reduces to Brownian motion—and FGN to white Gaussian noise—when $H = \frac{1}{2}$. Thus $X(t), t \in \mathbb{Z}_+$, becomes completely uncorrelated. Since Gaussian processes are characterized by their second-order structure, for each $H, 0 < H < 1$, there is a unique Gaussian process that is the stationary increment of a H-sssi process. FBM is the corresponding unique Gaussian H-sssi process. By the same token, for Gaussian processes, distributional self-similarity and second-order self-similarity yield equivalent definitions.

Now we examine why heavy tails are considered the root cause of long-range dependence in network traffic. We take a constructive approach by presenting input processes—in various guises—with probabilistic activity times, which then are shown to lead to long-range dependence if, and only if, they are heavy-tailed. We first present the on/off model by Willinger et al. [74] followed by a related model used by Likhanov et al. [43], which has a slightly different, but complementary, source arrival perspective.

The on/off model considers N independent traffic sources $X_i(t), i \in [1, N]$, where each is a 0/1 *reward renewal process* with i.i.d. on periods and i.i.d. off-periods. This just means that $X_i(t)$ takes on the values 1 ("on") and 0 ("off") on alternating, nonoverlapping time intervals called on and off periods, respectively. $X_i(t) = 1$ is interpreted as there being a packet transmission. Thus an on period can be viewed as constituting a "packet train" [36]. Three such on/off sources and their aggregation are depicted in Fig. 1.6. Let $S_N(t) = \sum_{i=1}^{N} X_i(t)$ denote the aggregate traffic at time t. Consider the *cumulative* process $Y_N(Tt)$ defined as

$$Y_N(Tt) = \int_0^{Tt} \left(\sum_{i=1}^{N} X_i(s) \right) ds, \tag{1.12}$$

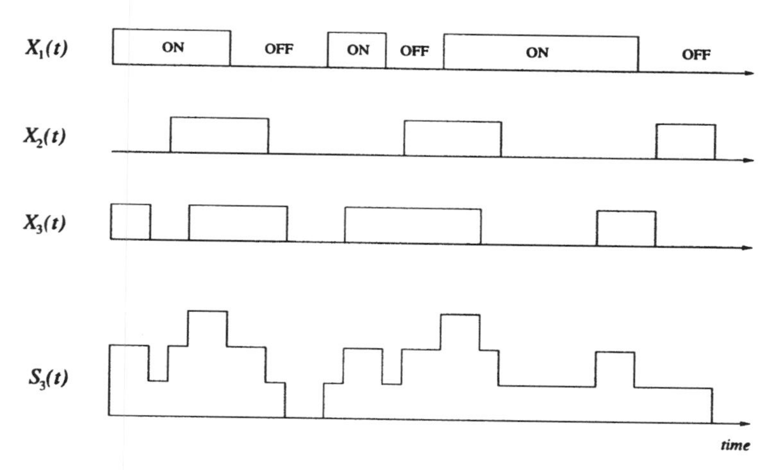

Fig. 1.6 $N = 3$ on/off sources, $X_1(t), X_2(t)$, and $X_3(t)$, and their aggregation $S_3(t) = X_1(t) + X_2(t) + X_3(t)$.

where $T > 0$ is a scale factor that is explicitly incorporated. Thus $Y_N(Tt)$ measures the total traffic up to time Tt. What is the behavior of $Y_N(Tt)$ for large T and N? We will simplify the discussion so as to concentrate on the single salient feature of how heavy-tailedness influences long-range dependence. Let τ_{on} be the random variable describing the duration of the on periods and let τ_{off} be the random variable associated with the durations of the off periods. Let

$$\Pr\{\tau_{\text{on}} > x\} \sim cx^{-\alpha}, \qquad x \to \infty,$$

where $1 < \alpha < 2$ and $c > 0$ is a constant. As to τ_{off}, it can be either heavy tailed or light tailed with finite variance. It can be shown [62, 74] that $Y_N(Tt)$ behaves like FBM in the following sense.

Theorem 1.4.5 (On/Off Model and FBM) *$Y_N(Tt)$ behaves statistically as*

$$\frac{E(\tau_{\text{on}})}{E(\tau_{\text{on}}) + E(\tau_{\text{off}})} NTt + CN^{1/2}T^H B_H(t) \tag{1.13}$$

for large T, N, where $H = (3 - \alpha)/2$, $B_H(t)$ is FBM with parameter H, and $C > 0$ is a quantity depending only on the distributions of τ_{on} and τ_{off}.

Thus $Y_N(Tt)$ asymptotically behaves as fractional Brownian motion fluctuating around $NTtE(\tau_{\text{on}})/(E(\tau_{\text{on}}) + E(\tau_{\text{off}}))$ when suitably normalized. It is long-range dependent ($\frac{1}{2} < H < 1$) iff $1 < \alpha < 2$; that is, τ_{on}'s distribution is heavy-tailed. If neither τ_{on} nor τ_{off} is heavy-tailed, then $Y_N(Tt)$ is short-range dependent. It is in this sense that heavy-tailedness (in this case, of the on or off periods) is an essential component to inducing long-range dependence in the aggregated time series. Of less practical import in the networking context is the case when the off period is heavy tailed but the on period is not, which nonetheless also yields long-range dependence.

A related but slightly different source model is obtained when viewing each source $i \in \mathbb{Z}_+$ as emitting a *singular* packet train but being otherwise silent [43]. Thus, a single on/off source in the on/off model can be construed to be the output behavior of a network host, which may service multiple TCP connections, whereas in the singular packet train case, the source corresponds to a single TCP connection transporting a byte stream such as a file. To each source i $\in \mathbb{Z}_+$, we associate a time interval $[t_i, t_i + \tau_i)$, $t_i, \tau_i \in \mathbb{Z}_+$, where $X_i(t) = 1$ if $t \in [t_i, t_i + \tau_i)$, and 0 otherwise. We assume that the $\tau_i, i \in \mathbb{Z}_+$, are i.i.d. and t_i is determined by a Poisson process $\xi(t)$, which indicates how many *new* connections arrive at time t.

$$X(t) = \sum_{i \in \mathbb{Z}_+} X_i(t) < \infty$$

then counts how many connections are active at time t. Alternatively, $X(t)$ can be viewed as the aggregate (over flows) traffic rate emitted at time instance t. The behavior of $X(t)$ and its generalized brethren can be analyzed directly [43, 65, 66], but a more succinct and elegant approach that reveals the influence of heavy-tailedness on long-range dependence can be found in a result due to Cox involving the $M/G/\infty$ queueing system [12]. An $M/G/\infty$ queue is defined to be the *busy server process* where connection arrivals are Poisson and each connection is serviced by a server—there are infinitely many—with a general service time. Thus, at any instant of time, we count how many servers are busy servicing requests. If the i.i.d. service times are given by $\tau = \tau_i, i \in \mathbb{Z}_+$, then it is easy to see that the busy server process in the $M/G/\infty$ queue corresponds to the aggregate traffic rate $X(t)$ in the Poisson source model with a single on period. Let τ be heavy-tailed with tail index $1 < \alpha < 2$.

Theorem 1.4.6 **(**$M/G/\infty$ **and LRD)** *$X(t), t \in \mathbb{Z}_+$, is asymptotically second-order self-similar with parameter $H = (3 - \alpha)/2$.*

Thus $1 < \alpha < 2$, via $\frac{1}{2} < H = (3 - \alpha)/2 < 1$, is directly tied to long-range dependence. Theorem 1.4.6, in turn, implies by the previous correspondence that when connections with a single heavy-tailed on period arrive in a Poisson manner, then the resulting aggregate traffic is long-range dependent. In its raw form, $X(t)$ has Poisson marginals [57], but it can be shown that FBM arises naturally as a limiting process by appropriately scaling the Poisson arrival rate and service times [39]. The $M/G/\infty$ approach to modeling network traffic has proved useful in analyzing queueing behavior fed by long-range dependent input [54, 55].

We remark that from a purely *mathematical modeling* point of view, heavy-tailedness is not necessary to generate long-range dependence in aggregate traffic. As pointed out by Beran [9] (and further explored in Chapter 11), an *infinite* aggregation of short-range dependent sources—in particular, heterogeneous on/off sources with exponential on/off times—can produce long-range dependence when suitably calibrated. *Finite* aggregations of short-range dependent sources, however, cannot induce long-range dependence; hence the assumption of infinite aggregation is crucial. Empirical traffic measurements provide strong evidence that file sizes and connection durations are heavy-tailed, and hence the *heavy-tailedness causes long-range dependence* rule-of-thumb is supported by physical modeling. The practical implications—if any—of the observation that short-range dependent flows can produce long-range dependence are not clear, and we include them for completeness.

1.5 ORGANIZATION OF THE BOOK

This book is a collection of chapter contributions that bring together relevant works spanning a cross section of topics covering traffic measurement, modeling, performance analysis, and traffic control for self-similar network traffic.

The first part of the book deals with traffic characterization, estimation, and simulation issues. Wavelet analysis is introduced as a powerful technique for both modeling and estimation of self-similar traffic. The wavelet-based approach naturally lends itself to a multifractal view of network traffic, where a shift in traffic properties at long and short time scales is captured using cascade constructions superimposed on heavy-tailed renewal processes. This is further discussed in Chapter 20. Complementing the theme of traffic modeling is the issue surrounding simulation, such as in the generation of synthetic workloads and self-similar traffic, which entails, in many instances, sampling from heavy-tailed distributions. Due to the slow convergence of sample statistics to population statistics, special care needs to be exercised when performing simulations that involve sampling from heavy-tailed distributions so as not to advance erroneous conclusions attributable to sampling effects including underestimation.

The second part of the book focuses on performance evaluation issues, in particular, queueing behavior of finite and infinite buffer systems when fed with long-range dependent input. Due to the breakdown of Markovian assumptions, which are key to tractability in traditional queueing analysis, the technical challenges encountered with self-similar input are great. This part of the book gives an exposition of what is known about queueing with self-similar input, starting with the trademark phenomenon that the queue length distribution decays subexponentially—as opposed to exponentially—and advancing to packet scheduling, transient analysis, tight buffer asymptotics, and impact of resource boundedness and finite time horizons. Queueing based performance analysis also forms the foundation of traffic control based on resource provisioning and dimensioning. The traffic models considered can be viewed as variants of on/off renewal reward processes where session arrivals are allowed to be Poisson, however, on or off periods are assumed to be heavy-tailed. Some of the input processes are intimately related to fractional Brownian motion and its increment process, fractional Gaussian noise, which in turn can be analyzed by various techniques including large deviations theory.

The third part of the book covers traffic control issues that arise under self-similar traffic conditions. There are two main facets to the question, one centered on the problem of resource provisioning and dimensioning—a form of open-loop control—and ensuing trade-off relations, and the other based on the traditional traffic control framework of feedback control and its realization in network protocols including TCP. With respect to resource provisioning, due to the amplified queueing delay incurred when employing buffer dimensioning, an alternative resource provisioning strategy based on bandwidth dimensioning as the principal control variable has been advanced. In this "bufferless" traffic engineering regime, by reserving sufficient resources to meet the peak rate of multiplexed input traffic—that is, over-provisioning—a desired level of quality of service in the form of statistical guarantees can be achieved. Feedback traffic control, on the other hand, represents a more subtle challenge where the central idea revolves around exploiting correlation structure at multiple time scales—in particular, "large" time scales exceeding the round-trip time associated with the feedback loop—as afforded by long-range dependence and self-similarity, to affect traffic control decisions executed at smaller time scales. When effectively facilitated, this can result in significant performance improvements including mitigation of the delay–bandwidth product problem in broadband wide–area networks due to proactivity.

1.6 CHAPTER CONTRIBUTIONS

In the following, we give a brief outline of the various chapter contributions organized into three parts: (i) estimation and simulation, (ii) queueing with self-similar input, and (iii) traffic control and resource provisioning. We describe how each chapter fits into the overall picture and comment on the potential role and relevance of each chapter for future advances in these areas. The threefold categorization is not strict in the sense that some chapters encompass subject matters that cross the set boundaries. Also, part (ii) may be more generally characterized as performance evaluation with self-similar input as queueing is the predominant, but not exclusive, theme contained therein.

Chapter 1 by Park and Willinger serves as an introductory chapter that provides the necessary technical background including definitions for following the rest of the book. The chapter is self-contained and thus can also be read as a modern introduction to the topic of self-similar network traffic. It gives an overview of the various research activities surrounding self-similar traffic and outlines the principal issues in the areas of traffic modeling, statistical and scientific inference, performance analysis, and traffic control. Chapter 1 concludes with an overview of the book including the present section describing each chapter contribution.

1.6.1 Estimation and Simulation

The chapter by Abry, Flandrin, Taqqu, and Veitch (Chapter 2) discusses the state of the art in identification of scaling phenomena in traffic series—the crucial component of self-similarity—using the framework of wavelets. Due to their ability to localize a given signal or time series both in time and scale (or frequency), wavelets provide a powerful and refined technique for detecting and quantifying scaling behavior in measured traffic. Since wavelets are, in part, parameterized by scaling parameters, this lends itself naturally to a multiscale representation and analysis of time series, which, in turn, allows a qualitatively more informative and quantitatively more accurate estimation of underlying scaling properties. Abry and co-authors present a comprehensive overview of the fundamentals of wavelet analysis and its application to estimating scaling behavior in self-similar traffic, focusing on properties related to self-similar scaling. The chapter concludes with a discussion of the "inverse" operation, that is, *generating* synthetic self-similar time series using wavelet expansions.

Chapter 20 by Riedi and Willinger describes an even further refined modeling based, in part, on the wavelet framework where a notion of large-time-scale and small-time-scale behavior and their observed empirical differences are captured. The novel aspect here is that the resulting representation combines in a natural manner self-similar and *multifractal* scaling behaviors. In one interpretation, multifractals can be viewed as being composed of heavy-tailed renewal processes, which collectively determine the large-time-scale behavior leading to long-range dependence, and a more fine granular "within-connection" structure that reveals itself in a locally highly irregular scaling behavior and conforms to multifractality over fine time scales. The fact that long-range dependence of such processes does not depend on the finer details underlying variations within on periods or connection times is shown by a result of Kurtz [39]. Riedi and Willinger, however, show that the fine granular structure at smaller time scales deviates significantly from the self-similar scaling behavior over large time scales and may therefore be relevant for traffic modeling purposes. They argue that multifractals in the form of certain cascade models or multiplicative processes provide a natural modeling framework and also give initial evidence to suggest that the deviation from self-similar scaling observed in the short-correlation structure may be due to protocol stack effects such as those stemming from TCP's feedback congestion control.

Chapter 20 also provides a high-level overview of recent developments in traffic modeling with foundations in physical modeling. In this capacity, it serves to delineate a future program for traffic characterization and modeling with emphasis on achieving a comprehensive understanding of network traffic and workloads and their scaling behavior across multiple time scales.

Chapter 3 by Crovella and Lipsky serves to draw attention to issues surrounding simulation under self-similar traffic conditions, which, in many instances, involve sampling from heavy-tailed distributions such as those arising in the context of generating long-range dependent traffic series as well as generating heavy-tailed workloads in related contexts (e.g., WWW). Slowly decaying tails lead to slow convergence of sample statistics to their corresponding population statistics, in fact, leading to biasedness in terms of underestimation. In other words, the sample mean is consistent but biased, which has ramifications when performing simulations whose sampling frequency may not be sufficiently large. Crovella and Lipsky discuss various issues and possible remedies associated with this important practical problem.

1.6.2 Queueing with Self-Similar Input

The chapter by Norros (Chapter 4) gives an updated overview of the fundamental queueing results associated with fractional Brownian motion (FBM) input processes, also called fractional Brownian storage. Due to the fundamental importance played by FBM and its increment process fractional Gaussian noise (FGN) as a model of *aggregate* traffic, understanding the queueing behavior under self-similar input as captured by FBM is of importance to many other results, serving as a reference point. Norros derives the Weibullian tail behavior of the queue length distribution arising in fractional Brownian storage models with Hurst parameter in the range $\frac{1}{2} < H < 1$ and discusses the importance of the Gaussian property of the input process and the role that it plays in the analysis.

In Chapter 5, Brichet, Massoulié, Simonian, and Veitch give a more refined extension of Gaussian input processes stemming from superposition of on/off processes in the heavy-load case where the arrival rate is close to the service rate. As the number of i.i.d. on/off sources increases, they are able to derive limit results that characterize the queue length distribution asymptotics and show it to be Weibullian, consistent with the result in the FBM case.

The chapter by Boxma and Cohen (Chapter 6) discusses queueing behavior as a function of the service discipline, in particular, first-come-first-served (FCFS), processor sharing (PS), and last-come-first-served preemptive resume (LCFS-PR), in the context of the $M/G/1$ queueing model. Due to the close connection between heavy-tailed on/off processes and the $M/G/1$ system when the service time is heavy-tailed, the latter represents a natural queueing model in which to incorporate the impact of long-range dependence. Their main contribution is toward deriving heavy-traffic—utilization approaches 1—limit theorems for the $M/G/1$ queue under the aforementioned service disciplines and showing that the tail of the queue length distribution can indeed be significantly less heavy when using PS and LCFC-PR in place of FCFS. The conclusions advanced have potential applications to packet and workload scheduling in router and Web server design.

Chapter 7 by Resnick and Samorodnitsky investigates performance issues for three classes of input models where heavy tails induce long-range dependence on the input side and a single server works at constant rate. In particular, they consider a single on-off renewal process with heavy-tailed on periods, a finite number of i.i.d. heavy-tailed renewal processes, and, lastly, an infinite number of sources whose transmission times or on periods are heavy-tailed. In all three cases, the input process is long-range dependent and queueing-affected performance leads to polynomial (versus exponential) dependence, which agrees with related queueing results.

The chapter by Likhanov (Chapter 8) derives results for the queueing behavior for a broad class of asymptotic second-order self-similar processes—also related to the $M/G/\infty$ model—which can be viewed as superposition of sessions that arrive in a Poisson fashion and whose session durations are heavy-tailed. Each session or connection, however, can be viewed as having a finite lifetime—after a burst of activity, it is silent forever thereafter—which distinguishes it from the on/off model where connections alternate between active and idle periods ad infinitum. The author establishes asymptotic bounds for the queue length distribution, which relate the various parameters of the asymptotic second-order self-similar arrival process model to performance, giving rise to polynomially decaying tails of the queue length distribution.

In Chapter 9, Makowski and Parulekar present a comprehensive treatment of the large buffer asymptotics for the $M/G/\infty$ input model—a constant rate server fed with $M/G/\infty$ inputs—using large deviation techniques to approximate the tail behavior. They show that there exist compact relationships between buffer occupancy asymptotics and the service time distribution of $M/G/\infty$, but the tightness of buffer asymptotics is influenced by the shape of the service time distribution—exponential or subexponential—the latter leading to upper and lower bounds that collapse only under certain restrictions. In addition to its technical content, the chapter provides a detailed discussion on why the $M/G/\infty$ model is a useful queueing model under which to study performance issues relating to long-range dependence and explicates their performance results with respect to other well-known results in buffer asymptotics for long-range dependent input. In particular, the authors point out that the same buffer asymptotics can be induced by vastly different input streams—long-range and short-range dependent.

Chapter 10 by Jelenković investigates the subexponential queueing behavior of very general queueing systems when subject to subexponential arrival processes. In particular, he considers the class of $GI/GI/1$ queueing systems and some of its variations, including finite buffers and truncated heavy-tailed arrivals. It is shown that the asymptotic approximation for the loss rate in a finite buffer $GI/GI/1$ queue is independent of the service process, and buffer capacity dimensioning—in certain buffer size regimes—can exert a significant impact on performance. The chapter concludes with a study of multiplexing behavior under long-range dependent input for fluid queues. It is shown that dominance is at play, which allows simplified reasoning of multiplexing effects by suitable replacement of dominated input processes by more simple ones (e.g., constant bit rate).

The chapter by Jacquet (Chapter 11) is concerned with long-range dependent processes that are the superposition of an infinite number of suitably calibrated on/off sources with *exponential* on/off times. While in the networking context such constructions are artificial and have little in common with empirical information gained from measured traffic, it serves to show that mathematically—under *infinite* aggregation—light-tailed on and off periods can induce long-range dependence, and due to the inherently Markovian nature of the individual components, they may be useful when studying the behavior of queueing systems. For example, Jacquet uses Mellin transforms, a transform technique used in traditional teletraffic theory, to track the polynomial tails of the ensuing queue length distribution in a simple queueing system.

Chapter 12 by Heyman and Lakshman discusses the relative import of short-range and long-range dependence for performance analysis in certain networking contexts—for example, where buffer capacities are sufficiently "small," for certain types of applications such as variable bit rate video, and for certain first-order performance measures such as long-term packet loss rate—where short-range dependence can dominate queueing. The existence of such regimes is not surprising but worthwhile remarking as with other manifold qualifications on the impact and role of self-similarity and heavy-tailedness discussed in the book. In these cases, given the choice between equally well-fitting short-range dependent "black box" processes and long-range dependent "physical" models, the short-range dependent ones can be effectively employed for performance analysis purposes. However, a change in the underlying assumptions—different networking context, application, or performance measures—renders this approach inflexible in contrast with physical models, which are more robust and parsimonious. The limitation of short-range dependent black box models is most pronounced when studying "what if" questions, in particular, those involving traffic control with feedback.

Chapter 13 by Li and Li is the last chapter of the second part of the book and presents a transient analysis of queueing behavior under long-range dependent input. The analysis is "transient" in the sense that loss probabilities are derived *conditioned* on the state of the system at a previous time instance that facilitates tractability while utilizing the correlation structure present in the stationary i.i.d. increments of the long-range dependent input considered. This work can be viewed as an effort toward understanding genuinely transient phenomena in queueing systems with self-similar input, noting that complete transient analysis is a difficult task even for simple Markovian systems such as $M/M/1$, which involves modified Bessel functions. The practical relevance of transient results becomes obvious in the presence of heavy-tailed distributions, which can slow down convergence to the steady state to the point where its value for engineering purposes is greatly diminished. Chapter 17—in the third part of the book—gives a complementary and more sophisticated form of transient analysis geared toward an application to traffic control.

1.6.3 Traffic Control and Resource Provisioning

Chapter 14 by Park, Kim, and Crovella discusses how the causality of self-similar network traffic can be traced back to a high-level structural property of the underlying networked system, namely, the heavy-tailed nature of file size or Web document distributions at the application layer. The authors show that when objects sampled from such distributions are exchanged via the mediation of a "typical" protocol stack—application layer (e. g., FTP, HTTP), transport layer (e.g., TCP, flow-controlled UDP), network layer (e.g., IP)—with focus on the transport layer which governs congestion control and reliability (if so desired), then the transfer and manifestation of the application layer causal seed of self-similarity at multiplexing points in the network layer in the form of self-similar traffic is robust, being largely impervious to the details of the actions carried out in the protocol stack and network configuration. In conjunction with evidence of heavy-tailed file size distribution observed in UNIX file systems of the past—that is, in the 1980s well before the onset of the World Wide Web and its constituent traffic—the chapter provides a causal, physical explanation for why self-similar traffic may be so ubiquitous and is a phenomenon likely to persist in the future Internet. The chapter also shows that if the transport protocol behaves in "extreme" ways—for example, minimal or no congestion control—it is possible for the protocol stack to exert sufficient influence such that the transfer mechanism of causality is significantly impeded.

Chapter 20 and its discussion of multifractal IP traffic with different scaling behavior at small and large time scales, where the multiplicative scaling at small time scales—at roughly sub-RTT times—is taken to stem from the actions associated with TCP's feedback congestion control, can be viewed as a refined characterization of the influence of the protocol stack, in this case, for short-term correlation structure of network traffic.

In Chapter 15, Feldmann presents an empirical study of the characteristics of TCP connection arrivals and shows that in today's Internet, in addition to self-similarity at the packet level, self-similar scaling is already encountered at the session or application layer when analyzing time series of the number of TCP connections per time unit. To this end, Feldmann relies on wavelet-based inference techniques and uncovers that various facets of TCP connection arrival characteristics conform to Weibullian-type distributions. This detailed workload characterization is relevant from both traffic modeling and control perspectives since knowing the structure of TCP connection arrivals and their durations can help in devising improved traffic control mechanisms.

The chapter by Roberts (Chapter 16) gives a high-level discussion of traffic control and resource provisioning issues under long-range dependent traffic conditions. The basic premise is predicated on segmenting traffic into two broad classes—stream and elastic traffic—where the former are subjected to open-loop control, that is, resource reservation, and the latter are handled using closed-loop control. Due to the heavy queueing cost associated with provisioning resources using buffer sizing, it is explicitly proposed that for stream traffic, bandwidth allocation with small buffer capacity be the default resource allocation policy employed. Roberts sketches the components of a multiservice network architecture advocating measurement-based admission control for stream traffic considered effective for self-similar traffic, and end-to-end feedback control for elastic traffic, with pricing applicable to both. The influence of self-similarity and heavy tailedness on architectural considerations and traditional traffic control are discussed throughout.

In Chapter 17, Duffield and Whitt adopt a bufferless model for performance analysis and traffic control where instantaneous offered load is given by a long-term level process (i.e., "DC" component shifts across a range of traffic levels) with "within-level" fluctuations. They investigate the problem of approximating the conditional mean of aggregate traffic—conditioned on past traffic profile or demand parameterized by level and age (or duration)—using numerical transform inversion. They show that the age variable plays an important role in facilitating prediction. Duffield and Whitt show applications of the "transient analysis framework" by estimating the probability of high levels of congestion in steady state using a large deviation principle approximation. They also analyze the converse situation captured by the time to recover—that is, return to a traffic level corresponding to a given resource capacity—after the excursion. The approach advanced in Chapter 17 is interesting due to its focus on the long-range dependent aggregate input process, dispensing with its impact on queueing, and directly analyzing the transient or dynamic variability structure based on the predictability inherent to long-range dependent processes. A similar "on-line" framework is adopted in Chapter 18 where long-term predictability structure is exploited for feedback congestion control.

Tuan and Park (Chapter 18) show that in spite of the "bad news" associated with scale-invariant burstiness, there is "good news" in the sense of there being the potential of exploiting long-term correlation structure present in long-range dependent traffic for traffic control purposes. They advance the *multiple time scale congestion control* framework and show that nonnegligible correlations at large time scales can be effectively detected on-line and engaged to improve the performance of feedback congestion controls in rate-based settings. The central idea underlying the technology is *selective aggressiveness control*, which allows explicit prediction of large time scale network state to be used to modulate the aggressiveness of bandwidth consumption behavior exhibited by feedback congestion control acting at small time scales (i.e., time scale of RTT). An important consequence is the mitigation of the delay–bandwidth product problem of reactive controls in broadband wide-area networks.

Finally, Chapter 19 by Adas and Mukherjee addresses the problem of how resource reservation in a time-division multiplexing set up—per-VC framing—can be used to facilitate end-to-end quality of service (QoS) under long-range dependent traffic conditions. The asynchronous framing approach described follows the resource provisioning paradigm espoused for long-range dependent traffic, namely, that of bufferless queueing, which then allows computation of QoS guarantees by appealing to the central limit theorem and equivalent bandwidth computations.

REFERENCES

1. P. Abry and D. Veitch. Wavelet analysis of long-range dependent traffic. *IEEE Trans. Information Theory*, **44**(1):2–15, 1998.
2. A. Adas and A. Mukherjee. On resource management and QoS guarantees for long range dependent traffic. In *Proc. IEEE INFOCOM '95*, pp. 779–787, 1995.
3. R. Addie, M. Zukerman, and T. Neame. Fractal traffic: measurements, modelling and performance evaluation. In *Proc. IEEE INFOCOM '95*, pp. 977–984, 1995.
4. G. Amdahl. Validity of the single-processor approach to achieving large scale computing capabilities. In *AFIPS Conf. Proc.*, pp. 483–485, 1967.
5. V. Anantharam. Queueing analysis with traffic models based on deterministic dynamical systems. In *Proc. 25th Allerton Conference on Communication, Control and Computing*, pp. 233–241, 1996.
6. M. Arlitt and C. Williamson. Internet Web servers: workload characterization and performance implications. *IEEE/ACM Trans. Networking*, **5**(5):631–645, 1997.
7. P. Barford and M. Crovella. Generating representative workloads for network and server performance evaluation. In *Proc. ACM SIGMETRICS '98*, pp. 151–160, 1998.
8. M. Barnsley. *Fractals Everywhere*. Academic Press, New York, 1988.

9. J. Beran. *Statistics for Long-Memory Processes*. Monographs on Statistics and Applied Probability. Chapman and Hall, New York, 1994.
10. J. Beran, R. Sherman, M. S. Taqqu, and W. Willinger. Long-range dependence in variable-bit-rate video traffic. *IEEE Trans. Commun.,* **43**:1566–1579, 1995.
11. E. Biersack. Performance evaluation of forward error correction in ATM networks. In *Proc. ACM SIGCOMM '92*, pp. 248–257, 1992.
12. D. R. Cox. Long-range dependence: a review. In H. A. David and H. T. David, eds., *Statistics: An Appraisal*, pp. 55–74. Iowa State University. Press, Ames, 1984.
13. M. Crovella and A. Bestavros. Self-similarity in World Wide Web traffic: evidence and possible causes. In *Proceedings of the 1996 ACM SIGMETRICS International Conference on Measurement and Modeling of Computer Systems*, May 1996.
14. M. E. Crovella and A. Bestavros. Self-similarity in World Wide Web traffic: evidence and possible causes. *IEEE/ACM Trans. Networking* **5**, 835–846,1997.
15. M. Crovella and L. Lipsky. Long-lasting transient conditions in simulations with heavy-tailed workloads. In *Proc. 1997 Winter Simulation Conference*, 1997.
16. N. G. Duffield, J. T. Lewis, N. O'Connel, R. Russell, and F. Toomey. Statistical issues raised by the Bellcore data. In *Proc. 11th IEE Teletraffic Symposium*, 1994.
17. N. G. Duffield and N. O'Connell. Large deviations and overflow probabilities for the general single server queue, with applications. *Math. Proc. Cambridge Philos. Soc.* **118**: 363–374, 1995.
18. N. Duffield and W. Whitt. Control and recovery from rare congestion events in a large multi-server system. *Queueing Syst.*, **26**:69–104, 1997.
19. N. Duffield and W. Whitt. A source traffic model and its transient analysis for network control. *Stochastic Models*, **14**:51–78, 1998.
20. A. Erramilli, O. Narayan, and W. Willinger. Experimental queueing analysis with long-range dependent packet traffic. *IEEE/ACM Trans. Networking*, **4**:209–223, 1996.
21. A. Erramilli, R. Singh, and P. Pruthi. An application of deterministic chaotic maps to model packet traffic. *Queueing Syst.*, **20**:171–206, 1995.
22. A. Feldmann, A. C. Gilbert, P. Huang, and W. Willinger. Dynamics of IP traffic: a study of the role of variability and the impact of control. In *Proc. ACM SIGCOMM '99*, pp. 301–313, 1999.
23. A. Feldmann, A. C. Gilbert, and W. Willinger. Data networks as cascades: investigating the multifractal nature of Internet WAN traffic. In *Proc. ACM SIGCOMM '98*, 42–55, 1998.
24. A. Feldmann, A. C. Gilbert, W. Willinger, and T. G. Kurtz. The changing nature of network traffic: scaling phenomena. *Comput. Commun. Rev.*, **28**: 5–29, 1998.
25. M. R. Frater, P. Tan, and J. F. Arnold. Variable bit rate video traffic on the broadband ISDN: modelling and verification. In J. Labetoulle and J. W. Roberts, eds, *The Fundamental Role of Teletraffic in the Evolution of Telecommunications Networks*, pp. 1351–1360, Elsevier, Amsterdam, The Netherlands, 1994.
26. M. Garret and W. Willinger. Analysis, modeling and generation of self-similar VBR video traffic. In *Proc. ACM SIGCOMM '94*, pp. 269–280, 1994.
27. R. J. Gibbens. Traffic characterization and effective bandwidths for broadband network traces. In F. P. Kelly, S. Zachary, and I. Ziedins, eds, *Stochastic Networks: Theory and Applications*, pp. 169–179. Clarendon Press, Oxford, 1996.
28. A. C. Gilbert, W. Willinger, and A. Feldmann. Scaling analysis of conservative cascades,

with applications to network traffic. *IEEE Trans. Information Theory,* **45**(3):971–991, 1999.

29. M. Grossglauser and J.-C. Bolot. On the relevance of long-range dependence in network traffic. In *Proc. ACM SIGCOMM '96*, pp. 15–24, 1996.
30. M. Harchol-Balter. Process lifetimes are not exponential, more like $1/t$: implications on dynamic load balancing. Technical report, EECS, University of California, Berkeley, 1996. CSD-94-826.
31. M. Harchol-Balter and A. Downey. Exploiting process lifetime distributions for dynamic load balancing. In *Proc. SIGMETRICS '96*, pp. 13–24, 1996.
32. D. Heyman and T. Lakshman. Source models for VBR broadcast video traffic. *IEEE/ACM Trans. Networking*, **5**(1): 40–48, Feb 1996.
33. D. Heyman and T. Lakshman. What are the implications of long-range dependence for VBR-video traffic engineering? *IEEE/ACM Trans. Networking*, **4**(3):301–317, June 1996.
34. C. Huang, M. Devetsikiotis, I. Lambadaris, and A. Kaye. Modeling and simulation of self-similar variable bit rate compressed video: a unified approach. In *Proc. ACM SIGCOMM '95*, pp. 114–125, 1995.
35. P. Jacquet. Analytic information theory in service of queueing with aggregated exponential on/off arrivals. In *Proc. 25th Allerton Conference on Communication, Control and Computing*, p. 242–251, 1996.
36. R. Jain and S. Routhier. Packet trains—measurements and a new model for computer network traffic. *IEEE J. Select. Areas Commun.*, **4**(6):986–995, 1986.
37. P. Jelenkovic and B. Melamed. Automated TES modeling of compressed video. In *Proc. IEEE INFOCOM '95*, pp. 746–752, 1995.
38. Leonard Kleinrock. *Queueing Systems, Volume 1: Theory.* Wiley-Interscience, New York, 1975.
39. T. G. Kurtz. Limit theorems for workload input models. In F. P. Kelly, S. Zachary, and I. Ziedins, eds, *Stochastic Networks: Theory and Applications*. Clarendon Press, Oxford, 1996.
40. W. E. Leland and T. J. Ott. UNIX process behavior and load balancing among loosely-coupled computers. In O. J. Boxma, J. W. Cohen, and H. C. Tijms, eds, *Teletraffic Analysis and Computer Performance Evaluation*, pp. 191–208, Elsevier, Amsterdam, The Netherlands, 1986.
41. W. Leland, M. Taqqu, W. Willinger, and D. Wilson. On the self-similar nature of Ethernet traffic. In *Proc. ACM SIGCOMM '93*, pp. 183–193, 1993.
42. W. E. Leland, M.S. Taqqu, W. Willinger, and D.V. Wilson. On the self-similar nature of Ethernet traffic (extended version). *IEEE/ACM Trans. Networking*, **2**:1–15, 1994.
43. N. Likhanov, B. Tsybakov, and N. Georganas. Analysis of an ATM buffer with self-similar ("fractal") input traffic. In *Proc. IEEE INFOCOM '95*, pp. 985–992, 1995.
44. R. Lukose and B. Huberman. Surfing as a real option. In *Proc. 1st International Conference on Information and Computation Economies*, pp. 45–51, 1998.
45. B. Mandelbrot and J. Van Ness. Fractional Brownian motions, fractional noises and applications. *SIAM Rev.* , **10**:422–437, 1968.
46. B. B. Mandelbrot. Long-run linearity, locally gaussian processes, h-spectra and infinite variances. *Int. Econom. Rev.*, **10**:82–113, 1969.

47. B. B. Mandelbrot. *The Fractal Geometry of Nature*. W. H. Freeman, New York, 1982.
48. D. Menasce and V. Almeida. *Capacity Planning for Web Performance: Metrics, Models, and Methods*. Prentice Hall, Englewood Cliffs, NJ, 1998.
49. I. Norros. A storage model with self-similar input. *Queueing Syst.*, 1**6**:387–396, 1994.
50. K. Park, G. Kim, and M. Crovella. On the relationship between file sizes, transport protocols, and self-similar network traffic. In *Proc. IEEE International Conference on Network Protocols*, pp. 171–180, 1996.
51. K. Park, G. Kim, and M. Crovella. On the effect of traffic self-similarity on network performance. In *Proc. SPIE International Conference on Performance and Control of Network Systems*, pp. 296–310, 1997.
52. K. Park and W. Wang. QoS-sensitive transport of real-time MPEG video using adaptive forward error correction. In *Proc. IEEE Multimedia Systems '99*, pp. 426–432, 1999.
53. M. Parulekar and A. Makowski. Tail probabilities for a multiplexer with self-similar traffic. In *Proc. IEEE INFOCOM '96*, pp. 1452–1459, 1996.
54. M. Parulekar and A. Makowski. $M/G/\infty$ input processes: a versatile class of models for traffic network. In *Proc. IEEE INFOCOM '97*, 1997.
55. M. Parulekar and A. Makowski. Tail probabilities for a multiplexer driven by $M/G/\infty$ input processes (i): preliminary asymptotics. *Queueing Syst.*, **27**:271–296, 1997.
56. V. Paxson and S. Floyd. Wide-area traffic: the failure of Poisson modeling. In *Proc. ACM SIGCOMM '94*, pp. 257–268, 1994.
57. V. Paxson and S. Floyd. Wide-area traffic: the failure of Poisson modeling. *IEEE/ACM Trans. Networking* **3**:226–244, 1995.
58. B. Ryu and A. Elwalid. The importance of long-range dependence of VBR video traffic in ATM traffic engineering: myths and realities. In *Proc. ACM SIGCOMM '96*, pp. 3–14, 1996.
59. B. Ryu. Fractal network traffic modeling: past, present, and future. In *Proc. 25th Allerton Conference on Communication, Control and Computing*, pp. 252–260, 1996.
60. G. Samorodnitsky and M. Taqqu. *Stable Non-Gaussian Random Processes: Stochastic Models with Infinite Variance*. Chapman and Hall, New York, 1994.
61. A. Shaikh, J. Rexford, and K. Shin. Load-sensitive routing of long-lived IP flows. In *Proc. ACM SIGCOMM '99*, pp. 215–226, 1999.
62. M. Taqqu, W. Willinger, and R. Sherman. Proof of a fundamental result in self-similar traffic modeling. *Comput. Commun. Rev.*, **26**:5–23, 1997.
63. M. S. Taqqu and J. B. Levy. Using renewal processes to generate long-range dependence and high variability. In E. Eberlein and M. S. Taqqu, eds, *Progress in Probability and Statistics, Vol. 11*. Birkhauser, Boston, 1996.
64. M. S. Taqqu, V. Teverovsky, and W. Willinger. Estimators for long-range dependence: an empirical study, 1995. Preprint.
65. B. Tsybakov and N. D. Georganas. On self-similar traffic in ATM queues: definitions, overflow probability bound and cell delay distribution. *IEEE/ACM Trans. Networking*, **5**(3):379–409, 1997.
66. B. Tsybakov and N.D. Georganas. Self-similar traffic and upper bounds to buffer overflow in an ATM queue. *Performance Evaluation*, **36**(1):57–80, 1998.

67. T. Tuan and K. Park. Multiple time scale congestion control for self-similar network traffic. *Performance Evaluation*, **36**:359–386, 1999.
68. T. Tuan and K. Park. Multiple time scale redundancy control for QoS-sensitive transport of real-time traffic. To appear in *Proc. IEEE INFOCOM '00*, 2000.
69. T. Tuan and K. Park. Performance evaluation of multiple time scale TCP under self-similar traffic conditions. Technical report, Dept. of Computer Sciences, Purdue University, 1999. CSD-TR-99-040.
70. A. Shwartz and A. Weiss. *Large Deviations for Performance Analysis*. Chapman and Hall, London, 1995.
71. W. Willinger. The discovery of self-similar traffic. In G. Haring, C. Lindemann and M. Reiser, eds, *Performance Evaluation: Origins and Directions*, LNCS. Springer-Verlag, New York (to appear).
72. W. Willinger and V. Paxson. Discussion of 'Heavy tail modeling and teletraffic data' by S. I. Resnick. *Ann. Statistics* **25**,1856–1866, 1998.
73. W. Willinger and V. Paxson. Where mathematics meets the Internet. *Notices of the AMS* **45**:961–970, 1998.
74. W. Willinger, M. Taqqu, R. Sherman, and D. Wilson. Self-similarity through high-variability: statistical analysis of Ethernet LAN traffic at the source level. In *Proc. ACM SIGCOMM '95*, pp. 100–113, 1995.
75. W. Willinger, M. Taqqu, and A. Erramilli. A bibliographical guide to self-similar traffic and performance modeling for modern high-speed networks. In F. P. Kelly, S. Zachary, and I. Ziedins, eds., *Stochastic Networks: Theory and Applications*, pp. 339–366, Clarendon Press, Oxford, UK, 1996.

2

WAVELETS FOR THE ANALYSIS, ESTIMATION, AND SYNTHESIS OF SCALING DATA

P. ABRY AND P. FLANDRIN

CNRS UMR 5672, École Normale Supérieure de Lyon, Laboratoire de Physique, 69 364 Lyon Cedex 07, France

M. S. TAQQU

Department of Mathematics, Boston University, Boston, MA 02215-2411

D. VEITCH

Software Engineering Research Centre, Carlton, Victoria 3053, Australia

2.1 THE SCALING PHENOMENA

2.1.1 Scaling Issues in Traffic

The presence of scaling behavior in telecommunications traffic is striking not only in its ubiquity, appearing in almost every kind of packet data, but also in the wide range of scales over which the scaling holds (e.g., see Beran et al. [18], Leland et al. [43], and Willinger et al. [78]). It is rare indeed that a physical phenomenon obeys a consistent law over so many orders of magnitude. This may well extend further, as increases in network bandwidth over time progressively "reveal" higher scales.

While the presence of scaling is now well established, its impact on teletraffic issues and network performance is still the subject of some confusion and uncertainty. Why is scaling in traffic important for networking? It is clear, as far as modeling of the traffic itself is concerned, that a feature as prominent as scaling

Self-Similar Network Traffic and Performance Evaluation, Edited by Kihong Park and Walter Willinger
ISBN 0-471-31974-0

should be built into models at a fundamental level, if these are to be both accurate and parsimonious. Scaling, therefore, has immediate implications for the choice of classes of traffic models, and consequently on the choice, and subsequent estimation, of model parameters. Such estimation is required for initial model verification, for fitting purposes, as well as for traffic monitoring.

Traffic modeling, however, does not occur in isolation but in the context of performance issues. Depending on the performance metric of interest, and the model of the network element in question, the impact and therefore the relevance of scaling behavior will vary. As a simple example, it is known that, in certain infinite buffer fluid queues fed by long-range-dependent (LRD) on/off sources, the stationary queueing distribution has infinite mean, a radically nonclassical result. Such infinite moments disappear, however, if the buffer is finite, intuitively because a finite reservoir cannot "hold" long memory. The long-range dependence of the input stream will strongly affect the overflow loss process but cannot seriously exacerbate the conditional delay experienced by packets that are *not* lost, as this is bounded by the size of the buffer. The importance of scaling in the performance sense, apart from being as yet unknown in a great many cases, is therefore context dependent.

We focus here on the fundamental issues of detection, identification, and measurement of scaling behavior. These cannot be ignored even if one is interested in *performance questions that are not directly related to scaling*. This is because scaling induces nonclassical statistical properties that affect the estimation of *all* parameters, not merely those that describe scaling. This, in turn, affects the predictive abilities of performance models and therefore their usefulness in practice.

The reliable detection of scaling should thus be our first concern. By detecting the absence or presence of scaling, one will know whether the data need be analyzed by using traditional statistics or by using special statistical techniques that take the presence of scaling into account. Here it is vital to be able to distinguish artifacts due to nonstationarities, with the appearance of scaling, from true scaling behavior. Identification is necessary since more than one kind of scaling exists, with differing interpretations and implications for model choice. Finally, should scaling of a given kind be present, an accurate determination of the parameters that describe it must be made. These parameters will control the statistical properties of estimates made of all other quantities, such as the parameters needed in traffic modeling or quality of service metrics.

As a simple yet powerful example of the above, consider a second-order process $X(t)$, which we know to be stationary, and whose mean μ_X we wish to estimate from a given data set of length n. For this purpose the simple sample mean estimator is a reasonable choice. The classical result is that asymptotically for large n the sample mean follows a normal distribution, with expectation equal to μ_X, and variance σ_X^2/n, where σ_X^2 is the variance of X. In the case where X is LRD the sample mean is also asymptotically normally distributed with mean μ_X; however, the variance is given by $[2c_r n^\alpha/(1+\alpha)\alpha](1/n)$, where $\alpha \in [0, 1)$ and $c_r \in (0, \infty)$ are the parameters describing the long-range dependence [17, p. 160]. This expression reveals that the variance of the sample mean decreases with the sample size n at a rate that is slower than in the classical case. Noting that the ratio of the size of the LRD-based variance to the classical one grows to infinity with n, it becomes apparent that confidence intervals based on traditional assumptions, even for a quantity as simple as the sample mean, can lead to serious errors when in fact the data are LRD.

We focus here on how a wavelet-based approach allows the threefold objective of the detection, identification, and measurement of scaling to be efficiently achieved. Fundamentally, this is due to the nontrivial fact that the analyzing wavelet family itself possesses a scale-invariant feature, a property not shared by other analysis methods. A key advantage is that quite different kinds of scaling can be analyzed by the same technique, indeed by the same set of computations. The semiparametric estimators of the scaling parameters that follow from the approach have excellent properties—negligible bias and low variance—and in many cases compare well even against parametric alternatives. The computational advantages, based on the use of the discrete wavelet transform (DWT), are very substantial and allow the analysis of data of arbitrary length. Finally, there are very valuable robustness advantages inherent in the method, particularly with respect to the elimination of superposed smooth trends (deterministic functions).

Another important issue connected with modeling and performance studies concerns the generation of time series for use in simulations. Such simulations can be particularly time consuming for long memory processes where the past exerts a strong influence on the future, disallowing simple approximations based on truncation. Wavelets offer in principle a parsimonious and natural way to generate good approximations to sample paths of scaling processes, which benefit from the same DWT-based computational advantages enjoyed by the analysis method. This area is less well developed than is the case for analysis, however.

2.1.2 Mapping the Land of Scaling and Wavelets

The remainder of the chapter is organized as follows.

Section 2.2, Wavelets and Scaling: Theory, discusses in detail the key properties of the wavelet coefficients of scaling processes. It starts with a brief, yet precise, introduction to the continuous and discrete wavelet transforms, to the multiresolution analysis theory underlying the latter, and the low complexity decomposition algorithm made possible by it. It recalls concisely the definitions of two of the main paradigms of scaling—self-similarity and long-range dependence. The properties of the wavelet coefficients of self-similar, long-range-dependent, and fractal processes are then given, and it is shown how the analysis of these various kinds of scaling can be gathered into a single framework within the wavelet representation. Extensions to more general classes of scaling processes requiring a collection of scaling exponents, such as multifractals, are also discussed.

The aim of Section 2.3, Wavelets and Scaling: Estimation, is to indicate how and why this wavelet framework enables the efficient analysis of scaling processes. This is achieved through the introduction of the *logscale diagram*, where the key analysis tasks of the detection of scaling—interpretation of the nature of scaling and estimation of scaling parameters—can be performed. Practical issues in the use of the logscale diagram are addressed, with references to examples from real traffic data and artificially generated traces. Definitions, statistical performance, and pertinent features of the estimators for scaling parameters are then studied in detail. The logscale diagram, first defined with respect to second-order statistical quantities, is then extended to statistics of other orders. It is also indicated how the tool allows for and deals with situations/processes departing from pure scaling, such as superimposed deterministic nonstationarities. Finally, clear connections between the wavelet tool and a number of more classical statistical tools dedicated to the analysis of scaling are drawn, showing how the latter can be profitably generalized in their wavelet incarnations.

Section 2.4, Wavelet and Scaling: Synthesis, proposes a wavelet-based synthesis of the fractional Brownian motion. It shows how this process can be naturally and efficiently expanded in a wavelet basis, allowing, provided that the wavelets are suitably designed, its accurate and computationally efficient implementation.

Finally, in Section 2.5, Wavelets and Scaling: Perspectives, a brief indication is given of what may lay ahead in the broad land of scaling and wavelets.

2.2 WAVELET AND SCALING: THEORY

2.2.1 Wavelet Analysis: A Brief Introduction

2.2.1.1 The (Continuous) Wavelet Transform The continuous wavelet decomposition (CWT) consists of the collection of coefficients

$$\{T_X(a, t) = \langle X, \psi_{a,t}\rangle, a \in \mathbb{R}^+, t \in \mathbb{R}\}$$

that compares (by means of inner products) the signal X to be analyzed with a set of analyzing functions

$$\left\{\psi_{a,t}(u) \equiv \frac{1}{\sqrt{a}}\psi_0\left(\frac{u-t}{a}\right), a \in \mathbb{R}^+, t \in \mathbb{R}\right\}.$$

This set of analyzing functions is constructed from a reference pattern ψ_0, called the mother wavelet, by the action of a time-shift operator $(\mathscr{T}_\tau\psi_0)(t) \equiv \psi_0(t-\tau)$ and a dilation (change of scale) operator

$$(\mathscr{D}_a\psi_0)(t) \equiv 1/\sqrt{a}\psi_0(t/a).$$

ψ_0 is chosen such that both its spread in time and frequency are relatively limited. It consists of a small wave defined on a support, which is almost limited in time and having most of its energy within a limited frequency band. While the time support and frequency band cannot both be finite, there is an interval on which they are *effectively* limited. The time-shift operator enables the selection of the time instant around which one wishes to analyze the signal, while the dilation operator defines the scale of time (or, equivalently, the range of frequencies) over which it will be observed. The quantity $|T_X(a, t)|^2$, referred to as a "scalogram," can therefore be interpreted as the energy content of X around time t within a given range of frequencies controlled by a. In addition to being well localized in both time and frequency, the mother wavelet is required to satisfy the *admissibility condition*, whose weak form is

$$\int \psi_0(u)du = 0, \tag{2.1}$$

which shows it is a bandpass or oscillating function, hence the name "wavelet."

Wavelets that are often used in practice include the Haar wavelet, the Daubechies wavelets, indexed by a parameter $N = 1, 2, \ldots,$ and the Meyer wavelets. The Haar wavelet $\psi_0(u)$ is discontinuous; it equals 1 at $0 \le u < \frac{1}{2}$, -1 at $\frac{1}{2} \le u \le 1$, and 0 otherwise. The Daubechies wavelet with $N = 1$ is in fact that the Haar wavelet, but the other Daubechies wavelets with $N > 1$ are continuous with bounded support and have N vanishing moments (i.e., they satisfy Eq. (2.5)). The Meyer wavelets do not have bounded support, in neither the time nor frequency domain, but all their moments vanish and they belong to the Schwartz space; that is, they are infinitely differentiable and decrease very rapidly to 0 as u tends to $\pm\infty$.

On the condition that the wavelet be admissible, the transform can be inverted:

$$X(t) = C_\psi \int\int T_X(a, \tau)\psi_{a,\tau}(t)\frac{da\, d\tau}{a^2}$$

where C_ψ is a constant depending on ψ_0. This reconstruction formula expresses X in terms of a weighted integral of wavelets (acting as elementary atoms) located around given times and frequencies, thereby constituting quanta of information in the time–frequency plane. For a more general presentation of the wavelet analysis see, for example, Daubechies [24].

Because the wavelet transform represents in a plane (i.e., a two-dimensional (2D) space) the information contained in a signal (i.e., one-dimensional (1D) space), it is a redundant transform, which means that neighboring coefficients in the time–scale plane share a certain amount of information. A mathematical theory, the *multiresolution analysis* (MRA), proves that it is possible to critically sample the time–scale plane, that is, to keep, among the $\{T_X(a, t), a \in \mathbb{R}^+, t \in \mathbb{R}\}$, only a discrete set of coefficients while still retaining the total information in X. That procedure defines the so-called discrete (or nonredundant) wavelet transform.

2.2.1.2 Multiresolution Analysis and Discrete Wavelet Transform A multiresolution analysis (MRA) consists of a collection of nested subspaces $\{V_j\}_{j\in\mathbb{Z}}$, satisfying the following set of properties [24]:

1. $\bigcap_{j\in\mathbb{Z}} V_j = \{0\}$, $\bigcup_{j\in\mathbb{Z}} V_j$ is dense in $L^2(\mathbb{R})$.
2. $V_j \subset V_{j-1}$.

3. $X(t) \in V_j \Leftrightarrow X(2^j t) \in V_0$.
4. There exists a function $\phi_0(t)$ in V_0, called the *scaling function*, such that the collection $\{\phi_0(t-k),\ k \in \mathbb{Z}\}$ is an unconditional Riesz basis for V_0.

To understand the significance of these properties, observe that, from Property 1, the V_j are approximation subspaces of the space of square integrable functions $L^2(\mathbb{R})$. Property 4 expresses the fact that the set of shifted scaling functions $\{\phi_0(t-k),\ k \in \mathbb{Z}\}$ form a "Riesz basis" for V_0; that is, they are linearly independent and span the space V_0, but they are not necessarily orthogonal nor do they have to be of unit length. Finding such a function $\phi_0(t)$ is hard, but many candidates for $\phi_0(t)$ are known in the literature.

Similarly, Properties 3 and 4 together imply that the scaled and shifted functions

$$\{\phi_{j,k}(t) = 2^{-j/2}\phi_0(2^{-j}t-k),\ k \in \mathbb{Z}\}$$

constitute a Riesz basis for the space V_j. The multiresolution analysis involves successively projecting the signal X to be studied into each of the approximation subspaces V_j:

$$\text{approx}_j(t) = (\text{Proj}_{V_j}X)(t) = \sum_k a_X(j,k)\phi_{j,k}(t).$$

Since, from Property 2, $V_j \subset V_{j-1}$, approx_j is a coarser approximation of X than is approx_{j-1}. (Note that some authors use the opposite convention and set $V_j \subset V_{j+1}$.) Property 1 moreover indicates that in the limit of $j \to +\infty$, all information is removed from the signal. The key idea of the MRA, therefore, consists in studying a signal by examining its coarser and coarser approximations, by canceling more and more high frequencies or details from the data.

The information that is removed when going from one approximation to the next, coarser one is called the detail:

$$\text{detail}_j(t) = \text{approx}_{j-1}(t) - \text{approx}_j(t).$$

The MRA shows that the detail signals detail_j can be obtained directly from projections of X onto a collection of subspaces, the $W_j = V_j \ominus V_{j-1}$, called the wavelet subspaces. Moreover, the MRA theory shows that there exists a function ψ_0, called the mother wavelet, to be derived from ϕ_0, such that its templates

$$\{\psi_{j,k}(t) = 2^{-j/2}\psi_0(2^{-j}t-k),\ k \in \mathbb{Z}\}$$

constitute a Riesz basis for W_j:

$$\text{detail}_j(t) = (\text{Proj}_{W_j}X)(t) = \sum_k d_X(j,k)\psi_{j,k}(t).$$

For example, if the scaling function $\phi_0(t)$ is the function that equals 1 if $0 \le t \le 1$ and 0 otherwise, then the corresponding mother wavelet $\psi_0(u)$ is the Haar wavelet.

Theoretically, this projection procedure can be performed from $j \to -\infty$ up to $j \to +\infty$. *In practice*, one limits the range of indices j to $j = 0, \ldots, J$ and thus only considers

$$V_J \subset V_{J-1} \subset \cdots \subset V_0.$$

This means that we restrict the analysis of X to that of its (orthogonal) projection $\text{approx}_0(t)$ onto the reference space V_0, labeled as zero by convention, and rewrite this fine scale approximation as a collection of details at different resolutions together with a final low-resolution approximation that belongs to V_J:

$$\begin{aligned}\text{approx}_0(t) &= \text{approx}_J(t) + \sum_{j=1}^{J} \text{detail}_j(t) \\ &= \sum_k a_X(J, k)\phi_{J,k}(t) + \sum_{j=1}^{J} \sum_k d_X(j, k)\psi_{j,k}(t). \end{aligned} \tag{2.2}$$

If X is in V_0, one can obviously replace approx_0 by X in the above relation.

Except in the case where X actually belongs to V_0, selecting V_0 implies some unavoidable information loss [11]. This is entirely analogous to the loss induced by the necessary prefiltering operation involved in Shannon–Whittaker sampling theory to band-limit a process prior to sampling. Note, however, that there is no additional information loss after the initial projection. Varying J simply means deciding if more or less information is written in details as opposed to the final approximation approx_J.

Since the approx_j are essentially coarser and coarser approximations of X, ϕ_0 needs to be a lowpass function. The detail_j, being an information "differential," indicates rather that ψ_0 is a bandpass function, and therefore a small wave, a *wavelet*. More precisely, the MRA shows that the mother wavelet must satisfy $\int \psi_0(t)\,dt = 0$ [24].

Given a scaling function ϕ_0 and a mother wavelet ψ_0, the discrete (or non-redundant) wavelet transform (DWT) consists of the collection of coefficients

$$X(t) \to \{\{a_X(J, k), k \in \mathbb{Z}\}, \{d_X(j, k), j = 1, \ldots, J, k \in \mathbb{Z}\}\}. \tag{2.3}$$

These coefficients are defined through inner products of X with two sets of functions:

$$\begin{aligned} a_X(j, k) &= \langle X, \mathring{\phi}_{j,k} \rangle, \\ d_X(j, k) &= \langle X, \mathring{\psi}_{j,k} \rangle, \end{aligned} \tag{2.4}$$

where $\mathring{\psi}_{j,k}$ (resp., $\mathring{\psi}_{j,k}$) are shifted and dilated templates of $\{\mathring{\psi}$ (resp., $\mathring{\psi}_0$), called the dual mother wavelet (resp., the dual scaling function), and whose definition depends on whether one chooses to use an orthogonal, semiorthogonal, or biorthogonal DWT (e.g., see Daubechies [24]). In Eqs. (2.2) and (2.4), the role of the wavelet and its dual can arbitrarily be exchanged, and similarly for the scaling function and its dual. In what follows this exchange is performed for simplicity of notation. The $d_X(j, k)$ constitute a subsample of the $\{T_X(a, t),\ a \in \mathbb{R}^+,\ t \in \mathbb{R}\}$, located on the so-called dyadic grid,

$$d_X(j, k) = T_X(2^j, 2^j k).$$

The logarithm (base 2) of the scale $\log_2(a = 2^j) = j$ is called the *octave* j, and a scale will often be referred to by its corresponding octave. For the sake of clarity, we henceforth restrict our presentation to the DWT (characterized by the $d_X(j, k)$), which brings with it considerable computational advantages. However, the fundamental results based on the wavelet approach hold for the CWT; see Abry et al. [3, 4].

2.2.1.3 Key Features of the Wavelet Transform In the study of the scaling processes analyzed below, the following two features of the wavelet transform play key roles:

- **F1:** The wavelet basis is constructed from the dilation (change of scale) operator, so that the analyzing family itself exhibits a scale-in-variance feature.
- **F2:** ψ_0 has a number $N \geq 1$ of *vanishing moments*:

$$\int t^k \psi_0(t)\, dt \equiv 0, \quad k = 0, 1, 2, \ldots, N-1. \tag{2.5}$$

The value of N can freely be chosen by selecting the mother wavelet ψ_0 accordingly. The Fourier transform $\Psi_0(\nu)$ of ψ_0 satisfies $|\Psi_0(\nu)| \approx |\nu|^N$, $|\nu| \to 0$ [24].

2.2.1.4 Fast Pyramidal Algorithm In all of what follows, we always assume that we are dealing with continuous time stochastic processes, and therefore that the wavelet (and approximation) coefficients are defined through continuous time inner products (Eq. (2.4)). One major consequence of the nested structure of the MRA consists in the fact that the $d_X(j, k)$ and the $a_X(j, k)$ can actually be computed through a discrete time convolution involving the sequence $a_X(j-1, k)$ and two discrete time filters h_1 and g_1. The DWT can therefore be implemented using a recursive filter-bank-based pyramidal algorithm, as sketched on Fig. 2.1, which has a lower computational cost than that of a fast Fourier transform (FFT) [24]. The coefficients of the filters h_1 and g_1 are to be derived from ϕ_0 and ψ_0 [24]. The use of the discrete time algorithm to compute the continuous time inner products $d_X(j, k) = \langle X, \psi_{j,k} \rangle$ requires an initialization procedure. It amounts to computing an initial discrete time sequence to feed the algorithm (see Fig. 2.1): $a_X(0, k) = \langle X, \phi_{0,k} \rangle$, which corresponds to the coefficients of the expansion of the projection of X on V_0. From a practical point of view, one deals with sampled versions of X, which implies that the initialization stage has to be approximated. More details can be found in Delbeke and Abry [27] and Veitch and Abry [75]. The fast pyramidal algorithm is not only scalable because of its linear complexity, $O(n)$ for data of length n, but is simple enough to implement on-line and in real time in high-speed packet networks. An on-line wavelet-based estimation method for the scaling parameter with small memory requirements is given by Roughan et al. [62].

2.2.2 Scaling Processes: Self-Similarity and Long-Range Dependence

We can define *scaling behavior* broadly as a property of scale invariance, that is, when there is no controlling characteristic scale or, equivalently, when all scales have equal importance. There is no one simple definition that can capture all systems or processes with this property; rather there are a set of known classes open to

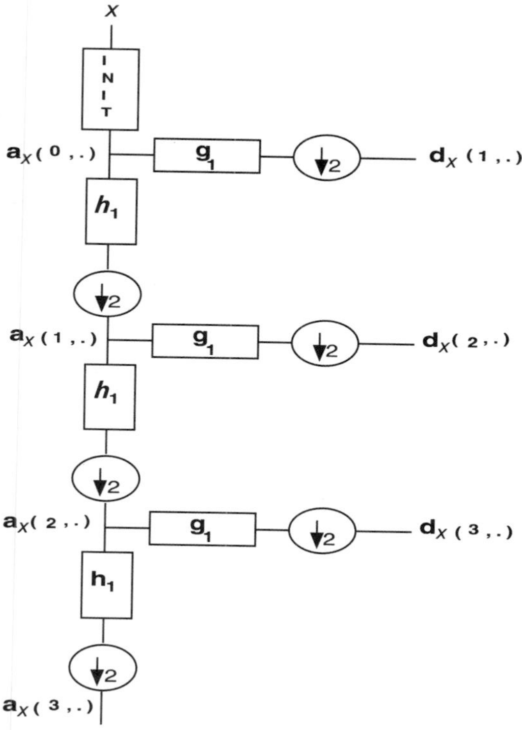

Fig. 2.1 Fast filter-bank-based pyramidal algorithm. The DWT can be computed using a fast pyramidal algorithm: that is, given that we have approximation $a_X(j-1,k)$ at level $j-1$, we obtain approximation $a_X(j,k)$ and detail $d_X(j,k)$ at level j by convolving with h_1 and g_1, respectively, and decimating. The coefficients of the filters h_1 and g_1 are derived from the chosen scaling function and wavelet ϕ_0 and ψ_0. The downarrow stands for a decimation by a factor of 2 operation: one drops the odd coefficients. An initialization step is required to go from the process X to the approximation of order 0: $a_X(0,k)$.

expansion. In this section we briefly introduce the most well known of these, namely, *self-similar*, self-similar with stationary increments, and *long-range-dependent* processes. Please note that throughout this chapter we will use the following convention: $f(x) \sim g(x)$ as $x \to a$ means that $\lim_{x\to a} f(x)/g(x) = 1$, and $f(x) \approx g(x)$ as $x \to a$ means that $\lim_{x\to a} f(x)/g(x) = C$, where C is some finite constant.

Recall that a process $X = \{X(t),\ t \in \mathbb{R}\}$ is *self-similar with parameter* $H > 0$ (H-ss) if $X(0) = 0$ and $\{X(ct),\ t \in \mathbb{R}\}$ and $\{c^H X(t),\ t \in \mathbb{R}\}$ have the same finite-dimensional distributions. Such a process, obviously, cannot be stationary. The process X is H-sssi if it is H-ss and if, in addition, it has stationary increments, that is, if the finite-dimensional distributions of its increments $\{X(t+h) - X(t),\ t \in \mathbb{R}\}$ do not depend on t. An H-sssi process with $H < 1$ has zero mean and a variance that behaves as $\mathbb{E}X^2(t) = \sigma^2|t|^{2H}$. The *fractional Brownian motion* (FBM), for example, is the (unique) Gaussian H-sssi process, which is simply Brownian motion for $H = \frac{1}{2}$.

Long-range dependence,[1] on the other hand, is associated with *stationary* processes. A stationary finite-variance process X displays *long-range dependence* if its spectral density $\Gamma_X(\nu)$ satisfies

$$\Gamma_X(\nu) \sim c_f |\nu|^{-\alpha} \quad \text{as } \nu \to 0, \tag{2.6}$$

where $0 < \alpha < 1$ and where c_f is a nonzero constant.[2] Equation (2.6) implies that the autocovariance $r(k) = \mathbb{E}Z(j)Z(j+k)$ satisfies

$$r(k) \sim c_r k^{\alpha-1} \quad \text{as } k \to \infty, \tag{2.7}$$

where $c_r = c_f 2\Gamma(1-\alpha)\sin(\pi\alpha/2)$, Γ being (here) the Gamma function [17, p. 43]. Equation (2.7) and (2.6) imply that the covariances $r(k)$ decay so slowly, that $\sum_{k=-\infty}^{\infty} r(k) = \infty$, or equivalently, $\Gamma_Z(0) = \infty$.

There is a close relationship between long-range dependence and self-similar processes. Indeed, the increments of any finite variance H-sssi process have long-range dependence, as long as $\frac{1}{2} < H < 1$, with H and α related through

$$\alpha = 2H - 1. \tag{2.8}$$

In particular, fractional Gaussian noise (FGN), which is the increment process of fractional Brownian motion[3] (FBM) [50] with $\frac{1}{2} < H < 1$, has long-range depen-

[1]Long-range dependence is sometimes referred to as "long memory" or "second-order asymptotic self-similarity."

[2]The index f indicates that this constant is in force in the frequency domain. The corresponding constant appearing in the autocovariance is denoted c_r. One can also replace these constants by slowly varying functions but for the sake of simplicity, we will not do this here.

[3]Discrete standard FGN is the time series $X(j) = B_H(j+1) - B_H(j), j = 0, 1, \ldots,$ where B_H is FBM. Its spectral density satisfies $\Gamma_X(-\nu) = \Gamma_X(\nu)$, and because it is a discrete-time sequence, $\Gamma_X(\nu)$ is concentrated on the interval $[-\frac{1}{2}, \frac{1}{2}]$.

dence. FGN is close to an "ideal" model because its spectral density is-close to $\nu^{1-2H} \equiv \nu^{-\alpha}$ for a large range of frequencies ν in the interval $[0, \frac{1}{2}]$, and because its correlation function,

$$r(k) = \tfrac{1}{2}\{(k+1)^{2H} - 2k^{2H} + |k-1|^{2H}\}, \tag{2.9}$$

is invariant under *aggregation* (see Section 2.3.5.1).

We now recall the properties of the wavelet coefficients of H-sssi processes (such as FBM) and LRD processes (such as FGN) and show that they can be gathered into a unified framework. We subsequently show that other stochastic processes exhibiting scaling behavior also fit into this framework, opening up the prospect of a single approach covering diverse forms of scaling.

2.2.3 Wavelet Transform of Scaling Processes

2.2.3.1 Discrete Wavelet Transform of Stochastic Processes Whereas the wavelet theory was first established for deterministic finite-energy processes, it has clearly been demonstrated in the literature that the wavelet transform can be applied to stochastic processes; for example, see Cambanis and Houdré [20] and Masry [49]. More specifically, for the second-order random processes of interest here, it is well known that the wavelet transform is a second-order random field, on the condition that the scaling function ϕ_0 (and hence the wavelet ψ_0) satisfy certain mild conditions [20, 49] related to the covariance structure of the analyzed process. We will assume hereafter that the scaling functions and wavelets decay at least exponentially fast in the time domain, so that the second-order statistics of the wavelet transform exist for all of the random processes we discuss here.

2.2.3.2 Wavelet Transform (WT) of H-ss and H-sssi Processes Let X be an H-ss process. Its wavelet coefficients $d_X(j,k)$ exactly reproduce the self-similarity through the following central scaling property; see Delbeke [25] and Delbeke and Abry [26] or Pesquet-Popescu [57]:

- **P0 SS:** For the DWT, $d_X(j,k) = \langle X, \psi_{j,k}\rangle$, so that

$$\begin{aligned}(d_X(j,0), d_X(j,1), \ldots, d_X(j,N_j-1))&\\ \stackrel{d}{=} 2^{j(H+1/2)}(d_X(0,0), d_X(0,1), \ldots, d_X(0,N_j-1)).&\end{aligned} \tag{2.10}$$

  For the CWT, $T_X(a,t) = \langle X, \psi_{a,t}\rangle$, and hence

$$(T_X(ca, ct_1), \ldots, T_X(ca, ct_n)) \stackrel{d}{=} c^{H+1/2}(T_X(a,t_1), \ldots, T_X(a,t_n)), \quad \forall c > 0.$$

These equations mimic the self-similarity of the process. Let us emphasize that this, nontrivially, results from the fact that the analyzing wavelet basis is designed from the dilation operator and is therefore, by nature, scale invariant (**F1**). For second-order processes, a direct consequence of Eq. (2.10) is

$$\mathbb{E}d_X(j,k)^2 = 2^{j(2H+1)}\mathbb{E}d_X(0,k)^2. \tag{2.11}$$

Moreover, if we add the requirement that X has stationary increments (i.e., X is H-sssi), ingredients **F1** and **F2** combine, resulting in:

- **P1 SS:** The wavelet coefficients with fixed scale index $\{d_X(j,k), k \in \mathbb{Z}\}$ form a stationary process.

  This follows from the stationary increments property of the analyzed processes [20, 25, 49]. This property is not trivial, given that self-similar processes are nonstationary processes, and is a consequence of $N \geq 1$ (**F2**). In this case, Eq. (2.11) reduces to the fundamental result:

$$\mathbb{E}d_X(j,k)^2 = 2^{j(2H+1)}C(H,\psi_0)\sigma^2, \quad \forall k, \tag{2.12}$$

  with $C(H,\psi_0) = \int |t|^{2H}(\int \psi_0(u)\psi_0(u-t)\,du)\,dt$ and $\sigma^2 = \mathbb{E}X(1)^2$.
- **P2 SS:** Using the specific covariance structure of an H-sssi process $X(t)$, namely,

$$\mathbb{E}X(t)X(s) = \frac{\sigma^2}{2}\{|t|^{2H} + |s|^{2H} - |t-s|^{2H}\}, \tag{2.13}$$

  it can be shown [32, 73] that the correlations between wavelet coefficients located at different positions is extremely small as soon as $N \geq H + \frac{1}{2}$ and their decay can be controlled by increasing N:

$$\mathbb{E}d_X(j,k)\,d_X(j',k') \approx |2^j k - 2^{j'}k'|^{2H-2N}, \; |2^j k - 2^{j'}k'| \to +\infty. \tag{2.14}$$

These two results have been obtained and illustrated originally in the case of the FBM [31–34] (see also Tewfik and Kim [73]) and have been stated in more general contexts [20, 25, 26, 49].

2.2.3.3 *WT of LRD Processes* Let X be a second order *stationary* process, its wavelet coefficients $d_X(j,k)$ satisfy the following:

- **P0 LRD:**

$$\mathbb{E}d_X(j,k)^2 = \int \Gamma_X(\nu)2^j|\Psi_0(2^j\nu)|^2\,d\nu \tag{2.15}$$

where $\Gamma_X(\nu)$ and $\Psi_0(\nu)$ stand for the power spectrum of X and the Fourier transform of ψ_0, respectively. This can be understood as the classical *interference* formula of the linear filter theory and receives a spectral estimation interpretation: $\mathbb{E}d_X(j,k)^2$ is a measure of $\Gamma_X(\cdot)$ at frequency $\nu_j = 2^{-j}\nu_0$ (ν_0 depends on ψ_0) through the constant relative bandwidth wavelet filter [1–3, 34].

In the specific context of LRD processes, **F1** and **F2** together yield the two following key properties:

- **P1 LRD:** Using $\Gamma_X(\nu) \sim c_f|\nu|^{-\alpha}$, $\nu \to 0$ (2.15), we obtain

$$\mathbb{E}d_X(j,k)^2 \sim 2^{j\alpha}c_f C(\alpha,\psi_0), \quad j \to +\infty, \tag{2.16}$$

  where $C(\alpha,\psi_0) = \int |\nu|^{-\alpha}|\Psi_0(\nu)|^2\, d\nu$, $\alpha \in (0,1)$. The case of $\alpha = 0$ is well defined, corresponding to trivial scaling at large scales, leaving only short-range dependence at small scales. Again, this asymptotic recovering of the underlying power law is not a trivial result. It would not, for instance, be obtained with periodogram-based estimates [3] and is due to **F1**.
- **P2 LRD:** It can also be shown [3] that the covariance function of any two wavelet coefficients is controlled by N and therefore can decay much faster than that of the LRD process itself and is no longer LRD as soon as $N \geq \alpha/2$. Since $\alpha \in [0,1)$, this is in fact always satisfied.

$$\mathbb{E}d_X(j,k)\, d_X(j',k') \approx |2^j k - 2^{j'}k'|^{\alpha-1-2N},\, |2^j k - 2^{j'}k'| \to +\infty. \tag{2.17}$$

Observe that the exponents in **P1 LRD** and **P2 LRD** are different from those in **P1 SS** and **P2 SS**, respectively.

2.2.3.4 WT of Generalized Scaling Processes The results above can be generalized in a straightforward manner to processes that are neither strictly H-sssi nor LRD but whose wavelet coefficients share *equivalent scaling* properties. Some important cases are detailed here.

- Start with a H-sssi process X, and define Y as

$$Y(t) = \underbrace{\int_0^t dt_{p-1} \int_0^{t_{p-1}} \cdots dt_1 \int_0^{t_1} du}_{p\text{-integrals}}\ X(u).$$

Then Y is a H_Y-ss process with self-similarity parameter $H_Y = H + p$ and with stationary increments of order $p+1$. We say that Z is the pth-order ($p > 0$) increment process of Y if $Z(t) = Y^{(p-1)}(t+1) - Y^{(p-1)}(t)$ and $Y^{(p-1)}(t) = d^{(p-1)}Y/dt^{(p-1)}$ (note that we use such a "mixed" definition because an H-sssi process (i.e., with $0 < H < 1$) is not differentiable, whereas its integrals are). Then, properties **P1 SS** and **P2 SS** still hold replacing H by H_Y. The condition for **P1 SS** becomes $N \geq p + 1$ [10] and can be rewritten as $N \geq H_Y$ [10]. We hereafter say that X is an H-sssi(p) process if it is H-ss and has stationary increments of order $p + 1$. Note that with this definition H-sssi($p = 0$) and H-sssi are equivalent.

- Let X be a second-order stationary $1/f$*-type process*; that is, $\Gamma_X(\nu) = c_f|\nu|^{-\alpha}$, $\nu_1 \leq |\nu| \leq \nu_2$, $\alpha \geq 0$. Note that the term $1/f$ implicitly implies the physicist point of view, where the power-law behavior is supposed to hold for a wide range of frequencies, that is, $\nu_1 \ll \nu_2$. Recall that the mother wavelet is a bandpass function whose frequency content is essentially concentrated between ν_A and ν_B and negligible elsewhere, if nonzero. In the case of $1/f$ processes, it is therefore assumed that $|\nu_2 - \nu_1| \gg |\nu_B - \nu_A|$. We henceforth have

$$\mathbb{E}d_X(j,k)^2 \simeq \int_{2^{-j}\nu_A < |\nu| < 2^{-j}\nu_B} \Gamma_X(\nu)2^j|\Psi(2^j\nu)|^2\, d\nu.$$

This means that for all j's such that $\nu_1 \leq 2^{-j}\nu_A \leq 2^{-j}\nu_B \leq \nu_2$, the wavelet coefficients of X will reproduce the power law: $\mathbb{E}d_X(j,k)^2 \simeq 2^{j\alpha}c_f C(\alpha, \psi_0)$. Strictly speaking, this last relation holds for wavelets whose frequency support is finite, but it is generally valid to an excellent approximation. $1/f$-type processes with $\alpha < 1$ and $\nu_1 \equiv 0$ can be seen as the special case of LRD processes. Note that the definition of $1/f$ processes naturally extends to include $\alpha < 0$.

- Let X be such that $\Gamma_X(\nu) \sim c_f|\nu|^{-\alpha}$, $\nu \to 0$, $\alpha \geq 0$. For $\alpha \geq 1$, the variance does not exist (the integral of the spectrum diverges). X can, however, be seen as a generalized second-order stationary $1/f$-type process, in the sense that the variance of the wavelet coefficients remains finite,

$$\mathbb{E}d_X(j,k)^2 = \int \Gamma_X(\nu)2^j|\Psi_0(2^j\nu)|^2\, d\nu = 2^{j\alpha}c_f \int |\nu|^{-\alpha}|\Psi_0(\nu)|^2\, d\nu < \infty,$$

on condition that $N > (\alpha - 1)/2$. This is possible as the power-law decrease of the spectrum of the wavelet at the origin $|\Psi_0(\nu)| \approx \nu^N$, $|\nu| \to 0$ balances the divergence of $\Gamma_X(\nu)$ (see Abry et al. [3, 4] for details). Then, just as before, we have $\mathbb{E}d_X(j,k)^2 \sim 2^{j\alpha}c_f C(\alpha, \psi_0)$, $j \to +\infty$.

- Let X be such that $\Gamma_X(\nu) \sim c_f|\nu|^{-\alpha}$, $\nu \to \infty$, $\alpha \geq 1$, (i.e., $\nu_2 = \infty$). Its autocovariance function reads $\mathbb{E}X(t)X(t+\tau) \sim \sigma^2(1 - C|\tau|^{2h})$, $\tau \to 0$, with $h = (\alpha - 1)/2$. Equivalently, it implies that $\mathbb{E}(X(t+\tau) = X(t))^2 \approx |\tau|^{2h}$, $\tau \to 0$. If X is moreover Gaussian, this implies that the sample path of each realization of the process is fractal, with fractal dimension (strictly speaking Hausdorff dimension) $D = (5 - \alpha)/2$ [28]. This means that the local regularity of the sample path of the process or, equivalently, its local correlation structure exhibits scaling behavior. Such processes are called *fractal*.

Fractality is reproduced in the wavelet domain (generalization of **P1**) through $\mathbb{E}d_X(j,k)^2 \approx 2^{j(2h+1)}$, $j \to -\infty$, or equivalently for the CWT: $\mathbb{E}|T_X(a,t)|^2 \approx a^{2h+1}$, $a \to 0$ [35, 26], which allows an estimation of the fractal dimension through that of the scaling exponent $\alpha = 2h + 1 = 5 - 2D$.

2.2.3.5 Summary for Scaling Processes Let X be either an H-sssi(p) process, or a LRD process, or a (possibly generalized) second-order stationary $1/f$-type process or a fractal process. Then the wavelet coefficient, due to the combined effects of **F1** and **F2**, will exhibit the two following properties, which will play a key role in the estimation of the scaling exponent presented below:

- **P1:** The $\{d_X(j,k), k \in \mathbb{Z}\}$ is a stationary process if $N \geq (\alpha - 1)/2$ and the variance of the $d_X(j,k)$ accurately reproduces, *within a given range of octaves* $j_1 \leq j \leq j_2$, the underlying scaling behavior of the data:

$$\mathbb{E}d_X(j,k)^2 = 2^{j\alpha} c_f C(\alpha, \psi_0), \tag{2.18}$$

  where

  (i) in the case of an H-sssi(p) process, $\alpha = 2H + 1$, $C(\alpha, \psi_0)$ is to be identified from Eq. (2.12), and $j_1 = -\infty$ and $j_2 = +\infty$;

  (ii) in the case of an LRD process, α is defined as in Eq. (2.6), $C(\alpha, \psi_0)$ is to be identified from Eq. (2.16), and $j_2 = +\infty$ and j_1 is to be identified from the data;

  (iii) in the case of a (generalized) second-order stationary $1/f$-type process, α is defined from $\Gamma_X(\nu) = c_f|\nu|^{-\alpha}$, $\nu_1 \leq |\nu| \leq \nu_2$, $C(\alpha, \psi_0) = \int |\nu|^{-\alpha} |\Psi_0(\nu)|^2 \, d\nu$, and (j_1, j_2) are to be derived from (ν_1, ν_2);

  (iv) in the case of a fractal process, $\alpha = 2h + 1$, expressions for $C(\alpha, \psi_0)$ can be found in Flandrin and Gonçalvès [35, 36] and $j_1 = 1$ and j_2 is to be identified from the data.

- **P2:** $\{d_X(j,k), k \in \mathbb{Z}\}$ is stationary and no longer exhibits long-range statistical dependences but only short-term residual correlations; that is, it is short-range dependent (SRD) and not LRD, on condition that $N \geq \alpha/2$. Moreover, the higher N the shorter the correlation:

$$\mathbb{E}d_X(j,k)\, d_X(j,k') \approx |k - k'|^{\alpha - 1 - 2N}, \; |k - k'| \to +\infty.$$

Note that these two properties of the wavelet coefficients do not rely on an assumption of Gaussianity. In **P2** above, we used only weak reformulations (setting $j = j'$) of **P2 SS** and **P2 LRD**. Their general versions (j not necessarily equal to j') can be used to formulate a stronger idealization of strict decorrelation:

ID1: $\mathbb{E}d_X(j,k)\, d_X(j',k') = 0$ if $(j',k') \neq (j,k)$.

The relevance of this idealization has already been illustrated by, for instance Abry et al. [3], Abry and Veitch [5], and Flandrin [32, 33], and will play a key role in the next section.

2.2.3.6 Multiple Exponents, Multifractal Processes

Property **P1** (wavelet reproduction of the power law) extends further to classes of generalized scaling processes whose behavior cannot be described by a single scaling exponent, but which requires a collection, even an infinite collection, of exponents. We briefly describe three classes of examples.

The first example is in the spirit of the simple fractal processes described in Section 2.2.3.4. Consider a generalization where the exponent h, which describes the statistics of local scaling properties, is no longer constant in time: $\mathbb{E}(X(t+\tau)-X(t))^2 \approx |\tau|^{2h(t)}$, $\tau \to 0$. One consequence is that the local regularity of sample paths is no longer uniform but depends on t. A class of processes called *multifractional Brownian motion* has been proposed [56], which satisfies such a property, with h being a continuous function of t. As detailed in Flandrin and Gonçalvès [35, 36] the time evolution of h can be traced through an analysis of the continuous wavelet transform coefficients at small scales: $\mathbb{E}|T_X(a,t)|^2 \approx a^{2h(t)+1}$, $a \to 0$. This relation is to be understood as a time-dependent generalization of **P1**.

The second class, *multifractal processes*, is one that allows an extremely rich scaling structure at small scales, far richer than simply fractal in general. There is not the space here to give precise definitions of such processes, nor of the related *multifractal formalism*. We aim rather to give some intuition of their relation to wavelets and refer the reader to Riedi [59] and Riedi et al. [60] and to Chapter 20 of the present volume, and references therein, for a thorough presentation. For multifractal processes, the local regularity of almost every (i.e., with probability one) sample path, which we write as $|X(\omega,t+\tau)-X(\omega,t)| \approx |\tau|^{h(\omega,t)}$, $\tau \to 0$ (where ω denotes an element of the probability space underlying the process), exhibits an extraordinary variability over time; indeed, it is itself fractal-like. One therefore abandons the idea of *following* the time variations of h, since this is realization dependent and in any case is too complex, and instead studies it statistically. Classically this has been done through the *Hausdorff multifractal spectrum* $D(h)$, which consists of the *Hausdorff dimension* of the set of points where $h(\omega,t)=h$. The same multifractal spectrum is obtained for almost all realizations and is therefore a useful invariant describing the scaling properties of the process. A classical tool to obtain the multifractal spectrum is to calculate, from any typical sample path, the *structure functions* or *partion functions*: $S_q(\tau) = \int |X(\omega,t+\tau)-X(\omega,t)|^q \, dt$. It is known that for given classes of multifractal processes [42], such $S_q(\tau)$ exhibit power-law behavior $S_q(\tau) \approx |\tau|^{\zeta(q)}$, $\tau \to 0$, $q \in \mathbb{R}$, which is deeply related to their multifractal nature. Another multifractal spectrum, namely, the *Legendre multifractal spectrum*, can then be obtained by taking the Legendre transform of $\zeta(q)$. Although it is possible that the Legendre spectrum is different and in fact less rich than the Hausdorff spectrum, it is used because it is far more numerically accessible. The connection between multifractals and wavelets arises from the fact that the increments involved in the study of the local regularity of a sample path can be seen as simple examples of wavelet coefficients [52]. It has therefore been proposed heuristically [52] to replace increments by wavelet coefficients in the partition functions and shown theoretically that, in some cases, the multifractal formalism can be based directly on wavelet coefficients [16, 42, 60]. For the Legendre multifractal spectrum, this amounts to using wavelet-based partition functions that exhibit, for small scales, power-law behavior: $\int |T_{X(\omega)}(a, t)|^q \, dt \approx a^{\zeta(\omega,q)+q/2}$, $a \to 0$. This last relation can be thought of as a generalization of **P1** to statistics of order both above and below 2. In addition, it is important to understand that even though the relation describes a property of a single (typical) realization, it deals directly with the object $\zeta(\omega, q)$ central to the description of the scaling, and *not* to an estimator of it. This is in contrast to self-similar processes, for example, and the fractal class of the previous paragraph, where the fundamental scaling relations and exponents are defined at the level of the ensemble. Such a change of perspective is meaningful for multifractals as almost all realizations yield a *common* function $\zeta(q)$. Finally, let us note that more refined wavelet-based partition functions have been proposed to overcome various difficulties arising in signal processing; the reader is referred to Bacry et al. [16] and Muzy et al. [52].

The third example is that of *multiplicative cascades*, a paradigm introduced by Mandelbrot [51] in 1974. It involves a recursive procedure whereby an initial mass is progressively subdivided according to a geometric rule and assigned to subsets of an initial set, typically an interval. It provides a powerful tool to define multifractal processes and was originally considered as a natural synthesis procedure for them. Indeed, cascade-based methods of generating multifractals have been the preferred option thus far in teletraffic applications (see Chapter 15). However, the infinitely divisible model proposed by Castaing et al. [21] shows that multiplicative cascade processes can also very effectively model scaling phenomena in other cases, even where the scaling is barely observable in the time domain. Again, the wavelet tool has proved useful for the analysis of such situations, as comprehensively detailed by Arnéodo et al. [14, 15]. This tool has been applied, for instance, in the study of turbulence [22, 63].

2.2.3.7 Processes with Infinite Second-Order Statistics: α-Stable Processes The existence of the wavelet coefficients, the extensions of **P0 SS**, **P1 SS**, and **P2 SS**, to H-sssi processes without second-order statistics, such as α-stable processes, for instance, have recently been obtained [25, 26, 58] (see also, Pesquet-Popescu [57]) but will not be detailed here.

2.3 WAVELETS AND SCALING: ESTIMATION

In this section it is shown in detail how the statistical properties of the wavelet detail coefficients, summarized in the previous section in the form of properties **P1** and **P2**, can be applied to the related tasks of the detection, identification, and measurement of scaling. The estimation of scaling exponents, "magnitude of scaling" parameters, and the multifractal spectrum are discussed. Practical issues in the use of the estimators are addressed and comparisons are made with other estimation methods. Robustness of different kinds is also discussed. It is shown how wavelet methods allow statistics other than second order to be analyzed, with applications in the identification of self-similar and *multifractal* processes. It is explained how the wavelet framework allows a reinterpretation and a fruitful extension of the natural idea of *aggregation* in the study of scaling. It is shown how the *Allan variance*, an effective time domain estimator of scaling, belongs in fact to this framework. Finally, it is shown how the same analysis methods can be applied to the measurement of generalized forms of the *Fano factor*, a well-known descriptor of the burstiness of point processes.

2.3.1 An Analysis Tool: The Logscale Diagram

2.3.1.1 The Legacy of P1 and P2 Property **P2** is the key to the statistical advantages of analysis in the wavelet domain. In sharp contrast to the problematic statistical environment in the time domain due to the long-range dependence, non-stationarity, or fractality of the original process $X(t)$, in the wavelet domain we need only deal with the stationary, short-range-dependent (SRD) processes $d_X(j, \cdot)$ for each j. (Due to the admissibility condition of the mother wavelet these processes each have zero mean.) The stationarity allows us to meaningfully average across "time" within each process to reduce variability. The short-range dependence results in these average statistics having small variance. An example of central importance here is given by

$$\mu_j = \frac{1}{n_j} \sum_{k=1}^{n_j} |d_X(j, k)|^2, \tag{2.19}$$

where n_j is the number of coefficients at octave j available to be analyzed. The random variable μ_j is a nonparametric, unbiased estimator of the variance of the process $d_X(j, \cdot)$. Despite its simplicity, because of the short-range dependence the variance of μ_j decreases as $1/n_j$ and it is in fact asymptotically efficient (of minimal variance). The variable μ_j can therefore be thought of as a near-optimal way of concentrating the gross second-order behavior of X at octave j. Furthermore, again from **P2**, the μ_j are themselves only weakly dependent, so the analysis of each scale is largely decoupled from that at other scales. To analyze the second-order dependence of $X(t)$ on scale, therefore, we are naturally led to study μ_j as a function of j.

Property **P1** now enters by showing explicitly, in the case of scaling, the underlying power-law dependence in j of the variance (second moment) of the processes at each scale, of which the μ_j are estimates. The importance of **P1** is that its pure power-law form suggests that the scaling exponent α could be extracted simply by considering the slope in a plot of $\log_2(\mu_j)$ against j. Here it is essential to understand that, although log–log plots are a natural and familiar tool whenever exponents of power laws are at issue, using them as a basis for semiparametric estimation of the exponent is only effective statistically if properties equivalent to **P1**–**P2** hold. This is typically not the case. For example, for the *correlogram*—a time domain semiparametric estimator [17] based on direct estimation of the covariance function—covariance estimates at fixed lag are biased, resulting in bias in the exponent estimate. Furthermore, across lags the covariance estimates are strongly correlated, resulting in misleadingly impressive "straight lines" in the log–log plot, which in reality are symptomatic of high variance in the resulting estimates. In addition to these issues, the complication that in general $\mathbb{E}[\log(\cdot)] \neq \log(\mathbb{E}[\cdot])$ is overlooked in the correlogram and in many other estimators based on log–log plots. For simplicity of presentation we set $y_j = \log(\mu_j)$ for the moment but address this refinement in the estimation section below. We now introduce a wavelet-based anlaysis tool, the *logscale diagram*, which exploits the key properties **P1** and **P2** and serves as an effective and intuitive central starting point for the analysis of scaling.

Definition 2.3.1. The (second-order) *logscale diagram* (LD) consists of the graph of y_j against j, together with confidence intervals about the y_j.

Examples of logscale diagrams analyzing synthesized scaling data are given in Fig. 2.2, where the plot on the left is of a LRD series, and that on the right side of a self-similar series. It follows from the nature of the dilation operator generating the wavelet basis that the number n_j of detail coefficients at octave j halves with each increase in j (in practice the presence of border effects results in slightly lower values). Confidence intervals about the y_j therefore increase monotonically with j as one moves to larger and larger scales, as seen in each of the diagrams in Fig. 2.2. The exact sizes of these intervals depend on details of the process and in practice are calculated using additional distributional and quasi-decorrelation assumptions. If necessary they could also be estimated from data.

Generalizations to the qth-order logscale diagrams can be defined, $q > 0$, where the second moment of the details in Eq. (2.19) is replaced by the qth. Here we mainly concentrate on the second-order logscale diagram or simply "logscale diagram," both as an illustrative example and because it is the most important special case, being central for LRD and $1/f$ processes by definition, definitive for Gaussian processes, and sufficient for exactly self-similar processes. Like any second-order approach, it is of course insufficient for processes whose second moments do not determine all the properties of interest. We discuss this further in Section 2.4.3 in the particular context of multifractals.

The logscale diagram is first of all a means to visualize the scale dependence of data with a minimum of preconceptions. Scaling behavior is not assumed but *detected*, through the region(s) of *alignment*, if any, observed in the log–log plot. By an alignment region we mean a range of scales where, up to statistical variation, the y_j fall on a straight line. *Estimation* of scaling parameters, if relevant, can then be effectively performed through weighted linear regression over the region(s). Finally,

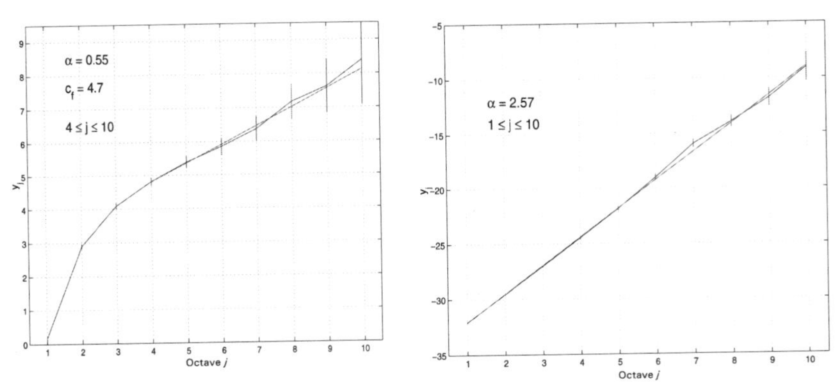

Fig. 2.2 Logscale diagrams. *Left*: An example of the y_j against j plot and regression line for a LRD process with strong short-range dependence. The vertical bars at each octave give 95% confidence intervals for the y_j. The series is simulated FARIMA $(0, d, 2)$ with $d = 0.25$ and second-order moving average operator $\Psi(B) = 1 + 2B + B^2$, implying $(\alpha, c_f) = (0.50, 6.38)$. Alignment is observed over scales $[j_1, j_2] = (4, 10)$, and a weighted regression over this range allows an accurate estimation despite the strong short-range dependence: $\hat{\alpha} = 0.55 \pm 0.07$, $\hat{c}_f = 6.0$ with $4.5 < \hat{c}_f < 7.8$. The scaling can be identified as LRD as the value is in the correct range, $\hat{\alpha} \in (0, 1)$, and the alignment region includes the largest scales in the data. *Right*: Alignment is observed over the full range of scales with $\hat{\alpha} = 2.57$, corresponding to $\hat{H} = 0.79$, consistent with the self-similarity of the simulated FBM ($H = 0.8$) series analyzed.

the *identification* of the kind of scaling is made by interpreting the estimated value in the context of the observed range. These different aspects of the aims and use of the logscale diagram are expanded upon next.

2.3.1.2 The Detection of Scaling A priori it is not known over which scales, if any, a scale-invariant property may exist. By the *detection* of scaling in the logscale diagram we mean the identification of region(s) of *alignment* and the determination of their lower and upper cutoff octaves, j_1 and j_2, respectively, which are taken to correspond to scaling regimes. In a sense this is an insoluble problem, as scaling often occurs asymptotically or has an asymptotic definition, with no clear way to define how a scaling range begins or ends. Nonetheless experience shows that good estimates are possible. Note the semantic difference between the term *scaling region* or range, a theoretical concept that refers to where scaling is truly present (an unknown in real data), and *alignment region* or range, an estimation concept corresponding to what is actually observed in the logscale diagram for a given set of data.

The first essential point here is that the concept of alignment is relative to the confidence intervals for the y_j, and *not* to a close alignment of the y_j themselves. Indeed, an undue alignment of the actual estimates y_j indicates strong correlations between them, a highly undesirable feature typical of time domain log–log based methods such as variance–time plots. As mentioned earlier the μ_j, and hence the y_j, are weakly dependent, resulting in a natural and desirable variation around the calculated regression line as seen, for example, in Fig. 2.2. Using weighted regression incorporates the varying confidence intervals into the estimation phase; however, the selection of the *range of scales* defining the alignment region is prior to this, and great care is required to avoid poor decisions.

We now discuss the selection, in practice, of the cutoff scales j_1 and j_2. A preliminary comment is that for the regression to be well defined at least two scales are required, for a Chi-squared goodness of fit test three, and in practice four, are needed before any estimate can be taken seriously: it is simply too easy for three points to align fortuitously if the confidence intervals are not very small. A useful heuristic in the selection of a range is that the regression line should cut, or nearly so, each of the confidence intervals within it. This can help avoid the following two errors: (1) the nondetection of an alignment region due to the apparently wild variation of the y_j, when in fact to within the confidence intervals the alignment is good (this typically occurs when the slope is small, such as in the right-hand plot in Fig. 2.4, as the vertical scale on plots is reduced, increasing the apparent size of variations), and (2) the erroneous inclusion of extra scales to the left of an alignment region, since to the eye they appear to accurately continue a linear trend, whereas in fact the small confidence intervals about the y_j for small j reveal that they depart significantly from it.

The above heuristic can be formalized somewhat by a Chi-squared goodness of fit test [9], where the critical level of the goodness of fit statistic is monitored as a function of the endpoints of the alignment range. At least in the case of the lower scale, this can make a very clear and relatively objective choice of cutoff possible, eliminating the error of type (2) above. An example of this is afforded by the left-hand plot of Fig. 2.2, where the octave $j = 3$, if included, results in a drop of the Chi-squared goodness of fit of several orders of magnitude! An even subtler example is that of the right-hand plot in Fig. 2.3, where $j = 2$ was excluded from the alignment region for the same reason, whereas in the left-hand plot in the same figure it is clear even to the eye that, given the small size of the confidence interval about octave $j = 8$, it should not be included. Further work is required to develop reliable automated methods of cutoff scale determination. This is especially true for upper cutoff scales, where the difficulties are compounded by a lack of data. On the other hand, at smaller scales the technical assumptions used in the calculation of the confidence intervals (see below) may be less reliable, whereas at large scales the data are highly aggregated and therefore Gaussian approximations are reasonable.

2.3.1.3 The Interpretation of Scaling By the *interpretation* of scaling we mean the identification of the kind of underlying scaling phenomenon—LRD, H-ss, and so on—generating the observed alignment in the logscale diagram. The task is the meaningful interpretation of the estimated value of α in the context of the range of scales defining the alignment region, informed where possible by other known or assumed properties of the time series such as stationarity. It is in fact partly a question of model choice, and there may be no unique solution. We now consider, nonexhaustively, a number of important cases.

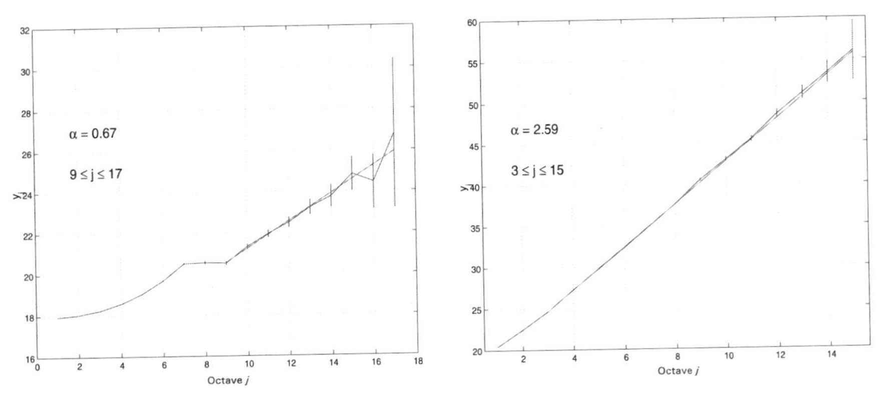

Fig. 2.3 LRD and H-sssi behavior in Ethernet traffic data. *Left*: Logscale diagram for the discrete series of successive interarrival times, showing a range of alignments and an α estimate consistent with long-range dependence. *Right*: Logscale diagram for the cumulative work process (bytes up to time t), consistent with an asymptotically self-similar (close to exactly self-similar) process with stationary increments.

If an estimate of the scaling exponent α is found to lie in (0, 1), and the range of scales is from some initial value j_1 up to the largest one present in the data, then the scaling could be said to correspond to long-range dependence with a scaling exponent that is simply the measured α. Examples are afforded by the left-hand plots in each of Figs. 2.2, and 2.3. If there were a priori physical reasons to believe that the data were stationary, then long-range dependence would be an especially relevant conclusion. This applies to the left-hand plots in Fig. 2.3, as the series corresponds to successive interarrival times of Ethernet packets, which under steady traffic conditions one would expect to be stationary (the Ethernet data in Fig. 2.3 is from the "pAug" Bellcore trace [43]).

Another key example, illustrated in the right-hand plots in Figs. 2.2 and 2.3, is a value of α *greater* than 1 but also measured over a range including the largest scales. Such a value precludes long-range dependence and may indicate that a self-similar or asymptotically self-similar model is required, implying that the data are nonstationary. The exponent would then be reexpressed as $H = (\alpha - 1)/2$, the Hurst parameter. Again conclusions should be compared with a priori physical reasoning. The right-hand plot in Fig. 2.3 is the analysis of a cumulative work process for Ethernet traffic, that is, the total number of bytes having arrived by time t. Such a series is intrinsically nonstationary, though under steady traffic conditions one would expect it to have stationary increments. Thus a conclusion of nonstationarity is a natural one, and the estimated value of $\hat{H} = 0.80$, being in (0,1), indeed corresponds to an H-sssi process. It would have been problematic, however, if underlying physical reasons had indicated that in fact stationarity was to be expected. Such an apparent paradox could be resolved in one of two ways. It may be that the underlying process is indeed stationary and exhibits $1/f$ noise over a wide range of scales, but that the data set is simply not long enough to include the upper cutoff scale. The alternative is to accept that empirical evidence has shown the physical reasoning concluding stationarity to be invalid.

If, on the other hand, the scaling was concentrated at the lowest scales (high frequencies), that is, $j_1 = 1$ with some upper cutoff j_2, then the scaling may best be understood as indicating the fractal nature of the sample path. The observed α should then be reexpressed as $h = (\alpha - 1)/2$, the local regularity parameter. Values of h in the range (0, 1], for example, would then be interpreted as indicating continuous but nondifferentiable sample paths (under Gaussian assumptions [28]), as observed in the leftmost alignment region in the Internet delay data in the left plot in Fig. 2.4. The stationarity or otherwise of the data in such a case may then not be relevant. Note that $j = 1$ has been excluded from the leftmost alignment region in each plot in Fig. 2.4. This is not in contradiction with interpretations of fractality, as it is known that the details at $j = 1$ can be considerably polluted due to errors in the initialization of the multiresolution algorithm (see Section 2.2.1.4).

If scaling with $\alpha > 1$ is found over all or almost all of the scales in the data, such as in the right-hand plot in Fig. 2.2, then exact self-similarity could be chosen as a model, again with $H = (\alpha - 1)/2$ being the relevant exponent. However, in this case one could equally well use the local regularity parameter $h = (\alpha - 1)/2$, with the interpretation that the fractal behavior at small scales is constant over time and happens to extend right up to the largest scales in the data.

Finally, more than one alignment region is certainly possible within a single logscale diagram, a phenomenon that we refer to as *biscaling*. One could imagine, for example, fractal characteristics leading to an alignment at small scales with one exponent, and long-range dependence resulting in alignment at large scales with a separate scaling exponent. Examples of this phenomenon are shown in Fig. 2.4 in

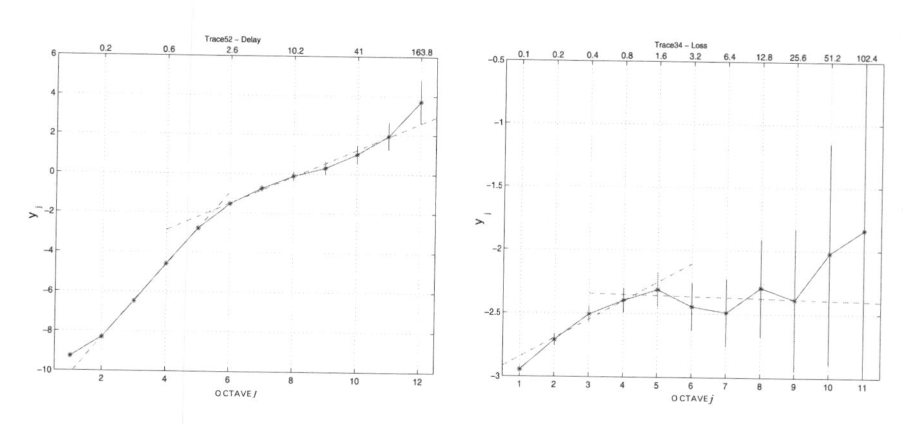

Fig. 2.4 Logscale diagrams with biscaling. Internet UDP packet data displaying two scaling regimes, examples of *biscaling*. *Left*: Delay series. Regime I on the left (small j's) is related to continuous but nondifferentiable sample paths ($h \in (0, 1)$) and regime II (large j's) to long-range dependence. *Right*: Loss series (1 for lost packets, else 0). Regime I corresponds to discontinuous sample paths ($h < 0$). In regime II there is trivial white noise scaling ($\alpha = 0$), indicating stationary SRD behavior.

the context of delays (left) and losses (right) experienced by consecutive user datagram protocol (UDP) packets sent over the Internet in a regular stream (see Andren et al. [12] for details of such data sets). In both figures the alignment at large scales corresponds to long-range dependence, whereas at small scales it is associated with highly irregular sample paths. Note that when second-order properties are insufficient to fully describe the scaling nature of the process (an extreme example is afforded by multifractals), then the correct interpretation of each branch of the bisecting will require the examination of logscale diagrams across a range of orders, as discussed in Section 2.3.3.

2.3.2 Estimation Within the Logscale Diagram

In this subsection it is assumed that a scaling range $j \in [j_1, j_2]$ has been correctly identified. Sums over j, and regressions, are always taken over this range. The estimators to be defined are semiparametric, as they depend on the range of scales $j \in [j_1, j_2]$, where scaling is deemed to be present, and the scaling property **P1** valid there, but not on any tightly specified parametric model.

2.3.2.1 Estimating the Scaling Exponent α Because of property **P1**, the measurement of α is reduced to the determination of the slope over the alignment region in the logscale diagram. A natural way to achieve this in a statistical estimation context is through linear regression, the defining hypothesis of which is $\mathbb{E}y_j = \alpha j + a$, where a is a real constant. Because in general $\mathbb{E}[\log(\cdot)] \neq \log(\mathbb{E}[\cdot])$, this condition is not exactly satisfied, however. We therefore introduce small corrective deterministic factors $g(j)$, discussed below, and redefine the y_j as $y_j = \log(\mu_j) - g(j)$, so that $\mathbb{E}y_j = \alpha j + a$ by definition.

Any kind of linear regression of y_j on j constitutes an unbiased estimator of α, as the lack of bias does not require decorrelation between the y_j, nor knowledge of their variances or distributions. A weighted regression where the weights are related to the variances σ_j^2 of the y_j is preferable, however, as this is the minimum variance unbiased estimator (MVUE) for the regression problem [48]. Intuitively this refinement is significant as we know that the σ_j^2 are far from equal. To exploit the optimality, however, the correlation factors $g(j)$ and the variances σ_j^2 must be calculated—a difficult task. They can nonetheless be well approximated, provided simplifying idealized properties are adopted. The particular idealizations chosen here are the following:

ID1a: For each fixed j the $d_X(j, \cdot)$ are stationary sequences of uncorrelated variables.

IDIb: The processes $d_X(j, \cdot)$ and $d_X(j', \cdot), j \neq j'$, are uncorrelated.

ID2: The process X and, hence, the processes $d_X(j, \cdot)$ are Gaussian.

The above conditions may appear unduly restrictive at first; however, the underlying effectiveness of the method is based on **P1** and **P2**, **ID1–ID2** being added mainly to extend the quantitative analysis. Robustness with respect to departures from these idealizations is discussed in detail below. Note that **ID1a** and **ID1b** together make **ID1**, the idealization of complete decorrelation. It is split here to highlight the fact that **ID1b**, independence between scales, is not needed for the key results.

Under **ID1a** and **ID2** it can be shown [75] that $g(j)$ is a negative, increasing function of n_j only, given by

$$g(j) = \psi(n_j/2)/\ln 2 - \log_2(n_j/2), \tag{2.20}$$

where $\psi(z) = \Gamma'(z)/\Gamma(z)$ is the so-called Psi function and $\Gamma(z)$ the Gamma function. This function can easily be calculated for all values of n_j.

Under **ID1a** and **ID2**, with $g(j)$ as above, the variables $y_j = \log(\mu_j) - g(j)$ are scaled and shifted logarithms of Chi-squared variables, satisfying

$$\mathbb{E}y_j = j\alpha + \log_2 c_f C, \tag{2.21}$$

$$\mathrm{Var}(y_j) \equiv \sigma_j^2 = \zeta(2, n_j/2)/\ln^2 2, \tag{2.22}$$

where $\zeta(z, v)$ is a generalized Riemann Zeta function (see Gradshteyn and Ryzhik [38, p.1072]).

There may be other sets of idealizations under which the $g(j)$ and σ_j^2, if they cannot be expressed in closed form, may nonetheless be accurately calculated. In such cases the estimator would have the same properties including a variance that is known, although no longer expressible in terms of standard functions.

The estimator $\hat{\alpha}$ of α is the slope of a weighted linear regression of y_j on j given by

$$\hat{\alpha} = \frac{\sum y_j(Sj - S_j)/\sigma_j^2}{SS_{jj} - S_j^2} \equiv \sum w_j y_j, \tag{2.23}$$

where $S = \sum 1/\sigma_j^2$, $S_j = \sum j/\sigma_j^2$, and $S_{jj} = \sum j^2/\sigma_j^2$.

2.3.2.2 Properties of the Estimator By construction, $\hat{\alpha}$ is *unbiased* under **ID1a** and **ID2**, and if, in addition, we assume **ID1b** its variance is simply

$$\mathrm{Var}(\hat{\alpha}) = \sum \sigma_j^2 w_j^2. \tag{2.24}$$

Note that the variance is a function of the *amount* of data, the n_j, but is independent both of the data itself, Eq. (2.22), and of the actual (unknown) value of α. It is also independent of the precise choice of mother wavelet, except indirectly through the choice of N, the number of vanishing moments. A quantitative study of this dependence is given by Delbeke and Abry [27].

It has been shown [75], in the limit of n_j large for each j in $[j_1, j_2]$, that the Cramér–Rao bound for the full problem is attained, showing that $\hat{\alpha}$ is asymptotically the minimum variance unbiased estimator under **ID1–ID2**. The decrease in variance of $\hat{\alpha}$ as a function of the size of the data is then explicitly seen to be $1/n$, a remarkable result, being the rate of decay typical of SRD problems, yet appearing in a difficult scaling context. Numerical comparisons [75] show that away from the limit the variance of $\hat{\alpha}$ remains extremely close to the Cramér–Rao bound. This is not surprising as the assumption that n_j is large is a very good one, except possibly for the n_j corresponding to the largest j, since $n_{j+1} \approx n_j/2$. Examining the limit in more detail, for large n_j we have [75]

$$g(j) \to \frac{\log_2 e}{n_j} \tag{2.25}$$

$$\sigma_j^2 \to \frac{2(\log_2 e)^2}{n_j}. \tag{2.26}$$

The first of these relations indicates that for n_j large, y_j can be identified with $\log_2(\mu_j)$. It has moreover been shown [3] that, under **ID1** and **ID2**, $\log_2(\mu_j)$ is asymptotically normally distributed:

$$\log_2(\mu_j) \overset{d}{\sim} N\left(j\alpha + \log_2(c_f C), \frac{2(\log_2 e)^2}{n_j}\right).$$

Since $\hat{\alpha}$ consists of a sum of the y_j, most of which are approximately Gaussian and weighted according to their (known) variances, $\hat{\alpha}$, can be considered as approximately Gaussian distributed. Confidence intervals for the y_j and $\hat{\alpha}$ have been calculated using these arguments.

2.3.2.3 Robustness with Respect to ID1 and ID2 Simulation studies show [5, 75] that the above properties hold to an excellent approximation, even for small size data, upon the mild departures from **ID1** characteristic of the FGN series used.

Numerical simulations presented by Abry and Veitch [5], as well as those described below, show that the above properties also hold to an excellent approximation when the Gaussian hypothesis **ID2**, as well as **ID1**, is dropped. The robustness with respect to **ID2** can be justified using the following asymptotic arguments. Let $Y = f(X)$ and σ_X^2 and σ_Y^2 be the variances of X and Y, respectively. Standard approximation formulas for a change of variable [54] are $\mathbb{E}f(X) \simeq f(\mathbb{E}X) + f''(\mathbb{E}X)\sigma_X^2/2$ and $\sigma_Y^2 \simeq |f'(\mathbb{E}X)|^2\sigma_X^2$. Because Var μ_j decreases as $1/n_j$ in the limit of large n_j, we can apply these formulas to $\log_2(\mu_j)$. Using **ID1** we obtain

$$\begin{aligned}\mathbb{E}\log_2\mu_j &\simeq \log_2\mathbb{E}\mu_j - \frac{(\log_2 e)}{2}\frac{\mathrm{Var}\,\mu_j}{(\mathbb{E}\mu_j)^2}\\ &= \log_2\mathbb{E}\mu_j - \frac{(\log_2 e)}{2}\frac{\mathrm{Var}\,d_X^2(j,\cdot)}{n_j(\mathbb{E}_X^2(j,\cdot))^2}\\ &= \log_2\mathbb{E}\mu_j - (\log_2 e)\frac{1+C_4(j)/2}{n_j},\end{aligned} \tag{2.27}$$

where $C_4(j)$ is the (normalized) fourth-order cumulant of the random process $d_X(j,\cdot)$ given by $C_4(j) = (\mathbb{E}d_X^4(j,k) - 3(\mathbb{E}d_X^2(j,k))^2)/(\mathbb{E}d_X^2(j,k))^2$. From Eq. (2.25) it can be seen that the term $(\log_2 e)(1 + C_4(j)/2)/n_j$ plays the role of $g(j)$ from the Gaussian case and, up to the $C_4(j)$ term, has the same form. Performing the regression of $y_j = \mathbb{E}\log_2\mu_j$ on j, we obtain, using $n_j = 2^{-j}n$ and under **ID1**,

$$\mathbb{E}\hat{\alpha} = \mathbb{E}\sum_j w_j\log_2\mu_j \simeq \alpha - \left((\log_2 e)\sum_j(1 + C_4(j)/2)w_j2^j\right)/n, \tag{2.28}$$

which (1) shows that the estimate is asymptotically unbiased irrespective of the Gaussian hypothesis and (2) allows us to subtract a first-order approximation of that bias. Similarly,

$$\mathrm{Var}\log_2\mu_j \simeq 2(\log_2 e)^2\frac{1 + C_4(j)/2}{n_j}, \tag{2.29}$$

which again shows the similarity of form with the corresponding Gaussian case, Eq. (2.26), in the asymptotic limit. It follows that

$$\mathrm{Var}\,\hat{\alpha} = \sum_j w_j\log_2\mu_j \simeq \left(2(\log_2 e)^2\sum_j(1 + C_4(j)/2)w_j^2 2^j\right)/n, \tag{2.30}$$

which again is identical, up to the C_4 term, to what would be obtained asymptotically from Eq. (2.24) under **ID2**.

The C_4 term can be estimated for each octave using the sample moment estimators of the fourth and second moments of the details, combined as per the definition. Due to the quasi-decorrelation in the wavelet plane, this simple estimator will have low bias.

The above arguments clearly show that, in the limit of a large number of samples, the key statistical features of the estimator—namely lack of bias and low variance—are retained and are therefore not **ID2** dependent. Numerical simulations, summarized in Table 2.1, show that this asymptotic behavior is moreover reached for a relatively small number of samples. The simulations were performed using 1000

TABLE 2.1 Robustness with Respect to Gaussianity (ID2)[a]

RV	Bias	Variance	MSE
Gaussian	0.015	0.039	0.039
Uniform	0.017	0.036	0.036
Exponential	0.016	0.039	0.039
Lognormal	0.017	0.045	0.045
Pareto $\alpha = 1.75$	0.015	0.149	0.150
Pareto $\alpha = 1.25$	0.019	0.131	0.132
α-stable $\alpha = 1.75$	0.013	0.122	0.122
α-stable $\alpha = 1.25$	−0.004	0.228	0.228

[a]It can be seen that the residual bias and variance of the wavelet-based estimator of the scaling exponent α are not very sensitive to the form of the marginal of the process $X(t)$. They are, moreover, very close to the theoretical performance derived assuming exact decorrelation of the wavelet coefficients and Gaussianity of $X(t)$.

realizations of FARIMA$(0, d, 0)$ processes (of length $n = 2^{14} = 16{,}384$) with a variety of probability density functions for the marginals. We used a Daubechies3 [24] wavelet and j_1 was set to $j_1 = 4$ from a preliminary analysis involving the Chi-squared goodness of fit test described above. Table 2.1 shows that the performance obtained with the non-Gaussian processes is quite close to that obtained from Gaussian processes. Note that the four last chosen (Pareto and α-stable) processes are infinite variance processes. The estimates, however, remain unbiased and the variances, though larger, remain controlled as explained in Delbeke [25] and Pesquet-Popescu and Abry [58].

2.3.2.4 Estimating the Second Parameter of Scaling The scaling exponent α is a dimensionless parameter, which can be thought of as characterizing the *qualitative* nature of the scaling phenomena in question. Although α is clearly the key, defining the parameter of scaling, it is not sufficient to fully characterize a given scaling phenomenon, nor therefore the effect that scaling may have on the distributions of various statistics, nor the impact of scaling on performance issues in applications. At the very least, there is a need for a second parameter to describe the *quantitative* aspect of the scaling, a magnitude or “volume” of scaling parameter. This was illustrated in the introduction in the context of the variance of the sample mean of a LRD process. There c_r was introduced as a second parameter with the dimensions of variance describing the relative role that long-range dependence plays. Similarly, for self-similar processes the variance σ^2 of the marginal at $t = 1$ is a free parameter that also requires estimation. These “magnitude” parameters are also problematic to estimate using traditional methods; however, as with α they can be simply and effectively estimated from the logscale diagram. For simplicity we will continue the discussion for the LRD case only. However, an essentially identical procedure, estimator definition, and properties hold in the self-similar case. For other kinds of scaling also, magnitude parameters can be defined and estimated in a similar way; however, this will not be discussed here.

We briefly summarize the results of Veitch and Abry [74, 75] for a two-dimensional joint estimator $(\hat{\alpha}, \hat{c}_f)$ of long-range dependence. One of the powerful properties of $\hat{\alpha}$ is that its statistics are entirely independent of the specific form of the mother wavelet, depending on the coefficients of the linear regression and the amount of data n_j at each scale. It is this feature that allows an explicit expression for its variance to be obtained independently of the wavelet basis. It is clear that this property is shared by $\hat{a} = \sum v_j y_j$, the unbiased estimator of the intercept a of the same linear regression defining α. (The coefficients v_j are given by $v_j = \sum(S_{jj} - S_j j)/(\sigma_j^2(SS_{jj} - S_j^2))$, cf. Eq. (2.23).) From Eq. (2.16) it is apparent that whereas α is simply the slope in the logscale diagram, the magnitude parameter c_f is related to the intercept, being essentially proportional to $2^{\hat{a}}$, a quantity that retains the wavelet independence advantages of $\hat{a}$. Unfortunately, $2^{\hat{a}}$ does not correspond exactly to c_f but rather to the dimensionless quantity $c_f C(\alpha, \psi_0)$, and an attempt to isolate the former necessarily brings in a wavelet dependence. It is advantageous, however, to study $c_f C$ as the largest "wavelet-independent part" c_f, and to define subsequently the estimator of c_f as $\hat{c}_f = \widehat{c_f C}/\hat{C}$, where $\hat{C}$ is an estimator of the integral $C(\alpha, \psi_0)$, which will not be detailed here. It can be shown [75] that $\hat{C}$ has small variance, so that the properties of $\hat{c}_f$ closely resemble those of $\widehat{c_f C}$. More specifically, define $\widehat{c_f C} = p2^{\hat{a}}$, where p is a wavelet-independent bias correcting (and variance reducing) factor given by

$$p = \prod \frac{\Gamma(n_j/2)\exp(\psi(n_j/2)v_j)}{\Gamma(v_j + n_j/2)},$$

which is typically close to 1 (here ψ is the Psi function, not the wavelet!). This estimator is unbiased and complex, but explicit expressions for the variance of $\widehat{c_f C}$ (and covariance of $(\hat{\alpha}, \widehat{c_f C})$) are obtainable, and again the Cramér–Rao bound is attained in the limit n_j large, provided the additional condition $v_j/n_j \to 0$ is also satisfied for each j in $[j_1, j_2]$. Based on these explicit expressions for $(\hat{\alpha}, \widehat{c_f C})$, very accurate approximate expressions (which to first order are also wavelet independent) can be derived for the covariance matrix of $(\hat{\alpha}, c_f)$.

The estimator $\hat{c}_f$ is asymptotically unbiased and efficient, and approximately lognormally distributed. The correlation coefficient of $(\hat{\alpha}, \hat{c}_f)$ is negative and large in magnitude: typically around -0.9. As before, simulation studies show [75] that the above properties hold to an excellent approximation, even for small size data, upon mild departures from **ID1**. An example of the joint estimation is given in the left-hand plot of Fig. 2.2.

2.3.2.5 ***Comparisons with Other Estimators*** In the evaluation of an estimator both statistical and computational aspects must be considered. The logscale diagram based estimators are essentially optimal computationally speaking, as they have a complexity of only $O(n)$, and a direct, nonproblematic implementation that can even be performed in real time [62]. Simple estimators of scaling such as the variogram [17] also have excellent computational properties; however, they suffer from significant bias and high variance [68]. At the other end of the spectrum, fully parametric maximum likelihood estimators require the inversion of an $n \times n$ autocovariance matrix, an $O(n^3)$ operation, which is unsuitable for anything but small data sets. Even approximate forms such as the Whittle estimator, or discrete versions of it [5, 17], involve numerical minimization and are prohibitively slow for the large ($n > 2^{14}$) data sets now routinely encountered in teletraffic studies.

Statistically, the best performing estimators are fully parametric, such as those based on maximum likelihood, which offer zero bias and optimal variance *provided that the data fit the chosen parametric model.* As mentioned above, to avoid extreme computational difficulties encountered for all but small data sets, approximate forms are used in practice, which retain these desirable statistical properties asymptotically [5, 17]. Taqqu et al. [68, 71], give a comparative discussion of the statistical properties of a variety of estimators of long-range dependence. It is shown that approximate maximum likelihood-based estimators such as the Whittle, aggregated Whittle, and local Whittle methods still offer the best statistical performance when compared against alternatives such as the absolute value method, the variance method, the variance of residuals method, the R/S method, and the periodogram method (see also Teverovsky et al. [66] and Taqqu and Teverovsky [69, 72]). We therefore compare against such parametric alternatives. Being parametric based, such estimators can make full use of the data and will therefore outperform logscale diagram based estimators, which are constrained to use only those scales where the scaling is both present and apparent. Although it is possible that all the data may be accessible to a logscale diagram based estimator—that is, that $[j_1, j_2]$ can be chosen correctly as $[j_1, j_2] = [1, \log_2(n)]$, this is unlikely in general. It is even possible that a data set is too short to contain scales in the scaling range, in which case the logscale diagram based estimators, or indeed any semiparametric estimator of scaling, will be useless. On the other hand, in practice typically one cannot know the "true" model for the data, and parametric estimators based on the wrong model can yield meaningless results. In contrast, logscale diagram based estimation is not sensitive to nonscaling details of the data, provided the scaling range is correctly identified, and is also more robust in other ways, as discussed below. Another key advantage is the ability to measure in a uniform framework both stationary and nonstationary forms of scaling.

A detailed comparison of the performance of $\hat{\alpha}$ against that of the discrete Whittle estimator for the FGN and Gaussian FARIMA$(0, d, 0)$ processes is given by Abry and Veitch [5]. Veitch and Abry [75] compared $(\hat{\alpha}, \hat{c}_f)$ against a joint discrete Whittle estimator for the Gaussian FARIMA$(0, d, 0)$ process, and $(\hat{\alpha}, \widehat{c_f C})$ against a joint maximum likelihood estimator for a "pure" scaling process defined in the wavelet domain [79]. The main conclusion is that the logscale diagram based estimators offer almost unbiased estimates, even for data of small length, with far greater robustness, for the price of a small to moderate increase in variance compared to that of the parametric alternatives. They can also be used to treat data of arbitrary length both from the computational complexity and memory requirement points of view [5, 62, 75]. Comparison against parametric estimators for processes that are further from such "ideal" processes will appear elsewhere.

2.3.3 The Multiscale Diagram and Multifractals

2.3.3.1 The Need for Statistics Other than Second Order It is natural and straightforward to generalize the logscale diagram to the study of statistics other than second order, by replacing Eq. (2.19) with $\mu_j^{(q)} = 1/n_j \sum_k |d_X(j,k)|^q$, $q \in \mathbb{R}$. The resulting qth-*order logscale diagrams* are, naturally enough, of interest in situations where information relevant to the analysis of scaling is beyond the reach of second-order statistics. Let us concentrate on two important examples: self-similarity and multifractality.

Self-Similarity From the definition of self-similarity, the moments of $X(t)$ satisfy $\mathbb{E}|X(t)|^q = \mathbb{E}|X(1)|^q \cdot |t|^{qH}$, $\forall t$. As for the wavelet coefficients, it follows from Eq. (2.10) that $\mathbb{E}|d_X(j,k)|^q = \mathbb{E}|d_X(0,k)|^q \cdot 2^{j(qH+q/2)}$, implying that $\mathbb{E}\mu_j^{(q)} = C_q 2^{j(\zeta(q)+q/2)}$, $\forall j$, with $\zeta(q) = qH$. This relation suggests that self-similarity can be detected by testing the linearity of $\zeta(q)$ with q.

Multifractal Processes For the class of multifractal processes, assuming that $\int |T_{X(\omega)}(a,t)|^q \, dt \approx a^{\zeta(q)+q/2}$, $a \to 0$, can be related to the multifractal properties of the process, $\mu_j^{(q)}$ is expected to behave according to $\mu_j^{(q)} \approx 2^{j(\zeta(q)+q/2)}$ for small j. From these relations one can measure $\zeta(q)$ in practice and therefore estimate the Legendre multifractal spectrum. A particularly interesting question in the multifractal formalism is to test whether, in the range of q where $\int |T_{X(\omega)}(a,t)|^q \, dt$ is finite, $\zeta(q)$ takes a simple linear form: $\zeta(q) = qh$, or not. Clearly in such cases the Legendre spectrum is somewhat degenerate as it is entirely determined by h and the range of q where $\zeta(q)$ is defined. For instance, self-similar processes, for which $\mu_j^{(q)} \approx 2^{j(qH+q/2)}$ for *all* scales, satisfy $\zeta(q) = qH$ and are therefore fractal processes with $h = H$. More specifically the self-similar Lévy processes have infinite variance and are multifractal [41], yet their spectra are parameterized by H and are therefore derivative of the strong self-similar property. The FBM is another, even simpler example, often referred to as "monofractal," confirming the intuition that a single scaling parameter controls all of its scaling properties.

2.3.3.2 The Multiscale Diagram and Its Use In both the self-similar and multifractal cases, therefore, accurately measuring the deviation of $\zeta(q)$ from a simple linear form is a crucial issue. To test this, and to investigate the form of $\zeta(q)$ in general, the qth-order scaling exponent $\alpha_q = \zeta(q) + q/2$ can be estimated in the qth-order logscale diagram for a variety of q values, and then the q dependence examined in the following tool:

Definition 2.3.2 The *multiscale diagram* (MD) consists of the graph of $\hat{\zeta}(q) = \hat{\alpha}_q - q/2$ against q, together with confidence intervals about the $\hat{\zeta}(q)$.

A lack of alignment in the multiscale diagram strongly suggests multifractal scaling that is not of the simple type where $\zeta(q) = qH$. Alignment in the multiscale diagram may indicate the presence of the simple type of multifractal scaling, or self-similarity, or both simultaneously as in the self-similar Lévy processes above. It is important to note that the question of alignment, just as in the logscale diagram, is relative to confidence intervals, and that conclusions cannot be drawn in their absence. The confidence intervals shown are derived through approximations presented by Delbeke and Abry [27].

To better examine the presence of alignment, it is also of interest to use an alternative *linear* multiscale diagram (LMD), where $h_q = \alpha_q/q - \frac{1}{2}$ is plotted against q. In the LMD the linear form is detected as a horizontal alignment, and the value of h is obtained by estimating its level. Although the LMD is more convenient for visual inspection of alignment, the two forms of multiscale diagram are statistically equivalent in terms of the determination of the linearity or otherwise of $\zeta(q)$.

Figure 2.5 illustrates the use of the multiscale diagrams by superimposing an analysis of a synthesized FBM and that of actual Internet data (delays of UDP packets as described in Section 2.3.1.2, see Fig. 2.4). The regressions in the underlying logscale diagrams were performed for small scales: $[j_1, j_2] = (2, 6)$, for each of 11 different q values (as discussed in Section 2.3.1.3, the lowest scale is problematic due to initialization issues). The multiscale diagram in the left plot

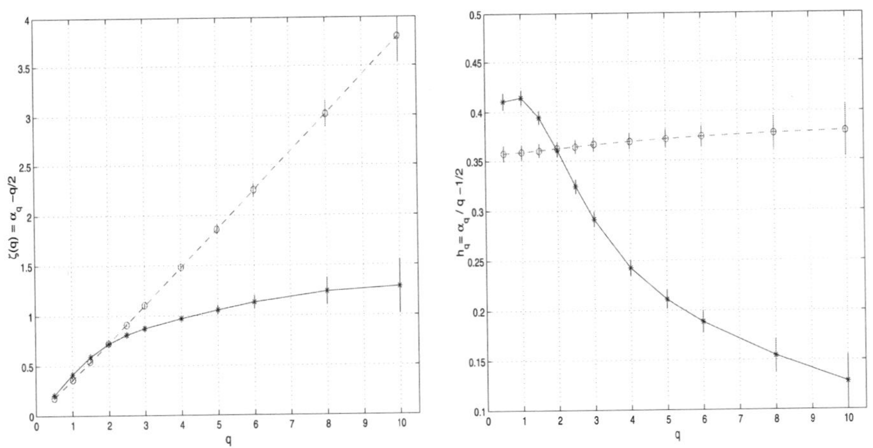

Fig. 2.5 Testing for multifractality. Each plot superposes results for a synthesized FBM (dashed line) and actual Internet data (solid line). *Left*: Multiscale diagram. Alignment for the FBM series is consistent with its known self-similar nature, and no alignment for the Internet data suggests nondegenerate multifractal behavior. *Right*: Linear MD. Horizontal alignment for the FBM series suggests even more clearly that $\zeta(q) = qH$, whereas it is even more clearly not the case for the Internet data.

shows a clear alignment for the FBM series, consistent with its self-similar nature, and a clear absence of alignment for the Internet data, indicating that it is not monofractal and that a multifractal model may be appropriate. These conclusions are confirmed in the right-hand plot, where the same data are analyzed in the linear multiscale diagram. A clear horizontal alignment of the FBM data is observed while it is confirmed that the Internet data cannot obey $\zeta(q) = qH$.

The Internet data example in the figure is given to illustrate the use of the multiscale diagrams, although the presence of possible multifractal scaling behavior in such traffic is of interest in its own right [76]. More generally, however, multifractals have recently attracted considerable attention in traffic modeling because of their ability to flexibly model highly variable local behavior in a natural way. Multifractal processes can also have positive increments that are approximately lognormally distributed, in agreement with a number of empirical traffic studies. For more details on these empirical findings the reader is referred to Lévy Véhel and Riedi [45, 61], where such analyses were first performed, or to Feldmann et al. [29, 30], Gilbert et al. [37], Riedi et al. [60], Taqqu et al. [70], and Veitch et al. [76] for further investigations, and where connections to multiplicative cascade models are presented. Chapter 20 of this volume and references therein can also be consulted and Chapter 15 for details of lognormal distributions in traffic.

2.3.4 Departures from Scaling

There are two broad categories of departures from models possessing a pure scaling feature. One is the presence of nonstationarities that are not in themselves of a scaling nature, and in the presence of which a scaling exponent continues to be well defined. Here the issue is one of the correct measurement of the constant underlying scaling parameter(s), despite the nonstationary "noise." The other category is the possibilty that the scaling is itself changing in time: that is, the parameters of constant-in-time scaling are not well defined. A central issue in that case is the reliable detection of such a variation. In the following, special properties of the wavelet approach are exploited to yield partial, but nonetheless significant and useful, replies to these two different challenges.

2.3.4.1 Superimposed Trends In many situations of practical interest, the assumption that the data are fully described by a scaling model—be it self-similar, fractal, or LRD—is much too restrictive to be realistic. This is especially the case when an observed time series is in fact the result of an additive contamination of a scaling process $X(t)$ by some extraneous contribution $T(t)$, that is, $Y(t) = X(t) + T(t)$. We will not consider here the case where $T(t)$ is random (this situation of a scaling process corrupted by some "observation noise" is considered, e.g., by Wornell and Oppenheim [79]) but will comment on the case when $T(t)$ is deterministic and can be thought of as a "trend."

A simple model for a trend amounts to choosing $T(t)$ as a polynomial, say, of order p. If no care is taken when analyzing a scaling process corrupted by such a trend, important features of interest (e.g., stationarity of the increments) are likely to be lost, thus impairing the estimation of the relevant scaling parameters. Power-law trends can even mimic LRD correlations when added to a stationary short-range process, leading to entirely erroneous conclusions [5, 17, 65]. It is therefore desirable, prior to any analysis, to eliminate possible trends or, at least, to be able to evaluate and control their effects on the final estimates: wavelets offer a versatile and easy way of doing so.

In order to understand where the effectiveness of wavelets comes from in the context of trend removal, it is worthwhile to start from the admissibility condition, Eq. (2.1), satisfied by any wavelet ψ_0: saying that a wavelet is zero-mean is in fact equivalent to saying that it is orthogonal, and therefore "blind," to nonzero mean values. A natural generalization consists of considering wavelets with more than one vanishing moment, in the sense of Eq. (2.5), since assuming that the number of vanishing moments is N allows the analysis to be blind to polynomials of order up to $N - 1$. In other words, the removal of a polynomial trend of order p is guaranteed by a wavelet with $N \geq p + 1$. From a practical point of view, when p is the unknown order of a polynomial approximation of a trend, trend removal amounts to analyzing the data with different wavelets such that $N = 1, 2, \ldots$. Until the effective value $N = p + 1$ is reached, the analysis is governed by the trend and gives N-dependent results, whereas, as soon as $N \geq p + 1$, stabilized results are obtained and reveal relevant features of the detrended data. Exact detrending is expected to occur in the case of polynomial trends, but it is worthwhile to remark that the procedure still remains effective in the case of nonpolynomial trends, including many power-law trends, and oscillatory functions. In Fig. 2.6 an example is given of an FGN series contaminated with linear and sinusoidal trends. Although the logscale diagram with $N = 2$ is free from the effects of the linear trend (top right), higher values of N are required to effectively remove the sinusoidal trend (bottom right). Note that increasing N unfortunately also decreases the number of scales available for the analysis. The usefulness of an ability to effectively remove smooth nonpolynomial trends is again illustrated in Fig. 2.7, where, using a value of N as low as 2, a change in mean level in Ethernet byte data is shown to not affect the estimation of the exponent α of long-range dependence. See Veitch and Abry [74], Fig. 9, for further discussion. The data set is derived from the "pOct" Bellcore trace [43].

The versatility of wavelets with respect to the freedom of choice of N makes them an easy and efficient tool for trend removal. This contrasts with other methods, such as the Whittle estimator and parametric estimators in general, whose performance is heavily affected by trends (see Abry and Veitch [5, Fig. 2 and Table 1]). An important advantage is hat one can choose N to eliminate preselected trends, without having to know if they are actually present a priori nor to jointly estimate their characteristics.

2.3.4.2 Time Varying Scaling Exponents Because of the high variability inherent in scaling processes, instances of scaling behavior can incorrectly be judged as "nonstationarity," in the broad sense of unstructured time variation, or conversely, variability due to nonstationarity may be erroneously taken to be scaling in nature.

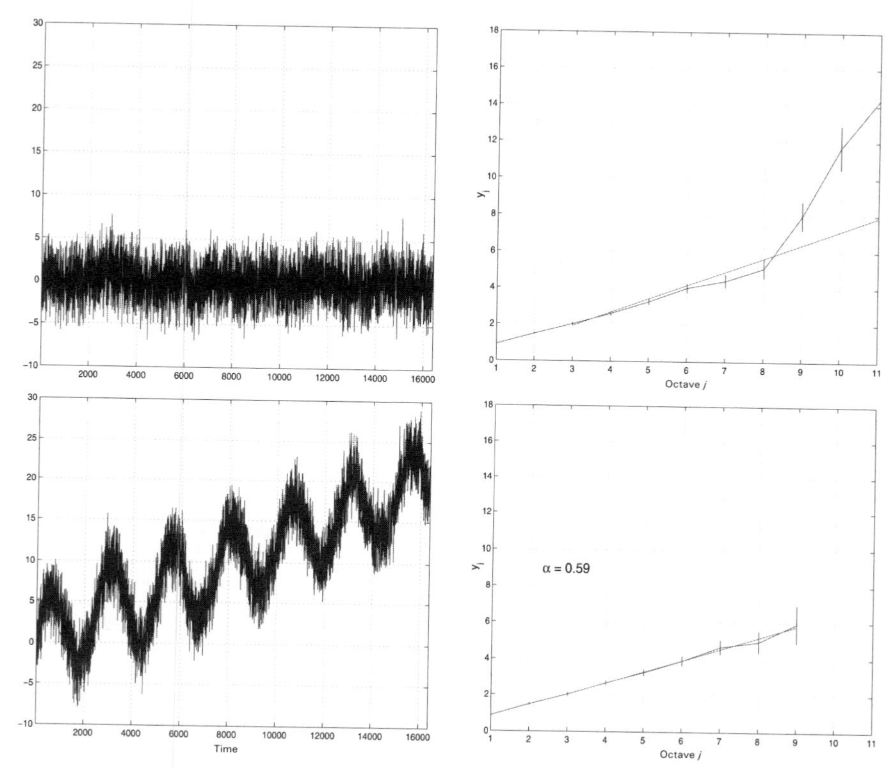

Fig. 2.6 Robustness with respect to trends. *Left*: Synthesized FGN with $H = 0.80$ (top), and with superimposed sinusoidal and linear trends (bottom). *Right*: Logscale diagrams of the data *with trends* using a Daubechies2 (i.e., $N = 2$) wavelet (top) and Daubechies9 wavelet (bottom). It is seen that increasing N allows the effects of superimposed trends to be removed from the logscale diagram, enabling uncontaminated estimates of the scaling exponent. Here, with $N = 9$, $\hat{\alpha} = 0.59$ and therefore $\hat{H} = 0.795$.

There is a strong need therefore to monitor the value of scaling parameters over time to distinguish between these possibilities, and to test if constancy of scaling can be concluded or not. A basic approach is to split a data set into a number m of adjacent blocks, and to examine the corresponding logscale diagrams. If alignment is found in each over roughly the same range of scales, then separate estimations $\hat{\alpha}^{(m)}$ of the scaling exponent could be performed for each block. A null hypothesis might be that the (unknown) values $\alpha^{(m)}$ share a common value. The difficulty is how to combine these estimates in a well-defined statistical test, given that the problematic statistical nature of scaling processes would in general imply strong correlations between the $\hat{\alpha}^{(m)}$. Because of the quasi-decorrelation in the wavelet plane, however, it can be shown [77] that the $\hat{\alpha}^{(m)}$ can be treated as almost independent, and moreover that they are approximately normally distributed with known variances. The difficult problem of the constancy of the scaling exponent can thereby be reduced to a simple

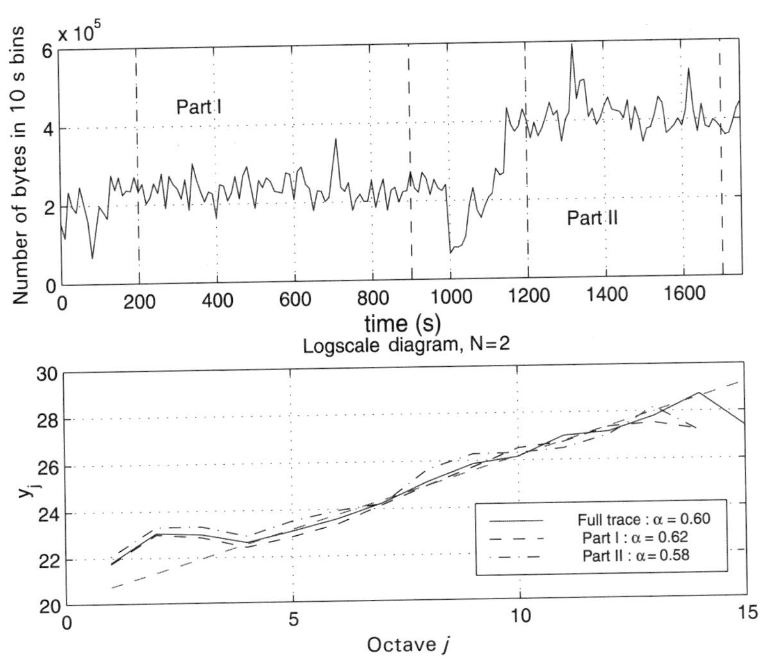

Fig. 2.7 Eliminating mean level shifts. *Top*: Ethernet byte data aggregated over 10 s intervals. A marked, but smooth, level shift seems to occur around 1050 seconds. *Bottom*: Logscale diagram of the byte data aggregated over 10 ms intervals. With $N = 2$, estimates to the left and the right of the shift, and over the whole series, are consistent, showing that smooth level shifts can effectively be eliminated in practice.

model inference problem, to which an optimal (uniformly most powerful invariant) test exists [77] (see also the discussion in Abry and Veitch [5]).

2.3.5 Relations to Other Tools

2.3.5.1 Aggregation Procedure Let $X(t)$ be a centered second-order stationary process with finite variance and form the discrete-time process $X^{(T)}(n)$, $n \in \mathbb{Z}$, defined by

$$X^{(T)}(n) = \frac{1}{T} \int_{(n-1)T}^{nT} X(t)\, dt.$$

Such a quantity, referred to as the *aggregated* process of X with aggregation level T, turns out to play a key role with respect to self-similarity and long-range dependence. In fact, if we assume that X is LRD in the sense that its autocovariance decreases asymptotically as in Eq. (2.7), it can be shown [23] that, in the limit of arbitrarily large aggregation levels T, the normalized autocovariance

$$r^{(T)}(k) = \frac{\mathbb{E}X^{(T)}(n)X^{(T)}(n+k)}{\operatorname{Var} X^{(T)}(n)}, \quad k = 1, 2, \ldots$$

of the aggregated process tends to a well-defined limiting form that coincides with that of an FGN with unit variance (see Eq. (2.9)), while its variance satisfies

$$\operatorname{Var} X^{(T)}(n) \approx T^{\alpha-1}, \quad T \to +\infty. \tag{2.31}$$

Aggregation appears therefore as a natural renormalization tool that reveals, in a well-defined manner, the possible occurrence of long-range dependence in an observed process. By construction, aggregation amounts to averaging the analyzed process over adjacent blocks of larger and larger support. This is of course reminiscent of the way the Haar MRA, whose scaling function is simply the indicator function of [0, 1], is built. More precisely, we have the exact equivalence [6]

$$X^{(2^j)}(k) = 2^{-j/2}\langle X, \phi_{j,k}^{(\mathrm{Haar})}\rangle,$$

from which we can deduce that

$$\operatorname{Var} a_X^{(\mathrm{Haar})}(j, k) \approx 2^{j\alpha}, \quad j \to +\infty.$$

Revisiting the aggregation procedure from the Haar perspective allows for two levels of generalization [6]. First, we can relax the choice of the Haar system and move toward any MRA. Second, we can replace approximation coefficients by details, since we also have, for any admissible wavelet ψ,

$$\operatorname{Var} d_X(j, k) \approx 2^{j\alpha}, \quad J \to +\infty.$$

2.3.5.2 Allan Variance Although well rooted in wavelet theory, this idea of using details rather than approximations can itself be viewed as a generalization of a method put forward by Allan in 1966 [13]. Specifically, whereas Eq. (2.31) shows the theoretical link between variance and the scaling behavior associated with long-range dependence, Allan showed that the estimation of α is greatly improved when the standard variance estimate $(1/N)\sum_n (X^{(T)}(n))^2$ is replaced by the "Allan variance," defined as $(1/N)\sum_n (X^{(T)}(n+1) - X^{(T)}(n))^2$. In wavelet words, this just amounts—when dyadic intervals of the form $T = 2^j$ are considered—to computing Haar details [32]. As explained previously, the reason for the improved efficiency of such an approach is to be found in the ability of wavelets to almost decorrelate LRD processes. Moreover, revisiting the Allan variance in the light of the

Haar system suggests that more versatile generalizations could be considered based on wavelets with more vanishing moments. This indeed not only guarantees smaller residual correlations but also an increased robustness of the analysis to polynomial trends.

2.3.5.3 ***Fano Factor*** Apart from aggregation and the Allan variance, wavelets also offer a way of generalizing standard approaches in other areas, for example, in the case of point processes [4]. Let us consider, for instance, processes of the form

$$P(t) = \sum_{k=-\infty}^{+\infty} G(t - t_k),$$

where the t_k are Poisson distributed, with a stationary (but possibly time varying) density $\lambda(t)$, and $G(t)$ is the impulse response of some filter. With such a model, it turns out that scaling features or long-range dependence may result from the joint properties of $\lambda(t)$ and $G(t)$. This is especially the case when $G(t) = \delta(t)$ and $\lambda(t)$ is FGN with a nonzero mean. A classical way of revealing the scaling structure of such a "fractal point process" amounts to looking at how it departs from an ordinary Poisson process when observed over larger and larger scales. With the ordinary Poisson process, the associated counting process $N(T)$ is such that $\mathrm{Var}\, N(T) = \mathbb{E}N(T)$, no matter how large T is, thus guaranteeing the "Fano factor," defined as

$$F(T) = \frac{\mathrm{Var}\, N(T)}{\mathbb{E}N(T)},$$

to be constant. This situation contrasts with the fractal case for which we have

$$F(T) \approx T^{2H-1}, \quad T \to +\infty. \tag{2.32}$$

We use H here rather than α as the process $N(t)$, which has stationary increments, is not LRD but rather asymptotically self-similar.

By definition, the Fano factor is a ratio between fluctuations and averages at a given observation scale. As such, it naturally admits a wavelet-based alternative definition:

$$WF(j) = 2^{j/2} \frac{\mathrm{Var}\, d_P(j, k)}{\mathbb{E}a_P(j, k)},$$

in which details (resp. approximations) play the role of fluctuations (resp. averages). Given this definition, the scaling relation (2.32) now reads

$$WF(j) \approx 2^{j(2H-1)}, \quad j \to +\infty,$$

and moreover we have the explicit equivalence $WF^{(\text{Haar})}(j) = F^{(\text{Allan})}(2^j)$. Again, this wavelet revisiting of the Fano factor allows for versatile generalizations with increased performance, in which the Haar wavelet is to be replaced by a wavelet with a larger number of vanishing moments.

From the above examples, it is worthwhile stressing the fact that wavelets provide the user with a unified framework that (1) possesses increased versatility and performance as compared to the earlier methods it generalizes and (2) is equally applicable to continuous processes and point processes.

2.4 WAVELETS AND SCALING: SYNTHESIS

After having allocated the greater part of this chapter to the wavelet *analysis* of scaling processes, we now turn to the question of their synthesis. To do so, we focus here on a wavelet-based synthesis of the FBM. This may seem restrictive; however, unfortunately generation is more difficult than analysis and the area is less well developed. Although it is not difficult to generate time series with approximate scaling properties using wavelets, there are very few examples where a specific process can be reproduced accurately. We discuss this further at the end of this section.

There exist a number of methods dedicated to the synthesis of the FBM that are not based on wavelets. Among these, the so-called Choleski or Durbin–Levinson methods [19, 39, 68] are notable as they are exact. They have severe computational drawbacks, however, which make them unsuitable to the generation of long traces, particularly the enormous series required in teletraffic simulations. Various approximate synthesis methods with reasonable computational loads are known, for example, the so-called spectral synthesis method (for a recent variant, see Paxson [55]), the on/off superposition method [78], a method based on the discretization of the integral representation, and recently a refined random midpoint displacement method [53], which has a number of advantages. These methods yield tractable practical implementations (sometimes on-the-fly implementations) but often suffer from the drawback that the errors (due to approximations made) cannot be controlled and that it is not clear which property of the FBM has been lost. Wavelet-based methods have also already been proposed [79, 80] to produce stationary $1/f$-type processes. We present here a wavelet-based synthesis of the FBM, introduced by Sellan [64], which is developed and implemented in Abry and Sellan [7] and Meyer et al. [47]. The method relies on an exact wavelet expansion of the FBM. This allows the generation of approximate sample paths, but with controlled errors, in a practical implementation framework using a fast pyramidal algorithm similar to that underlying the inversion of DWT [24, 46]. This method reproduces accurately the key features that characterize the FBM—stationarity of the increments, self-similarity, and long-range dependence—and allows new and clearer insights into them, as described by Abry and Sellan [7] and Meyer et al. [47]. We detail below the key steps of this synthesis.

2.4.1 Wavelet Expansions for the FBM

When $0 < H < 1$, and $H \neq \frac{1}{2}$, FBM has an integral representation [50, 67],

$$B_H(t) = \frac{\sigma}{C}\left\{\int_{-\infty}^{0} \left[(t-x)^{H-1/2} - (-x)^{H-1/2}\right] dB(x) - \int_{0}^{t} (t-x)^{H-1/2}\, dB(x)\right\}, \quad (2.33)$$

where c is a positive normalization constant which depends on H.

From this integral representation, one can see FBM as a fractionally integrated version of a Gaussian white noise (denoted dB). The starting point of the wavelet-based representation of FBM lies in the fact that, in an orthonormal wavelet basis expansion of a Gaussian white noise process (in the sense of distributions), the weights of the expansion are simply Gaussian independent, identically distributed (i.i.d.) random variables. To obtain the wavelet expansion of the FBM, one can fractionally integrate the wavelet Gaussian white noise expansion, which basically amounts to fractionally integrating the wavelets. One may therefore conjecture that it can be represented as

$$B_H(t) \stackrel{(?)}{=} \sum_{j=-\infty}^{\infty} \sum_{k=-\infty}^{\infty} 2^{jH} \psi_H(2^{-j}t - k)\epsilon_{j,k}, \quad (2.34)$$

where ψ_H is a suitable wavelet and the $\epsilon_{j,k}$ are i.i.d. $N(0, \sigma_0^2)$ random variables. One can check that the right-hand side (RHS) of (2.34) scales correctly at dyadic points, namely, $B_H(at) \stackrel{d}{=} a^H B_H(t)$ for $a = 2^{-l}$, $l = -1, 0, 1, \ldots$, as can easily be verified by making the change of variables $j \to j' = j + l$ and using the fact that the $\epsilon_{j,k}$ are i.i.d. However, the RHS of (2.34), as written, does not converge because of the growth of 2^{jH} as $j \to \infty$, that is, because of its behavior at low frequencies. The representation (2.34) is almost correct, however. It is just necessary to add an "infrared correction," namely, to write

$$B_H(t) = \sum_{j=-\infty}^{\infty} \sum_{k=-\infty}^{\infty} 2^{jH}[\psi_H(2^{-j}(t-k)) - \psi_H(-k)]\epsilon_{j,k}. \quad (2.35)$$

It is shown by Meyer et al. [47] that, for a suitable function ψ_H (see below), the RHS of (2.35) converges to fractional Brownian motion. Not only does the scaling by $a = 2^{-l}$ hold, but in fact (2.35) scales (in the sense of the finite-dimensional distributions) for any $a > 0$. Observe that the infrared correction also ensures that the resulting process has stationary increments and equals 0 at $t = 0$. We are, in fact, representing the increment $B_H(t) - B_H(0)$, which equals $B_H(t)$ because $B_H(0) = 0$, a similar idea is behind the integral representation (2.33).

Whereas the representation (2.33) involves *integration* with respect to "continuous" white noise dB, the representation (2.35) involves a summation over discrete white noise (i.i.d. $\epsilon_{j,k}$), and, as such, is very close to the intuition of a "Karhunen–Loève representation" of the process. It is based on translations (by k) and dilations (by j) of the function ψ_H. A value $k > 0$ corresponds to translation to the right, $k < 0$

to the left, $j > 0$ to dilation, and $j < 0$ to compression. Thus the limit $j \to -\infty$ captures the high frequencies and the limit $j \to \infty$ captures the low frequencies. Note that Meyer et al. [47] use $j' = -j$ instead of j. The function ψ_H is continuous, and, in fact, the RHS of (2.35) converges to $B_H(t)$ uniformly on every closed bounded interval $t \in [-T, T]$.

There exist different wavelet representations that differ (1) by the way the low frequencies are represented and (2) by the choice of wavelets. Instead of using ψ_H for both the low and high frequencies, it is possible to generate the low frequencies by using a different function. As shown by Meyer et al. [47], one also has

$$B_H(t, \omega) = \sum_{k=-\infty}^{\infty} \tilde{\phi}_H(t-k) B_H(k, \omega') + \sum_{j=-\infty}^{0} \sum_{k=-\infty}^{\infty} 2^{jH} \psi_H(2^{-j}t - k)\epsilon_{j,k}(\omega'') - \tilde{b}_0(\omega) \tag{2.36}$$

for some suitably chosen function $\tilde{\phi}_H$. The function ψ_H is as before and the random variable $\tilde{b}_0(\omega)$ is a random level shift, independent of t, necessary to ensure that $B_H(0, \omega) = 0$. We write here $B_H(t, \omega)$ instead of $B_H(t)$ not only to underline the fact that B_H is random (ω is an element of the probability space), but also because $\{B_H(k, \omega'),\ k \in \mathbb{Z}\}$ and $\{\epsilon_{j,k}(\omega''),\ j \le 0,\ k \in \mathbb{Z}\}$ are assumed independent and $\omega = (\omega', \omega'')$. Thus, to generate *continuous* time $\{B_H(t),\ t \ge 0\}$, one generates first a *discrete* sequence $B_H(k, \omega')$ in addition to the i.i.d. $\epsilon_{j,k}$. It is the sum $\sum_{k=-\infty}^{\infty} \tilde{\phi}_H(t-k) B_H(k, \omega')$ that generates the low-frequency components and hence the corresponding long-range dependence. Observe that $B_H(k, \omega)$ is not equal to $B_H(k, \omega')$ because $B_H(k, \omega)$ depends also on the $\epsilon_{j,k}$. Here again, convergence to $B_H(t)$ holds uniformly for any $t \in [-T, T]$.

The sequence $\{B_H(k, \omega'),\ k \in \mathbb{Z}\}$ is a discrete time series whose increments $\{\Delta B_H(k) = B_H(k+1, \omega') - B_H(k, \omega'),\ k \in \mathbb{Z}\}$ are stationary and have long-range dependence. It is possible to replace $\{B_H(k, \omega'),\ k \in \mathbb{Z}\}$ by a sum of other stationary time series with long-range dependence, in particular, by a Gaussian FARIMA$(0, d, 0)$ time series with $d = H - \frac{1}{2}$ [40].

Let then $\{Z_i^{(H)},\ i \in \mathbb{Z}\}$ represent this mean zero FARIMA time series independent of the $\epsilon_{j,k}$ and let

$$S_0^{(H)} = 0, \quad S_{-k}^{(H)} = -\sum_{i=-k+1}^{0} Z_i^{(H)}, \quad S_k^{(H)} = \sum_{i=1}^{k} Z_i^{(H)} \quad \text{for } k \ge 1,$$

denote the partial sums. Then fractional Brownian motion can also be represented as

$$B_H(t) = \sum_{k=-\infty}^{\infty} \phi_H(t-k) S_k^{(H)} + \sum_{j=-\infty}^{0} \sum_{k=-\infty}^{\infty} 2^{jH} \psi_H(2^{-j}t - k)\epsilon_{j,k} - b_0 \tag{2.37}$$

for some suitably chosen function ϕ_H. The function ψ_H is as before and the random variable b_0 is such that $B_H(0) = 0$, that is,

$$b_0 = \sum_{k=-\infty}^{\infty} \phi_H(-k) S_k^{(H)} + \sum_{j=-\infty}^{0} \sum_{k=-\infty}^{\infty} 2^{jH} \psi_H(-k) \epsilon_{j,k}.$$

Once again, convergence holds uniformly for $t \in [-T, T]$.

Equation (2.37) is a corrected version of a representation in Sellan [64]. It describes the FBM as a trend $\sum_{k=-\infty}^{\infty} \phi_H(t-k) S_k^{(H)}$ over which are superimposed (high-frequency) details $\sum_{j=-\infty}^{0} \sum_{k=-\infty}^{\infty} 2^{jH} \psi_H(2^{-j}t - k)\epsilon_{j,k}$. The nonstationarity property of the FBM is already apparent in the expression of the trend, which involves $S^{(H)}$. The self-similarity is carried in the actual variance of the wavelet coefficients of the expansion $d_{B_H}(j, k) = 2^{j(H+1/2)}\epsilon_{j,k}$, which follows a power law of the scale 2^j. Moreover, ψ_H is tailored in such a way that the wavelet basis it generates exactly catches the correlation of the FBM so that the wavelet coefficients are strictly uncorrelated. This specific wavelet basis therefore acts as a Karhunen–Loève basis for the high frequencies of the process.

2.4.2 Wavelet Design

It remains to indicate what are the functions ψ_H, ϕ_H, and $\tilde{\phi}_H$. They are derived from a scaling function $\phi_0(t)$ and a corresponding wavelet function $\psi_0(t)$, yielding an orthonormal wavelet basis. Using the Fourier transform notation $U(v) = \int_{-\infty}^{\infty} u(t)e^{-itv}\, dt$, start with the Fourier transforms Ψ_0 and Φ_0 of ψ_0 and ϕ_0 and let

$$\Psi_{0,1/2-H}(v) = (iv)^{1/2-H}\Psi_0(v), \quad \Phi_{0,1/2-H}(v) = \left(\frac{iv}{1-e^{-iv}}\right)^{1/2-H} \Phi_0(v).$$

Going back to the time domain, get the functions ψ_H and ϕ_H by integration:

$$\psi_H(t) = \int_{-\infty}^{t} \psi_{0,1/2-H}(u)\, du = \psi_{0,-(H+1/2)}(t)$$

and

$$\phi_H(t) = \int_{t-1}^{t} \phi_{0,1/2-H}(u)\, du = \phi_{0,-(H+1/2)}(t).$$

The expansion for $\tilde{\Phi}_H$ is more involved and we refer the reader to Meyer et al. [47].

There are many possible choices for the pair (ψ_0, ϕ_0). Because the spectral density [50, 67] of the increments of FBM may diverge at the frequency $v = 0$, it is necessary that the Fourier transform $\Psi_0(\nu)$ of the wavelet ψ_0 tend to 0 as $\nu \to 0$ relatively fast. In Meyer et al. [47], the Meyer (or Lemarié–Meyer) wavelets are used. Their Fourier transforms vanish not just at the origin but in a whole neighborhood of the origin. The Meyer scaling and wavelet functions $\phi_0(t)$ and $\psi_0(t)$ as well as $\phi_H(t)$ and $\psi_H(t)$ are very smooth: they are infinitely differentiable and, while they do not have bounded support, they tend to 0 as $t \to \pm\infty$ faster than any polynomial.

Instead of the Meyer wavelets, one can use, to obtain ϕ_H and ψ_H, any orthonormal wavelet basis with enough vanishing moments, such as the Daubechies wavelets with $N \geq H + \frac{1}{2}$, that is, $N \geq 2$. One can also use the so-called Lemarié–Battle orthonormal spline wavelets [24].

2.4.3 Fast Implementation

It is well known from the wavelet literature [24] that the synthesis in Eq. (2.37), which we rewrite for convenience,

$$\begin{aligned} B_H(t) + b_0 = & \sum_{k=-\infty}^{\infty} S_k^{(H)} \phi_H(t-k) \\ & + \sum_{j=-\infty}^{0} \sum_{k=-\infty}^{\infty} (2^{j(H+1/2)} \epsilon_{j,k}) 2^{-j/2} \psi_H(2^{-j}t - k), \end{aligned}$$

can be implemented using a discrete-time fast pyramidal algorithm on condition that $\phi_H(t)$ and $\psi_H(t)$ are designed from a multiresolution analysis (see Section 2.3). The implementation we present here is based on a slightly modified version of the one developed by Abry and Sellan [7].

The coefficients of the discrete-time filters h_2 and g_2 involved in the synthesis depend on ϕ_H and ψ_H through

$$\begin{aligned} h_2 &= u_H, \\ g_2 &= v_H, \end{aligned} \tag{2.38}$$

where u_H and v_H are the generating sequences of ϕ_H and ψ_H, respectively:

$$\begin{aligned} \phi_H(t/2) &= \sqrt{2} \sum_k u_H(k) \phi_H(t-k), \\ \psi_H(t/2) &= \sqrt{2} \sum_k v_H(k) \phi_H(t-k). \end{aligned} \tag{2.39}$$

This means that we do not necessarily need to provide explicit expressions for these functions. Let us denote by $\phi_0(t)$ and $\psi_0(t)$ a scaling function and a mother wavelet giving birth to an orthonormal wavelet basis with sufficient regularity and by u and v their generating sequences:

$$\begin{aligned}\phi_0(t/2) &= \sqrt{2}\sum_k u(k)\phi_0(t-k),\\ \psi_0(t/2) &= \sqrt{2}\sum_k v(k)\phi_0(t-k).\end{aligned} \tag{2.40}$$

We have shown [7] that u_H and v_H can be obtained from u and v by

$$\begin{aligned}u_H &= f^{(s)} * u, \quad F^{(s)}(z) = 2^{-s}(1+z^{-1})^{s},\\ v_H &= g^{(s)} * v, \quad G^{(s)}(z) = 2^{-s}(1-z^{-1})^{-s},\end{aligned} \tag{2.41}$$

where the $*$ denotes the discrete-time convolution, $s = H + \frac{1}{2}$, and $f^{(s)}$ and $g^{(s)}$ are infinite length sequences whose z-transforms are labeled $F^{(s)}(z)$ and $G^{(s)}(z)$. From a practical point of view, the discrete-time convolutions above are computed using an approximation technique described in Abry and Sellan [7].

By using the pyramidal algorithm sketched in Fig. 2.8, we end up with an approximation of fractional Brownian motion that is computationally efficient and conceptually simple. The "trend" part involves the cumulated sum of a FARIMA(0, d, 0) time series, which can be obtained from i.i.d. Gaussian random variables, with zero mean and variance σ_0^2. The "details" (successive high frequencies) are labeled here by the indices $j = 0, -1, \ldots$. We include them up to some level $-J$. Observe that the details at levels $j = 0, \ldots, -J$ involve independent Gaussian random variables $\{\eta_{j,k} = 2^{j(H+1/2)}\epsilon_{j,k}, j \le 0,\ k \in \mathbb{Z}\}$, i.i.d. in k, with zero mean and variances that follow a power law in j: Var $2^{j(H+1/2)}\epsilon_{j,k} = \sigma_0^2 2^{j(2H+1)}$.

2.4.4 Synthesis of Other Scaling Processes

Although the FBM is an important scaling process, it is of course very specific and unsuitable for many modeling purposes. It is of great interest to be able to rapidly and accurately synthesize other scaling processes that are not strictly FBM but which may have looser constraints (e.g., $1/f$-type processes) or richer scaling properties (e.g., multifractal processes), or more flexible nonscaling features together with scaling features (e.g., LRD processes with flexible SRD structure). Despite these broader needs, the highly focused understanding of the wavelet synthesis described above remains of interest, because it shows where to modify the synthesis scheme either to give up, or to maintain, certain properties of the FBM. For instance, Wornell [80] presents a wavelet-based synthesis that is in the same spirit as that given here, but which produces *only* $1/f$-type processes. In fact, there exists in the teletraffic literature a variety of wavelet-based syntheses of scaling processes, which typically amounts to producing $1/f$-type processes, though in a loose and poorly controlled manner. However, wavelets can also be be used to implement multiplicative cascades, which are themselves capable of generating processes with rich scaling behavior. This has been exploited in recent work for the synthesis of multifractal

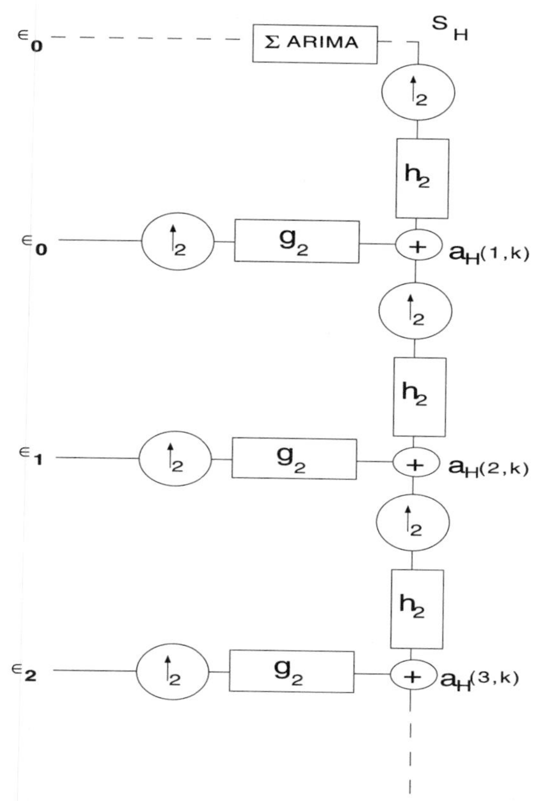

Fig. 2.8 Wavelet-based synthesis for FBM. The FBM can be synthesized numerically using a fast pyramidal algorithm based on discrete-time filters. The coefficients of the filters h_2 and g_2 depend on functions ϕ_H and ψ_H obtained by the fractional integration of order $H + \frac{1}{2}$ of an orthonomal wavelet basis. The inputs of the filters, denoted $\{\epsilon_j\}$ in the figure, stand for independent random vectors $\{\eta_{j,k}, k \in \mathbb{Z}\}$, labeled by j, whose components are i.i.d. Gaussian random variables with zero mean and variance $\sigma_0^2 2^{j(2H+1)}$. The uparrow operator indicates an upsample by a factor of 2 operation obtained by inserting a zero between each sample. One gets, for example, $a_H(2, k)$ by adding the upsampled $a_H(1, k)$ *convolved with* h_2 to the upsampled ϵ_1 convolved with g_2.

processes [60] and for the synthesis of a large variety of multiplicative cascades processes [15].

2.5 WAVELETS AND SCALING: PERSPECTIVES

Wavelets have been shown to present many advantages when dealing with scaling phenomena. A first point is that, whereas "scaling" may refer to different models (e.g., self-similar processes, fractal sample paths, LRD), depending on the range of scales over which it is actually observed and the value of scaling exponents, wavelets offer a unified framework that applies equally well to any of these situations. A second point of interest is that wavelets allow for a well-controlled splitting of the analyzed process in a number of sub-processes at different scales, each of which being much better behaved than the original process considered as a whole. This is especially true of long-range dependence, a phenomenon that prohibits the use of most classical statistical tools: when properly translated to the wavelet domain, it turns out that the situation is much simpler, with short-range dependencies at each scale, thus permitting the design of simple, yet efficient, estimators based on the usual empirical variance estimates. A third point is that wavelets are naturally equipped with a number of degrees of freedom in their design (e.g., the number of their vanishing moments, which both controls decorrelation properties of the coefficients and increases the robustness of the analysis with respect to trends), as well as with fast pyramidal (filter-bank) implementations, which give rise to computationally efficient schems for analysis and synthesis. Generally speaking, using wavelets when dealing with scaling processes is in some sense "natural," because of the structural affinity that exists between the *mathematical* framework they offer (multiresolution) and the *physical* nature (scaling) of the processes under study.

Regarding directions for future work, a number of important and active areas have already been referred to in the text. These include the need to be able to objectively and automatically determine the range of scales for a given scaling phenomenon, how to judge if scaling exponents are constant or not in time, and how to generate accurately in a wavelet framework more flexible classes of scaling processes. Looking further afield, from the particular perspective of teletraffic research, an avenue that is as yet quite unexplored is that of going beyond merely describing scaling traffic processes, to instead defining entire queueing systems with scaling traffic input, reexpressed and hopefully resolved in a wavelet world.

ACKNOWLEDGEMENTS

P. Abry and P. Flandrin were partially supported by the CNRS grant TL 97035, Programme Télécommunications. M. S. Taqqu was partially supported by the NSF grant DMS-94044093 and ANI-9805623 at Boston University.

REFERENCES

1. P. Abry. *Ondelettes et Turbulences—Mutirésolutions, Algorithmes de Décompositions, Invariance d'Echelle et Signaux de Pression*. Diderot, Editeur des Sciences et des Arts, Paris, 1997.
2. P. Abry, P. Gonçalvès, and P. Flandrin. Wavelet-based spectral analysis of $1/f$ processes. In *Proc. IEEE–ICASSP'93*, pp. III.237–III.240, Minneapolis, MN, 1993.
3. P. Abry, P. Gonçalvès, and P. Flandrin. Wavelets, spectrum estimation and $1/f$ processes. In A. Antoniadis and G. Oppenheim, eds., *Wavelets and Statistics, Lectures Note in Statistics*, **103**, pp. 15–30. Springer-Verlag, New York, 1995.

4. P. Abry and P. Flandrin, Point processes, long-range dependence and wavelets. In A. Aldroubi and M. Unser, eds. *Wavelets in Medicine and Biology*, pp. 413–438. CRC Press, Boca Raton, FL, 1996.
5. P. Abry and D. Veitch. Wavelet analysis of long-range dependent traffic. *IEEE Trans. Inf. Theory*, **44**(1):2–15, 1998.
6. P. Abry, D. Veitch, and P. Flandrin. Long-range dependence: revisiting aggregation with wavelets. *J. Time Series Anal.*, **19**(3):253–266, 1998.
7. P. Abry and F. Sellan. The wavelet based synthesis for fractional Brownian motion proposed by F. Sellan and Y. Meyer: remarks and implementation. *Appl. Comput. Harmonic Anal.*, **3**: 377–383, 1996.
8. P. Abry, L. Delbeke, and P. Flandrin. Wavelet-based estimator for the self-similarity parameter of α-stable processes. In *ICASSP-99*, Phoenix, USA, 1999.
9. P. Abry, M. S. Taqqu, D. Veitch, and P. Flandrin. On the automatic selection of scaling range in the semi-parametric estimation of scaling exponents. In preparation.
10. P. Abry, P. Flandrin, M. S. Taqqu, and D. Veitch. Unpublished note.
11. A. Aldroubi and M. Unser. Sampling procedure in function spaces and asymptotic equivalence with Shannon's sampling theory. *Numer. Funct. Anal. Optim.*, **15**:1–21, 1994.
12. J. Andren, M. Hilding, and D. Veitch. Understanding end-to-end Internet traffic dynamics. In *Proc. Globecom '98*, Sydney, pp. 1118–1122, 1998.
13. W. Allan. Statistics of atomic frequency standards. *Proc. IEEE*, **54**:221–230, 1966.
14. A. Arnéodo, J. F. Muzy, and S. G. Roux. Experimental analysis of self-similar random cascade processes: application to fully developed turbulence. *J. Phys. II France*, **7**:363–370, 1997.
15. A. Arnéodo, E. Bacry, and J. F. Muzy. Random cascades on wavelet dyadic trees. *J. Math. Phys.*, **39**(8):4142–4164, August 1998.
16. E. Bacry, J. F. Muzy, and A. Arnéodo. Singularity spectrum of fractal signals from wavelet analysis: exact results, *J. Statist. Phys.*, **70**:635–674, 1994.
17. J. Beran. *Statistics for Long-Memory Processes*. Chapman and Hall, New York, 1994.
18. J. Beran, R. Sherman, M. S. Taqqu, and W. Willinger. Long-range dependence in variable-bit-rate video traffic. *IEEE Trans. Commun.*, **43**:1566–1579, 1995.
19. P. J. Brockwell and R. A. Davis. *Time Series:Theory and Methods*, 2nd ed. Springer-Verlag, New York, 1991.
20. S. Cambanis and C. Houdré. On the continuous wavelet transform of second-order random processes. *IEEE Trans. Inf. Theory*, **41**(3):628–642, 1995.
21. B. Castaing, Y. Gagne, and M. Marchand. Log-similarity for turbulent flows? *Physica D*, **68**:387–400, 1993.
22. P. Chainais, P. Abry, and J. F. Pinton. Intermittency and coherent structures in a swirling flow: a wavelet analysis of joint pressure and velocity measurements. *Phys. of Fluids*, **11**(11): 3524–3539, 1999.
23. D. R. Cox. Long-range dependence: a review. In H. A. David and H. T. David, eds., *Statistics: An Appraisal*, pp. 55–74. Iowa State University Press, Ames, 1984.
24. I. Daubechies. *Ten Lectures on Wavelets*. SIAM, Philadelphia, 1992.
25. L. Delbeke. *Wavelet Based Estimators for the Scaling Index of a Self-Similar Process with Stationary Increments*. Ph.D. Thesis, KU Leuven, Belgium, 1998.

26. L. Delbeke and P. Abry. Stochastic integral representation and properties of the wavelet coefficients of linear fractional stable motion. To appear in *Stochastic Processes and Their Applications*.
27. L. Delbeke and P. Abry. Wavelet based estimators for the self-similarity parameter of the fractional Brownian motion. Internal report, 1998.
28. K. Falconer. *Fractal Geometry—Mathematical Foundations and Applications*. Wiley, New York, 1990.
29. A. Feldmann, A. C. Gilbert, and W. Willinger. Data networks as cascades: investigating the multifractal nature of Internet WAN traffic. In *Proc. ACM SIGCOMM '98*, Vancouver, Canada, Sept. 1998.
30. A. Feldmann, A. C. Gilbert, W. Willinger, and T. G. Kurtz. Looking behind and beyond self-similarity: on scaling phenomena in measured WAN traffic. Preprint.
31. P. Flandrin. On the spectrum of fractional Brownian motions. *IEEE Trans. Inf. Theory*, **35**:197–199, 1989.
32. P. Flandrin. Wavelet analysis and synthesis of fractional Brownian motion. *IEEE Trans. Inf. Theory*, **38**:910–917, 1992.
33. P. Flandrin. Fractional Brownian motion and wavelets. In M. Farge, J. C. R. Hunt, and J. C. Vassilicos, eds., *Wavelets, Fractals, and Fourier Transforms*, pp. 109–122. Clarendon Press, Oxford, 1993.
34. P. Flandrin. *Time-Frequency/Time-Scale Analysis*. Academic Press, San Diego, 1999.
35. P. Flandrin and P. Gonçalvès. From wavelets to time-scale energy distributions In L. L. Schumaker and G. Webb, eds., *Recent Advances in Wavelet Analysis*, pp. 309–334. Academic Press, San Diego, 1994.
36. P. Gonçalvès and P. Flandrin. Scaling exponents estimation from time-scale energy distributions. In *Proc. IEEE Int. Conf. on Acoust., Speech and Signal Proc. ICASSP-92*, San Francisco, pp. V.157–V.160, 1992.
37. A. C. Gilbert, W. Willinger, and A. Feldman. Scaling analysis of random cascades, with applications to network traffic. *IEEE Trans. Inf. Theory*, Special Issue on Multiscale Statistical Signal Analysis and its Applications, **45**(3) 971–991. April 1999.
38. I. S. Gradshteyn and I. M. Ryzhik. *Table of Integrals, Series and Products*. Academic Press, New York, corrected and enlarged edition, 1980.
39. M. A. Hauser, W. Hörmann, R. M. Kunst, and J. Lenneis. A note on generation estimation and prediction of stationary processes. In R. Dutter and W. Grossmann, eds., *COMPSTAT*, pp. 323–329. Physica Verlag, Berlin, 1994.
40. J. R. M. Hosking. Fractional differencing. *Biometrika*, **68**(1):165–176, 1981.
41. S. Jaffard. Sur la nature multifractale des processus de Lévy. *C. R. Acad. Sci. Paris,Ser. 1*, **323**:1059–1064, 1996.
42. S. Jaffard. Multifractal formalism for functions, Part II: Self-similar functions. *SIAM J. Math. Anal.*, **28**(4):971–988, 1997.
43. W. E. Leland, M. S. Taqqu, W. Willinger, and D. V. Wilson. On the self-similar nature of Ethernet traffic (extended version). *IEEE/ACM Trans. Networking*, **2**:1–15, 1994.
44. J. Lévy Véhel. Fractal approaches in signal processing. In C. J. G. Evertsz, H.-O. Peitgen, and R. F. Voss, eds., *Fractal Geometry and Analysis*, The Mandelbrot Festschrift, Curacao, 1995. World Scientific Publishing, Singapore, 1996.

45. J. Lévy Véhel and R. Riedi. Fractional Brownian Motion and data traffic modeling: The other end of the spectrum. J. Lévy Véhel, E. Lutton, and C. Tricot, eds., *Fractals in Engineering*, pp., 185–203. Springer, London, 1997.

46. S. Mallat. *A Wavelet Tour of Signal Processing*. Academic Press, Boston, 1997.

47. Y. Meyer, F. Sellan, and M. S. Taqqu. Wavelets, generalized white noise and fractional integration: the synthesis of fractional Brownian motion. Preprint, 1998.

48. S. M. Kay. *Fundamentals of Statistical Signal Processing*. Prentice-Hall, Englewood Cliffs, NJ, 1993.

49. E. Masry. The wavelet transform of stochastic processes with stationary increments and its application to fractional Brownian motion. *IEEE Trans. Inf. Theory*, **39**(1):260–264, 1993.

50. B. B. Mandelbrot and J. W. Van Ness. Fractional Brownian motions, fractional noises and applications. *SIAM Rev.*, **10**:422–437, 1968.

51. B. B. Mandelbrot. Intermittent turbulence in self-similar cascades: divergence of high moments and dimension of the carrier. *J. Fluid Mech.*, **62**(2): 331–358, 1974.

52. J. F. Muzy, E. Bacry, and A. Arnéodo. The multifractal formalism revisited with wavelets. *Int. J. Bifurc. Chaos*, **4**(2):245–301, 1994.

53. I. Norros, P. Mannersalo, and J. L. Wang. Simulation of fractional Brownian motion with conditionalized random midpoint displacement. Preprint, Oct. 1998.

54. A. Papoulis. *Probability, Random Variables, and Stochastic Processes*, 2nd ed. McGraw-Hill, New York, 1984.

55. V. Paxson. Fast approximation of self-similar network traffic. Preprint, 1995.

56. R. Peltier and J. Lévy-Véhel. Multifractional Brownian motion: definition and preliminary results. Submitted to *Stochastic Processes and Their Applications*, 1997.

57. B. Pesquet-Popescu. Statistical properties of the wavelet decomposition of some non-Gaussian self-similar processes. Invited paper, *Signal Processing* **75**(3): pp. xx–xx, 1999.

58. B. Pesquet-Popescu, and P. Abry. Wavelet based estimators for the self-similarity parameter of α-stable processes. In preparation, 1999.

59. R. H. Riedi. An improved multifractal formalism and self-similar measures. *J. Math. Anal. Appl.*, **189**:462–490, 1995.

60. R. Riedi, M. S. Crouse, V. J. Ribeiro, and R. G. Baraniuk. A multifractal wavelet model with application to network traffic. *IEEE Trans. Inf. Theory*, Special issue on Multiscale Statistical Signal Analysis and Its Applications, **45**(3): 992–1018, April 1999.

61. R. Riedi and J. Lévy Véhel. TCP traffic in multifractal: a numerical study. Submitted to *IEEE Trans. Networking*.

62. M. Roughan, D. Veitoch, and P. Abry. On-line estimation of LRD parameters. In *Proc. Globecom '98*, Sydney, pp. 3716–3721, 1998.

63. S. Roux, J. F. Muzy, and A. Arnéodo. Detecting vorticity filaments using wavelet analysis: about the statistical contribution of vorticity filaments to intermittency in swirling turbulent flows. *Eur. Phys. J.* **B8**: 301–322, 1998.

64. F. Sellan. Synthèse de mouvements browniens fractionnaires à l'aide de la transformation par ondelettes. *C.R. Acad. Sci. Ser. I*, **321**:351–358, 1995.

65. V. Teverovsky and M. S. Taqqu. Testing for long-range dependence in the presence of shifting means or a slowly declining trend using a variance-type estimator. *J. Time Series Anal.*, **18**:279–304, 1997.

66. V. Teverovsky, M. S. Taqqu, and W. Willinger. A critical look at Lo's modified *R/S* statistics. *J. Statist. Planning Inference*, **80**: 211–227, 1999.

67. G. Samorodnitsky and M. S. Taqqu. *Stable Non-Gaussian Processes: Stochastic Models with Infinite Variance*. Chapman and Hall, New York, 1994.

68. M. S. Taqqu, V. Teverovsky, and W. Willinger. Estimators for long-range dependence: an empirical study. *Fractals*, **3**(4):785–798, 1995. Reprinted in C. J. G. Evertsz, H.-O. Peitgen, and R. F. Voss, eds., *Fractal Geometry and Analysis*. World Scientific Publishing, Singapore, 1996.

69. M. S. Taqqu and V. Teverovsky. Semi-parametric graphical estimation techniques for long-memory data. In P. M. Robinson and M. Rosenblatt, eds., *Athens Conference on Applied Probability and Time Series Analysis. Volume II: Time Series Analysis in Memory of E. J. Hannan*, Lecture Notes in Statistics **115**, pp. 420–432. Springer-Verlag, New York, 1996.

70. M. S. Taqqu, V. Teverovsky, and W. Willinger. Is network traffic self-similar or multi-fractal? *Fractals*, **5**:63–74, 1997.

71. M. S. Taqqu and V. Teverovsky. On estimating the intensity of long-range dependence in finite and infinite variance series. In R. Adler, R. Feldman, and M. S. Taqqu, eds., *A Practical Guide to Heavy Tails: Statistical Techniques and Applications*. Birkhäuser, Boston, 1998.

72. M. S. Taqqu and V. Teverovsky. Robustness of Whittle-type estimates for time series with long-range dependence. *Stoch. Models*, **13**:723–757, 1997.

73. A. H. Tewfik and M. Kim. Correlation structure of the discrete wavelet coefficients of fractional Brownian motion. *IEEE Trans. Inf. Theory*, **38**:904–909, 1992.

74. D. Veitch and P. Abry. Estimation conjointe en ondelettes des paramètres du phénonène de dépendance longue. In *Proc. 16ième Colloque GRETSI*, Grenoble, France, pp. 1451–1454, 1997.

75. D. Veitch and P. Abry. A wavelet-based joint estimator of the parameters of long-range dependence. *IEEE Trans. Inf. Theory*, Special Issue on Multiscale Statistical Signal Analysis and its Applications, **45**(3): 878–897, April 1999.

76. D. Veitch, P. Abry, and J. Bolot. Wavelet tools for the analysis of scaling phenomena in traffic. Preprint, 1999.

77. D. Veitch and P. Abry. A statistical test for the constancy of scaling exponents. Preprint, 1998.

78. W. Willinger, M. S. Taqqu, R. Sherman, and D. V. Wilson. Self-similarity through high-variability: statistical analysis of Ethernet LAN traffic at the source leve. *IEEE/ACM Trans. Networking*, **5**(1):71–86, 1997. Extended version of the paper with the same title that appeared in *Comput. Commun. Rev.*, **25**:100–113, 1995.

79. G. W. Wornell and A. V. Oppenheim. Estimation of fractal signals from noisy measurements using wavelets. *IEEE Trans. Signal Process.*, **40**(3):611–623, 1992.

80. G. W. Wornell. *Signal Processing with Fractals: A Wavelet-Based Approach*. Prentice-Hall, Englewood Cliffs, NJ, 1995.

3

SIMULATIONS WITH HEAVY-TAILED WORKLOADS

MARK E. CROVELLA

Department of Computer Science, Boston University, Boston, MA 02215

LESTER LIPSKY

Department of Computer Science and Engineering, University of Connecticut, Storrs, CT 06268

3.1 INTRODUCTION

Recently the phenomenon of network traffic *self-similarity* has received significant attention in the networking community [10]. Asymptotic self-similarity refers to the condition in which a time series's autocorrelation function declines like a power law, leading to positive correlations among widely separated observations. Thus the fact that network traffic often shows self-similarity means that it shows noticeable bursts at a wide range of time scales—typically at least four or five orders of magnitude. A related observation is that file sizes in some systems have been shown to be well described using distributions that are *heavy-tailed*—distributions whose tails follow a power law—meaning that file sizes also often span many orders of magnitude [3].

Heavy-tailed distributions behave quite differently from the distributions more commonly used to describe characteristics of computing systems, such as the normal distribution and the exponential distribution, which have tails that decline exponentially (or faster). In contrast, because their tails decline relatively slowly, the proabability of very large observations occurring when sampling random variables that follow heavy-tailed distributions is nonnegligible. In fact, the distributions we discuss in this chapter have *infinite variance*, reflecting the extremely high variability that they capture.

Self-Similar Network Traffic and Performance Evaluation, Edited by Kihong Park and Walter Willinger
ISBN 0-471-31974-0

As a result, designers of computing and telecommunication systems are increasingly interested in employing heavy-tailed distributions to generate workloads for use in simulation. See, for example, Chapters 14 and 18 in this volume. However, simulations employing such workloads may show unusual characteristics; in particular, they may be much less stable than simulations with less variable inputs. In this chapter we discuss the kind of instability that may be expected in simulations with heavy-tailed inputs and show that they may exhibit two features: first, they will be very slow to converge to steady state; and second, they will show highly variable performance at steady state. To explain and quantify these observations we rely on the theory of *stable* distributions [4, 15].

These problems are not unique to simulation of telecommunications systems, arising also in risk and insurance modeling [2]. Solutions to certain aspects of these problems have been proposed, drawing on rare event simulation and variance reduction techniques [8, 14].

In general, however, many of the problems associated with the simulations using heavy-tailed workloads seem quite difficult to solve. This chapter does not primarily suggest solutions but rather draws attention to these problems, both to yield insight for researchers using simulation and to suggest areas in which more research is needed.

3.2 HEAVY-TAILED DISTRIBUTIONS

3.2.1 Background

Let X be a random variable with cdf $F(x) = P[X \leq x]$ and complementary cdf (ccdf) $\bar{F}(x) = 1 - F(x) = P[X > x]$. We say here that a distribution $F(x)$ is *heavy-tailed* if

$$\bar{F}(x) \sim cx^{-\alpha}, \quad 0 < \alpha < 2, \tag{3.1}$$

for some positive constant c, where $a(x) \sim b(x)$ means $\lim_{x\to\infty} a(x)/b(x) = 1$. (We note that more general definitions of heavy tails are common; see, for example, Goldie and Klüppelberg [6].) If $F(x)$ is heavy tailed then X shows very high variability. In particular, X has infinite variance, and, if $\alpha \leq 1, X$ has infinite mean. Section 3.2.2 will explore the implications of infinite moments in practice; here we note simply that if $\{X_i, i = 1, 2, \ldots\}$ is a sequence of observations of X, then the sample variance of $\{X_i\}$ as a function of i will tend to grow without limit, as will the sample mean if $\alpha \leq 1$.

The simplest heavy-tailed distribution is the *Pareto* distribution, which is power law over its entire range. The Pareto distribution has pmf

$$p(x) = \alpha k^{\alpha} x^{-\alpha-1}, \quad 0 < k \leq x,$$

and cdf

$$F(x) = P[X \leq x] = 1 - (k/x)^{\alpha}, \tag{3.2}$$

in which the positive constant k represents the smallest possible value of the random variable.

In practice, random variables that follow heavy-tailed distributions are characterized as exhibiting many small observations mixed in with a few large observations. In such data sets, most of the observations are small, but most of the contribution to the sample mean or variance comes from the few large observations.

This effect can be seen in Fig. 3.1, which shows 10,000 synthetically generated observations drawn from a Pareto distribution with $\alpha = 1.2$ and mean $\mu = 6$. In Fig. 3.1(a) the scale allows all observations to be shown; in Fig. 3.1(b) the y axis is expanded to show the region from 0 to 200. These figures show the characteristic, visually striking behavior of heavy-tailed random variables. From plot (a) it is clear that a few large observations are present, some on the order of hundreds to one thousand; while from plot (b) it is clear that most observations are quite small, typically on the order of tens or less.

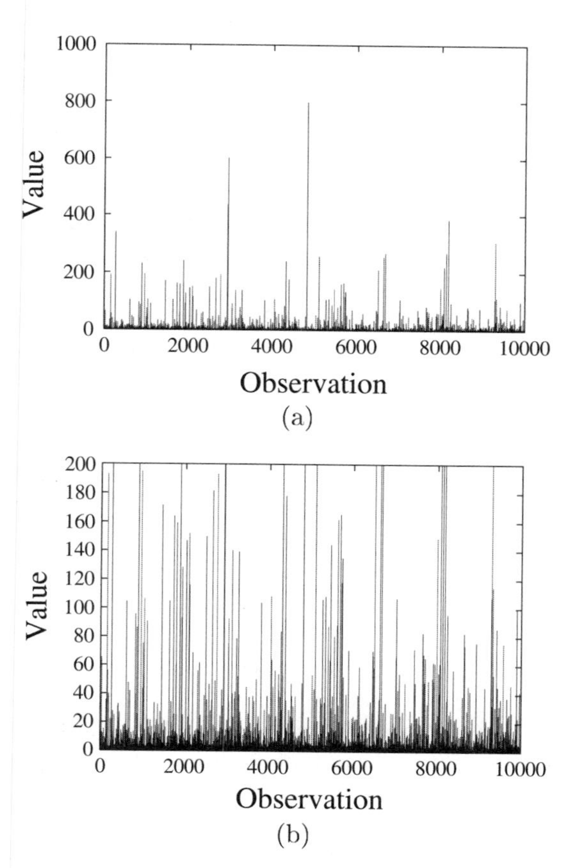

Fig. 3.1 Sample data from heavy-tailed distribution with $\alpha = 1.2$.

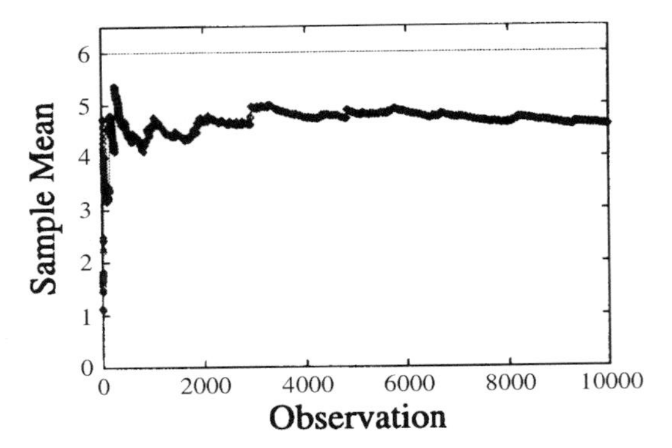

Fig. 3.2 Running mean of data from Fig. 3.1.

An example of the effect of this variability on sample statistics is shown in Fig. 3.2. This figure shows the running sample mean of the data points from Fig. 3.1, as well as a level line showing the mean of the underlying distribution (6). Note that the sample mean starts out well below the distributional mean, and that even after 10,000 observations it is not close in relative terms to the distributional mean.

3.2.2 Heavy Tails in Computing Systems

A number of recent studies have shown evidence indicating that aspects of computing and telecommunication systems can show heavy-tailed distributions. Measurements of computer network traffic have shown that autocorrelations are often related to heavy tails; this is the phenomenon of *self-similarity* [5, 10]. Measurements of file sizes in the Web [1, 3] and in I/O patterns [13] have shown evidence that file sizes can show heavy-tailed distributions. In addition, the CPU time demands of UNIX processes have also been shown to follow heavy-tailed distributions [7, 9].

The presence of heavy-tailed distributions in measured data can be assessed in a number of ways. The simplest is to plot the ccdf on log–log axes and visually inspect the resulting curve for linearity over a wide range (several orders of magnitude). This is based on Eq. (3.1), which can be recast as

$$\lim_{x \to \infty} \frac{d \log \bar{F}(x)}{d \log x} = -\alpha,$$

so that for large x, the ccdf of a heavy-tailed distribution should appear to be a straight line on log–log axes with slope $-\alpha$.

An example empirical data set is shown in Fig. 3.3, which is taken from Crovella and Bestavros [3]. This figure is the ccdf of file sizes transferred through the network due to the Web, plotted on log–log axes. The figure shows that the file size distribution appears to show power-law behavior over approximately three orders of magnitude. The slope of the line fit to the upper tail is approximately -1.2, yielding $\hat{\alpha} \approx 1.2$.

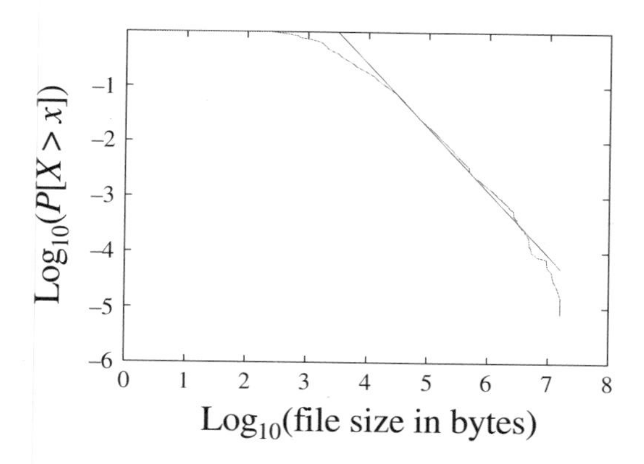

Fig. 3.3 Log–log complementary distribution of sizes of files transferred through the Web.

3.3 STABILITY IN SYSTEMS WITH HEAVY-TAILED WORKLOADS

As heavy-tailed distributions are increasingly used to represent workload characteristics of computing systems, researchers interested in simulating such systems are beginning to use heavy-tailed inputs to simulations. For example, Paxson [12] describes methods for generating self-similar time series for use in simulating network traffic and Park et al. [11] use heavy-tailed file sizes as inputs to a network simulation. However, an important question arises: How stable are such simulations? This can be broken down into two questions:

1. How long until such simulations reach steady state?
2. How variable is system performance at steady state?

In this section we will show that if simulation outputs are dependent on all the moments of the distribution F, then the answers to the above questions can be surprising. Essentially, we show that such simulations can take a very long time to reach steady state; and that such simulations can be much more variable at steady state than is typical for traditional systems.

Note that some simulation statistics may not be affected directly by all the moments of the distribution F, and our conclusions do not necesssarily apply to those cases. For example, the mean number of customers in an $M/G/\infty$ queueing system may not show unusual behavior even if the service time distribution F is heavy tailed because that statistic only depends on the mean of F.

Since not all simulation statistics will be affected by heavy-tailed workloads, we choose a simple statistic to show the generality of our observations: the sample mean of the heavy-tailed inputs. Since our results apply to the sample mean of the input, we expect that any system property that behaves like the sample mean should show similar behavior. For example, assume we want to achieve steady state in a particular simulation. This implies that the measured system utilization $\lambda\bar{x}$ (where λ^{-1} is the average interarrival time and $\bar{x}$ is the sample mean of service times over some period) should be close to the desired system utilization ρ. For this to be the case, $\bar{x}$ must be close to its desired mean μ.

To analyze the behavior of the sample mean, we are concerned with the convergence properties of sums of random variables. The normal starting point for such discussions would be the *central limit theorem* (CLT). Unfortunately, the CLT applies only to sums of random variables with finite variance, and so does not apply in this case. In the place of the CLT we instead have limit theorems for heavy-tailed random variables first formulated by Lévy [4, 5].

To introduce these results we need to define the notation $A \xrightarrow{d} B$, which means that the random variable A converges in distribution to B (roughly, has distribution B for large n). Then the usual CLT can be stated as follows. For X_i i.i.d. and drawn from some distribution F with mean μ and variance $\sigma^2 < \infty$, define

$$A_n = \frac{1}{n}\sum_{i=1}^{n} X_i$$

and

$$Z_n = n^{1/2}(A_n - \mu); \tag{3.3}$$

then

$$Z_n \xrightarrow{d} \mathcal{N}(0, \sigma^2), \tag{3.4}$$

where $\mathcal{N}(0, \sigma^2)$ is a normal distribution.

However, when X_i are i.i.d. and drawn from some distribution F that is heavy tailed with tail index $1 < \alpha < 2$, then if we define

$$Z_n = n^{1-1/\alpha}(A_n - \mu) \tag{3.5}$$

we find that

$$Z_n \xrightarrow{d} \mathcal{S}_\alpha, \tag{3.6}$$

where $\mathcal{S}_\alpha$ is an α-*stable* distribution. The α-stable distribution has four parameters: α, a location parameter (analogous to the mean), a scale parameter (analogous to the standard deviation), and a skewness parameter. Based on the value of the last parameter, the distribution can be either skewed or symmetric. A plot of the symmetric α-stable distribution with $\alpha = 1.2$ and location zero is shown in Fig. 3.4. From the figure it can be seen that this distribution has a bell-shaped body much like the normal distribution but that it has much heavier tails. In fact, the α-stable distribution has power-law tails that follow *the same* α as that of the distribution F from which the original observations were drawn.

From Eqs. (3.5) and (3.6) we can make two observations about the behavior of sums of heavy-tailed random variables. First, Eq. (3.5) states that such sums may converge much more slowly than is typical in the finite variance case. Second, Eq. (3.6) states that, even after convergence, the sample mean will show high variability—it follows a heavy-tailed distribution.

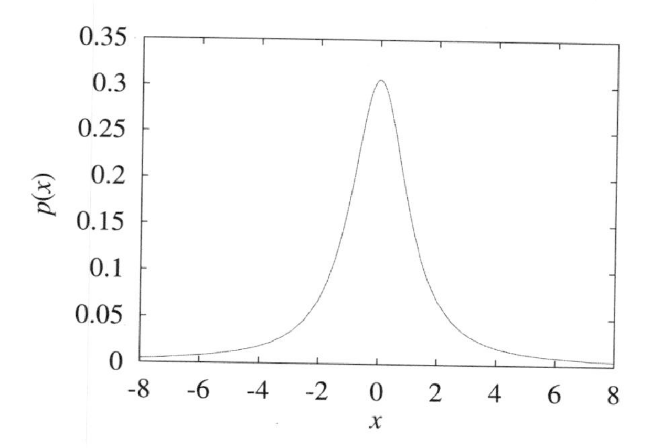

Fig. 3.4 The pmf of an α-stable distribution.

These effects can be seen graphically in Fig. 3.5. This figure shows histograms of A_n for varying values n. In plot (a) we show the case in which the X_i were drawn from an exponential distribution; in plot (b) we show the case in which the X_i were drawn from a strictly positive heavy-tailed distribution with $\alpha = 1.4$. In both cases the mean of the underlying distribution was 1. Plot (a) shows that the most likely value of the sample mean is equal to the true mean, even when summing only a small number of samples. In addition, it shows that as one sums larger numbers of samples, the sample mean converges quickly to the true mean. However, neither of these observations are true for the case of the heavy-tailed distribution in plot (b). When summing small numbers of samples, the most likely value of the sample mean is far from the true mean, and the distribution progresses to its final shape rather slowly.

Thus we have seen that the convergence properties of sums of heavy-tailed random variables are quite different from those of finite variance random variables. We relate this to steady state in simulation as follows: presumably for a simulation to reach steady state, it must at a minimum have seen enough of the input workload to observe its mean. Of course, it may be necessary for much more of the input to be consumed before the simulation reaches steady state, so this condition is a relatively weak one. Still, we show in the next two subsections that this condition has surprising implications for simulations.

3.3.1 Slow Convergence to Steady State

Equation (3.6) states that for large n, Z_n converges in distribution. Thus another way of formulating Eq. (3.5) is

$$|A_n - \mu| \sim n^{1/\alpha - 1}.$$

In this form it is more clear how slowly A_n converges to μ. If α is close to 1, then the rate of convergence, measured as the difference between A_n and μ, is very slow—

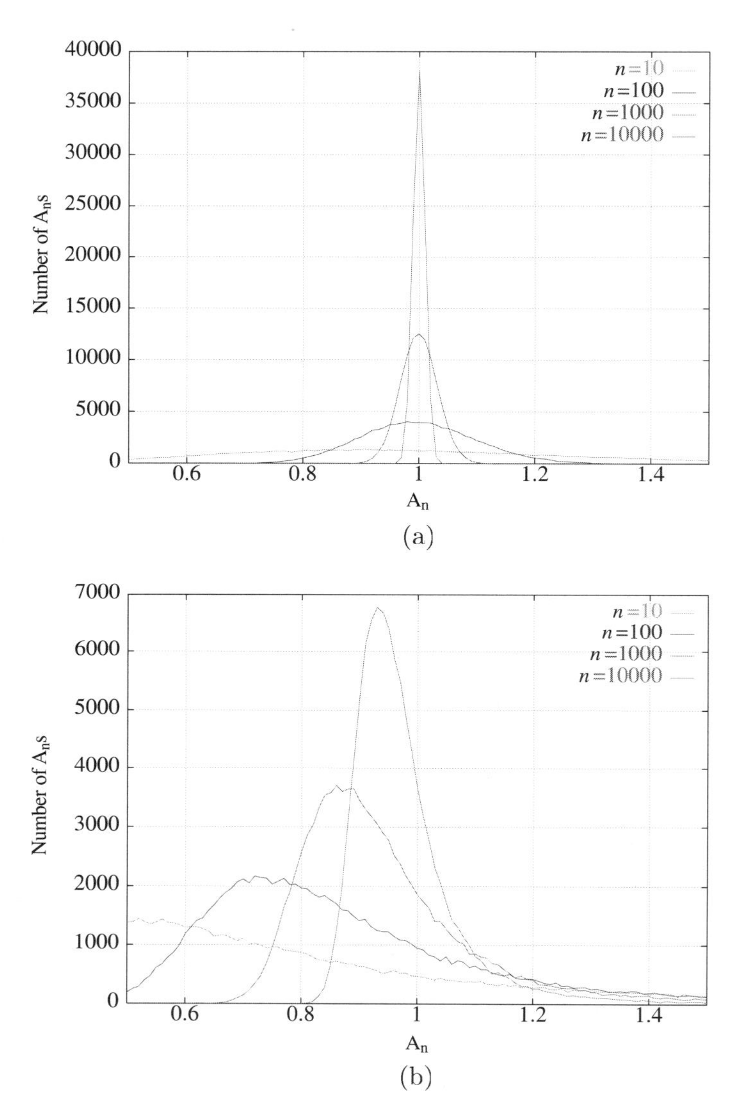

Fig. 3.5 Histogram of A_n as n varies for (a) exponential and (b) heavy-tailed random variables.

until, for $\alpha = 1$, the average does not converge at all, reflecting the fact that the mean is infinite.

Suppose one would like to use A_n to form a estimate of the mean μ that is accurate to k digits. Alternatively, one might state that a simulation has reached steady state when the observed mean of the input A_n agrees with μ to k digits. Then we would like

$$|A_n - \mu|/\mu \leq 10^{-k}.$$

Now, as a rough approximation,

$$|A_n - \mu| = c_1 n^{1/\alpha - 1}$$

for some positive constant c_1. Then we find that

$$n \geq c_2 10^{k/(1-1/\alpha)}.$$

We can say that given this many samples, k digit accuracy is "highly likely."

For example, assume we would like two-digit accuracy in A_n, and suppose $c_2 \approx 1$. Then the number of samples n necessary to achieve this accuracy is shown in Table 3.1. This table shows that as $\alpha \to 1$, the number of samples necessary to obtain convergence in the sample mean explodes. Thus it is *not feasible* in any reasonable amount of time to observe steady state in such a simulation as we have defined it. Over any reasonable time scale, such a simulation is *always in transient state.*

3.3.2 High Variability at Steady-State

Equation (3.6) shows that, even at steady state, the sample mean will be distributed according to a heavy-tailed distribution, and hence will show high variability. Thus the likelihood of an erroneous measurement of μ is still nonnegligible. Equivalently, the simulation still behave erratically.

To see this more clearly, let us define a *swamping* observation as one whose presence causes the estimate of μ to be at least twice as large as it should be. That is, if we happen to encounter a swamping observation in our simulation, the observed mean of the input will have a relative error of at least 100%.

In a simulation consisting of n inputs, a swamping observation must have value at least $n\mu$. Let us assume that the inputs are drawn from a Pareto distribution. Such a distribution has $\mu = k\alpha/(\alpha - 1)$. Then the probability $p_{n\mu}$ of observing a value of $n\mu$ or greater is

$$p_{n\mu} = P[X > n\mu] = \left(\frac{k}{nk\alpha/(\alpha - 1)}\right)^{\alpha} = \left(\frac{\alpha - 1}{n\alpha}\right)^{\alpha},$$

TABLE 3.1 Number of Samples Necessary to Achieve Two-Digit Accuracy in Mean as a Function of α

α	n
2.0	10,000
1.7	72,000
1.5	1,000,000
1.2	10^{12}
1.1	10^{22}

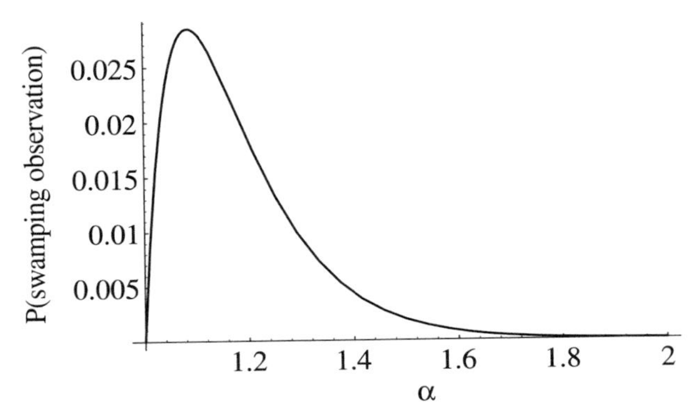

Fig. 3.6 Probability of a swamping observation in 10^5 inputs as a function of α.

and the probability p of observing such a value at least once in n trials is

$$p = 1 - (1 - p_{n\mu})^n.$$

Figure 3.6 shows a plot of p as a function of α for $n = 10^5$. (The figure is not significantly different for other values of n, say 10^6 or 10^7.) It shows that even in a relatively long simulation, the probability of a swamping observation is not negligible; when α is below about 1.3, such an observation could occur more often than once in a hundred simulations. The probability declines very rapidly for $\alpha < 1.1$ not because the variability of the simulation is declining, but because of the way we have defined the swamping observation: in terms of the distributional mean. When $\alpha = 1$, the mean is infinite, and so it becomes impossible to observe a value greater than the mean.

Taken together, Table 3.1 and Fig. 3.6 also provide some insight into the value of α above which it may be possible to obtain convergent, consistent simulations. The table shows that simulation convergence becomes impractical when α is somewhere in the region between 1.7 and 1.5; and the figure shows that simulations become erratic at steady state in approximately the same region. As a result, we can conclude that the difficulties inherent in simulations with heavy-tailed inputs are likely to be particularly great when α is less than about 1.7; and that when α is greater than or equal to about 1.7 it may be feasible (given sufficient computing effort) to obtain consistent steady state in simulation.

3.4 CONCLUSIONS

We have shown that a difficult problem arises when simulating systems with heavy-tailed workloads. In such systems, steady-state behavior can be elusive, because average-case behavior depends on the presence of many small observations as well as a few large observations.

This problem has two implications. First, since a number of large but rare observations must occur before average-case behavior is evident, convergence of a simulation to steady state may be slow. It may not be possible in any reasonable time to achieve steady state. Second, since many small observations must occur to balance the presence of large observations, large observations can have a dominating effect on performance results even at steady state. Simulations may still behave erratically even at steady state.

ACKNOWLEDGMENTS

This research was supported in part by NSF grant CCR-9501822 and by a grant from Hewlett-Packard Company.

REFERENCES

1. M. F. Arlitt and C. L. Williamson. Web server workload characterization: the search for invariants. *IEEE/ACM Trans. Networking*, **5**(5):631–645, 1997.
2. S. Asmussen and K. Binswanger. Simulation of ruin probabilities for subexponential claims. Preprint, 1995.
3. M. E. Crovella and A. Bestavros. Self-similarity in World Wide Web traffic: evidence and possible causes. *IEEE/ACM Trans. Networking*, **5**(6):835–846, December 1997.
4. W. Feller. *An Introduction to Probability Theory and Its Applications*, Vol. II, 2nd ed. Wiley, New York, 1971.
5. M. Garrett and W. Willinger. Analysis, modeling, and generation of self-similar VBR video traffic. In *Proc. ACM SIGCOMM '94*, September 1994.
6. C. M. Goldie and C. Klüppelberg. Subexponential distributions. In R. J. Adler, R. E. Feldman, and M. S. Taqqu, eds., *A Practical Guide to Heavy Tails*, pp. 435–460. Chapman and Hall, New York, 1998.
7. M. Harchol-Balter and A. Downey. Exploiting process lifetime distributions for dynamic load balancing. *ACM Trans. Comput. Syst.*, **15**(3):253–285, 1997.
8. P. Heidelberger. Fast simulation of rare events in queueing and reliability models. *Lecture Notes Comput. Sci.*, **729**:165–202, 1993.
9. W. E. Leland and T. J. Ott. Load-balancing heuristics and process behavior. In *Proceedings of Performance and ACM Sigmetrics*, pp. 54–69, 1986.
10. W. E. Leland, M. S. Taqqu, W. Willinger, and D. V. Wilson. On the self-similar nature of Ethernet traffic (extended version). *IEEE/ACM Trans. Networking*, **2**:1–15, 1994.
11. K. Park, G. T. Kim, and M. E. Crovella. On the relationship between file sizes, transport protocols, and self-similar network traffic. In *Proceedings of the Fourth International Conference on Network Protocols (ICNP'96)*, pp. 171–180, October 1996.
12. V. Paxson. Fast approximation of self-similar network traffic. Technical Report LBL-36750, Lawrence Berkely National Laboratory, April 30 1995.

13. D. L. Peterson, Data center I/O patterns and power laws. In *CMG Proceedings*, December 1996.
14. R. Y. Rubinstein and B. Melamed. *Efficient Simulation and Monte Carlo Methods*, Wiley, New York, 1997.
15. G. Samorodnitsky and M. S. Taqqu. *Stable Non-Gaussian Random Processes*. Stochastic Modeling. Chapman and Hall, New York, 1994.

4

QUEUEING BEHAVIOR UNDER FRACTIONAL BROWNIAN TRAFFIC

ILKKA NORROS

VTT Information Technology, Espoo, Finland

4.1 INTRODUCTION

This chapter gives an overview of some properties of the storage occupancy process in a buffer fed with "fractional Brownian traffic," a Gaussian self-similar process. This model, called here "fractional Brownian storage," is the logically simplest long-range-dependent (LRD) storage system having strictly self-similar input variation. The impact of the self-similarity parameter H can be very clearly illustrated with this model. Even in this case, all the known explicitly calculable formulas for quantities like the storage occupancy distribution are only limit results, for example, large deviation asymptotics. Scaling formulas, on the other hand, hold exactly for this model.

The simplicity is won at the price that the input model is not meaningful at smallest time scales, where half of the "traffic" is negative. The model can be justified by rigorous limit theorems, but it should be emphasized that this involves not only a central limit theorem (CLT) argument for Gaussianity but also a heavy traffic limit—see Chapter 5. From a less rigorous, practical viewpoint one can say that fractional Brownian storage gives usable results when, at time scales relevant for queueing phenomena, the traffic consists of independent streams such that a large number of them are simultaneously active, and second-order self-similarity (see Chapter 1) holds.

Chapter 7 describes many features of storage processes with finitely aggregated on/off input traffic, which differ qualitatively from those of fractional Brownian

---

Self-Similar Network Traffic and Performance Evaluation, Edited by Kihong Park and Walter Willinger
ISBN 0-471-31974-0

storage. For example, the correlation function of the input process remains unchanged if one changes the distributions of on and off periods to each other, whereas the queueing process usually becomes entirely different. Thus second-order self-similarity alone tells *nothing* about queueing behavior, if the traffic variation cannot be considered Gaussian. Another important difference can be seen in how a queue builds up. Assume that the individual sources are relatively powerful so that a queue starts to build up when a small number, say, k of them are simultaneously active. Then a small increase in link speed can result in a considerable performance improvement, if it happens to increase the small integer k by one. A related observation about a possibly big effect of a small increase in buffer size is discussed in Chapter 10 in $GI/GI/1$ context. In fractional Brownian traffic, in contrast, individual source streams are thought of as being infinitely thin compared with the link speed, and phenomena caused by the granularity of traffic are not reflected in the model.

On the other hand, if the Gaussian property is sufficiently well satisfied, but second-order self-similarity holds only asymptotically, then some of the techniques presented in this chapter are still usable in a modified form. See Addie et al. [1].

As a special case with $H = \frac{1}{2}$, the model includes the "Brownian storage," which is often used in the heavy traffic limit as a diffusion approximation for short-range-dependent (SRD) queueing systems. Many problems have explicit solutions in the Brownian case, and one method to study the fractional case is to analyze their derivations and identify the steps that are not justified when H does not equal to $\frac{1}{2}$. Regrettably, almost all methodological cornerstones of Brownian motion are based on independence and thus unavailable when $H \neq \frac{1}{2}$: independent increments, Markov property, margingale property, renewal times.

Besides self-similarity, the general methods available for fractional Brownian storage come from the literature on Gaussian processes. In particular, the beautiful theory of their large deviations in path space turns out to be well suited for the needs of performance analysis.

This chapter is structured as follows. The definitions are given in Section 4.2. Some basic scaling formulas are derived in Section 4.3. Results based on large deviations in path space are presented in Section 4.4. Finally, some other approaches are outlined in Section 4.5.

4.2 INPUT, OUTPUT, AND STORAGE PROCESSES

We consider in continuous time an unlimited fluid storage that is fed by *fractional Brownian traffic*, defined below, and emptied at constant service rate c.

4.2.1 The Input Process, "Fractional Brownian Traffic"

The fluid input in time interval $(s, t]$ is denoted by $A(s, t)$ and it has the form

$$A(s, t) = m(t - s) + \sigma(Z_t - Z_s), \quad s, t \in (-\infty, \infty), s \leq t,$$

where m and σ are nonnegative parameters, $m < c$, and the process $(Z_t)_{t \in \Re}$ is a *normalized fractional Brownian motion* (FBM), defined as a centered Gaussian process with stationary increments, continuous paths, and variance $EZ_t^2 = |t|^{2H}$ (see Chapter 1). Z is a self-similar process:

$$(Z(\alpha t), t \in \Re) \stackrel{\mathscr{D}}{=} (\alpha^H Z(t), t \in \Re) \tag{4.1}$$

for every $\alpha > 0$, where $\stackrel{\mathscr{D}}{=}$ denotes that the processes have the same finite-dimensional distributions (considered as random elements of some appropriate space of continuous functions, the whole processes then have the same distribution). H is called the self-similarity parameter, and it is a number from the interval (0, 1). Thus the traffic model has three parameters—m, σ^2, and H—and the storage model has in addition a fourth parameter—c. The parameter m is the mean input rate, and σ^2 is the variance of traffic in a time unit. It is often useful to write

$$\sigma^2 = ma,$$

where a, the index of dispersion at unit time, has sometimes been called "peakedness." The point in using ma instead of σ^2 is that varying m can now be interpreted as varying the number of traffic sources alone, without changing their characteristics.

The parameter H characterizes dependence in the input process. For $H \in (\frac{1}{2}, 1)$, all the random variables $A(s, t)$ with $s < t$ are strictly positively correlated. For $H = \frac{1}{2}$, the input process is a Brownian motion, and the storage model is a classical diffusion approximation for a queueing process. For $H \in (0, \frac{1}{2})$, inputs on disjoint intervals are negatively correlated. It is possible that this case has no natural applications in teletraffic contexts, but including it comes usually for free, so we do not exclude it.

We also write

$$A_t = A(0, t) \quad \text{for } t \geq 0, \qquad A_t = -A(t, 0) \quad \text{for } t \leq 0. \tag{4.2}$$

Then $A(s, t) = A_t - A_s$.

This model was identified by Leland et al. [8] and by Norros [12, 13] as a simple way to include the observed self-similarity features of data traffic in mathematical performance analysis.

4.2.2 The Storage Process

The storage occupancy process with fractional Brownian traffic as input is defined by Reich's formula:

$$V_t = \sup_{s \leq t}(A(s, t) - c(t - s)), \quad t \in (-\infty, \infty). \tag{4.3}$$

In words: if $A(s, t) > c(t - s)$, the buffer must contain at time t at least the difference; a short reasoning yields that, on the other hand, it does not contain more than the maximum over s of such differences.

Since Z has stationary increments, V is a stationary process. Z is invertible in time, so V_0 is distributed like $\sup_{t \geq 0}(A(0, t) - ct)$. An application of Kolmogorov's continuity criterion to the process $z_t/(1 + |z|)$ yields $\lim_{t \to \infty} A(0, t)/t = m$ with probability 1. Since we have assumed $m < c$, it follows that V_0 is a.s. finite. Note that V is nonnegative, although the input process has (regrettably!) negative increments also.

The ruggedness (nondifferentiability) of the fractional Brownian path implies a paradoxical property of V: the storage is almost always nonempty. Indeed, it can be shown that the supremum in Eq. (4.3) is positive with probability one, and by stationarity, the positivity must also hold for almost every time point in almost every realization of the process. The set of times t with $V_t = 0$ is uncountable a.s., with almost every point being an accumulation point, so that between any two distinct busy periods there are a.s. infinitely many tiny busy periods. This is, of course, an anomaly of the continuous-time model only, it has no counterpart in the teletraffic reality being modeled. Note, on the other hand, that it is a natural feature of a heavy traffic limit process (cf. Chapter 5), and that the case $H = \frac{1}{2}$ is no exception here.

4.2.3 The Output Process

It is natural to define the output within an interval $(s, t]$ as

$$U(s, t) = A(s, t) - V_t + V_s, \quad U_t = U(0, t) \quad \text{for } t \geq 0. \tag{4.4}$$

We then have from Eqs. (4.3) and (4.2) that

$$U_t = V_0 + ct - \sup_{s \leq t}(cs - A_s),$$

so that U is the difference of two increasing processes and thus has paths that are differentiable almost everywhere. Thus the microscale behavior of the output process is entirely different from that of the input process—one more nonpleasant anomaly of our model!

Since the storage is almost never empty, the output proceeds almost always with full rate c. However, the output within the set of time points where the storage is empty is negative, and the mean rate is still m as can be seen from (1.4) dividing by t and letting $t \to \infty$.

4.3 SCALING RULES

The self-similarity property (4.1) allows for deriving some useful relations. First, we observe that from a mathematical point of view, H is the only "real parameter" of the storage system, since the effect of the others reduces to scaling.

Proposition 4.3.1. *Let the self-similarity parameter H be fixed, and denote by $V^{(m,\sigma^2,c)}$ the storage occupany process of a fractional Brownian storage with parameters m, σ^2, and c. Then*

$$\left(V_t^{(m,\sigma^2,c)}\right)_{t\in\Re} \overset{\mathcal{D}}{=} \left(\frac{c-m}{\alpha^*} V_{\alpha^* t}^{(0,1,1)}\right)_{t\in\Re}, \quad \textit{where} \quad \alpha^* = \left(\frac{c-m}{\sigma}\right)^{1/(1-H)}.$$

Proof. For any $\alpha > 0$, we have by Eq. (4.1) that

$$\begin{aligned}
V_t &= \sup_{s\le t}(A(s,t) - c(t-s)) \\
&= \sup_{s\le t}(\sigma(Z_t - Z_s) - (c-m)(t-s)) \\
&\overset{\mathcal{D}}{=} \sup_{s\le t}(\sigma\alpha^{-H}(Z_{\alpha t} - Z_{\alpha s}) - (c-m)(t-s)) \\
&= \sup_{s\le t}(\sigma\alpha^{-H}(Z_{\alpha t} - Z_{\alpha s}) - (c-m)\alpha^{-1}(\alpha t - \alpha s)),
\end{aligned}$$

where the similarity in distribution holds for the whole processes, not just for a single t. Now, choose $\alpha = \alpha^*$ by requiring that

$$\sigma\alpha^{-H} = (c-m)\alpha^{-1}.$$

■

In particular, Proposition 4.3.1 has the following consequences.

Corollary 4.3.2. *The storage occupancy distribution obeys the scaling law*

$$\begin{aligned}
P(V^{(m,\sigma^2,c)} > x) &= P\left(V^{(0,1,1)} > \frac{\alpha^*}{c-m}x\right) \\
&= P\left(V^{(0,1,1)} > \frac{(c-m)^{H/(1-H)}}{\sigma^{1/(1-H)}}x\right).
\end{aligned}$$

Corollary 4.3.3. *Denote by $B^{(m,\sigma^2,c)}$ the length of the busy period containing time zero in fractional Brownian storage with parameters m, σ^2, c. Its distribution obeys the scaling law*

$$P(B^{(m,\sigma^2,c)} > x) = P(B^{(0,1,1)} > \alpha^* x).$$

Another type of scaling law is obtained from Corollary 4.3.2 by fixing a "quality of service criterion," requiring that the probability of exceeding a certain storage level x equals a given small number ϵ.

Corollary 4.3.4. *Denote*

$$f(y) = P\left(\sup_{t \geq 0}(Z(t) - yt) > 1\right).$$

Then the condition $P(V^{(m,\sigma^2,c)} > x) = \epsilon$ is equivalent to the "buffer dimensioning formula"

$$x = f^{-1}(\epsilon) a^{1/(2(1-H))} c^{(2H-1)/(2(1-H))} \frac{\rho^{1/(2(1-H))}}{(1-\rho)^{H/(1-H)}}, \tag{4.5}$$

where $\rho = m/c$, and to the "bandwidth allocation rule"

$$c = m + f^{-1}(\epsilon) a^{1/(2H)} x^{-(1-H)/H} m^{1/(2H)}, \tag{4.6}$$

where $a = \sigma^2/m$.

Except for the unknown constant $f^{-1}(\epsilon)$, these formulas fully describe the effect of the system parameters on buffer and bandwidth allocation. The qualitative difference between the classical case $H = \frac{1}{2}$ and the LRD case $H > \frac{1}{2}$ is clearly seen. For example, the buffer requirement depends on the utilization factor ρ as $(1-\rho)^{-H/(1-H)}$; with $H = 0.8$, a typical value for data traffic, this is $(1-\rho)^{-4}$, which means very fast increase with ρ. This kind of scaling analysis can be elaborated further and applied to many types of problems—see Krishnan et al. [7].

For the rest of this chapter, we assume always that the parameters are

$$m = 0, \quad \sigma^2 = 1, \quad c = 1.$$

4.4 LARGE DEVIATIONS IN PATH SPACE

4.4.1 Large Deviation Principle for Fractional Brownian Motion

The large deviation asymptotics of the complementary storage occupancy distribution function

$$Q(x) = P\left(\sup_{t \geq 0}(Z_t - t) > x\right)$$

was obtained by Duffield and O'Connell [5] (the easy lower bound part was observed already by Norros [12]). Here we outline the derivation of this result within a richer framework, applying the theory of large deviations of Gaussian processes in path space. This allows simple heuristic derivations that can also be made accurate, in the spirit of Shwartz and Weiss [20]. This type of result has recently been obtained also by Chang et al. [2] and by O'Connell and Procissi [18].

Denote by $(\Omega, \|\cdot\|_\Omega)$ the function space

$$\Omega = \left\{ \omega\colon \omega \text{ is continuous } \Re \to \Re,\ \omega(0) = 0,\ \lim_{t\to\infty} \frac{\omega(t)}{1+|t|} = \lim_{t\to-\infty} \frac{\omega(t)}{1+|t|} = 0 \right\},$$

equipped with the norm

$$\|\omega\|_\Omega = \sup\left\{ \frac{|\omega(t)|}{1+|t|} : t \in \Re \right\}.$$

Ω is a separable Banach space. Let P be the unique probability measure on the Borel sets of Ω such that the coordinate process $Z_t(\omega) = \omega(t)$ is a normalized fractional Brownian motion with some fixed self-similarity parameter H. Denote the covariance function of Z by

$$\Gamma(s, t) = \tfrac{1}{2}(|s|^{2H} + |t|^{2H} - |s-t|^{2H}), \quad s, t \in \Re.$$

The *reproducing kernel Hilbert space* R of the Gaussian process Z is a space of functions $\Re \to \Re$, which is defined by the condition that the mapping

$$Z_t \mapsto \Gamma(t, \cdot)$$

extends to an isometry between R and the Gaussian space of Z, that is, the smallest closed linear subspace of $L^2(\Omega, \mathscr{B}_\Omega, P)$ containing all the Z_t. It is not difficult to check that R is a subset of Ω. However, the topology of R is finer than that of Ω. The name of R comes from the important *reproducing kernel property*

$$\langle f, \Gamma(t, \cdot)\rangle_R = f(t), \quad f \in R, \tag{4.7}$$

which is an extension of the relation

$$\langle \Gamma(s, \cdot), \Gamma(t, \cdot)\rangle_R = \Gamma(s, t)$$

defining the inner product for the functions generating R.

Let us now turn to the large deviations. The general result we need is called generalized Schilder's theorem—see, for example, Theorem 3.4.12 in Deuschel and Stroock [4]. Here is a formulation appropriate in our case:

Theorem 4.4.1. *Denote*

$$I(f) = \begin{cases} \frac{1}{2}\|f\|_R^2, & \text{if } f \in R, \\ \infty, & \text{otherwise,} \end{cases}$$

and, for any $A \subset \Omega$, $I(A) = \inf_{\omega \in A} I(f)$. *The function* I *is a good rate function for the centered Gaussian measure* P, *and the following large deviation principle holds:*

$$\textit{for } F \textit{ closed in } \Omega\text{:} \quad \limsup_{n\to\infty} \frac{1}{n} \log P\left(\frac{Z}{\sqrt{n}} \in F\right) \le -I(F);$$

$$\textit{for } G \textit{ open in } \Omega\text{:} \quad \liminf_{n\to\infty} \frac{1}{n} \log P\left(\frac{Z}{\sqrt{n}} \in G\right) \ge -I(G).$$

4.4.2 Large Queues

In order to use Theorem 4.4.1 for studying the asymptotics of $Q(x)$, we have to make a scaling transformation first:

$$\begin{aligned} Q(x) &= P\left(\sup_{t\ge 0}(Z_t - t) > x\right) \\ &= P\left(\sup_{t\ge 0}(x^H Z_{t/x} - t) > x\right) \\ &= P\left(\sup_{t\ge 0}(x^{H-1} Z_{t/x} - t/x) > 1\right) \\ &= P\left(\sup_{T\ge 0}(x^{H-1} Z_T - T) > 1\right) \\ &= P\left(\frac{Z}{x^{1-H}} \in G\right), \end{aligned}$$

where G is the union over $T \ge 0$ of the open sets

$$G_T = \{\omega: Z_T(\omega) - T > 1\}.$$

Now, the lower bound part of Theorem 4.4.1 yields that

$$\liminf_{x\to\infty} \frac{1}{x^{2-2H}} \log P\left(\frac{Z}{x^{1-H}} \in G\right) \ge -I(G).$$

It remains to determine $I(G)$. First, note that $I(G) = \inf_{T\ge 0} I(G_T)$. Thus, we have to minimize $\| f \|_R$ over all paths $f \in R \cap G_T$. In general, such questions lead to difficult variational problems. For the particular set G_T, however, the solution is extremely simple. Indeed, by the reproducing kernel property (4.7),

$$f \in G_T \Leftrightarrow \langle f, \Gamma(T, \cdot)\rangle_R > 1 + T. \tag{4.8}$$

For any $a > 0$, the path f with smallest R-norm satisfying $\langle f, \Gamma(T, \cdot)\rangle_R = a$ is, of course,

$$f = f^{T,a} = \frac{a}{\|\Gamma(T, \cdot)\|_R^2}\Gamma(T, \cdot) = \frac{a}{T^{2H}}\Gamma(T, \cdot),$$

and it follows that

$$I(G_T) = \tfrac{1}{2}\| f^{T,1+T}\|_R^2 = \frac{1}{2}\left(\frac{1+T}{T^H}\right)^2.$$

The last step is to minimize this over T. It is quickly seen that the minimum is obtained at $T^* = H/(1-H)$. Substituting this, we obtain the final result

$$I(G) = \frac{1}{2\kappa(H)^2}, \quad \text{where} \quad \kappa(H) = H^H(1-H)^{1-H}.$$

It is not difficult to show that $\overline{G} = \{\omega\colon \sup_{t\geq 0}(Z_t(\omega) - t)1\}$, and it follows that $I(\overline{G}) = I(f^{T^*,1+T^*}) = I(G)$. Thus we have shown that the storage occupancy distribution has Weibullian tail behavior:

Proposition 4.4.2. *Denoting $\kappa(H) = H^H(1-H)^{1-H}$, we have*

$$\lim_{x\to\infty} x^{-(2-2H)}\log Q(x) = -1/(2\kappa(H)^2);$$

that is,

$$Q(x) \sim \exp\left(-\frac{x^{2-2H}}{2\kappa(H)^2}\right) \tag{4.9}$$

in the sense of logarithmic asymptotics.

Note that Eq. (4.9) reduces to the exact expression e^{-2x} in the Brownian case $H = \frac{1}{2}$.

Moreover, the path space approach gives the additional information that the "most typical path" of Z to obtain a high buffer occupancy level x at time 0 is

$$f(t) = \frac{x}{1-H}\left(\Gamma\left(1, \frac{t+T^*x}{T^*x}\right) - 1\right), \quad \text{where} \quad T^* = \frac{H}{1-H}. \tag{4.10}$$

We used quotes, since this smooth path is in fact very different from really typical paths of fractional Brownian motion. The storage is empty until time $-T^*x$, when it starts to fill, reaching x at time 0. The "typical storage occupancy path with $V_0 = 1$" is plotted in Fig. 4.1 in the three cases $H = 0.5$, $H = 0.9$, and $H = 0.2$. Note that for $H > \frac{1}{2}$, the derivative of f is 1 at $-T^*x$ and at 0, whereas it is infinite at both points for $H < \frac{1}{2}$. Note also that the "typical queue-mountain" is not symmetric when $H \neq \frac{1}{2}$. For $H > \frac{1}{2}$, the buildup phase takes more time than coming down, whereas the opposite happens with $H < \frac{1}{2}$.

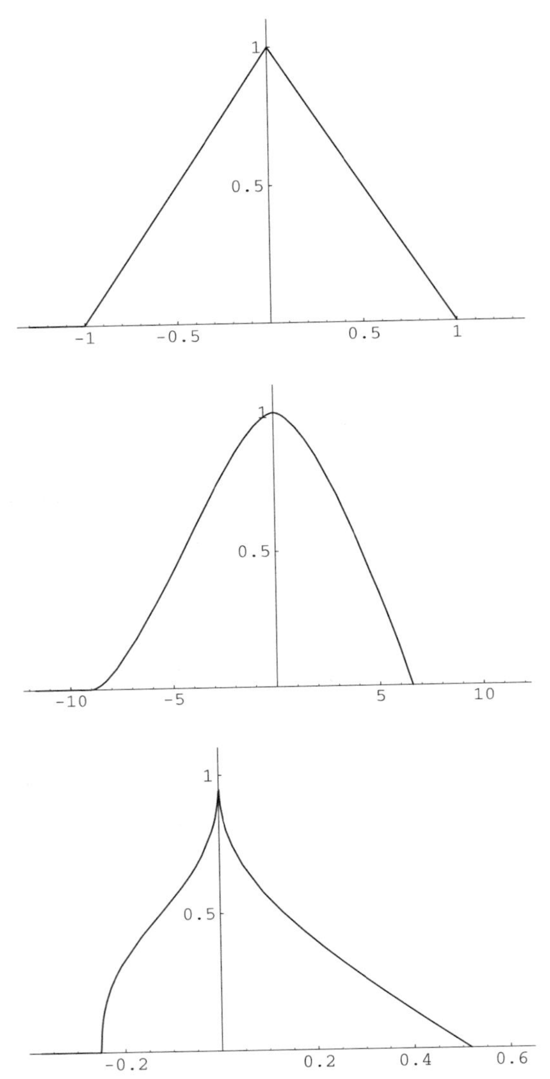

Fig. 4.1 The "most typical" storage occupancy path that reaches level 1 at time 0. From top: $H = 0.5$, $H = 0.9$, and $H = 0.2$.

4.4.3 Long Busy Periods

In a similar way, we can estimate the distribution of long busy periods (the peculiarities of busy periods of fractional Brownian storage were briefly explained in Section 4.2.2). There is no such thing as an arbitrary busy period, but the busy period containing the origin (or any other given time point) is a well-defined object. Let us have a look at the main steps in applying Schilder's theorem to the asymptotics of the busy period containing the origin (for a detailed presentation, see Norros [15]).

It can be deduced from the stationarity of the system that, as regards large deviation asymptotics, instead of considering any busy period of length greater than T that covers 0, we can consider one that extends from 0 to T. Denote by $Q^*(s, t)$ the set of paths $\omega \in \Omega$ such that $\omega(u) - \omega(s) > u - s$ for all $u \in (s, t]$. If $\omega \in Q^*(s, t)$, then the storage process $V(\omega)$ has a busy period that contains the interval $(s, t]$. We apply once again the scaling trick to obtain

$$\begin{aligned} P(Z \in Q^*(0, T)) &= P(Z_T > t\ \forall t \in (0, T]) \\ &= P(Z_{Tt} > Tt\ \forall t \in (0, 1]) \\ &= P(T^H Z_t > Tt\ \forall t \in (0, 1]) \\ &= P\left(\frac{Z}{T^{1-H}} \in Q^*(0, 1)\right). \end{aligned}$$

Now, Theorem 4.4.1 yields (we skip all details concerning open and closed sets) that the distribution tail of the length B of the busy period containing the origin is

$$\lim_{T \to \infty} \frac{1}{T^{2-2H}} \log P(B > T) = -I(Q^*(0, 1)), \tag{4.11}$$

or

$$P(B > T) \sim \exp(-I(Q^*(0, 1))x^{2-2H}).$$

We have $I(Q^*(0, 1)) = \|f\|_R^2/2$, where f has the meaning of being "the most typical path that gives a busy period from 0 to 1." The function f seems to be unknown, as is its R-norm. The latter is obviously larger than 1, since all functions $g \in Q^*(0, 1)$ satisfy $g(1) \geq 1$, and the reasoning in Eq. (4.8) shows that their norms are larger than $\|\Gamma(1, \cdot)\|_R = 1$. On the other hand, the path χ, which is between 0 and 1 a straight line from 0 to 1, and "most typical" elsewhere, is in the closure of $Q^*(0, 1)$ (it produces a nongenuine "busy period" where the storage remains empty). The squared norm of χ is known to be

$$\|\chi\|_R^2 = c_2^2 \doteq (H(2H - 1)(2 - 2H)B(H - \tfrac{1}{2}, 2 - 2H))^{-1}, \tag{4.12}$$

which is very close to 1 for $H > \frac{1}{2}$ but diverges as $H \to 0$ (see Norros [16]). Thus, for practical purposes in teletraffic context, one can use the approximation

$$P(B > T) \approx e^{-T^{2-2H}/2}. \tag{4.13}$$

4.5 OTHER APPROACHES

The tail behavior of $P(V > x)$ has recently been identified more accurately. Massoulie and Simonian [9] applied the theory of extremal values of Gaussian processes and discovered that

$$P(V > x) \approx K x^{-\gamma} \exp(-x^{2-2H}/2\kappa(H)^2), \tag{4.14}$$

where $\gamma = (1-H)(2H-1)/H$ and K is some constant independent of x. Such a result is derived by using specific metric properties of the centered Gaussian process $G_t = Z_t/(1+t)$ for the canonical distance d defined by $d(s,t) = E(G_t - G_s)^2$. The same result was obtained almost simultaneously by Narayan [11] with a completely original method based on the Fourier decomposition of FBM, which allows for numerically estimating the prefactor K, and by Hüsler and Piterbarg [6].

Two other approaches are described in Norros [14]. An attempt [16] to apply the so-called Beneš method (see Roberts et al. [19]) resulted in the presentation of the storage occupancy distribution as

$$P(V > x) = 2E(Z_1^*) \int_0^\infty L_H(x,t) P(Z_t \in x + t + dt), \tag{4.15}$$

where $Z_1^* = \sup_{t \in [0,1]} Z_t$ and

$$L_H(x,t) = \lim_{u \downarrow t} \frac{1}{(u-t)^{1-H}} P[Z_v < x + v, \forall v > u | \exists s \in [t,u]\colon Z_s = x + s]$$

is a finite positive function. When L_H is replaced simply by a normalizing constant, Eq. (4.15) is slightly better than the Weibull approximation (4.9). However, it has turned out to be difficult to obtain more accurate estimates with the Beneš presentation.

Finally, an approach based on the Girsanov formula for fractional Brownian motion [3, 10, 17] was presented by Norros [14]. The process

$$M_t \doteq c_1 \int_0^1 s^{1/2-H}(t-s)^{1/2-H}\, dZ_s,$$

where

$$c_1 = \left[H(2H-1)B(\tfrac{3}{2}-H, H-\tfrac{1}{2})\right]^{-1}$$

and $B(\mu, \nu)$ denotes the beta function

$$B(\mu,\nu) = \int_0^1 x^{\mu-1}(1-x)^{\nu-1}\, dx = \frac{\Gamma(\mu)\Gamma(\nu)}{\Gamma(\mu+\nu)},$$

turns out to be a Gaussian martingale with zero mean and variance $EM_t^2 = c_2^2 t^{2-2H}$, where c_2 is the constant given in Eq. (4.12). Moreover, $\exp(aM_t - a^2 c_2^2 t^{2-2H})$ is the Radon–Nikodym derivative between probability measures corresponding to fractional Brownian motions on time interval $[0, t]$ with drifts a and 0, respectively. Now, the storage occupancy distribution has the presentation

$$P(V > x) = E \exp\left(-M_{T_x} - \frac{c_2^2}{2} T_x^{2-2H} \right),$$

where T_x is the hitting time $T_x = \inf\{t \geq 0\colon Z_t = x\}$. Good approximations can be obtained in the case $H > \frac{1}{2}$ by replacing c_2 by 1. This leads to

$$\begin{aligned} P(V > x) &\approx E \exp\left(-xT_x^{1-2H} - \tfrac{1}{2} T_x^{2-2H}\right) \\ &= E\left(\exp\left[-\left(\frac{x}{Z_1^*}\right)^{(1-H)/H} Z_1^* - \frac{1}{2}\left(\frac{x}{Z_1^*}\right)^{(2-2H)/H} \right] \right), \end{aligned}$$

where the last step follows from the fact that $T_x \stackrel{\mathscr{D}}{=} (x/Z_1^*)^{1/H}$. This can be expected to be rather accurate, but we still have to use some approximation for the unknown distribution of Z_1^*. It is interesting to note that the rough substitution $Z_1^* \stackrel{\mathscr{D}}{\approx} |Z_1|$ gives in fact the same approximation as the simple Beneš approximation mentioned after Eq. (4.15).

Both the Beneš and Girsanov approaches have the appealing feature that they are based on nontrivial exact formulas. However, attempts to use them to obtain inequalities, or alternative derivations of the exact asymptotics (4.14), have not succeeded so far.

4.6 CONCLUSION

The fractional Brownian storage is a nice mathematical object that offers many interesting challenges for further work. For example, little seems to be known about the finite buffer case, priority queueing, or dependence between busy periods. On the other hand, new and powerful analytical techniques for working with FBM, based on Malliavian calculus, are already waiting for applications—see Decreusefond and Üstünel [3].

The problems connected with the motivation of fractional Brownian storage by limit theorems were briefly discussed in Section 4.1. Describing the applicability of this model or, more generally, any Gaussian storage model in quantitative terms is a task that clearly requires further work and innovations.

REFERENCES

1. R. G. Addie, P. Mannersalo, and I. Norros. Performance formulae for queues with Gaussian input. In P. Key and D. Smith, eds., *Teletraffic Engineering in a Competitive World.* Proceedings of the International Teletraffic Congress-ITC 16.

2. C. S. Chang, D. D. Yao, and T. Zajic. Large deviations, moderate deviations, and queues with long-range dependent input. *Adv. Appl. Probab.* **31**(1):254–278, 1999.
3. L. Decreusefond and A. S. Üstünel. Stochastic analysis of the fractional Brownian motion. *Potential Anal.* **10**(2):177–214, 1999.
4. J.-D. Deuschel and D. W. Stroock. *Large Deviations*. Academic Press, Boston, 1989.
5. N. G. Duffield and N. O'Connell. Large deviations and overflow probabilities for the general single-server queue, with applications. *Math. Proc. Cambridge Philos. Soc.*, **118**(2):363–374, 1995.
6. J. Hüsler and V. Piterbarg. Extremes of a certain class of Gaussian processes. *Stoch. Proc. Appl.*, **83**(2): 257–271, 1999.
7. K. R. Krishnan, A. L. Neidhardt, and A. Erramilli. Scaling analysis of traffic management of self-similar processes. In V. Ramaswami and P. E. Wirth, eds., *Teletraffic Contributions for the Information Age. Proceedings of the 15th International Teletraffic Congress ITC 15*, Washington, DC, pp. 1087–1096. Elsevier, New York, 1997.
8. W. E. Leland, M. S. Taqqu, W. Willinger, and D. V. Wilson. On the self-similar nature of Ethernet traffic (extended version). *IEEE/ACM Trans. Networking*, **2**(1):1–15, February 1994.
9. L. Massoulie and A. Simonian. Large buffer asymptotics for the queue with FBM input. *J. Appl. Probab.*, **36**(3), xxx–xxx, 1999.
10. G. M. Molchan and Ju. I. Golosov. Gaussian stationary processes with asymptotic power spectrum. *Sov. Math. Dokl.*, **10**(1):134–137, 1969.
11. O. Narayan. Exact asymptotic queue length distribution for fractional Brownian traffic. *Adv. Perf. Anal.*, **1**:39–63, 1998.
12. I. Norros. A storage model with self-similar input. *Queueing Syst.*, **16**:387–396, 1994.
13. I. Norros. On the use of fractional Brownian motion in the theory of connectionless networks. *IEEE J. Selected Areas Commun.*, **13**(6), 953–962, August 1995.
14. I. Norros. Four approaches to the fractional Brownian storage. In J. Lévy Véhel, E. Lutton, and C. Tricot, eds., *Fractals in Engineering*. Springer, New York, 1997.
15. I. Norros. Busy periods of fractional Brownian storage: a large deviations approach. *Adv. Perf. Anal.*, **2**(7):1–19, 1999.
16. I. Norros, A. Simonian, D. Veitch, and J. Virtamo. A Beneš formula for a buffer with fractional Brownian input. In *9th ITC Specialists Seminar '95: Teletraffic Modelling and Measurement*, Leidschendam, The Netherlands, 1995.
17. I. Norros, E. Valkeila, and J. Virtamo. An elementary approach to a Girsanov formula and other analytical results on fractional Brownian motions. *Bernoulli* **5**(4):571–587, 1999.
18. N. O'Connell and G. Procissi, On the build-up of large queues in a queueing model with fractional Brownian motion input. Technical Report HPL-BRIMS-98-18, Hewlett-Packard's Basic Research Institute in the Mathematical Sciences, 1998.
19. J. Roberts, U. Mocci, and J. Virtamo, eds. *Broadband Network Teletraffic*. Number 1155 in Lecture Notes in Computer Science. Springer, New York, 1996.
20. A. Shwartz and A. Weiss. *Large Deviations for Performance Analysis*. Chapman and Hall, London, 1995.

5

HEAVY LOAD QUEUEING ANALYSIS WITH LRD ON/OFF SOURCES

F. BRICHET AND A. SIMONIAN
France Télécom, CNET, 92794 Issy-Moulineaux Cédex 9, France
L. MASSOULIÉ
Microsoft Research Ltd., Cambridge, CB2 3NH, United Kingdom
D. VEITCH
Software Engineering Research Centre, Carlton, Victoria 3053, Australia

5.1 INTRODUCTION

In this chapter, we evaluate the impact of long-range dependence on a single network element (multiplexer or router) of a data network, this element being modeled as a fluid queueing system. Such a system has constant output rate C and is fed by a number N of i.i.d. on/off traffic sources. The case where the number of sources N is fixed is treated, for instance, in Boxma and Dumas [4]. For N sufficiently large, C exceeds the peak rate of an individual source. If the ratio "source peak rate/total output rate C" decreases with N (e.g., is proportional to $1/N$), then we are in the realm of "small" sources. In the meantime, we consider the "heavy load" case when the output rate C is slightly larger than the total mean input rate so that the queue is almost always nonempty. This "heavy load" situation associated with "small sources" in a fluid queueing system motivates the derivation of limit theorems for both input traffic and queue occupancy processes when the number N increases to infinity, as detailed below.

More precisely, represent an on/off source by mutually independent, alternating silence periods A and activity periods B. When active, the source emits data at

Self-Similar Network Traffic and Performance Evaluation, Edited by Kihong Park and Walter Willinger
ISBN 0-471-31974-0

constant rate, its peak rate, taken as unity. Given $\mathbf{E}(A) = 1/\alpha$ and $\mathbf{E}(B) = 1/\beta$, the activity probability of a source is then $v = \alpha/(\alpha + \beta)$ and we require that $C > Nv$ so that the queue has a stationary regime. Provided that the probability density of duration $A + B$ satisfies a simple regularity condition, we first show in this chapter that, once properly centered and normalized, the aggregate input rate and traffic processes produced by the superposition of N i.i.d. on/off sources converge as $N \uparrow +\infty$ toward continuous Gaussian processes. This is the content of our "functional central limit" theorems. In contrast with the classical heavy load approximation, note that the limit input rate process obtained this way is generally non-Markovian for arbitrary distributions of on and off durations A and B.

Denote further by $V_0^{(N)}$ the corresponding queue content in stationary conditions. Our essential aim is to obtain estimates for the limiting distribution of $V_0^{(N)}$ as $N \uparrow +\infty$, the capacity C scaling with N as

$$C = Nv + \gamma\sqrt{N} \tag{5.1}$$

for some positive constant γ. Equation (5.1) is clearly identified as a heavy load condition since the queue load Nv/C tends to 1 for increasing N (with speed $1/\sqrt{N}$). Now, by considering so-called heavy-tailed distributions for on and/or off periods, the input process becomes long-range dependent (LRD). Based on the above convergence results for input processes, our analysis then subsequently shows that the tail of the limiting distribution of the scaled queue content $V_0^{(N)}/\sqrt{N}$ is not exponential but Weibullian. Specifically, defining

$$\lim_{N\uparrow+\infty} \mathbf{P}(V_0^{(N)} > \xi\sqrt{N}) = \mathscr{H}(\xi)$$

for any fixed $\xi \geq 0$, we have

$$\lim_{\xi\uparrow+\infty} \frac{1}{\xi^{2(1-H)}} \cdot \log \mathscr{H}(\xi) = -\frac{\kappa^2}{2}$$

where both constants κ and $H \in]\frac{1}{2}, 1[$ depend on the distributions of on and off durations. Weibullian tails imply finite moments of all orders but, nonetheless, necessitate buffer sizes that grow much faster with load than in the classical Markovian case since the exponent $2(1 - H)$ is strictly less than 1 for $H > \frac{1}{2}$. Such a result reminds one of the Weibullian nature of the distribution of a fluid queue fed by the so-called fractional Brownian motion (FBM), as discussed in Norros [13] (see Chapter 4 in this volume for a full account). We also show in this chapter how the limit input processes obtained as $N \uparrow +\infty$ can easily be related to the FBM by means of proper time and space scale changes.

Other heavy load queueing analysis with heavy-tailed distributions has been considered in related contexts such as the $M/G/1$ discrete queue (see Chapter 6 in this volume) and the fluid queue with $M/G/\infty$ input rate process (refer to Chapter 9 in this volume and references contained therein). In comparison to the analysis performed in Chapters 6, 9 and 10, we stress the fact that the limit considered here, with many small sources and heavy load queueing regime, enables the derivation of continuous limit processes together with the Weibullian queueing behavior, as opposed to the heavy-tailed queueing behavior obtained when fixed "source peak rate/total output rate C" ratio and a finite number of sources are considered.

The rest of the chapter is organized as follows. Section 5.2 contains some preliminary notation and basic properties related to the on/off source model. In Section 5.3, we state and prove functional central limit theorems (CLTs) for the properly scaled input processes as the number of sources tends to infinity. In Section 5.4, these CLTs are used to derive general limiting upper and lower bounds for the stationary distribution of the queue content $V_0^{(N)}$ under the heavy load condition (5.1). Section 5.5 addresses the impact of long-range dependence on these bounds, demonstrating their Weibullian behavior in this case and relating the present model to FBM as considered in Chapter 4. Much of the material presented in this chapter was first published in Brichet et al. [5]. Alternative arguments for convergence results can also be found in Kurtz [10]

5.2 PRELIMINARY PROPERTIES OF THE ON/OFF SOURCE MODEL

Let $\{\lambda_t\}_{t\geq 0}$ be the stationary process representing the input rate of a single on/off source at time t and $\omega_t = \int_0^t \lambda_s \, ds$ the input it generates in the interval $[0, t]$. We denote $\{\Lambda_t^{(N)}\}$ and $\{W_t^{(N)}\}$ the sum of N i.i.d. copies of the processes $\{\lambda_t\}$ and $\{\omega_t\}$, respectively. For fixed t, λ_t is a Bernoulli variable with mean v and variance

$$\sigma^2 = v(1 - v), \tag{5.2}$$

while ω_t is a random variable taking values in $[0, t]$, with mean vt and variance denoted by

$$D(t)^2 = \mathrm{Var}(\omega_t) \tag{5.3}$$

It is also useful to note from the independence and homogeneity of all input sources that, given $\Lambda_0^{(N)} = r$, $r \in \{0, 1, \ldots, N\}$, the expectation of $W_t^{(N)}$ can be expressed as

$$\mathbf{E}(W_t^{(N)} | \Lambda_0^{(N)} = r) = (N - r)m_a(t) + rm_b(t), \tag{5.4}$$

where $m_a(t)$ and $m_b(t)$ are the first conditional moments

$$m_a(t) = \mathbf{E}(\omega_t | \lambda_0 = 0), \quad m_b(t) = \mathbf{E}(\omega_t | \lambda_0 = 1) \tag{5.5}$$

associated with ω_t. Finally, we denote by $\mathscr{R}$ the covariance function of process $\{\lambda_t\}$ defined for $t \geq 0$ by

$$\mathscr{R}(t) = \mathbf{E}(\lambda_0 - v)(\lambda_t - v) = \mathbf{E}(\lambda_0 \lambda_t) - v^2. \tag{5.6}$$

Using basic renewal properties of the on/off rate processes, it can easily be shown that the Laplace transform of covariance function $\mathscr{R}$ is given by

$$\mathscr{R}^*(p) = \frac{v(1-v)}{p}\left[1 - \frac{\alpha+\beta}{p} \cdot \frac{(1-a^*(p))(1-b^*(p))}{1-a^*(p)b^*(p)}\right] \tag{5.7}$$

for $p > 0$, where a^* (resp. b^*) denotes the Laplace transform of the off (resp. on) duration A (resp. B).

5.3 FUNCTIONAL CENTRAL LIMIT THEOREMS

The aim of this section is to find sufficient conditions for the rate and input processes associated with the on/off sources to satisfy functional CLTs. To this end, we first establish (Theorem 5.3.1) a CLT for renewal processes, using standard results on weak convergence [3]. The desired results are then obtained in Section 5.3.2.

5.3.1 CLT for Renewal Processes

For any interval I of $\mathbf{R}$, let $\mathscr{D}(I)$ denote the space of right-continuous functions on I with left limits. In the sequel, real-valued processes on some interval I are considered as $\mathscr{D}(I)$-valued random elements. For bounded I, $\mathscr{D}(I)$ is endowed with the J_1-Skorokhod topology. The space $\mathscr{D}(\mathbf{R}^+)$, denoted by $\mathscr{D}$ for short, is endowed with the topology of convergence in the J_1 topology on bounded intervals.

Theorem 5.3.1. *Let L be a stationary renewal process, with $L(0, t)$ representing the number of renewal points within time interval $]0, t]$, $t \geq 0$, and with finite and nonzero mean intensity l. Denoting by Π the distribution function of the inter-renewal times, we assume that:*

$$\text{the density } \Pi' \text{ exists and is bounded in some neighborhood of } 0. \tag{5.8}$$

Then, given a sequence $\{L_j\}_{j\geq 0}$ of i.i.d. copies of L, the sequence of processes $\{L^{(N)}\}_{N>0}$ defined by

$$L^{(N)}(t) = \frac{1}{\sqrt{N}}\sum_{j=1}^{N}(L_j(0,t) - lt), \quad t \in \mathbf{R}^+, \tag{5.9}$$

is tight and converges weakly to a limiting Gaussian process with a.s. continuous paths as $N \uparrow +\infty$.

In order to prove Theorem 5.3.1, the following results are needed. The first one is taken from Billingsley [3].

Theorem 5.3.2. [3, Theorem 15.6, p. 128]. *Let $I = [0, T]$ be some bounded interval of* **R**. *For a sequence $\{G^{(N)}\}_{N>0}$ of $\mathscr{D}(I)$-valued processes to converge weakly to the $\mathscr{D}(I)$-valued process G, it is sufficient that:*

- *for all $t_1, \ldots, t_n \in I$, the vector $(G^{(N)}(t_1), \ldots, G^{(N)}(t_n))$ converges weakly to $(G(t_1), \ldots, G(t_n))$, as $N \uparrow +\infty$;*
- *there exists $p > 0$, $q > 1$ and a continuous nondecreasing function F such that*

$$\mathbf{E}[|G^{(N)}(t) - G^{(N)}(t_1)|^p \cdot |G^{(N)}(t_2) - G^{(N)}(t)|^p] \leq [F(t_2) - F(t_1)]^q \tag{5.10}$$

for all N and $0 \leq t_1 \leq t \leq t_2 \leq T$.

The proof of Lemmas 5.3.3 and 5.3.4 below are deferred to Sections 5.7.1 and 5.7.2.

Lemma 5.3.3. *Let L be a stationary renewal process satisfying (5.8). Let $a_0 > 0$ denote an upper bound of the density Π' in some neighborhood of 0. Then for small enough $\epsilon > 0$, process L can be constructed on the same probability space as a homogeneous Poisson process M with intensity $m = l/(1 - l\epsilon) \vee a_0/(1 - a_0\epsilon)$ so that with probability one, the paths of M dominate those of L on $[0, \epsilon]$; that is,*

$$L(t) \leq M(t), \quad t \in [0, \epsilon].$$

Lemma 5.3.4. *In the setting of Theorem 5.3.1, let*

$$P^{(N)}(t_1, t, t_2) = \mathbf{E}[|L^{(N)}(t) - L^{(N)}(t_1)|^2 \cdot |L^{(N)}(t_2) - L^{(N)}(t)|^2] \tag{5.11}$$

for $0 \leq t_1 \leq t \leq t_2$, where the process $L^{(N)}$ is defined by Eq. (5.9). It holds that

$$P^{(N)}(t_1, t, t_2) \leq 2\{\mathbf{E}(\xi^2\eta^2) + \mathbf{E}(\xi^2)\mathbf{E}(\eta^2) + [\mathbf{E}(\xi\eta)]^2\}, \tag{5.12}$$

where the random pair (ξ, η) is distributed as $(L(t_1, t) - l(t - t_1),\ L(t, t_2) - l(t_2 - t))$.

Moreover, for $\epsilon > 0$ as in Lemma 5.3.3, there exists a constant $\delta > 0$ such that

$$t_2 - t_1 \leq \epsilon \Rightarrow P^{(N)}(t_1, t, t_2) \leq \delta(t_2 - t_1)^2. \tag{5.13}$$

Proof of Theorem 5.3.1. The plan is to apply Theorem 5.3.2. Note first that, by Lemma 5.3.3, the variance of $L(0, t)$ is finite for all $t > 0$. Thus, for any finite collection $t_1, \ldots, t_n$, the finite-dimensional vector $(L^{(N)}(t_1), \ldots, L^{(N)}(t_n))$ converges weakly to a multidimensional Gaussian distribution, as a consequence of the classical CLT. These finite-dimensional Gaussian distributions satisfy Kolmogorov's consistency criterion, hence the existence of some Gaussian process G with the corresponding finite-dimensional distributions. Lemma 5.3.3 provides bounds on the quantities $\mathbf{E}|G(t) - G(s)|^2$ or order $|t - s|$, which enables us to deduce, by

Kolmogorov's regularity criterion, that a version of G exists with almost surely continuous paths. The version is thus a $\mathscr{D}$-valued random element.

We now show that for any interval $I = [0, T]$, condition (5.10) of Theorem 5.3.2 is met. Let ϵ be fixed as in Lemma 5.3.3. Let $t_1 \leq t \leq t_2$ lie in I. If $t_2 - t_1 \leq \epsilon$, Condition (5.10) is met with $p = q = 2$ and $F(x) = \sqrt{\delta x}$, in view of (5.13). It remains to find a suitable bound on $P^{(N)}(t_1, t, t_2)$ when $t_2 - t_1 > \epsilon$. Using the basic inequality $2|ab| \leq a^2 + b^2$ in conjunction with the Cauchy–Schwarz inequality, (5.12) entails that

$$P^{(N)}(t_1, t, t_2) \leq 3(\mathbf{E}\xi^4 + \mathbf{E}\eta^4).$$

This bound is in turn less than

$$3\{\mathbf{E}L(t_1, t)^4 + \mathbf{E}L(t, t_2)^4 + [l(t - t_1)]^4 + [l(t_2 - t)]^4\}.$$

In order to bound the means in the last display expression, note that, by Lemma 5.3.3, both $L(t_1, t)$ and $L(t, t_2)$ are less than the sum of $\lceil T/\epsilon \rceil$ random variables, each of which is less than a Poisson random variable with mean $m\epsilon$. Let m_4 denote the fourth moment of such a Poisson random variable. In view of the straightforward inequality $(\sum_{i=1}^k a_i)^n \leq k^{n-1} \sum_{i=1}^k |a_i|^n$, each of the means in the preceding bound is less than $\lceil T/\epsilon \rceil \times \lceil T/\epsilon \rceil^4 m_4 = \lceil T/\epsilon \rceil^5 m_4$. One thus obtains

$$P^{(N)}(t_1, t, t_2) \leq 6\{\lceil T/\epsilon \rceil^5 m_4 + (lT)^4\}.$$

Denoting the latter upper bound by K, $t_2 - t_1 \geq \epsilon$ then entails that

$$P^{(N)}(t_1, t, t_2) \leq \frac{K}{\epsilon^2}(t_2 - t_1)^2.$$

Thus condition (5.10) of Theorem 5.3.2 is met, with $p = q = 2$ and $F(x) = \sqrt{\delta \vee (k/\epsilon^2)}x$. ■

5.3.2 CLT for Rate and Input Processes

The propositions to follow state the functional CLT for the two sequences of processes $\{\Lambda_t^{(N)}\}$ and $\{W_t^{(N)}\}$ introduced in Section 5.2.

Proposition 5.3.5. *Let $A + B$ denote the total duration of two successive on and off periods of rate process $\{\lambda_t\}$ and assume that the distribution of $A + B$ has a bounded density in the neighborhood of 0. The sequence of processes $Y^{(N)}$ defined by*

$$Y_t^{(N)} = \frac{\Lambda_t^{(N)} - Nv}{\sqrt{N}}, \quad t \geq 0, \tag{5.14}$$

then converges weakly in the space $\mathscr{D}$ toward a zero-mean stationary Gaussian process Y with almost surely continuous paths.

Proof. Write first Eq. (5.14) as

$$Y_t^{(N)} = \frac{1}{\sqrt{N}} \sum_{j=1}^{N} (\lambda_{j,t} - v), \tag{5.15}$$

where the $\{\lambda_{j,t}\}$ are i.i.d. copies of the process $\{\lambda_t\}$. In view of the ordinary CLT, the finite-dimensional distributions of the process $Y^{(N)}$ converge weakly to limiting Gaussian distributions. Weak convergence of the sequence $\{Y^{(N)}\}_{N>0}$ will then hold provided it is tight. Since $\Lambda^{(N)}$ is the superposition of N i.i.d. on/off rate processes λ_j with peak rate 1, Eq. (5.15) entails that

$$Y_t^{(N)} = Y_0^{(N)} + \frac{1}{\sqrt{N}} \sum_{j=1}^{N} \{L_j^+(0,t) - L_j^-(0,t)\} \tag{5.16}$$

for $t \geq 0$, where $L_j^+(0,t)$ (resp. $L_j^-(0,t)$) is the number of upward (resp. downward) jumps of process λ_j over time interval $[0,t]$. Each point process L_j^+ (resp. L_j^-) is a stationary renewal process with intensity l, where $1/l = 1/\alpha + 1/\beta$ is the mean cycle period of each process λ_j. We therefore deduce from Eq. (5.16) that

$$Y_t^{(N)} = Y_0^{(N)} + L_+^{(N)}(t) - L_-^{(N)}(t), \tag{5.17}$$

where

$$L_\pm^{(N)}(t) = \frac{1}{\sqrt{N}} \sum_{j=1}^{N} \{L_j^\pm(0,t) - lt\}.$$

In view of decomposition (5.17), in order to prove the tightness of the sequence $\{Y^{(N)}\}$, it is sufficient to show that each sequence $\{Y_0^{(N)}\}$, $\{L_+^{(N)}\}$, and $\{L_-^{(N)}\}$ is tight. The sequence $\{Y_0^{(N)}\}$ converges in distribution and is therefore tight. Tightness of both sequences $L_+^{(N)}$ and $L_-^{(N)}$ follows from Theorem 5.3.1. Since the weak limits of both sequences have a.s. continuous paths, it results that the weak limit of $Y^{(N)}$ also has a.s. continuous paths. Finally, stationarity of each process $Y^{(N)}$ ensures stationarity of the limiting Gaussian process. ■

The following result on the convergence of the normalized input process can be deduced as a simple corollary of the previous proposition.

Proposition 5.3.6. *In the framework of Proposition 5.3.5, the sequence of processes* $\{\Omega_t^{(N)}\}$ *defined by*

$$\Omega_t^{(N)} = \frac{W_t^{(N)} - Nvt}{\sqrt{N}}, \quad t \geq 0, \tag{5.18}$$

converges weakly toward a zero-mean, continuous Gaussian process $\{\Omega_t\}$ *with stationary increments and* $\Omega_0 = 0$. *In addition, this process is such that*

$$\lim_{t \to +\infty} \frac{\Omega_t}{t} = 0 \quad a.s.$$

Proof. Note that

$$\Omega_t^{(N)} = \int_0^t Y_u^{(N)} \, du$$

with $Y^{(N)}$ introduced in Eq. (5.14). The mapping φ, which associates to some $f \in \mathscr{D}([0, T])$ the integrated function $\varphi(f): t \mapsto \int_0^t f(u) \, du$, is continuous on the subset $C([0, T])$ of $\mathscr{D}([0, T])$ consisting of continuous functions (indeed, this holds since the trace of the J_1-Skorokhod topology on $C([0, T])$ coincides with the topology of uniform convergence; see Billingsley [3]). As $Y^{(N)}$ converges weakly to a limiting process Y with a.s. continuous paths, $\Omega^{(N)} = \varphi(Y^{(N)})$ converges weakly to $\varphi(Y)$. Process Y being Gaussian and centered, the limit process Ω is clearly centered, continuous, and Gaussian with $\Omega_0 = 0$. The stationarity of increments of Ω is readily deduced from the stationarity of process Y. In view of such stationarity, the ergodic theorem can be used to yield the existence of an almost sure limit for Ω_t/t as $t \to +\infty$. This limit is necessarily a Gaussian random variable, and we will be done if its variance is equal to zero. This variance is equal to (see Eq. (5.32))

$$\lim_{t \to +\infty} \frac{2}{t^2} \int_0^t (t - u)\mathscr{R}(u) \, du,$$

where $\mathscr{R}(u)$ is as defined in Eq. (5.6). It can be shown that this limit equals zero provided $\mathscr{R}(u)$ tends to zero as $u \to +\infty$. Recall that the process $\{\lambda_t\}$ is regenerative, with cycle lengths distributed as $A + B$. Under the present assumptions, this cycle length distribution has a density in the neighborhood of zero, and hence is nonlattice. This ensures (e.g., see Asmussen [1]) that λ_t converges weakly to its stationary distribution, irrespective of its initial value, as $t \to +\infty$. Writing

$$\mathscr{R}(t) = v\mathbf{P}(\lambda_t = 1 | \lambda_0 = 1) - v^2,$$

we conclude that $\mathscr{R}(t)$ indeed goes to zero as $t \to +\infty$. ■

5.4 QUEUE CONTENT DISTRIBUTION IN HEAVY LOAD CONDITION

Using the above convergence results for both the normalized rate process and the input process created by the superposition of a large number of on/off sources, we can now assert general limit bounds for the distribution of the normalized queue content $V_0^{(N)}/\sqrt{N}$ when $N \uparrow +\infty$ and heavy load condition (5.1) is fulfilled. To derive such limits, first recall [2] that the stationary queue content $V_0^{(N)}$ can be defined as the supremum of the transient process $\{W_t^{(N)} - Ct\}$, namely,

$$V_0^{(N)} = \sup_{t \geq 0}[W_t^{(N)} - Ct]. \tag{5.19}$$

Second, given a fluid queue with output rate γ, it has been shown [16] that, provided the input process $\{\Omega_t\}_{t\geq 0}$ is such that (1) $\Omega_t = \int_0^t Y_u \, du$ for all t, where the rate process $\{Y_t\}_{t\geq 0}$ is stationary, and (2) $\Omega_t - \gamma t$ tends to $-\infty$ almost surely as $t \uparrow +\infty$, then, for any $\xi \geq 0$, the corresponding stationary distribution of the queue content U_0 verifies the bounds

$$\begin{aligned} \sup_{t>0} \mathbf{P}(\Omega_t > \xi + \gamma t) \leq \mathbf{P}(U_0 > \xi), \\ \mathbf{P}(U_0 > \xi) \leq \int_{-\infty}^{\gamma} (\gamma - y)\mathbf{P}(Y_0 \in dy) \int_0^{+\infty} f_y(t, \xi + \gamma t) \, dt, \end{aligned} \tag{5.20}$$

where $f_y(t, .)$ denotes the probability density of Ω_t with given $Y_0 = y$. In order to state the next proposition, let Φ denote the distribution function of a standard normal variable and introduce the notation

$$\begin{aligned} \mu(t) &= m_b(t) - m_a(t), \\ \Delta(t)^2 &= D(t)^2 - \mu(t)^2\sigma^2, \\ S(t, \xi) &= \frac{D(t)}{\sigma \cdot \Delta(t)} \left\{ \gamma - (\xi + \gamma t) \frac{\mu(t)\sigma^2}{D(t)^2} \right\}, \\ R(t, \xi) &= \frac{\sigma \cdot \Delta(t)}{D(t)} \left\{ S(t, \xi)\Phi[S(t, \xi)] + \frac{1}{\sqrt{2\pi}} e^{-S(t,\xi)^2/2} \right\} \end{aligned} \tag{5.21}$$

where σ, $D(t)$, $m_a(t)$, and $m_b(t)$ are defined by Eqs. (5.2), (5.3), and (5.5), respectively.

Proposition 5.4.1. *Assume that heavy load condition (5.1) holds for some positive constant γ. Then the distribution of $V_0^{(N)}/\sqrt{N}$ converges weakly as $N \uparrow +\infty$. Furthermore, defining*

$$\mathscr{H}(\xi) = \lim_{N \uparrow +\infty} \mathbf{P}(V_0^{(N)} > \xi\sqrt{N})$$

for any $\xi \geq 0$, the bounds

$$q(\xi) \leq \mathscr{H}(\xi) \leq Q(\xi)$$

hold with

$$q(\xi) = \sup_{t>0} \bar{\Phi}\left(\frac{\xi + \gamma t}{D(t)}\right), \tag{5.22}$$

where $\bar{\Phi} = 1 - \Phi$ *and*

$$Q(\xi) = \int_0^{+\infty} \exp\left(-\frac{(\xi + \gamma t)^2}{2D(t)^2}\right) \frac{R(t, \xi)}{D(t)\sqrt{2\pi}} dt \tag{5.23}$$

where $R(t, \xi)$ *is given by Eq. (5.21).*

Proof. First, using the scaling condition (5.1) for C together with notation (5.18), relation (5.19) for $V_0^{(N)}$ reads

$$\frac{V_0^{(N)}}{\sqrt{N}} = \sup_{t>0}(\Omega_t^{(N)} - \gamma t).$$

Let ϕ denote the map $d \in \mathscr{D} \mapsto \phi(d) = \sup_{t>0}(d_t - \gamma t)$. From the weak convergence of processes $\Omega^{(N)}$ toward process Ω, one can conclude that $V_0^{(N)}/\sqrt{N} = \phi(\Omega^{(N)})$ converges weakly to $U_0 = \phi(\Omega)$ provided the distribution of Ω puts no mass on the set of trajectories $d \in \mathscr{D}$ for which ϕ is discontinuous. The fact that Ω has a.s. continuous paths and is such that $\Omega_t/t \to 0$ a.s., as implied by Proposition 5.3.6, is sufficient to conclude that this is so (the detailed argument is left to the reader). It therefore follows that, for fixed $\xi \geq 0$, $\mathbf{P}(V_0^{(N)} > \xi\sqrt{N}) \to \mathscr{H}(\xi)$ as $N \uparrow +\infty$, where

$$\mathscr{H}(\xi) = \mathbf{P}\left(\sup_{t>0}(\Omega_t - \gamma t) > \xi\right). \tag{5.24}$$

The limit distribution of $V_0^{(N)}/\sqrt{N}$ is therefore identical to that of the stationary distribution of a queue fed by input $(\Omega_t\}$ and with output rate $\gamma > 0$.

Second, process $\{\Omega_t\}$ verifying the conditions recalled above at the beginning of this section, the lower bound $\mathscr{H}(\xi) \geq q(\xi)$ can readily be derived, where $q(\xi)$ denotes the left-hand side of lower bound (5.20). Each variable Ω_t being Gaussian and centered with variance $D(t)^2$ for fixed $t > 0$, expression (5.22) follows.

Third, we further obtain the upper bound $\mathscr{H}(\xi) \leq Q(\xi)$ for upper bound (5.24), where $Q(\xi)$ denotes the right-hand side of upper bound (5.20). To compute this upper bound, we note that both variables Y_0 and Ω_t are Gaussian so that $f_y(t, \cdot)$ is a Gaussian density. Its mean and variance, by use of standard relations [15, pp. 302, 305] for Gaussian correlation and with notation (5.2) and (5.3), are given by

$$\mu_y(t) = \mathrm{Cov}(Y_0, \Omega_t) \times \frac{y}{\mathrm{Var}(Y_0)} = \mathbf{E}(Y_0\Omega_t) \times \frac{y}{\sigma^2} \tag{5.25}$$

and

$$\Delta(t)^2 = \mathrm{Var}(\Omega_t) - \frac{\mathrm{Cov}(Y_0, \Omega_t)^2}{\mathrm{Var}(Y_0)} = D(t)^2 - \frac{\mathbf{E}(Y_0\Omega_t)^2}{\sigma^2}, \tag{5.26}$$

respectively. Now, using expression (5.4) with $r = Nv + y\sqrt{N}$ along with definitions (5.14) and (5.18), we have

$$\mathbf{E}(\Omega_t^{(N)} | Y_0^{(N)} = y) = \frac{1}{\sqrt{N}}[N(1-v)m_a(t) - y\sqrt{N}m_a(t) + Nvm_b(t)$$
$$+ y\sqrt{N}m_b(t) - Nvt] = y[m_b(t) - m_a(t)] = \mu(t)y$$

since $(1-v)m_a(t) + vm_b(t) = vt$. Letting N tend to infinity in the latter conditional expectation then implies that $\mathbf{E}(\Omega_t | Y_0 = y) = y\mu(t)$ as well and therefore $\mathbf{E}(Y_0\Omega_t) = \sigma^2\mu(t)$ by deconditioning with respect to the centered variable Y_0. It consequently follows from the above discussion and relations (5.25) and (5.26) that

$$\mu_y(t) = \mu(t)y, \quad \Delta(t)^2 = D(t)^2 - \mu(t)^2\sigma^2. \tag{5.27}$$

Using Eq. (5.27) in upper bound (5.20), we then obtain

$$Q(\xi) = \int_{-\infty}^{\gamma} (\gamma - y)\frac{e^{-y^2/2\sigma^2}}{\sigma\sqrt{2\pi}}\, dy$$
$$\times \int_0^{+\infty} \exp\left(-\frac{[\xi + \gamma t - \mu(t)y]^2}{2\Delta(t)^2}\right)\frac{1}{\Delta(t)\sqrt{2\pi}}\, dt.$$

Performing the integration with respect to variable y, it is verified after some standard manipulation of Gaussian integrals that the above expression reduces to $Q(\xi)$ as given in Eq. (5.23). ■

5.5 IMPACT OF LONG-RANGE DEPENDENCE

We now wish to introduce long-range dependence into the source model and derive explicit estimates of bounds $q(\xi)$ and $Q(\xi)$ of Proposition 5.4.1 for such a case. As shown below, a natural way to do this is to assume a heavy tail for the distribution of either the silence period, the activity period, or both. In the rest of this section, we first specify the LRD behavior of the rate process associated with heavy-tailed distributions for on/off durations, then provide the estimates for bounds $q(\xi)$ and $Q(\xi)$ in such a case. We conclude this section by providing a weak convergence theorem for the limiting input process $\{\Omega_t\}$ obtained in Section 5.3, which, once rescaled in space and time, converges toward the so-called fractional Brownian motion.

5.5.1 LRD Properties

Recall [6] that a stationary process $\{\lambda_t\}$ is said to exhibit short- (resp. long-) range dependence if its correlation function $\mathscr{R}$ defined in Eq. (5.6) is such that the integral $\int_0^{+\infty} \mathscr{R}(t)\, dt$ is convergent (resp. divergent). If the distributions of on and off periods A and B are Coxian, that is, their Laplace transforms a^* and b^* are rational functions, then standard results for Laplace transform inversion applied to expression (5.7) entail that the correlation function $t \longmapsto \mathscr{R}(t)$ tends exponentially fast to 0 as $t \uparrow +\infty$, thus corresponding to short-term dependence for process $\{\lambda_t\}$. Now, the situation appears totally different in the case when the distribution of either off duration A or on duration B is heavy tailed. A distribution function F is here defined to be heavy tailed if

$$1 - F(t) = \frac{h(t)}{t^r} \tag{5.28}$$

for $t > 0$ and $1 < r < 2$, where function h has a finite limit $\bar{h}$ at infinity. A typical example is the Pareto distribution, where

$$F(t) = 1 - \left(\frac{\theta}{t + \theta}\right)^r$$

for some $\theta > 0$. As a consequence of definition (5.28), the Laplace transform $f^*: p > 0 \longmapsto \int_0^{+\infty} e^{-pt}\, dF(t)$ of a heavy-tailed distribution F with finite mean m and power $r \in]1, 2[$ verifies

$$f^*(p) = 1 - mp + \frac{\bar{h} \cdot \Gamma(2 - r)}{r - 1} p^r + o(p^r) \tag{5.29}$$

for small positive p, where Γ is Euler's function (see Section 5.7.3). For $1 < r < 2$, such an expression for the Laplace transform near 0 in fact characterizes the heavy-tailed behavior of the corresponding distribution F (in the case when $r = 2$, Eq. (5.29) is identical to a classical Taylor expansion, corresponding to a regular distribution having a finite variance). We can then show the following.

Proposition 5.5.1. *Assume that the Laplace transforms a^* and b^* of the off and on periods have expansions*

$$\begin{aligned} a^*(p) &= 1 - \frac{p}{\alpha} + a_r p^r + o(p^r), \\ b^*(p) &= 1 - \frac{p}{\beta} + b_s p^s + o(p^s), \end{aligned} \tag{5.30}$$

respectively, with positive coefficients a_r and b_s and where powers $r, s \in]1, 2]$ are such that $r + s \neq 4$. The covariance function of rate $\{\lambda_t\}$ then verifies

$$\mathscr{R}(t) \sim \frac{\mathscr{R}_0}{t^{q-1}} \tag{5.31}$$

for large t, with $q = \min(r, s)$ and

$$\mathscr{R}_0 = \frac{\beta v^3 a_r}{\Gamma(2-r)} \quad \left(\text{resp. } \mathscr{R}_0 = \frac{\alpha(1-v)^3 b_s}{\Gamma(2-s)}\right)$$

if $r < s$ (resp. $s < r$) and

$$\mathscr{R}_0 = \frac{\beta v^3 a_r + \alpha(1-v)^3 b_s}{\Gamma(2-r)}$$

if $r = s$.

In other terms, assuming such heavy-tailed distributions for A and B with $r, s \in]1, 2]$ entails long-range dependence for process $\{\lambda_t\}$ since $\int_0^{+\infty} \mathscr{R}(t)\, dt$ diverges for $0 < q - 1 \leq 1$. The proof of Proposition 5.5.1 can be performed simply by inserting expansions (5.30) for a^* and b^* into expression (5.7) and then using Tauberian Theorem 5.7.1 recalled in Section 5.7.3 for deriving the asymptotics of $\mathscr{R}(t)$ for large t. Concerning input process $\{\omega_t\}$ with $\omega_t = \int_0^t \lambda_u\, du$, it is easy to express its variance $D^2(t)$ defined in Eq. (5.3) as

$$D^2(t) = 2\int_0^t (t-u)\mathscr{R}(u)\, du \tag{5.32}$$

in terms of the covariance function $\mathscr{R}$ of process $\{\lambda_t\}$. As a consequence, it readily follows that if

$$\int_0^{+\infty} \mathscr{R}(u)\, du < +\infty,$$

then $D^2(t) = O(t)$ as $t \uparrow +\infty$. On the other hand, if behavior (5.31) holds with $1 < q < 2$, we have instead

$$D^2(t) \sim \mathscr{L} t^{3-q} \tag{5.33}$$

as $t \uparrow +\infty$, where

$$\mathscr{L} = \frac{2\mathscr{R}_0}{(3-q)(2-q)} \tag{5.34}$$

$\mathscr{R}_0$ being expressed as in Proposition 5.5.1.

5.5.2 Impact on Queue Content Distribution

To make bounds $q(\xi)$ and $Q(\xi)$ of Proposition 5.4.1 explicit in the case of heavy-tailed distributions for on and off periods, we first need some preliminary estimation of intermediate quantities $\mu(t)$ and $\Delta(t)$ defined in expressions (5.21).

Lemma 5.5.2. *Assume that expansions (5.30) hold for the Laplace transforms a^* and b^*. Then quantity $\mu(t)$ defined in Eq. (5.21) verifies*

$$\mu(t) \sim \left(\frac{3-q}{2}\right)\frac{\mathscr{L}}{\sigma^2}\cdot t^{2-q} \tag{5.35}$$

for large t, with $q = \min(r, s)$ and constants σ^2 and $\mathscr{L}$ defined in Eqs. (5.2) and (5.34), respectively.

The proof is deferred to Section 5.7.4. Note that the average difference $\mu(t)$ of input due to different initial conditions $\lambda_0 = 0$ or $\lambda_0 = 1$ grows with increasing t rather than tending to a constant, an eloquent testimony to long-range dependence. It further follows from estimate (5.33) and the latter result that the variance $\Delta^2(t)$ introduced in Eq. (5.21) satisfies

$$\Delta^2(t) = D^2(t) - \mu^2(t)\sigma^2 \sim D^2(t) \tag{5.36}$$

for large t since $2(2-q) < 3-q$, when $q > 1$. These observations lead to the next proposition, the main result in this chapter.

Proposition 5.5.3. *Assume that the distributions of either on or off periods have heavy tails with respective power $r, s, \in]1, 2[$ and that expansions (5.30) hold. Then limiting distribution $\xi \longmapsto \mathscr{H}(\xi)$ of Proposition 5.4.1 has the logarithmic estimate*

$$\lim_{\xi\uparrow+\infty}\frac{1}{\xi^{2(1-H)}}\cdot\log\mathscr{H}(\xi) = -\frac{\kappa^2}{2} \tag{5.37}$$

where $H = (3-q)/2$,

$$\kappa = \frac{1}{H^H(1-H)^{1-H}}\frac{\gamma^H}{\sqrt{\mathscr{L}}}$$

and with $\mathscr{L}$ specified in Eq. (5.34)

Proof. First, consider lower bound (5.22). As expansions (5.30) hold, estimate (5.33) applies to that $D^2(t) \sim \mathscr{L}t^{2H}$ for large t and $2H = 3-q$. Use the variable change $t = \xi u$ and write $D(\mu\xi) = \sqrt{\mathscr{L}}(u\xi)^H[1+\delta(u\xi)]$, where $\delta(t) \to 0$ as $t \uparrow +\infty$. Lower bound (5.22) then reads

$$q(\xi) = \sup_{u>0}\bar{\Phi}\left(\frac{\xi^{1-H}}{\sqrt{\mathscr{L}}}\frac{r(u)}{1+\delta(u\xi)}\right), \tag{5.38}$$

where

$$r(u) = \frac{1 + \gamma u}{u^H}.$$

It is easily verified that function $u \longmapsto r(u)$ has a unique minimum at point $u^* = H/\gamma(1-H)$ with $r(u^*) = \gamma^H/H^H(1-H)^{1-H}$. Supremum (5.38) entails, in particular, that $q(\xi) \geq \tilde{q}(\xi)$ with

$$\tilde{q}(\xi) = \bar{\Phi}\left(\frac{\xi^{1-H}}{\sqrt{\mathscr{L}}} \frac{r(u^*)}{1 + \delta(u^*\xi)}\right).$$

Using the asymptotics $\bar{\Phi}(z) \sim \exp(-z^2/2)/z\sqrt{2\pi}$ for large z, we may write that $\log \bar{\Phi}(z) = -z^2/2 + o(z^2)$. Since $\delta(u^*\xi) \to 0$ as $\xi \uparrow +\infty$, lower bound $\tilde{q}(\xi)$ thus verifies

$$\begin{aligned}\log \tilde{q}(\xi) &= -\frac{1}{2}\left(\frac{\xi^{1-H}}{1 + \delta(u^*\xi)} \cdot \frac{r(u^*)}{\sqrt{\mathscr{L}}}\right)^2 + o\left(\frac{\xi^{1-H}}{1 + \delta(u^*\xi)}\right)^2 \\ &= -\frac{1}{2}\left(\xi^{1-H} \cdot \frac{r(u^*)}{\sqrt{\mathscr{L}}}\right)^2 + o(\xi^{1-H})^2.\end{aligned}$$

The latter consequently implies that

$$\begin{aligned}\liminf_{\xi \uparrow +\infty} \frac{1}{\xi^{2(1-H)}} \cdot \log q(\xi) &\geq \liminf_{\xi \uparrow +\infty} \frac{1}{\xi^{2(1-H)}} \cdot \log \tilde{q}(\xi) \\ &= -\frac{\kappa^2}{2}\end{aligned} \tag{5.39}$$

with $\kappa = r(u^*)/\sqrt{\mathscr{L}}$.

Second, consider upper bound (5.23). Performing again the variable change $t = \xi u$ in the integral gives

$$Q(\xi) = \frac{\xi^{1-H}}{\sqrt{2\pi\mathscr{L}}} \int_0^{+\infty} \exp\left(-\frac{\xi^{2(1-H)}}{2\mathscr{L}} \cdot J_\xi(u)\right) L_\xi(u)\, du, \tag{5.40}$$

where

$$J_\xi(u) = \frac{r(u)^2}{[1 + \delta(u\xi)]^2}, \quad L_\xi(u) = \frac{R(u\xi, \xi)}{u^H[1 + \delta(u\xi)]},$$

and with functions $u \longmapsto r(u)$ and $u \longmapsto \delta(u\xi)$ defined as in Eq. (5.38). To evaluate integral (5.40), we apply the Laplace method [12] for the estimation of integrals with exponential integrand. For fixed ξ, the continuous function $u \mapsto J_\xi(u)$ tends to $+\infty$ as $u \downarrow 0$ and $u \uparrow +\infty$ (in fact, it is $O(u^{-2H})$ for small positive u and $O(u^{2(1-H)})$ for large u). It therefore has a minimum at some point u_ξ^*. As $J_\xi(u) \to r^2(u)$ as $\xi \uparrow +\infty$ for bounded u, it is clear that $u_\xi^* \to u^*$ as $\xi \uparrow +\infty$, where u^* is the minimum of $u \mapsto r^2(u)$. The contribution of the exponential factor to the value of the integral is thus preponderant in the neighborhood of that minimum u_ξ^*. Concerning the nonexponential factor $L_\xi(u)$ in Eq. (5.40), we note from Lemma 5.5.2 with $H = (3-q)/2$ that

$$\xi(1+\gamma u)\frac{\mu(u\xi)\sigma^2}{D(u\xi)^2} \to \gamma$$

as $\xi \uparrow +\infty$ and $u \to u^*$. As $D(u\xi) \sim \Delta(u\xi)$ for large ξ and fixed u in view of estimate (5.36), we successively deduce from the above and expressions (5.21) that $S(u\xi, \xi) \to 0$ and therefore $R(u\xi, \xi) \to \sigma/\sqrt{2\pi}$ as $\xi \uparrow +\infty$ and $u \to u^*$. These limiting results entail that $L_\xi(u) \to L$ with $L = \sigma/(u^*)^H\sqrt{2\pi}$. Factor $L_\xi(u)$ tending to a nonzero limit, integral (5.40) is therefore, up to powers of ξ, of order

$$\exp\left(-\frac{\xi^{2(1-H)}}{2\mathscr{L}}J_\xi(u_\xi^*)\right).$$

We therefore deduce from the above that

$$\lim_{\xi\uparrow+\infty}\frac{1}{\xi^{2(1-H)}}\cdot\log Q(\xi) = -\lim_{\xi\uparrow+\infty}\frac{J_\xi(u_\xi^*)}{2\mathscr{L}}, \tag{5.41}$$

but the latter simply equals $r(u^*)^2/2\mathscr{L} = \kappa^2/2$ by continuity. The conjunction of (5.39) and (5.41) together with the bounds $q(\xi) \leq \mathscr{H}(\xi) \leq Q(\xi)$ provide the desired logarithmic estimate (5.37). ■

The latter result consequently entails that the distribution of the content of a queue at heavy load, when fed by the superposition of an infinite number of on/off sources with LRD characteristics, has a Weibullian tail at infinity, namely,

$$\mathscr{H}(\xi) = \exp\left(-\frac{\kappa^2}{2}\cdot\xi^{2(1-H)} + o[\xi^{2(1-H)}]\right) \tag{5.42}$$

for large ξ. As shown in the proof of Proposition 5.5.3, only the leading term $D^2(t) \sim \mathscr{L}t^{2H}$ is used to derive the logarithmic estimate for $\mathscr{H}(\xi)$. Refined asymptotics for bounds $q(\xi)$ and $Q(\xi)$, and thus estimates for the $o[\xi^{2(1-H)}]$ term in Eq. (5.42), can further be derived from the *asymptotic expansion* of the variance $D^2(t)$ for large t; that is,

$$D^2(t) = \mathscr{L}t^{2H} + \mathscr{L}'t^{2H'} + o(t^{2H'}) \tag{5.43}$$

as $t \uparrow +\infty$, with $H' < H$. Using generalized Tauberian theorems [7, p. 142] and identify (5.32), such an expansion can be derived from that of the Laplace transform $p \mapsto 2\mathcal{R}^*(p)/p^2$ of $t \mapsto D^2(t)$ near $p = 0$ and the use of expansions (5.30) in formula (5.7) for $\mathcal{R}^*(p)$. Second-order power H' proves, however, difficult to specify for arbitrary powers r and s in expansions (5.30). By way of illustration, we just mention without proof that if $s < \min[r, (r+1)/2]$, then a second-order expansion (5.43) for $D^2(t)$ can be written with

$$2H = 3 - s, \ \mathcal{L} = \frac{2\alpha(1-v)^3 b_s}{(3-s)(2-s)\Gamma(2-s)},$$
$$2H' = 4 - 2s, \ \mathcal{L}' = \frac{\alpha^2(1-v)^3 b_s^2}{(2-s)\Gamma(4-2s)}.$$

Using Eq. (5.43) in formulas (5.22) and (5.23) then enables one to derive complete asymptotics for $q(\xi)$ and $Q(\xi)$ in the form

$$\begin{aligned} q(\xi) &\sim \frac{e^{\kappa'}}{\kappa\sqrt{2\pi}\xi^{1-H}} \cdot \exp\left(-\frac{\kappa^2}{2}\xi^{2(1-H)}\right), \\ Q(\xi) &\sim \frac{\sigma e^{\kappa'}}{\gamma\sqrt{2\pi}}\sqrt{\frac{H}{1-H}} \cdot \exp\left(-\frac{\kappa^2}{2}\xi^{2(1-H)}\right), \end{aligned} \tag{5.44}$$

respectively, where

$$\kappa' = \frac{\mathcal{L}'}{2H^2\mathcal{L}^2} \cdot \gamma^2.$$

We illustrate these approximations in the following numerical examples. Consider $N = 100$ on/off sources, assuming that 100 sources is sufficient for a useful application of our asymptotic results. We take the mean burst volume as unity ($1/\beta = 1$). The mean activity probability $v = \alpha/(\alpha+\beta)$ is set to 0.1 ($\alpha = \frac{1}{9}$) and the multiplexer load $Nv/C = (1 + \gamma/v\sqrt{N})^{-1}$ to 0.9, implying $C \approx 11.11$ ($\gamma = \frac{1}{9}$). Figure 5.1 shows a log plot of the complementary distribution function (interpreted as the overflow probability against buffer size) measured in units of mean burst volume, when sources have activity periods distributed as Pareto random variables. We also consider two values for H ($H = 0.8$ and $H = 0.85$). Curve A (B, resp.) corresponds to the limiting upper bound Q (lower bound q, resp.) given by Eq. (5.44). We observe that the buffer size required for a loss probability of 10^{-9} is around 3×10^4 times the mean burst size for $H = 0.8$ and more than 10^6 times the mean burst size for $H = 0.85$. These values have to be compared to the case where both silent and activity periods are exponentially distributed. In the latter case, a buffer size of around 50 times the mean burst size guarantees a loss probability less than 10^{-9}.

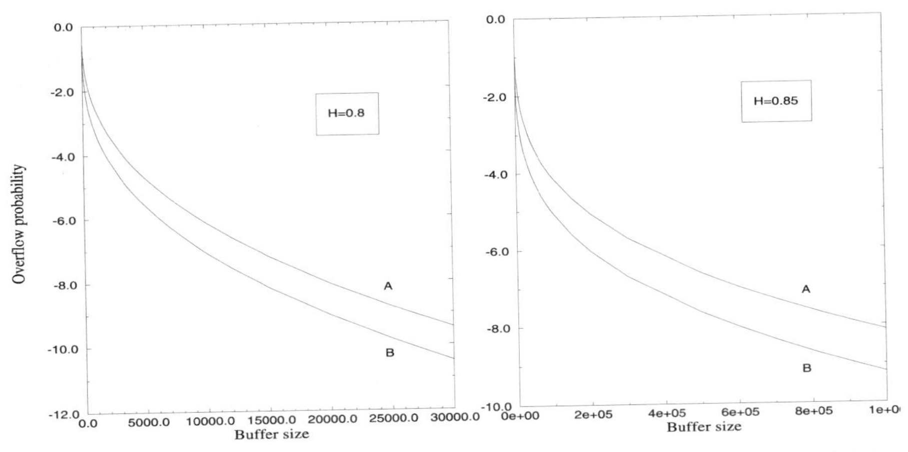

Fig. 5.1 Overflow probability as a function of the buffer size with Pareto activity distribution.

5.5.3 Convergence to the FBM

The limiting Gaussian process $\{\Omega_t\}$ defined in Proposition (5.3.6) of Section 5.3 has been derived as the integral of limiting rate process $\{Y_t\}$, that is,

$$\Omega_t = \int_0^t Y_u \, du, \quad t \geq 0.$$

Now, introduce the family of Gaussian centered processes $\{Z_t^{(\theta)}\}$ defined by

$$Z_t^{(\theta)} = \frac{1}{\theta^H} \cdot \Omega_{\theta t}, \quad t \geq 0, \tag{5.45}$$

for all parameter values $\theta > 0$ and with fixed constant H. Process $\{Z_t^{(\theta)}\}$ is thus deduced from $\{\Omega_t\}$ by the time and space scaling change $(t, x) \mapsto (\theta t, x\theta^H)$. Each process $\{Y_t^{(N)}\}$ defined by Eq. (5.15) being the superposition of i.i.d. rate processes distributed as $\{\lambda_t\}$, its covariance function is independent of N and is therefore identical, along with that of its limiting process $\{Y_t\}$, to the correlation function $\mathscr{R}$ of the single rate process $\{\lambda_t\}$. Similarly, in view of definition (5.18), the variance of Ω_t equals that of any $\Omega_t^{(N)}$ and is identical to $D^2(t) = \mathrm{Var}(\omega_t)$ defined in Eq. (5.3). The second-order properties of processes $\{Z_t^{(\theta)}\}$ can then be readily derived as follows. Formula (5.32) first entails that the variance of $Z_t^{(\theta)}$ can be written as

$$\mathrm{Var}(Z_t^{(\theta)}) = 2\theta^{2(1-H)} \int_0^t dy \int_0^y \mathscr{R}[\theta(y-x)] \, dx. \tag{5.46}$$

On the other hand, it is known that for any centered process $\{\Omega_t\}$ with *stationary increments*, the covariance $\mathbf{E}(\Omega_u \Omega_t)$ can be expressed in terms of the variance function $t \mapsto D(t)^2 = \mathrm{Var}(\Omega_t)$ only as

$$\mathbf{E}(\Omega_u \Omega_t) = \tfrac{1}{2}[D(t)^2 + D(u)^2 - D(t-u)^2] \tag{5.47}$$

with $u \leq t$. The covariance structure and, therefore, the distribution of each Gaussian process $\{Z_t^{(\theta)}\}$ are thus entirely defined through formulas (5.46) and (5.47).

From expression (5.46), in particular, we deduce that the convergence of the family of processes $\{Z_t^{(\theta)}\}$ as $\theta \uparrow +\infty$ toward some limiting process is governed by the behavior of the covariance function $\mathscr{R}$ in the neighborhood of infinity. In view of factor $\theta^{2(1-H)}$ in Eq. (5.46), it is then natural to expect that the variance $\mathrm{Var}(Z_t^{(\theta)})$ and, therefore, the sequence of processes $\{Z_t^{(\theta)}\}$ have a nontrivial limit as $\theta \uparrow +\infty$ in the case where $\mathscr{R}(t)$ behaves as a power of t for large t. Such a situation has been exemplified in Proposition 5.5.1 of this section, where it is shown that the covariance function $\mathscr{R}$ of rate process $\{\lambda_t\}$ is precisely $O(t^{1-q})$ for large t and some $q \in]1, 2[$, provided the distribution of either on or off periods is heavy tailed. A weak convergence result for the family of processes $\{Z_t^{(\theta)}\}$, $\theta > 0$, is therefore plausible in this framework.

Before stating the formal result, it is useful to recall here that [14, p. 318] there exists a unique centered Gaussian process $\{Z_t\}$ having

- continuous paths with $Z_0 = 0$,
- stationary increments,
- and with covariance function defined by

$$\mathbf{E}(Z_u Z_t) = \tfrac{1}{2}[|t|^{2H} + |u|^{2H} - |t-u|^{2H}] \tag{5.48}$$

for all $u, t \in \mathbf{R}$ and for some parameter $H \in [\frac{1}{2}, 1[$.

Note that, in view of the stationarity of increments of $\{Z_t\}$ and of general property (5.47), this definition is equivalent to stating that $\mathbf{E}(Z_t^2) = |t|^{2H}$ for all $t \in \mathbf{R}$. This process $\{Z_t\}$ is known as the fractional Brownian motion (FBM) with Hurst parameter H.

Proposition 5.5.4. *Assume that the covariance function $t \mapsto \mathscr{R}(t)$ of rate process $\{\lambda_t\}$ verifies asymptotics (5.31) for large t and some power $q \in]1, 2[$. The processes $\{Z_t^{(\theta)}\}$ defined in Eq. (5.45) with*

$$H = \frac{3-q}{2}$$

then converge weakly as $\theta \uparrow +\infty$ toward process $\sqrt{\mathscr{L}} \cdot \{Z_t\}$, where $\{Z_t\}$ is an FBM with Hurst parameter H and $\mathscr{L}$ is defined in Eq. (5.34).

Proof. To ensure the convergence of the sequence of centered Gaussian processes $\{Z_t^{(\theta)}\}$, one could, for instance, rely on Theorem 5.3.2. Here we content ourselves with the verification of the first half of this theorem's hypotheses, that is, convergence of the finite-dimensional distributions. The latter holds if the covariance functions converge toward the covariance function of some identified centered Gaussian process. In the present case, using estimate (5.31) in (5.46), we readily obtain

$$\mathrm{Var}(Z_t^{(\theta)}) \sim 2\theta^{2(1-H)} \int_0^t dy \int_0^y \frac{\mathscr{R}_0}{[\theta(y-x)]^{q-1}} dx, \tag{5.49}$$

where the condition $q \in]1, 2[$ ensures that the integral is finite. As

$$\int_0^t dy \int_0^y \frac{dx}{[y-x]^{q-1}} = \int_0^t \frac{y^{2-q}}{2-q} dy = \frac{t^{3-q}}{(3-q)(2-q)},$$

the right-hand side of Eq. (5.49), with $H = (3-q)/2$, is consequently asymptotic to

$$\theta^{q-1} \times \frac{2\mathscr{R}_0}{\theta^{q-1}} \frac{t^{3-q}}{(3-q)(2-q)} = \mathscr{L} \cdot t^{2H}$$

as $\theta \uparrow +\infty$. Using general property (5.47) for processes with stationary increments, we then deduce that the covariance function of $\{Z_t^{(\theta)}\}$ converges to

$$\tfrac{1}{2}[\mathscr{L}|t|^{2H} + \mathscr{L}|s|^{2H} - \mathscr{L}|t-s|^{2H}]$$

as $\theta \uparrow +\infty$. Up to multiplicative constant $\mathscr{L}$, the latter is equal to the covariance function of the FBM introduced in Eq. (5.48). We therefore conclude that the sequence of Gaussian centered processes $\{Z_t^{(\theta)}\}$ converges weakly toward the latter Gaussian process. ■

5.6 CONCLUSION

As noted by Boxma and Dumas [4] in the case of a finite number N of on/off sources, no matter how heavy the tail of the off duration distribution, the distribution of $V_0^{(N)}$ is light tailed (i.e., has finite moments) whenever the on duration distribution is light tailed. On the other hand, for an on duration distribution with Pareto tail, the distribution of $V_0^{(N)}$ also has a Pareto tail whatever the off duration distribution. In contrast, the Weibullian distribution derived in this chapter for $V_0^{(N)}$ as $N \uparrow +\infty$ in the heavy traffic limit can arise as long as at least one of the two distributions for the on and off periods is heavy tailed. Such an observation is in accordance with the usual fact that "small" sources on a link with large capacity perform better than "large" sources in terms of multiplexing efficiency. The amount of multiplexing gain has been derived here by means of powerful tools from weak convergence theory for random processes and some essential bounds for the distribution of the content of a fluid queue.

An important direction for future work is to consider the details of the convergence with N, so as to know when the results presented here will be relevant in practice. For example, for a given moderate to large but finite N, it is desirable to determine at which queue levels, up to some appropriate measure of precision, the Weibullian tail result holds approximately. Attempting to answer such questions through simulation would seem to be pointless, due to the small probabilities at issue, combined with the particular difficulties stemming from such LRD processes, such as the simulation of infinite variances. Analytical expressions are therefore required. This is an even more important question than one might first realize, as the following discussion makes clear. Let the output rate C be held constant and the source rates scaled down instead with increasing N. The distribution of the queue content will now converge to a limiting distribution as $N \uparrow +\infty$ and we can now meaningfully define upper and lower queue levels, $x_L(N)$ and $x_U(N)$, respectively, corresponding to where this distribution has an approximate Weibullian behavior. The lower level $x_L(N)$ will converge to a fixed value with increasing N, corresponding simply to the beginning of the limiting Weibullian tail (as judged by the precision criterion). Here the essential fact to know is simply this limiting value, although details on the rate of convergence would also be useful. The upper level $x_U(N)$, however, will diverge with N and it is essential to have good estimates of $x_U(N)$. As for the behavior for $x > x_U(N)$, it has recently been shown [8] that, with finite N, the input process will have regularly varying behavior, and likewise for the queue content. We are therefore faced with a situation whereby for content $x < x_L(N)$, the queueing distribution is unknown but undramatic (all moments are finite and upper bounds could be found); for contents satisfying $x_L(N) < x < x_U(N)$, it is approximately Weibullian; and for $x > x_U(N)$, the distribution is regularly varying with most moments infinite. Depending in which regime one deems oneself to be for a given range of x of practical interest, one could therefore conclude essentially anything one wished regarding the scaling of tail probabilities! It is therefore of crucial practical importance to be able to locate these different regimes as functions of N to avoid misleading conclusions.

5.7 APPENDIX

5.7.1 Proof of Lemma 5.3.3

Let $\mathcal{N}$ be a homogeneous Poisson process on $\mathbf{R}^2$, with Lebesgue measure as intensity measure. The renewal process L can then be constructed from $\mathcal{N}$ as follows (see Lindvall [11] for a more detailed description of this construction): plot against x the failure rate of the first renewal point on the right of the origin, which is known to equal $\bar{\Pi}(x)/\int_x^{+\infty} \bar{\Pi}(z)\, dz$, where Π is the distribution function of a typical inter-renewal time and $\bar{\Pi} = 1 - \Pi$. The first renewal point of L is then chosen as the smallest x-coordinate T_1 of points of $\mathcal{N}$ that fall between the plotted failure rate and the x-axis. Once T_1 is chosen, plot from T_1 on the failure rate of the inter-renewal time distribution: that is, plot for $x > T_1$ the function

$$x \longmapsto \frac{\Pi'(x - T_1)}{\bar{\Pi}(x - T_1)}$$

and select the second point T_2 as the smallest x-coordinate of points of $\mathcal{N}$ falling between the x-axis and this second curve; see Fig. 5.2. As proved by Lindvall [11], iterating this construction produces a stationary renewal process L with associated interarrival time distribution function Π. Now select $\epsilon > 0$ so that the density Π' is less than a_0 on $[0, \epsilon]$ (such ϵ exists by assumption), and verifying in addition $\epsilon < a_0^{-1} \wedge l^{-1}$. To conclude the proof of the lemma, it is enough to show that the plotted curves are smaller than $m = l/(1 - l\epsilon) \vee a_0/(1 - a_0\epsilon)$ for $x \in [0, \epsilon]$. Indeed, this will imply that the renewal process L is dominated by the point process M counting the x-coordinates of those points of $\mathcal{N}$ falling between the x-axis and the horizontal line with y-coordinate β, and M is clearly a Poisson process with intensity m. Consider the first function plotted to construct T_1, that is,

$$x \longmapsto \frac{\bar{\Pi}(x)}{\int_x^{+\infty} \bar{\Pi}(z)\, dz}.$$

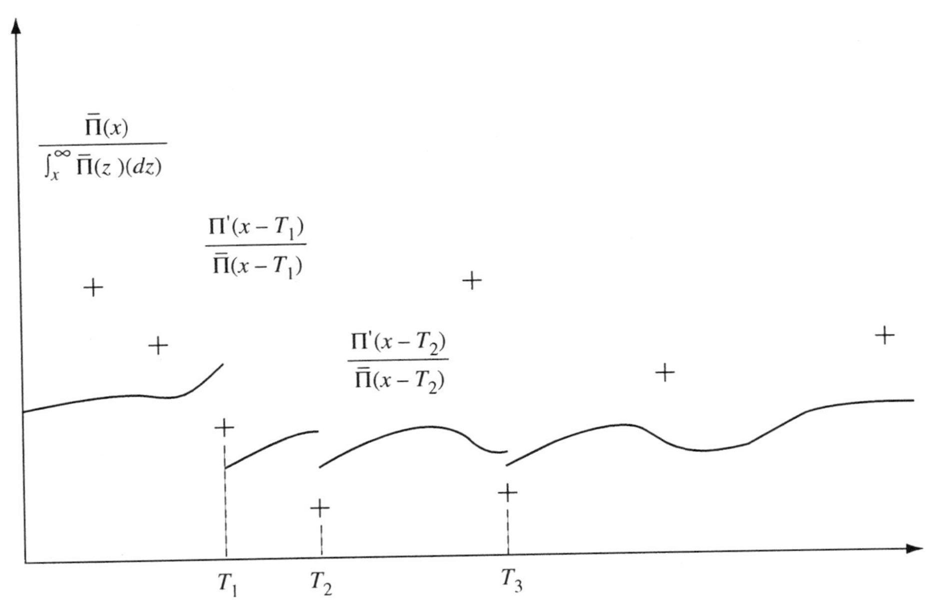

Fig. 5.2 Construction of the renewal process L.

Its numerator is less than 1, while its denominator is equal to $l^{-1} - \int_0^x \bar{\Pi}(z)\,dz$ and hence is larger than $l^{-1} - x$. This function is thus smaller than $l/(1 - l\epsilon)$ for all $x \in [0, \epsilon]$, and this bound is less than m.

Consider now the other functions, which are all of the form

$$x \longmapsto \frac{\Pi'(x - T_n)}{\bar{\Pi}(x - T_n)}, \qquad x \in (T_n, T_{n+1}].$$

Using the bounds a_0 on $\Pi'(z)$ for $z \in [0, \epsilon]$, it follows at once that the considered functions are smaller than $a_0/(1 - a_0\epsilon)$, itself smaller than m. ■

5.7.2 Proof of Lemma 5.3.4

Given $0 \le t_1 \le t \le t_2$ and $j \in \{1, \ldots, N\}$, define the centered variables

$$\xi_j = L_j(t_1, t) - l(t - t_1), \qquad \eta_j = L_j(t, t_2) - l(t_2 - t).$$

From definition (5.9), one obtains

$$P^{(N)}(t_1, t, t_2) = \mathbf{E}\left[\left(\frac{1}{\sqrt{N}}\sum_{j=1}^{N} \xi_j\right)^2 \left(\frac{1}{\sqrt{N}}\sum_{j=1}^{N} \eta_j\right)^2\right],$$

that is,

$$N^2 \cdot P^{(N)}(t_1, t, t^2) = \mathbf{E}\left[\left(\sum_j \xi_j^2\right)\left(\sum_j \eta_j^2\right)\right] + 4 \cdot \mathbf{E}\left[\sum_{j<k} \xi_j\xi_k \sum_{j<k} \eta_j\eta_k\right]$$
$$+ 2\mathbf{E}\left[\sum_j \xi_j^2 \sum_{j<k} \eta_j\eta_k + \sum_j \eta_j^2 \sum_{j<k} \xi_j\xi_k\right].$$

All processes L_j being mutually independent, all variables ξ_j (resp. η_k) are also mutually independent. Being all centered, we therefore deduce that for distinct values of j (resp. k), all terms $\mathbf{E}(\xi_j^2\eta_j\eta_k) = \mathbf{E}(\xi_j^2\eta_j)\mathbf{E}(\eta_k)$, $\mathbf{E}(\eta_j^2\xi_j\xi_k)$ for $j \neq k$ and $\mathbf{E}(\xi_j\xi_k\eta_m\eta_n)$ for $(j, k) \neq (m, n)$ are zero. The above sum then reduces to

$$N^2 P^{(N)}(t_1, t, t_2) = \mathbf{E}\left[\left(\sum_j \xi_j^2\right)\left(\sum_j \eta_j^2\right)\right] + 4\mathbf{E}\left[\sum_{j<k} \xi_j\eta_j\xi_k\eta_k\right];$$

hence

$$N^2 P^{(N)}(t_1, t, t_2) = N\mathbf{E}(\xi^2\eta^2) + N(N-1)\mathbf{E}(\xi^2)\mathbf{E}(\eta^2) + 2N(N-1)[\mathbf{E}(\xi\eta)]^2,$$

which entails inequality (5.12).

The terms on the right-hand side of Eq. (5.12) can now be upper bounded by applying the result of Lemma 5.3.3. For instance,

$$\begin{aligned}\mathbf{E}(\xi\eta) &= \mathbf{E}[L(0, t-t_1) - l(t-t_1)][L(0, t_2-t) - l(t_2-t)] \\ &= \mathbf{E}L(0, t-t_1)L(0, t_2-t) - l(t-t_1)l(t_2-t);\end{aligned}$$

hence,

$$\begin{aligned}[\mathbf{E}(\xi\eta)]^2 &= [\mathbf{E}L(0, t-t_1)L(0, t_2-t)]^2 \\ &\quad - 2l^2(t-t_1)(t_2-t)\mathbf{E}L(0, t-t_1)L(0, t_2-t) \\ &\quad + l(t-t_1)l(t_2-t)]^2 \\ &\leq [\mathbf{E}M(t_1, t)M(t, t_2)]^2 + [l(t-t_1)l(t_2-t)]^2 \\ &= [m^2(t-t_1)(t_2-t)]^2 + [l(t-t_1)l(t_2-t)]^2,\end{aligned}$$

where M is the majorizing Poisson process, the existence of which follows from Lemma 5.3.3. Similar computations can be done to bound the other terms on the right-hand side of inequality (5.12), all bounds being polynomial in $(t - t_1)$ and $(t_2 - t)$, with no terms of degree less than 2. Recalling that t_1, t, and t_2 lie in some bounded interval $[0, \epsilon]$, one eventually obtains bound (5.13). ■

5.7.3 Heavy-Tailed Distributions

From condition $r \in]1, 2]$, the function $g: t \mapsto t(1 - F(t))$ is bounded at infinity and its Laplace transform g^* is defined by

$$g^*(p) = \int_0^{+\infty} e^{-pt} g(t)\, dt$$

for $p > 0$. From definition (5.28), $g(t) \sim \bar{h}/t^{r-1}$ as $t \uparrow +\infty$ with $0 < \bar{h} < +\infty$. Therefore g is ultimately monotonic in the neighborhood of infinity. The following classical Tauberian theorem is then crucially needed in order to relate the behavior of function g at infinity to that of its Laplace transform near 0.

Theorem 5.7.1 [9, p. 446]. *Let function $g: [0, +\infty[\mapsto \mathbf{R}$ be an ultimately monotonic function in the neighborhood of infinity. It satisfies $g(t) \sim At^{s-1}$ as $t \uparrow +\infty$ for some constant A and some nonnegative s if* and only if *its Laplace transform verifies $g^*(p) \sim A\Gamma(s)p^{-s}$ as $p \downarrow 0$, where Γ denotes Euler's function.*

Applying the latter for $s = 2 - r > 0$ then enables us to assert that

$$g^*(p) \sim \frac{\bar{h} \cdot \Gamma(2-r)}{p^{2-r}} \tag{5.50}$$

as $p \downarrow 0$. On the other hand, the definition of g and standard properties of the Laplace transform entail

$$g^*(p) = \frac{1}{p^2}\left[1 - f^*(p) + p\frac{df^*}{dp}(p)\right], \tag{5.51}$$

where f^* is the Laplace transform of distribution F. Expression (5.51) together with asymptotics (5.50) consequently implies that f^* verifies the differential equation

$$\frac{df^*}{dp}(p) - \frac{f^*(p)}{p} = \bar{h} \cdot \Gamma(2-r)p^{r-1} - \frac{1}{p} + \epsilon(p),$$

where $\epsilon(p) = o(p^{r-1})$ for small p. Using the integration factor $1/p$, this equation solves to give

$$f^*(p) = 1 + kp + \frac{\bar{h} \cdot \Gamma(2-r)}{r-1}p^r + \epsilon_1(p),$$

where the integration constant k is identified as $-m$ and with

$$\epsilon_1(p) = p\int_0^p \frac{\epsilon(z)}{z}dz = o(p^r)$$

for $r < 2$. Result (5.29) hence follows.

5.7.4 Derivation of the Conditional Moments of ω_t

The first conditional moments of ω_t, namely,

$$m_a(t) = \mathbf{E}(\omega_t | \gamma_0 = 0), \quad m_b(t) = \mathbf{E}(\omega_t | \gamma_0 = 1),$$

can be calculated by taking successive derivatives at $z = 0$ of the Laplace transform $f_a^*(t, \cdot)$ (resp. $f_b^*(t, \cdot)$) for variable ω_t, given $\lambda_0 = 0$ (resp. variable ω_t, given $\lambda_0 = 1$). These transforms being difficult to calculate, we further consider their Laplace transform with respect to t, yielding

$$m_a^*(p) = -\partial_z^{(k)} f_a^{**}(p, z)_{|z=0}, \; m_b^*(p) = -\partial_z^{(k)} f_b^{**}(p, z)_{|z=0}.$$

As shown in Bensaou et al. [2, Eq. (17)], the double transforms f_a^{**} and f_b^{**} are given by

$$f_a^{**}(p, z) = \frac{1}{p}[1 - \alpha h(p, z)], \quad f_b^{**}(p, z) = \frac{1}{z+p}[1 + \beta h(p, z)],$$

where

$$h(p, z) = z \cdot \frac{(1 - a^*(p))(1 - b^*(p + z))}{p(z + p)(1 - a^*(p)b^*(p + z))}.$$

Note that $\partial_z h(p, z)_{|z=0}$, which is the key to the calculation of these moments, is symmetric with respect to $a^*(p)$ and $b^*(p)$. With the help of expansions (5.30) for $a^*(p)$ and $b^*(z + p)$, the above expressions for $m_a^*(p)$ and $m_b^*(p)$ can be expanded for small positive p. The appplication of Tauberian Theorem 5.7.1 recalled in Section 5.7.3 then provides us finally with the asymptotics for large t of the conditional moments $m_a(t)$ and $m_b(t)$ in the form $vt + O(t^{2-q})$ with $q = \min(r, s)$ (we omit here some easy but lengthy calculations). It can be verified, in particular, that

$$\mu(t) \sim \frac{a_r \alpha v}{\Gamma(3 - r)} t^{2-r}$$

if $r < s$ and

$$\mu(t) \sim \frac{b_s \beta(1 - v)}{\Gamma(3 - s)} t^{2-s}$$

if $s < r$, and $\mu(t)$ is asymptotic to the sum of the two latter equivalents if $r = s$. As a consequence, $\mu(t) \sim \mu_0 \cdot t^{2-q}$ for large t, where constant μ_0 is easily shown to equal $\mu_0 = (3 - q)\mathscr{L}/2\sigma^2$ in view of the above results and definition (5.34) of constant $\mathscr{L}$.

REFERENCES

1. S. Asumussen. *Applied Probability and Queues*. Wiley, New York, 1987.
2. B. Bensaou, J. Guibert, J. W. Roberts, and A. Simonian. Performance of an ATM multiplexer queue in the fluid approximation using the Beneš approach. *Ann. Oper. Res.* **49**:137–160, 1994.
3. P. Billingsley. *Convergence of Probability Measures*. Wiley, New York, 1968.
4. O. Boxma and V. Dumas. Fluid queues with long-tailed activity period distributions. *Comput. Commun.*, **21**:1509–1529, 1998.
5. F. Brichet, J. W. Roberts, A. Simonian, and D. Veitch. Heavy traffic analysis of a storage model with long-range dependent on/off sources. *Queueing Syst.*, **23**:197–215, 1996.
6. D. R. Cox. Long-range dependence. In H. A. David and H. T. David, eds., *A Review, Statistics: An Appraisal, Proceedings 50th Anniversary Conference, Iowa State Statistical Laboratory.* The Iowa State University Press, Ames, 1984.
7. G. Doetsch. *Handbuch der Laplace-transformation* (in German). Birkhäuser Verlag, Basel, 1955.
8. V. Dumas and A. Simonian. Asymptotic bounds for the fluid queue fed by sub-exponential on/off sources. To appear in *Adv. Appl. Prob.*, **32**(1), March 2000.

9. W. Feller. *An Introduction to Probability Theory and Its Applications, Volume II*. Wiley, New York, 1971.

10. T. Kurtz. Limit theorem for workload input models. In F. Kelly, S. Zachary, and I. Ziedins, eds., *Stochastic Networks: Theory and Applications*. Oxford University Press, New York, 1996.

11. T. Lindvall. *Lectures on the Coupling Method*. Wiley, New York, 1992.

12. J. D. Murray. *Asymptotic Analysis*. Springer-Verlag, New York, 1984.

13. I. Norros. A storage model with self-similar input. *Queueing Syst.*, **16**:387–396, 1994.

14. G. Samorodnitsky and M. S. Taqqu. *Stable Non Gaussian Random Processes*. Chapman and Hall, London, 1994.

15. A. N. Shiryayiev. *Probability*. Springer-Verlag, New York, 1984.

16. A. Simonian and J. Virtamo. Transient and stationary distributions for fluid queues and input processes with a density. *SIAM J. Appl. Math.*, **51**(6):1732–1739, December 1991.

6

THE SINGLE SERVER QUEUE: HEAVY TAILS AND HEAVY TRAFFIC

O. J. BOXMA

Department of Mathematics and Computing Science, Eindhoven University of Technology, 5600 MB Eindhoven, The Netherlands; and CWI, P.O. Box 94079, 1090 GB Amsterdam, The Netherlands

J. W. COHEN

CWI, P.O. Box 94079, 1090 GB Amsterdam, The Netherlands

6.1 INTRODUCTION

Recently, there has been much interest in the behavior of queues with heavy-tailed service time distributions. This interest has been triggered by a large number of traffic measurements on modern communication network traffic (e.g., see Willinger et al. [38] for Ethernet LAN traffic, Paxson and Floyd [30] for WAN traffic, and Beran et al. [3] for VBR video traffic; see also various chapters in this volume). These measurements and their statistical analysis (e.g., see Leland et al. [26]) suggest that modern communication traffic often possesses the properties of self-similarity and long-range dependence. A natural possibility to introduce long-range dependence in an input traffic process is to take a fluid queue and to assume that at least one of the input quantities I (on or off periods in the fluid queue fed by on/off sources) has the following "heavy-tail" behavior:

$$\mathbb{P}[I > t] \overset{t\to\infty}{\sim} h_v t^{-v}, \tag{6.1}$$

Self-Similar Network Traffic and Performance Evaluation, Edited by Kihong Park and Walter Willinger
ISBN 0-471-31974-0

with h_ν a positive constant and $1 < \nu < 2$ (here and later, $f(t) \overset{t\to\infty}{\sim} g(t)$ stands for $f(t)/g(t) \to 1$ with $t \to \infty$; and many-valued functions like $t^{-\nu}$ are defined by their principal value, so $t^{-\nu}$ is real for t positive). In this context, regularly varying and subexponential distributions [5] have received special attention. We refer to Boxma and Dumas [12] for a survey on *fluid* queues with heavy-tailed on-period distributions. In this chapter, we concentrate on the *ordinary* single server queue with regularly varying service and/or interarrival time distribution. The fluid queue is closely related to this ordinary queue (cf. Remarks 6.2.6 and 6.6.3), so that several of the results of the present chapter will also be relevant for fluid queues.

When $G(x)$ is the probability distribution of a nonnegative random variable G, then $1 - G(x) = \mathbb{P}[G > x]$ is said to be regularly varying at infinity of index $-\nu$ when

$$1 - G(x) = x^{-\nu}L(x), \quad x \geq 0, \tag{6.2}$$

with $L(x)$ a slowly varying function at infinity, that is,

$$\lim_{x\to\infty} \frac{L(tx)}{L(x)} = 1, \quad \forall t > 0.$$

$L(\cdot)$ could, for instance, be a constant or a logarithmic function. If Eq. (6.2) holds, then we write $\mathbb{P}[G > \cdot] \in \mathcal{R}(-\nu)$.

Regular variation is an important asymptotic concept in probabilistic analysis. The main reference text is the book by Bingham et al. [5].

An early result concerning regular variation in queueing theory is due to Cohen [13]. He proved the following result for the $GI/G/1$ queue under the first-come-first-served (FCFS) discipline: the tail of the waiting time distribution is regularly varying of index $1 - \nu$ if and only if the tail of the service time distribution is regularly varying of index $-\nu$, $\nu > 1$. This result has turned out to be very useful in relating the regularly varying tail behavior of on periods of on/off sources in fluid queues to the buffer content [7, 8]; see also Jelenković and Lazar [23] and Rolski et al. [32].

Cohen's result implies that, in the case of regular variation, the tail of the waiting time distribution is one degree heavier than that of the service time distribution. For $1 < \nu < 2$, the mean of the service time distribution is finite, but the mean of the waiting time distribution is infinite.

The first issue that we consider in this chapter is whether other service disciplines besides FCFS may lead to a less detrimental waiting time behavior. We consider the $M/G/1$ queue with the processor sharing (PS) discipline and the $M/G/1$ queue with the last-come-first-served preemptive resume (LCFS-PR) discipline. Note that PS and LCFS-PR are well-known and important disciplines, that both play a key role in product-form networks. For both disciplines, the waiting time tail behavior turns out to be regularly varying of index $-\nu$ iff the service time tail behavior is regularly varying of index $-\nu$. We refer to Anantharam [2] for a related investigation of the influence of the service discipline on the tail behavior of a single server queue.

The second issue under consideration in this chapter is the heavy traffic behavior of a queue with heavy-tailed service time distribution. A queueing system is said to be in heavy traffic when its traffic load $\rho \to 1$. This issue is of theoretical interest, since in the traditional heavy traffic limit theorems it is assumed that the second moments of service and interarrival times are finite, whereas Eq. (6.1) with $1 < \nu < 2$ leads to an infinite second moment. The issue is also of practical interest, since heavy traffic limit theorems may give rise to useful approximations in situations with a reasonably light traffic load. We first focus on the $M/G/1$ queue, again with the three service disciplines FCFS, PS, and LCFS-PR. Subsequently, we also allow interarrival times to be generally distributed and even heavy tailed. New heavy traffic limit results are presented for the waiting time in the $GI/G/1$ queue with heavy-tailed interarrival and/or service time distribution. The case in which both tails are "just as heavy" is particularly interesting from a mathematical point of view. We identify *coefficients of contraction* $\Delta(\rho)$ such that $\Delta(\rho)$ times the waiting time has a proper limiting distribution for $\rho \uparrow 1$.

The third issue under consideration is the convergence of the workload *process* $\{v_t, t \geq 0\}$ in the $GI/G/1$ queue with heavy-tailed interarrival and/or service time distribution. It is shown that $\Delta(\rho) v_{\tau/((1-\rho)\Delta(\rho))}$ converges in distribution for $\rho \uparrow 1$, for all $\tau \geq 0$. The thus scaled and contracted workload process converges weakly to the workload process of a queueing model of which the input process is described by ν-stable Lévy motion.

This chapter is organized in the following way. In Section 6.2 we discuss the relation between the tail behavior of service times and waiting times, for the $M/G/1$ queue with service disciplines FCFS, PS, and LCFS-PR. In Section 6.3, for the same three $M/G/1$ variants, we present heavy traffic limit theorems for waiting times, in the case of a regularly varying service time distribution with an infinite variance. In the LCFS-PR case, the sojourn time distribution coincides with the $M/G/1$ busy period distribution. The heavy traffic behavior of the latter distribution is investigated in detail; the tail behavior of the limiting distribution is studied in Section 6.4. Heavy traffic limit theorems for the waiting time in the $GI/G/1$ queue with the FCFS service discipline are presented in Section 6.5. A distinction is made between the cases in which the interarrival time distribution has a heavier tail, the service time distribution has a heavier tail, and both tails are "just as heavy." In Section 6.6 the input process and the workload process are studied for the $GI/G/1$ queue with heavy-tailed interarrival and/or service time distribution. The heavy traffic behavior of those processes is characterized. Section 6.7 contains conclusions and some topics for further research. (*Note*: Several of the results in this chapter have appeared in recent reports of the authors and their colleagues; in those cases, proofs are generally omitted. Also, several chapters in this volume are related to the present work. We mention in particular the contributions of Jelenković (Chapter 10), of Likhanov (Chapter 8), of Makowski and Parulekar (Chapter 9), of Norros (Chapter 4), of Brichet et al. (Chapter 5), and of Resnick and Samorodnitsky (Chapter 7).)

6.2 WAITING TIME TAIL BEHAVIOR

6.2.1 Introduction

In Section 6.1 we have already mentioned a result of Cohen [13] that relates the (regularly varying) tail behavior of service and waiting times in the $GI/G/1$ queue with the FCFS discipline; it shows that the waiting time is "one degree heavier" than the service time tail, in the case of regular variation. In Section 6.2.2 this result, and an extension, will be discussed in some more detail for the $M/G/1$ FCFS queue. A similar result for the $M/G/1$ queue with the PS discipline (due to Zwart and Boxma [42]) will be discussed in Section 6.2.3. That result shows that, under processor sharing, the waiting time tail is just as heavy as the service time tail. In Section 6.2.4 we prove that the latter phenomenon also occurs in the $M/G/1$ queue with the LCFS-PR discipline. In the present section we first introduce some notation, and we present a very useful lemma that relates the tail behavior of a regularly varying probability distribution and the behavior of its Laplace–Stieltjes transform (LST) near the origin.

Consider the $M/G/1$ queue. Customers arrive according to a Poisson process with rate λ; their service times $B_1, B_2, \ldots$ are i.i.d. (independent, identically distributed) random variables with finite mean β and LST $\beta\{s\}$. A generic service time is denoted by B. By B^* we denote a random variable of which the distribution is that of a residual service time:

$$\mathbb{P}[B^* > x] = \frac{1}{\beta}\int_x^\infty \mathbb{P}[B > u]\, du, \quad x \geq 0.$$

Its LST is given by $\beta^*\{s\} := (1 - \beta\{s\})/\beta s$. The traffic load $\rho := \lambda\beta$ of the $M/G/1$ queue is assumed to be less than one, so that the steady-state waiting time distribution exists.

A very useful property of probability distributions with regularly varying tails is a characterization of the behavior of its LST near the origin. Let $F(\cdot)$ be the distribution of a nonnegative random variable, with LST $\phi\{s\}$ and finite first n moments $\mu_1, \ldots, \mu_n$ (and $\mu_0 = 1$). Define

$$\phi_n\{s\} := (-1)^{n+1}\left[\phi\{s\} - \sum_{j=0}^{n} \mu_j \frac{(-s)^j}{j!}\right].$$

Lemma 6.2.1. *Let $n < v < n+1$, $C \geq 0$. The following statements are equivalent:*

$$\phi_n\{s\} = (C + o(1))s^v L(1/s), \quad s \downarrow 0,\ s \text{ real}, \tag{6.3}$$

$$1 - F(t) = (C + o(1))\frac{(-1)^n}{\Gamma(1-v)} t^{-v} L(t), \quad t \to \infty. \tag{6.4}$$

The case $C > 0$ is due to Bingham and Doney [4]. The case $C = 0$ is treated in Boxma and Dumas [12, Lemma 2.2]. The case of an integer v is more complicated; see Bingham et al. [5, Theorem 8.1.6 and Chap. 3].

6.2.2 The $M/G/1$ FCFS Queue

We first formulate the main result of Cohen [13] for the $GI/G/1$ queue with FCFS discipline in full generality. There is no need to specify the interarrival time distribution; the mean interarrival time and traffic load are denoted by $1/\lambda$ and $\rho = \lambda\beta$ (as before). In what follows, W denotes the steady-state waiting time.

Theorem 6.2.2. *For $\rho < 1$ and $v > 1$,*

$$\mathbb{P}[B > x] \overset{x\to\infty}{\sim} (v-1)\left(\frac{x}{\beta}\right)^{-v} L(x) \Longleftrightarrow \mathbb{P}[W > x] \overset{x\to\infty}{\sim} \frac{\rho}{1-\rho}\left(\frac{x}{\beta}\right)^{1-v} L(x). \tag{6.5}$$

Pakes [29] has extended this result to the larger class $\mathcal{S}$ of subexponential distributions (i.i.d. stochastic variables X_1 and X_2 have a subexponential tail if $\mathbb{P}(X_1 + X_2 > t)/\mathbb{P}(X_1 > t) \overset{t\to\infty}{\sim} 2$). His result states that $\mathbb{P}[W > \cdot] \in \mathcal{S}$ if and only if $\mathbb{P}[B^* > \cdot] \in \mathcal{S}$, and if either is the case then

$$\mathbb{P}[W > x] \overset{x\to\infty}{\sim} \frac{\rho}{1-\rho}\mathbb{P}[B^* > x]. \tag{6.6}$$

REMARK 6.2.3. Note that the interarrival time distribution has no influence on the above-mentioned waiting time tail behavior.

REMARK 6.2.4. In the case of Poisson arrivals, the Pollaczek–Khintchine formula reads:

$$\mathbb{E}[e^{-sW}] = \frac{1-\rho}{1-\rho\beta^*\{s\}}, \qquad \text{Re}\, s \geq 0, \tag{6.7}$$

and

$$\mathbb{P}[W > x] = \sum_{n=1}^{\infty}(1-\rho)\rho^n \mathbb{P}[B_1^* + \cdots + B_n^* > x], \quad x \geq 0. \tag{6.8}$$

Theorem 6.2.2 and Eq. (6.6) can be verified very easily in the $M/G/1$ case. For example, the above-mentioned definition of subexponentiality implies that $\mathbb{P}[B_1^* + \cdots + B_n^* > x] \overset{x\to\infty}{\sim} n\mathbb{P}[B_1^* > x]$; and combination of Lemma 6.2.1 and Eq. (6.7) (that relates the LSTs of W and B^*) readily yields the relation between the tail behavior of the distributions of W and B.

The possibility of having to wait at least a residual service time explains that the waiting time tail is one degree heavier than the service time tail: $\mathbb{P}[B > \cdot] \in \mathcal{R}(-\nu)$ implies that $\mathbb{P}[B^* > \cdot] \in \mathcal{R}(1-\nu)$ (cf. Bingham et al. [5]; obviously, integration increases the index of regular variation by one).

In the next subsections we shall consider two service disciplines in which an arriving customer does not first have to wait a residual service time, and where the waiting time tail is *just as heavy* as the service time tail.

REMARK 6.2.5. Denote the steady-state sojourn time by S; in the $GI/G/1$ FCFS queue, $S \simeq W + B$, where W and B are independent and where $\simeq$ denotes equality in distribution. Hence

$$\mathbb{P}[S > x] \overset{x\to\infty}{\sim} \frac{\rho}{1-\rho}\mathbb{P}[B^* > x]. \tag{6.9}$$

Denote the steady-state workload by V; in the $GI/G/1$ FCFS queue (cf. Cohen [15, p. 296]),

$$\mathbb{E}[e^{-sV}] = 1 - \rho + \rho\beta^*\{s\}\mathbb{E}[e^{-sW}], \qquad \text{Re}\, s \geq 0. \tag{6.10}$$

Hence, using Lemma 6.2.1, $\mathbb{P}[B > \cdot] \in \mathcal{R}(-\nu) \Longleftrightarrow \mathbb{P}[V > \cdot] \in \mathcal{R}(1-\nu)$.

REMARK 6.2.6. The goal of this remark is to indicate how results like Theorem 6.2.2 can readily be translated into results for fluid queues. Consider a fluid queue with an infinite buffer and an output rate equal to one. This queue is fed by a source that alternates between silence periods σ_n, $n \geq 1$, during which it generates no input, and activity periods α_n, $n \geq 1$, during which it generates fluid according to the rate process $(r_n(t))_{t\geq 0}$, $n \geq 1$ (this could be the aggregated input process due to several on/off sources). A crucial assumption is that

$$\forall n \geq 1, \forall t \geq 0 \colon 1 \leq r_n(t) \leq R \quad \text{(where } R \text{ is some given constant).} \tag{6.11}$$

In particular, the nth activity period results in a *net* input equal to

$$\hat{B}_n := \int_0^{\alpha_n} [r_n(t) - 1]\, dt. \tag{6.12}$$

Assume that $(\sigma_n)_{n\geq 1}$ and, similarly, $((\alpha_n, r_n(t)), t \geq 0)_{n\geq 1}$ are i.i.d., and that these sequences are independent. The buffer content W_n^b at the beginning of the nth activity period satisfies the recursion

$$W_{n+1}^b = \max[0, W_n^b + \hat{B}_n - \sigma_n], \qquad n \geq 1. \tag{6.13}$$

This is exactly the $GI/G/1$ Lindley waiting time recursion (identify W_n^b with W_n, $\hat{B}_n$ with B_n, and σ_n with the nth interarrival interval). Hence Theorem 6.2.2 also applies to this fluid queue. Similar results for the tail behavior of the *steady-state* buffer content can be proved, using Kella and Whitt [24]. The survey by Boxma and Dumas [12] contains an overview of the tail behavior of the buffer content in fluid queues fed by on/off sources, in which at least one on-period distribution is heavy tailed.

6.2.3 The $M/G/1$ PS Queue

In the (egalitarian) processor sharing service discipline, every customer is simultaneously being served with rate $1/X$, where X is the number of customers in the system. An extensive survey of processor sharing results is presented in Yashkov [39]. Let S_{PS} denote the steady-state sojourn time in the $M/G/1$ PS queue. In Zwart and Boxma [42] the following result is proved.

Theorem 6.2.7. *For $\nu > 1$ (ν noninteger),*

$$\mathbb{P}[B > x] \overset{x\to\infty}{\sim} x^{-\nu} L(x) \Longleftrightarrow \mathbb{P}[S_{\mathrm{PS}} > x] \overset{x\to\infty}{\sim} \frac{1}{(1-\rho)^{\nu}} x^{-\nu} L(x). \tag{6.14}$$

Both relations imply that, in the case of regular variation,

$$\mathbb{P}[S_{\mathrm{PS}} > x] \overset{x\to\infty}{\sim} \mathbb{P}[B > (1-\rho)x]. \tag{6.15}$$

Theorem 6.2.7 is proved by deriving a new expression for the LST of the sojourn time distribution, and applying Lemma 6.2.1. An interpretation of the factor $(1-\rho)$ in Eq. (6.15) is the following. When a tagged customer is in the system for a long time, then the distribution of the total number of customers is approximately equal to the steady-state distribution of the number of customers in a PS queue with one permanent customer. The latter model is a special case of the $M/G/1$ queue with the *generalized* processor sharing discipline, as studied in Cohen [14]. It follows from the results in Cohen [14] that, in steady state, the mean fraction of service given to the permanent customer in that $M/G/1$ queue is $1-\rho$. Hence, if a tagged customer has been in the $M/G/1$ PS queue during a large time x, one would expect that the amount of service received is approximately equal to $(1-\rho)x$.

6.2.4 The $M/G/1$ LCFS-PR Queue

In the LCFS preemptive resume discipline, an arriving customer K is immediately taken into service. However, this service is interrupted when another customer arrives, and it is only resumed when all customers who have arrived after K have left the system. Hence, the sojourn time of K has the same distribution as the busy period of this $M/G/1$ queue. The tail behavior of the busy period distribution in the $M/G/1$ queue (which distribution obviously is independent of the service discipline, when the discipline is work conserving) has been studied by De Meyer and Teugels [27] for the case of a regularly varying service time distribution. Their main theorem thus immediately leads to the following result. Let S_L denote the steady-state sojourn time in the $M/G/1$ LCFS-PR queue.

Theorem 6.2.8. *For $\nu > 1$,*

$$\mathbb{P}[B > x] \overset{x\to\infty}{\sim} x^{-\nu}L(x) \Longleftrightarrow \mathbb{P}[S_L > x] \overset{x\to\infty}{\sim} \frac{1}{(1-\rho)^{\nu+1}} x^{-\nu}L(x). \tag{6.16}$$

Both relations imply that, in the case of regular variation,

$$\mathbb{P}[S_L > x] \overset{x\to\infty}{\sim} \frac{1}{1-\rho}\mathbb{P}[B > (1-\rho)x]. \tag{6.17}$$

REMARK 6.2.9. Theorems 6.2.7 and 6.2.8 show that in both PS and LCFS-PR, regular variation of the service time distribution gives rise to regular variation of the sojourn time distribution, of the *same* index. In LCFS–*nonpreemptive*, it is easily seen that regular variation of the service time distribution gives rise to regular variation of the waiting time distribution of one *index higher* (just like FCFS). This result is obtained by applying Lemma 6.2.1 to the waiting time LST (for this LST, see, e.g., Cohen [15, (III.3.10)]). The increment of the index is not surprising, since—as in FCFS—a waiting time may include a residual service time.

6.3 HEAVY TRAFFIC LIMIT THEOREMS FOR THE $M/G/1$ QUEUE

6.3.1 Introduction

When the variance σ^2 of the service time distribution is finite, the standard heavy traffic limit theorem for the stationary waiting time W in the $M/G/1$ queue holds, that is,

$$\lim_{\rho\uparrow 1} \mathbb{P}[\Delta(\rho)W \leq t] = 1 - e^{-t}, \quad t \geq 0, \tag{6.18}$$

with $\Delta(\rho) := \lambda(1-\rho)/(1+\lambda^2\sigma^2/2)$. This exponential heavy traffic theorem was obtained by Kingman in the early 1960s; see Kingman [25] for an early survey, and Whitt [36] for an extensive overview of heavy traffic limit theorems for queues. In Boxma and Cohen [10] (see also Cohen [19]) a heavy traffic limit theorem has been proved for the $GI/G/1$ FCFS queue in which the service time and/or interarrival time distribution is regularly varying, with *infinite* second moment. In Section 6.3.2 we discuss that result for the case of the $M/G/1$ FCFS queue; in Section 6.5 the case of the $GI/G/1$ FCFS queue will be treated. The known heavy traffic result for the $M/G/1$ PS case is presented in Section 6.3.3. Section 6.3.4 contains a new heavy traffic limit theorem for the waiting time distribution in the $M/G/1$ LCFS-PR queue with a heavy-tailed regularly varying service time distribution.

6.3.2 The $M/G/1$ FCFS Queue

In Boxma and Cohen [10] the following result has been proved:

Theorem 6.3.1. *For the stable $M/G/1$ FCFS queue with regularly varying service time distribution of index $-\nu$, as specified in Eq. (6.2), the "contracted" waiting time $\Delta_W(\rho)W/\beta$ converges in distribution for $\rho \uparrow 1$. The limiting distribution $R_{\nu-1}(t)$ is specified by its LST $1/(1+r^{\nu-1})$, $\mathrm{Re}\, r \geq 0$, and the coefficient of contraction $\Delta_W(\rho)$ is the root of the equation*

$$\rho \frac{\Gamma(2-\nu)}{\nu-1} x^{\nu-1} L(x) = 1-\rho, \quad x > 0, \quad 0 < 1-\rho \ll 1, \tag{6.19}$$

with the property that $\Delta_W(\rho) \downarrow 0$ for $\rho \uparrow 1$.

Proof. We refer to Boxma and Cohen [10] for a detailed proof of the theorem, albeit under slightly weaker conditions on the service time distribution. Here we sketch a different proof. Using the Pollaczek–Khintchine formula (6.7) and Eq. (6.19) we can write, for $\mathrm{Re}\, s \geq 0$,

$$\begin{aligned} \mathbb{E}[e^{-sW}] &= \frac{1-\rho}{1-\rho\beta^*\{s\}} = \frac{1}{1+\dfrac{\rho}{1-\rho}(1-\beta^*\{s\})} \\ &= \frac{1}{1+\dfrac{1-\beta^*\{s\}}{\dfrac{\Gamma(2-\nu)}{\nu-1}\Delta_W(\rho)^{\nu-1}L(\Delta_W(\rho))}}. \end{aligned} \tag{6.20}$$

Now replace s by $r\Delta_W(\rho)$, $\mathrm{Re}\, r \geq 0$. Since $\Delta_W(\rho) \downarrow 0$ for $\rho \uparrow 1$, it follows from Lemma 6.2.1 that

$$1-\beta^*\{r\Delta_W(\rho)\} \sim \frac{\Gamma(2-\nu)}{\nu-1} r^{\nu-1} \Delta_W(\rho)^{\nu-1} L(r\Delta_W(\rho)) \quad \text{for } \rho \uparrow 1.$$

Hence, using the defining property of a slowly varying function,

$$\lim_{\rho\uparrow 1} \mathbb{E}[e^{-r\Delta_W(\rho)W}] = \lim_{\rho\uparrow 1} \frac{1}{1+\dfrac{1-\beta^*\{r\Delta_W(\rho)\}}{\dfrac{\Gamma(2-\nu)}{\nu-1}\Delta_W(\rho)^{\nu-1}L(\Delta_W(\rho))}} = \frac{1}{1+r^{\nu-1}}. \tag{6.21}$$

■

REMARK 6.3.2. In Boxma and Cohen [10] it is shown that Theorem 6.3.1 also holds for $\nu = 2$. Furthermore, $R_{\nu-1}(t)$ is discussed in detail in Boxma and Cohen

[10]. Its relation to the Mittag–Leffler function and (for $v = \frac{3}{2}$) to the complementary error function is pointed out. One can write, for $t \geq 0$,

$$1 - R_{v-1}(t) = \sum_{n=0}^{\infty}(-1)^n \frac{t^{n(v-1)}}{\Gamma(n(v-1)+1)}.$$

REMARK 6.3.3. Theorem 6.3.1 opens possibilities for approximating the waiting time distribution in the $M/G/1$ queue with heavy-tailed service time distribution; such possibilities are explored in Boxma and Cohen [9]. Replacing $\mathbb{P}[W > x]$ by $1 - R_{v-1}(\Delta_W(\rho)x/\beta)$ appears to yield remarkably accurate results, even if the traffic load ρ is much smaller than one (in particular when x is large).

6.3.3 The $M/G/1$ PS Queue

In view of the fact that, under the processor sharing discipline, no customer ever waits, it is natural to concentrate on the sojourn time distribution rather than the waiting time distribution. In Zwart and Boxma [42] a novel expression has been derived for the LST $v\{s, \tau\} = \mathbb{E}[e^{-sS_{PS}(\tau)}]$ in the $M/G/1$ PS queue; here $S_{PS}(\tau)$ is the sojourn time of a customer with service request τ. This expression leads to a new and easy proof [42] of the following $M/G/1$ PS heavy traffic limit theorem of Sengupta [34] and of Yashkov [40].

Theorem 6.3.4. *If $\beta < \infty$, then*

$$\lim_{\rho\uparrow 1} v\{s(1-\rho), \tau\} = \frac{1}{1+s\tau}, \quad \text{Re } s \geq 0, \ \tau \geq 0, \tag{6.22}$$

$$\lim_{\rho\uparrow 1} \mathbb{P}[(1-\rho)S_{PS}(\tau) \leq x] = 1 - e^{-x/\tau}, \quad x \geq 0, \ \tau \geq 0. \tag{6.23}$$

Since $v\{s(1-\rho), \tau\} \leq 1$, we have by dominated convergence and Theorem 6.3.4 the following heavy traffic limit for the unconditional sojourn time distribution with LST $v\{\cdot\}$ (the last statement of the theorem follows by observing that $1/(1+s\tau)$ is the LST of the negative exponential distribution).

Theorem 6.3.5. *For* Re $s \geq 0$,

$$\lim_{\rho\uparrow 1} v\{s(1-\rho)\} = \int_0^{\infty} \frac{1}{1+s\tau}\, d\mathbb{P}(B < \tau) = \int_0^{\infty} e^{-x}\beta\{sx\}\, dx. \tag{6.24}$$

In Zwart and Boxma [42] the convergence of the moments of the sojourn time in heavy traffic is also studied. The moments of the "contracted" sojourn time are shown to converge to the corresponding moments of the heavy traffic limiting distribution. In particular, we have the following theorem.

Theorem 6.3.6. *If $\beta < \infty$, then*

$$\lim_{\rho \uparrow 1} \mathbb{E}[((1-\rho)S_{\text{PS}}(\tau))^k] = k!\tau^k, \quad \tau \geq 0, \quad k \geq 1. \tag{6.25}$$

If $\beta_k < \infty$, $k \geq 1$, then

$$\lim_{\rho \uparrow 1} \mathbb{E}[((1-\rho)S_{\text{PS}})^k] = k!\beta_k. \tag{6.26}$$

The latter results can be used for the approximation of higher moments of the sojourn time distribution.

6.3.4 The $M/G/1$ LCFS-PR Queue

In Section 6.2.4 we have observed that the sojourn time S_L in the $M/G/1$ LCFS-PR queue has the same distribution as the $M/G/1$ busy period. We now concentrate on the latter quantity. Denote, for the $M/G/1$ queue, by $P(v)$ the first entrance time of the workload process into the zero state, starting with a workload v. $P(v)$ is usually called the busy period with initial workload v. Obviously $P(0) = 0$ and $P(v) > 0$ for $v > 0$. Let $P_1, P_2, \ldots$ be i.i.d. stochastic variables with distribution that of the busy period of the $M/G/1$ queue. Let $K(v)$ be the number of arrivals in an interval of length v, as defined above. Then $P(0) = 0$ and

$$P(v) = v + P_1 + \cdots + P_{K(v)} \quad \text{for } v > 0.$$

Hence for Re $s \geq 0$, with $I_{(\cdot)}$ an indicator function,

$$\begin{aligned}
\mathbb{E}[e^{-sP(v)}] &= I_{(v=0)} + I_{(v>0)} e^{-sv} \sum_{k=0}^{\infty} e^{-\lambda v} \frac{(\lambda v)^k}{k!} \mathbb{E}^k[e^{-sP}] \\
&= I_{(v=0)} + I_{(v>0)} \exp[-sv - \lambda v(1 - \mathbb{E}[e^{-sP}])] \\
&= I_{(v=0)} + I_{(v>0)} \exp\left[-\frac{sv}{1-\rho}(1 - \rho + \rho\mathbb{E}[e^{-sP^*}])\right].
\end{aligned} \tag{6.27}$$

Here P^* denotes a random variable with the distribution of a residual busy period, so its LST is $(1 - \mathbb{E}[e^{-sP}])/[s\beta/(1-\rho)]$, see Lemma 6.2.1. Integrating with respect to the distribution of the steady-state workload V in the $M/G/1$ queue (which is also the waiting time distribution because of PASTA), it follows for Re $s \geq 0$, with $\rho < 1$,

$$\mathbb{E}[e^{-sP(V)}] = \mathbb{E}\left[\exp\left(-\frac{sV}{1-\rho}\{1 - \rho + \rho\mathbb{E}[e^{-sP^*}]\}\right)\right]. \tag{6.28}$$

On the other hand, since $P(V)$ can be viewed as the time to empty the system when starting in steady state, we can write

$$\mathbb{E}[e^{-sP(V)}] = 1 - \rho + \rho\mathbb{E}[e^{-sP^*}], \quad \text{Re } s \geq 0. \tag{6.29}$$

Indeed, $P(V)$ is zero with probability $1 - \rho$, and it equals a residual busy period P^* with probability ρ.

Combination of Eqs. (6.28) and (6.29) yields

$$\mathbb{E}[e^{-sP(V)}] = \mathbb{E}\left[\exp\left(-\frac{s}{1-\rho}\mathbb{E}[e^{-sP(V)}]V\right)\right], \quad \text{Re } s \geq 0; \tag{6.30}$$

that is, $\mathbb{E}[e^{-sP(V)}]$ satisfies

$$k(s) = \mathbb{E}\left[\exp\left(-\frac{s}{1-\rho}k(s)V\right)\right] \quad \text{and} \quad |k(s)| \leq 1, \quad \text{Re } s \geq 0. \tag{6.31}$$

It may be shown [20] that these equations have one and only one solution for $\text{Re } s \geq 0$ with $k(0) = 1$; and $k(s)$ is regular for $\text{Re } s > 0$, continuous for $\text{Re } s \geq 0$, and decreasing in s for s real. Using Theorem 6.3.1, the heavy traffic limit theorem for the waiting time W in the $M/G/1$ FCFS queue (or equivalently for the workload V), one can now prove the following heavy traffic limit theorem for $P(V)$, or equivalently, see Eq. (6.29), for P^*.

Theorem 6.3.7. *For the stable $M/G/1$ queue with regularly varying service time distribution of index $-\nu$, as specified in Eq. (6.2), the "contracted" busy period $\Delta_P(\rho)P(V)$ converges in distribution for $\rho \uparrow 1$. The limiting distribution $P_{\nu-1}(t)$ is specified by the fact that its LST $p(r)$ is that zero of the equation*

$$y = \frac{1}{1 + (\beta r y)^{\nu-1}}, \quad \text{Re } r \geq 0, \tag{6.32}$$

with the property that $y(0) = 1$; and the coefficient of contraction $\Delta_P(\rho)$ equals $(1-\rho)\Delta_W(\rho)$. The same results hold for $\Delta_P(\rho)P^*$, the "contracted" residual busy period.

REMARK 6.3.8. In Cohen [20] it is shown that Theorem 6.3.7 also holds for $\nu = 2$.

REMARK 6.3.9. It follows from Eqs. (6.27) and (6.29), with $s = r\Delta_P(\rho)$, $\text{Re } r \geq 0$, and $v > 0$, that

$$\mathbb{E}[e^{-r\Delta_P(\rho)P(v)}] = \exp[-r\Delta_W(\rho)\mathbb{E}[e^{-r\Delta_P(\rho)P(V)}]v]. \tag{6.33}$$

In view of the just established limiting behavior of $\Delta_P(\rho)P(V)$ and the fact that $\Delta_W(\rho) \to 0$ for $\rho \uparrow 1$, it follows that the left-hand side of Eq. (6.33) tends to one when $\rho \uparrow 1$, for all finite v. However, taking $v = u/\Delta_W(\rho)$, apparently

$$\lim_{\rho \uparrow 1} \mathbb{E}[e^{-r\Delta_P(\rho)P(u/\Delta_W(\rho))}] = e^{-rp(r)u}, \quad \operatorname{Re} r \geq 0, \ u > 0. \tag{6.34}$$

Consequently, $\Delta_P(\rho)P(u/\Delta_W(\rho))$ converges in distribution for $\rho \uparrow 1$, and the LST of the (nondegenerate) limiting distribution equals $\exp(-rp(r)u)$. Note that its mean equals u. Expression (6.34) also occurs in a study of Prabhu [31], as entrance time from u in 0 of an infinitely divisible process with only positive jumps; its appearance as a heavy traffic limit result for $M/G/1$ is new to the best of the authors's knowledge.

REMARK 6.3.10. Since the sojourn time S_L in the $M/G/1$ LCFS-PR queue has the same distribution as the busy period, Theorem 6.3.7 also holds for the residual sojourn time S_L^*.

Noting that

$$\mathbb{P}[\Delta_P(\rho)P^* > t] = \int_{\Delta_P(\rho)t}^{\infty} \frac{\mathbb{P}[P > x]}{\mathbb{E}[P]}\, dx = \frac{1-\rho}{\beta} \int_{\Delta_P(\rho)t}^{\infty} \mathbb{P}[P > x]\, dx,$$

one can, via differentiation, obtain a heavy traffic limit theorem for P and S_L. We shall not go into detail here. See Ott [28] and Abate and Whitt [1] for such heavy traffic results in the case of *finite* service time variance.

6.4 LCFS-PR: ASYMPTOTICS FOR THE HEAVY TRAFFIC LIMITING DISTRIBUTION

The main goal of this section is to derive an asymptotic series, for $t \to \infty$, of the limiting distribution $P_{\nu-1}(t)$ that occurs in Theorem 6.3.7. In the previous section we have seen that $p(r)$, the LST of this limiting distribution, is the unique solution of Eq. (6.32) with $y(0) = 1$. A picture readily shows that, for $1 < \nu < 2$,

$$y = 1 - sy^{\nu} \tag{6.35}$$

has only one real root for $s > 0$, and that this root is positive for $s > 0$, equal to 1 for $s = 0$, and bounded by 1. Introduce $Y(s) := y((s/\beta)^{1/(\nu-1)})$. To investigate the singularities of this solution of Eq. (6.35), note that

$$\frac{dY(s)}{ds} = -\frac{Y^{\nu}(s)}{1 + \nu s Y^{\nu-1}(s)}. \tag{6.36}$$

The only pair $(s_\nu, Y(s_\nu))$ that satisfies Eq. (6.35) and for which the denominator on the right-hand side of Eq. (6.36) becomes zero is the pair

$$s_\nu = -\frac{1}{\nu}\left(\frac{\nu-1}{\nu}\right)^{\nu-1} \quad \text{and} \quad Y(s_\nu) = \frac{\nu}{\nu-1}. \tag{6.37}$$

To see this, write $Y(s) = 1 - sY^\nu(s) = 1 - (1/\nu)(\nu s Y^{\nu-1}(s))Y(s)$, which equals $1 + Y(s)/\nu$ for $s = s_\nu$. Hence, s_ν as specified in Eq. (6.37) (by its principal value) is the only singularity of $Y(s)$.

Note that for a solution $\hat{Y}(s)$ of Eq. (6.35) with $\hat{Y}(0) \neq 1$ holds that $\lim_{|s|\to 0} |\hat{Y}(s)| = \infty$.

Next, we derive a series expansion in powers of s for the solution $Y(s)$ with $Y(0) = 1$. It is obtained by using Lagrange's theorem (cf. Whittaker and Watson [37, Chap. 7]). Put

$$q(s) := Y(s) - 1, \tag{6.38}$$

so that

$$q(s) = (-s)(1 + q(s))^\nu, \quad q(0) = 0. \tag{6.39}$$

Let C be the circle with center at $q = 0$ and radius $q(s_\nu) = Y(s_\nu) - 1 = 1/(\nu - 1)$. Choosing $|s| < -s_\nu$, it is seen that, for $q \in C$,

$$|q| > |s||1 + q|^\nu. \tag{6.40}$$

Consider a circle C_ϵ that is concentric with C but has a radius that is ϵ smaller ($0 < \epsilon \ll 1$). Since q and $(1+q)^\nu$ are regular for q inside and on C_ϵ, while Eq. (6.40) holds on C, Eq. (6.39) seen as an equation in q has one root in the interior of C, and Lagrange's theorem gives a series development of the zero $q(s)$ of Eq. (6.39) in powers of s. For $|s| < -s_\nu$,

$$q(s) = \sum_{n=1}^{\infty} \frac{(-s)^n}{n!} \frac{d^{n-1}}{dt^{n-1}} (1+t)^{n\nu}|_{t=0} = \sum_{n=1}^{\infty} \frac{(-s)^n}{n!} \frac{\Gamma(n\nu + 1)}{\Gamma(n(\nu-1)+2)}. \tag{6.41}$$

Observe that $p(r) = Y((\beta r)^{\nu-1}) = 1 + q((\beta r)^{\nu-1})$, so that, for $0 < \text{Re}\,(\beta r)^{\nu-1} < -s_\nu$,

$$p(r) = 1 + \sum_{n=1}^{\infty} (-1)^n (\beta r)^{n(\nu-1)} \frac{\Gamma(n\nu+1)}{n!\Gamma(n(\nu-1)+2)}. \tag{6.42}$$

Hence

$$\frac{1-p(r)}{\beta r} = \sum_{n=1}^{\infty} (-1)^{n-1} (\beta r)^{n(\nu-1)-1} \frac{\Gamma(n\nu+1)}{n!\Gamma(n(\nu-1)+2)}. \tag{6.43}$$

Denote by $P_{\nu-1}(t)$ the probability distribution with LST $p(r)$, Re $r \geq 0$; then

$$\frac{1-p(r)}{\beta r} = \int_{t=0}^{\infty} e^{-rt}[1 - P_{\nu-1}(t)]\frac{dt}{\beta}. \tag{6.44}$$

By applying Theorem 2 of Doetsch [22, p. 159], an asymptotic series for $P_{\nu-1}(t)$ can be obtained from Eq. (6.43). We thus obtain the main result of this section. For $t \to \infty$, and for every finite positive integer H,

$$1 - P_{\nu-1}(t) = \sum_{n=1}^{H} (-1)^{n-1} \left(\frac{\beta}{t}\right)^{n(\nu-1)} \frac{\Gamma(n\nu+1)}{n!\Gamma(n(\nu-1)+2)} \frac{1}{\Gamma(1-n(\nu-1))} + O(t^{-(H+1)(\nu-1)}). \tag{6.45}$$

Using the identity $1/\Gamma(1-z) = \Gamma(z)(1/\pi)\sin(\pi z)$, we obtain the following. For $t \to \infty$, and for every finite positive integer H,

$$1 - P_{\nu-1}(t) = \frac{1}{\pi}\sum_{n=1}^{H} (-1)^{n-1} \left(\frac{\beta}{t}\right)^{n(\nu-1)} \frac{\Gamma(n\nu+1)\sin(n(\nu-1)\pi)}{n!n(\nu-1)(n(\nu-1)+1)} + O(t^{-(H+1)(\nu-1)}). \tag{6.46}$$

We close this section by comparing the limiting distributions $P_{\nu-1}(t)$ of $\Delta_P(\rho)P^*$ and $R_{\nu-1}(t)$ of $\Delta_W(\rho)W$ for $t \to \infty$. Remember that W is the waiting time in the $M/G/1$ FCFS queue, and that P^* has the same distribution as the residual sojourn time in the $M/G/1$ LCFS-PR queue. From Eq. (6.46), for $t \to \infty$,

$$1 - P_{\nu-1}(t) \sim \frac{1}{\pi} \frac{\sin(\nu-1)\pi}{(t/\beta)^{\nu-1}} \Gamma(\nu-1). \tag{6.47}$$

Interestingly, this is exactly the same behavior as that of the first term of the asymptotic series of $1 - R_{\nu-1}(t)$ (the latter behavior is studied in Boxma and Cohen [10]).

6.5 HEAVY TRAFFIC LIMIT THEOREMS FOR THE WAITING TIME IN THE $GI/G/1$ FCFS QUEUE

In Section 6.3.2 we have proved a heavy traffic limit theorem for the waiting time in the $M/G/1$ queue with FCFS service discipline. In the present section we summarize extensions of this result for the $GI/G/1$ FCFS queue, which were recently obtained by Boxma and Cohen [10] and Cohen [17]. We allow both the interarrival time distribution $A(\cdot)$ and the service time distribution $B(\cdot)$ to be regularly varying, with finite first moments α and β, respectively. We characterize the interarrival time distribution and the service time distribution by their respective LSTs $\alpha\{\cdot\}$ and $\beta\{\cdot\}$. For Re $s \geq 0$,

$$1 - \frac{1-\alpha\{s\}}{\alpha s} = g_a(\gamma s) + C_a(\gamma s)^{\nu_a - 1} L_a(\gamma s), \tag{6.48}$$

$$1 - \frac{1-\beta\{s\}}{\beta s} = g_b(\gamma s) + C_b(\gamma s)^{\nu_b - 1} L_b(\gamma s). \tag{6.49}$$

γ has been introduced in Cohen [17] as a constant that stands for the unit of time; below, we always take $\gamma = \alpha$, the mean interarrival time. Furthermore, for $j = a, b$:

C_j is a finite positive constant;
$1 < \nu_j \leq 2$;
$g_j(\gamma s)$ is regular for Re $s > -\delta$ for a positive δ, and $g_j(0) = 0$;
$L_j(\gamma s)$ is a regular function of s for Re $s > 0$, and continuous for Re $s \geq 0$ except possibly at $s = 0$, with
$L_j(\gamma s) \to b_j > 0$ for $|s| \to 0$, Re $s \geq 0$, and $b_j \leq \infty$ for $1 < \nu_j \leq 2$, $b_j = \infty$ for $\nu_j = 2$;

$$\lim_{x \downarrow 0} \frac{L_j(x\gamma s)}{L_j(\gamma s)} = 1 \quad \text{for every } s \text{ with Re } s \geq 0.$$

It is further assumed that the following limit exists:

$$f := \lim_{x \downarrow 0} \frac{L_b(x)}{L_a(x)} \geq 0. \tag{6.50}$$

Formulas (6.48) and (6.49), together with the above assumptions, imply according to Lemma 6.2.1 that $1 - A(t)$ and $1 - B(t)$ are regularly varying at infinity. The reason for introducing the explicit representation of the LSTs $\alpha\{s\}$ and $\beta\{s\}$ in the complex plane is that such a representation is needed to manipulate the expression for the waiting time LST in the complex plane. It is well known (cf. Cohen [16]) that for $\rho = \beta/\alpha < 1$ the following relation exists between the waiting time LST $\omega\{s\}$ and the LST $\chi\{s\} = (1 - \mathbb{E}[e^{-sI}])/s\mathbb{E}I$, with I the idle time of a busy cycle:

$$\chi\{\bar{s}\} = \frac{1 - \beta\{s\}\alpha\{\bar{s}\}}{(\beta - \alpha)s} \omega\{s\} \quad \text{for Re } s = 0, \tag{6.51}$$

where $\bar{s}$ denotes the complex conjugate of s, and where

(i) $\omega\{s\}$ is regular for Re $s > 0$, continuous and uniformly bounded by one for Re $s \geq 0$;

(ii) $\chi\{-s\}$ is regular for $\text{Re}\, s < 0$, continuous and uniformly bounded by one for $\text{Re}\, s \leq 0$, with $\chi\{\bar{s}\}$ its boundary value at $\text{Re}\, s = 0$;
(iii) $\chi\{0\} = \omega\{0\} = 1$.

These conditions constitute a Riemann boundary value problem for the functions $\omega\{s\}$ and $\chi\{s\}$. In Cohen [16] conditions on $\alpha\{\cdot\}$ and $\beta\{\cdot\}$ are provided, under which the solution of the boundary value problem is obtained. $\alpha\{\cdot\}$ and $\beta\{\cdot\}$ as specified in Eqs. (6.48) and (6.49) satisfy these conditions. The solution reads [16]

$$\omega\{s\} = e^{H(s)}, \quad \text{Re}\, s > 0, \tag{6.52}$$

$$\chi\{-s\} = e^{H(s)}, \quad \text{Re}\, s < 0, \tag{6.53}$$

where

$$H(s) = \frac{1}{2\pi i} \int_{\xi=-i\infty}^{i\infty} \left(\log \frac{1 - \beta\{\xi\}\alpha\{\bar{\xi}\}}{(\beta - \alpha)\xi} \right) \frac{s}{\xi - s} \frac{d\xi}{\xi}, \tag{6.54}$$

with the integral defined as a principal value integral at infinity and as a principal value singular Cauchy integral at s if $\text{Re}\, s = 0$ (cf. Cohen [16]). The above representation for $\omega\{s\}$ has been exploited in Cohen [17] (see also Cohen [19]) to obtain the following four heavy traffic limit theorems (the first two were already obtained in Boxma and Cohen [10], but under slightly stronger conditions). To present them, we need the following nomenclature.

Definition 6.5.1. The tail of $A(t)$ is said to be *heavier* than that of $B(t)$ whenever one of the following cases occurs:

(i) $v_a < v_b$;
(ii) $v_a = v_b$, $b_a = \infty$, $b_b < \infty$;
(iii) $v_a = v_b$, $b_a = b_b = \infty$, and $f = 0$.

Analogously, the tail of $B(t)$ is called heavier than that of $A(t)$ if above the a and b indices are interchanged and $f = 0$ is replaced by $f = \infty$.

$A(t)$ and $B(t)$ are said to have *similar tails* when the tail of $A(t)$ is not heavier than that of $B(t)$ and the tail of $B(t)$ is not heavier than that of $A(t)$.

Theorem 6.5.2. *Consider the stable GI/G/I FCFS queue with $A(t)$ and $B(t)$ specified by Eqs. (6.48) and (6.49). If the tail of $B(t)$ is heavier than that of $A(t)$, then the "contracted" waiting time $\Delta_B(\rho)W/\alpha$ converges in distribution for $\rho \uparrow 1$, the limiting distribution $R_{v_b-1}(t)$ is specified by its LST $1/(1 + r^{v_b-1})$, and the coefficient of contraction $\Delta_B(\rho)$ is that root of the equation*

$$C_b \rho x^{v_b-1} L_b(x) = 1 - \rho, \quad x > 0, \ 0 < 1 - \rho \ll 1, \tag{6.55}$$

with the property that $\Delta_B(\rho) \downarrow 0$ for $\rho \uparrow 1$.

Theorem 6.5.3. *Consider the stable $GI/G/I$ FCFS queue with $A(t)$ and $B(t)$ specified by Eqs. (6.48) and (6.49). If the tail of $A(t)$ is heavier than that of $B(t)$, then the "contracted" waiting time $\Delta_A(\rho)W/\alpha$ converges in distribution for $\rho \uparrow 1$, the limiting distribution is the negative exponential distribution with unit mean (and LST $1/(1+r)$), and the coefficient of contraction $\Delta_A(\rho)$ is that root of the equation*

$$C_a x^{v_a - 1} L_a(x) = 1 - \rho, \quad x > 0,\ 0 < 1 - \rho << 1, \tag{6.56}$$

with the property that $\Delta_A(\rho) \downarrow 0$ for $\rho \uparrow 1$.

Theorem 6.5.4. *Consider the stable $GI/G/1$ FCFS queue with $A(t)$ and $B(t)$ specified by Eqs. (6.48) and (6.49). If $A(t)$ and $B(t)$ have similar tails and $v_a = v_b = 2$, then the "contracted" waiting time $\Delta_{AB}(\rho)W/\alpha$ converges in distribution for $\rho \uparrow 1$, the limiting distribution is the negative exponential distribution with unit mean (and LST $1/(1+r)$), and the coefficient of contraction $\Delta_{AB}(\rho)$ is that root of the equation*

$$C_a x L_a(x) + C_b \rho x L_b(x) = 1 - \rho, \quad x > 0,\ 0 < 1 - \rho \ll 1, \tag{6.57}$$

with the property that $\Delta_{AB}(\rho) \downarrow 0$ for $\rho \uparrow 1$.

The previous three theorems give conditions under which ΔW converges in distribution. Δ appears to be of order (roughly) $(1-\rho)^{1/[\min(v_a, v_b)-1]}$. The limiting distribution is exponential when the interarrival time distribution is heavier, and is a heavy-tailed distribution $R_{v-1}(t)$ (which is related to the Mittag–Leffler function) when the service time distribution is heavier. It should be noted that $R_{1/2}(t) = 1 - e^t \operatorname{erfc}(\sqrt{t})$. The complementary error function $\operatorname{erfc}(x)$ is a special function that has been investigated in much detail; this has resulted in considerable insight into the behavior of the limiting distribution in this special case (cf. Boxma and Cohen [10]).

The most complicated case occurs when $\rho \uparrow 1$ while $A(t)$ and $B(t)$ have similar tails, with $1 < v := v_a = v_b < 2$. We introduce the following notation. Let

$$d_a := \frac{C_a}{D}, \quad d_b := \frac{C_b f}{D}, \tag{6.58}$$

with

$$D := \sqrt{C_a^2 + f^2 C_b^2 + 2(\cos v\pi) f C_a C_b}. \tag{6.59}$$

Let

$$\Phi\left(\frac{r}{\alpha}\right) := \frac{1}{2\pi i}\int_{-i\infty}^{i\infty} [\log\{1 + d_b \xi^{v-1} - d_a \bar{\xi}^{v-1}\}] \frac{r}{\xi - r}\frac{d\xi}{\xi} - \tfrac{1}{2}\log\{1 + d_b r^{v-1} - d_a \bar{r}^{v-1}\} I_{(\mathrm{Re}\, r = 0)}, \quad \mathrm{Re}\, r \geq 0, \tag{6.60}$$

where $I_{(\cdot)}$ denotes an indicator function. For real r, one can also write

$$\Phi\left(\frac{r}{\alpha}\right) = -\frac{1}{2\pi}\int_0^\infty \log\{1 + 2B(rx)^{\nu-1} + C(rx)^{2(\nu-1)}\}\frac{dx}{1+x^2}$$
$$-\frac{1}{\pi}\int_0^\infty \left(\arctan\frac{A(rx)^{\nu-1}}{1+B(rx)^{\nu-1}}\right)\frac{1}{1+x^2}\frac{dx}{x}, \quad r \geq 0, \tag{6.61}$$

where

$$\begin{aligned} A &:= (d_b + d_a)\sin\frac{\nu-1}{2}\pi > 0, \\ B &:= (d_b - d_a)\cos\frac{\nu-1}{2}\pi, \\ C &:= d_a^2 + d_b^2 + 2(d_b - d_a)\cos(\nu-1)\pi > 0. \end{aligned} \tag{6.62}$$

Theorem 6.5.5. *Consider the stable $GI/G/I$ FCFS queue with $A(t)$ and $B(t)$ specified by Eqs. (6.48) and (6.49). If $A(t)$ and $B(t)$ have similar tails and $1 < \nu_a = \nu_b < 2$, then the "contracted" waiting time $\Delta_{AB}(\rho)W/\alpha$ converges in distribution for $\rho \uparrow 1$, the limiting distribution is characterized by its LST $e^{\Phi(r/\alpha)}$, and the coefficient of contraction $\Delta_{AB}(\rho)$ is that root of the equation*

$$|(-1)^{\nu_a}C_a x^{\nu-1}L_a(x) + C_b\rho x^{\nu-1}L_b(x)| = 1-\rho, \quad x > 0, \; 0 < 1-\rho \ll 1, \tag{6.63}$$

with the property that $\Delta_{AB}(\rho) \downarrow 0$ for $\rho \uparrow 1$.

We refer to Cohen [17] for an extensive discussion of the tail behavior of the heavy traffic limiting distribution of the waiting time. In particular, for the case of Theorem 6.5.5,

$$\mathbb{P}[W > \alpha t] = \frac{\max(d_a, d_b)}{\Gamma(2-\nu)}t^{1-\nu}(1 + O(t^{1-\nu})), \quad t \to \infty.$$

REMARK 6.5.6. Relation (6.10) between the LSTs of the actual waiting time W and the virtual waiting time V readily implies that Theorems 6.5.2–6.5.5 also hold for the virtual waiting time.

6.6 A HEAVY TRAFFIC LIMIT THEOREM FOR THE WORKLOAD PROCESS IN THE $GI/G/1$ FCFS QUEUE

In the previous section we have considered the limiting distribution for $\rho \uparrow 1$ of the contracted waiting time, $\Delta(\rho)W$ (or equivalently $\Delta(\rho)V$, cf. Remark 6.5.6). In the present section we consider the workload *process* $\{v_t, t \geq 0\}$ of the $GI/G/1$ queue in heavy traffic. In order to get a proper limiting process, we not only apply the same coefficient of contraction $\Delta(\rho)$, but we also *scale time* by a factor $\Delta_1(\rho) := (1-\rho)\Delta(\rho)$. It can be shown that

$$w(\tau; \rho) := \Delta(\rho) v_{\tau/\Delta_1(\rho)} \tag{6.64}$$

converges in distribution for $\rho \uparrow 1$, for every $\tau > 0$. It can further be shown that the thus scaled and contracted workload process $\{w(\tau; \rho), \tau \geq 0\}$ converges weakly to the workload process of a queueing model of which the input is described by a stable Lévy motion if $1 < \nu < 2$ and by Brownian motion if $\nu = 2$ (with ν the index of the heaviest tail).

As a preparation for discussing these results in some detail, we consider the *noise traffic* $n_t = k_t - \rho t$ and the workload backlog $h_t = k_t - t$, with k_t the amount of traffic generated in $[0, t)$. It is shown in Theorem 4.1 of Cohen [21] that n_t and h_t, when scaled similarly as v_t in Eq. (6.64), have limiting distributions for $\rho \to 1$. In that theorem, the same cases are considered as in Theorems 6.5.2–6.5.5—and for the coefficient of contraction $\Delta(\rho)$ we take precisely the contraction factors of those theorems:

1. The tail of $B(t)$ is heavier than that of $A(t)$: $\Delta(\rho) = \Delta_B(\rho)$.
2. The tail of $A(t)$ is heavier than that of $B(t)$: $\Delta(\rho) = \Delta_A(\rho)$.
3. $A(t)$ and $B(t)$ have similar tails: $\Delta(\rho) = \Delta_{AB}(\rho)$.

Define

$$N(\tau; \rho) := \Delta(\rho) n_{\tau/\Delta_1(\rho)}, \quad H(\tau; \rho) := \Delta(\rho) h_{\tau/\Delta_1(\rho)}. \tag{6.65}$$

Theorem 6.6.1

(i) *The stochastic variables $N(\tau; \rho)$ and $H(\tau; \rho)$ converge in distribution for $\rho \uparrow 1$.*

(ii) *Let $N(\tau)$ and $H(\tau)$ be stochastic variables with as distribution the limiting distribution of $N(\tau; \rho)$ and $H(\tau; \rho)$, respectively; then for $\tau \geq 0$,* Re $r = 0$,

$$\begin{aligned} \mathbb{E}[e^{-rN(\tau)/\alpha}] &= e^{r^\nu \tau/\alpha} && \text{for } \nu = \nu_b < \nu_a \text{ or } C_a = 0, \\ &= e^{\bar{r}^\nu \tau/\alpha} && \text{for } \nu = \nu_a < \nu_b \text{ or } C_b = 0, \\ &= e^{[d_b r^\nu + d_a \bar{r}^\nu]\tau/\alpha} && \text{for } \nu = \nu_a = \nu_b,\ C_a > 0,\ C_b > 0; \end{aligned} \tag{6.66}$$

$$\mathbb{E}[e^{-rH(\tau)/\alpha}] = e^{\tau r/\alpha} \mathbb{E}[e^{-rN(\tau)/\alpha}]. \tag{6.67}$$

Proof. See Cohen [21]. To give the reader insight into the approach, we sketch the proof for the $M/G/1$ case with heavy-tailed service time distribution as given in

Eq. (6.49) (see also Cohen [20]). This is a case with $\nu = \nu_b < \nu_a$. In this $M/G/1$ case,

$$\mathbb{E}[e^{-sk_t}] = \sum_{n=0}^{\infty} e^{-t/\alpha} \frac{(t/\alpha)^n}{n!} \beta^n\{s\} = e^{-\rho st(1-\beta\{s\})/\beta s}, \tag{6.68}$$

so

$$\mathbb{E}[e^{-sn_t}] = e^{\rho st[1-(1-\beta\{s\})/\beta s]}. \tag{6.69}$$

Hence, putting

$$s = r\Delta(\rho), \quad t = \tau/\Delta_1(\rho), \tag{6.70}$$

it follows from Eqs. (6.65) and (6.69) that

$$\mathbb{E}[e^{-rN(\tau;\rho)}] = \exp\left(\frac{\rho}{1-\rho} r\tau\left[1 - \frac{1-\beta\{r\Delta(\rho)\}}{\beta r\Delta(\rho)}\right]\right). \tag{6.71}$$

For the term in square brackets, we have the representation (6.49). It can be verified that $g_b(\alpha r\Delta(\rho))/(1-\rho) \to 0$ for $\rho \uparrow 1$. The fact that $\Delta(\rho)$ is the unique zero of the contraction equation (6.55) with the property that $\Delta(\rho) \downarrow 0$ for $\rho \uparrow 1$ finally implies that

$$\lim_{\rho\uparrow 1} \mathbb{E}[e^{-rN(\tau;\rho)/\alpha}] = e^{r^\nu \tau/\alpha}. \tag{6.72}$$

The convergence in distribution of $N(\tau; \rho)$ follows after application of the convergence theorem for Laplace–Stieltjes transforms [20]. The statements concerning $H(\tau; \rho)$ follow immediately from those for $N(\tau; \rho)$. ■

REMARK 6.6.2. The distribution of the stochastic variable $N(\tau), \tau \geq 0$, and similarly of $H(\tau), \tau \geq 0$, is a ν-stable distribution (cf. Samorodnitsky and Taqqu [33, p. 5]). The process $\{N(\tau), \tau \geq 0\}$ is a process with stationary independent increments. As is evident from its LST representation, it is self-similar with index $1/\nu$, that is, $N(b\tau)$ and $b^{1/\nu}N(\tau), b > 0$, have the same distribution for every $\tau > 0$. Note that $\{H(\tau), \tau \geq 0\}$ is *not* self-similar. In Samorodnitsky and Taqqu [33] this process $\{N(\tau), \tau \geq 0\}$ with $1 < \nu < 2$ is called a ν-stable Lévy motion with independent self-similar increments; for $\nu = 2$, it is the Brownian motion.

In the $M/G/1$ case, it is verified in Cohen [20] (using the fact that the process $\{h_t, t \geq 0\}$ has stationary independent increments) that the finite-dimensional distributions of the $\{N(\tau; \rho), \tau \geq 0\}$ process converge to those of the $\{N(\tau), \tau \geq 0\}$ process.

The proof of Theorem 6.6.1 for the $GI/G/1$ case proceeds in principle along the same lines as sketched above for $M/G/1$. However, the proof of the weak convergence of $N(\tau;\rho)$ to $N(\tau)$ is more complicated.

Let us now turn to the workload processes $\{v_t, t \geq 0\}$ and the contracted and scaled $\{w(\tau;\rho), \tau \geq 0\}$. We again discuss the $M/G/1$ case.

It follows from formula (4.99) of Cohen [15, p. 262] that, for Re $s \geq 0$, Re $\theta > 0$,

$$\int_{t=0}^{\infty} e^{-\theta t}\mathbb{E}[e^{-sv_t}|v_0=0]\,dt = \frac{1}{\theta - s[1-\rho(1-\beta\{s\})/\beta s]} \times \left(1 - s\int_{t=0}^{\infty} e^{-\theta t}\mathbb{P}[v_t = 0|v_0=0]\,dt\right) \tag{6.73}$$

with (cf. Formula (4.94) of Cohen [15, p. 262] and Eq. (6.29))

$$\int_{t=0}^{\infty} e^{-\theta t}\mathbb{P}[v_t=0|v_0=0]\,dt = \frac{1-\rho}{\theta}\frac{1}{1-\rho+\rho\mathbb{E}[e^{-\theta P^*}]} = \frac{1-\rho}{\theta}\frac{1}{\mathbb{E}[e^{-\theta P(V)}]}. \tag{6.74}$$

Having established this relation with $P(V)$, one can now use Theorem 6.3.7 that specifies the heavy traffic behavior of $\Delta_P(\rho)P(V)$; note that $\Delta_P(\rho) = (1-\rho)\Delta(\rho) = \Delta_1(\rho)$. Again apply the transformation (6.70), with Re $r \geq 0$ and $0 < 1-\rho \ll 1$, and also

$$\theta = \omega\Delta_1(\rho), \qquad \text{Re}\,\omega \geq 0.$$

Hence,

$$\int_{t=0}^{\infty} e^{-\theta t}\mathbb{E}[e^{-sv_t}|v_0=0]\,dt = \frac{1}{\Delta_1(\rho)}\int_{\tau=0}^{\infty} e^{-\omega\tau}\mathbb{E}[e^{-rw(\tau;\rho)}|w(0;\rho)=0]\,d\tau. \tag{6.75}$$

It follows (cf. Cohen [20]) that $w(\tau;\rho)$ converges in distribution to a stochastic variable w_τ for $\rho\uparrow 1$ for every $\tau \geq 0$, and that the distribution of w_τ, $\tau \geq 0$ is specified as follows. For Re $\omega > 0$, Re $r \geq 0$,

$$\int_{\tau=0}^{\infty} e^{-\omega\tau}\mathbb{E}[e^{-rw_\tau/\alpha}|w_0=0]\,d\tau = \frac{1}{\omega - r(1+r^{\nu-1})}\left(1 - \frac{r}{\omega}\frac{1}{p(\omega)}\right). \tag{6.76}$$

As a by-product we find that

$$\lim_{\tau\to\infty}\mathbb{E}[e^{-rw_\tau/\alpha}] = \frac{1}{1+r^{\nu-1}}, \tag{6.77}$$

in agreement with Theorem 6.5.2.

For the $GI/G/1$ queue, the workload process $\{v_t, t \geq 0\}$ is described by Reich's formula ([15, p. 170]; in the sequel we assume that $v_0 = 0$, in order to make the following derivations somewhat simpler):

$$v_t = \max\left[h_t, \sup_{0<u<t} (h_t - h_u)\right], \quad t \geq 0. \tag{6.78}$$

It easily follows from Eqs. (6.64) and (6.78), with $w(0; \rho) = 0$, that

$$w(\tau; \rho) = \max\left[H(\tau; \rho), \sup_{0<u<\tau} (H(\tau; \rho) - H(u; \rho))\right], \quad \tau \geq 0. \tag{6.79}$$

Let us also consider the $\tilde{M}/\tilde{G}/1$ queue with input process $\{N(\tau), \tau \geq 0\}$ and workload $\tilde{w}_\tau$ at time τ defined by (cf. Reich's formula)

$$\tilde{w}_\tau = \max\left[H(\tau), \sup_{0<u<\tau} (H(\tau) - H(u))\right], \quad \tau \geq 0. \tag{6.80}$$

It is shown in Cohen [20] for $1 < \nu_b \leq 2$ that the process $\{w(\tau; \rho), \tau \geq 0\}$ indeed converges weakly (in the Skorokhod topology) for $\rho \to 1$ to the process $\{\tilde{w}_\tau, \tau \geq 0\}$. The proof is based on a theorem of Skorokhod for processes with independent increments [35]. Actually, the process $\{H(\tau; \rho), \tau \geq 0\}$ is a process with independent increments, of which for $\rho \uparrow 1$ all the finite-dimensional distributions of its increments converge weakly to those of the $\{H(\tau), \tau \geq 0\}$ process. The latter process is itself a process with independent increments. Because $\sup_{0<u<\tau}(H(\tau; \rho) - H(u; \rho))$ is a continuous functional of the $H(\tau; \rho)$ process, Skorokhod's theorem implies that the process $\{w(\tau; \rho), \tau > 0\}$ converges weakly to the process $\{\tilde{w}_\tau, \tau \geq 0\}$. In view of the conclusion below Eq. (6.75), $w_\tau \simeq \tilde{w}_\tau$ for $1 < \nu_b < 2$.

So for the $M/G/1$ queue with service time distribution specified by Eq. (6.49), we conclude the following. The workload $w(\tau; \rho)$, contracted and scaled as in Eq. (6.64), converges in distribution for $\rho \uparrow 1$ and every $\tau > 0$. The limiting distribution is specified by Eq. (6.76). The limiting process is also specified as the process that satisfies Reich's formula (6.80) for an $\tilde{M}/\tilde{G}/1$ queue with as input the ν-stable Lévy motion $\{N(\tau), \tau \geq 0\}$. These results are generalizations of the diffusion approximation of the $M/G/1$ queue with a finite service time variance. Similar results are obtained for the $GI/G/1$ queue, in particular, concerning the convergence of the input process $\{N(\tau; \rho), \tau \geq 0\}$ to ν-stable Lévy motion $\{N(\tau), \tau \geq 0\}$ and of the $\{w(\tau; \rho), \tau \geq 0\}$ process to the $\{\tilde{w}_\tau, \tau \geq 0\}$ process for $\rho \uparrow 1$ (see Cohen [21]); however, the proofs become more complicated.

REMARK 6.6.3. Although in this chapter only the $GI/G/1$ queue with instantaneous arrivals has been considered, many of the heavy traffic results should have a counterpart in fluid queues with regularly varying active and/or silent period distributions. Indeed, several authors [7, 18, 24] have established explicit distributional relations between buffer contents in fluid queues and (actual or virtual) waiting times in the $GI/G/1$ queue (cf. also Remark 6.2.6). And even if such results are not at hand, it is intuitively clear that in heavy traffic it does not play a crucial role whether work increases instantaneously or linearly.

6.7 CONCLUSION

This chapter has been devoted to the $GI/G/1$ queue with regularly varying interarrival and/or service time distribution.

First, we studied the tail behavior of the waiting time distribution in the $M/G/1$ case, for three different service disciplines: FCFS, LCFS preemptive resume, and processor sharing. The effect of the service discipline on the tail behavior was shown to be very pronounced, FCFS leading to heavier waiting time tails than the other two disciplines.

Second, we studied the heavy traffic behavior of the waiting time, for the cases in which the variance of the interarrival and/or service time is infinite. A coefficient of contraction $\Delta(\rho)$ was identified, such that the contracted waiting time $\Delta(\rho)W$ converges in distribution when the traffic load $\rho \uparrow 1$.

Third, the heavy traffic behavior of the workload process $\{v_t, t \geq 0\}$ was investigated. Both a coefficient of contraction $\Delta(\rho)$ and a time-scaling factor $\Delta_1(\rho)$ were identified, such that $\Delta(\rho)v_{\tau/\Delta_1(\rho)}$ converges in distribution for $\rho \uparrow 1$, for all $\tau \geq 0$. The thus scaled and contracted workload process converges weakly to the workload process of a queue of which the input process is described by a ν-stable Lévy motion.

Some topics of our recent and present research on these three subjects are:

- Investigation of the influence of long-tailed traffic characteristics of one type of customer on performance measures of other types of customer. Preliminary results are obtained in Boxma et al. [11] for the $M/G/1$ queue with (non)preemptive priority, in Zwart [41] for the $M/G/1$ queue with processor sharing and several customer classes, and in Borst et al. [6] for the generalized processor sharing discipline. The sometimes dramatic impact of queue scheduling disciplines on performance may have important implications for the choice of scheduling disciplines in designing switches in communication networks.
- We have extended the heavy traffic limit theorem 6.3.1 to the case of an $M/G/1$ queue with priority classes, and we have exploited the resulting limit theorem to derive a heavy traffic approximation for the waiting time distribution of low-priority customers [11]. In Boxma and Cohen [9] we had already obtained an approximation for the waiting time distribution for the $M/G/1$ queue with heavy-tailed service time distribution by using the heavy traffic limit theorem. In both studies, the resulting approximation is remarkably sharp, even when traffic is not heavy at all. It is conjectured that similar approximations can be developed for the $GI/G/1$ queue with heavy-tailed interarrival

and/or service time distribution. This is a point for further study. It is important to perform many more numerical experiments to get more insight into the effect of long-tailed traffic characteristics on performance measures, and to be able to develop useful approximations.

- The weak convergence results of Section 6.6 for the $GI/G/1$ workload in heavy traffic can probably be extended in several directions. For example, it is of interest to study *networks* of queues.

ACKNOWLEDGMENT

The authors are indebted to Dr. V. Dumas and Professor A. J. Stam for interesting suggestions and discussions.

REFERENCES

1. J. Abate and W. Whitt. Limits and approximations for the busy-period distribution in single-server queues. *Probab. Eng. Inf. Sci.*, **9**:581–602, 1995.
2. V. Anantharam. Scheduling strategies and long-range dependence. Report, Dept. of Electrical Engineering & Computer Sciences, University of California at Berkeley, 1997.
3. J. Beran, R. Sherman, M. S. Taqqu, and W. Willinger. Long-range dependence in variable-bit-rate video traffic. *IEEE Trans. Commun.*, **43**:1566–1579, 1995.
4. N. H. Bingham and R. A. Doney. Asymptotic properties of supercritical branching processes I: The Galton–Watson process. *Adv. Appl. Probab.*, **6**:711–731, 1974.
5. N. H. Bingham, C. M. Goldie, and J. L. Teugels. *Regular Variation*. Cambridge University Press, Cambridge, 1987.
6. S. C. Borst, O. J. Boxma, and P. R. Jelenković. Generalized processor sharing with long-tailed traffic sources. In P. Key and G. Smith, eds., *Proceedings of ITC-16*, pp. 345–354. North-Holland, Amsterdam, 1999.
7. O. J. Boxma. Fluid queues and regular variation. *Perf. Eval.*, **27–28**:699–712, 1996.
8. O. J. Boxma. Regular variation in a multi-source fluid queue. In V. Ramaswami and P. E. Wirth, eds., *Teletraffic Contributions for the Information Age (Proceedings of ITC-15)*, pp. 391–402. North- Holland, Amsterdam, 1997.
9. O. J. Boxma and J. W. Cohen. The $M/G/1$ queue with heavy-tailed service time distribution. *IEEE JSAC*, **16**:749–763, 1998.
10. O. J. Boxma and J. W. Cohen. Heavy-traffic analysis for the $GI/G/1$ queue with heavy-tailed distributions. *Queueing Syst.*, 1999.
11. O. J. Boxma, J. W. Cohen, and Q. Deng. Heavy-traffic analysis of the $M/G/1$ queue with priority classes. In P. Key and G. Smith, eds., *Proceedings of ITC-16*, pp. 1157–1167. North-Holland, Amsterdam, 1999.
12. O. J. Boxma and V. Dumas. Fluid queues with heavy-tailed activity period distributions. *Comput. Commun.*, **21**:509–529, 1998.

13. J. W. Cohen. Some results on regular variation for distributions in queueing and fluctuation theory. *J. Appl. Probab.*, **10**:343–353, 1973.
14. J. W. Cohen. The multiple phase service network with generalized processor sharing. *Acta Inf.*, **12**:245–284, 1979.
15. J. W. Cohen. *The Single Server Queue*, 2nd ed. North-Holland, Amsterdam, 1982.
16. J. W. Cohen. Complex functions in queueing theory. *Arch. Elektr. Uebertragung (Pollaczek Memorial Volume)*, **47**:300–310, 1993.
17. J. W. Cohen. Heavy-traffic limit theorems for the heavy-tailed $GI/G/1$ queue. Report PNA-R9719, CWI, 1997.
18. J. W. Cohen. The $M/G/1$ fluid model with heavy-tailed message length distributions. Report PNA-R9714, CWI, 1997.
19. J. W. Cohen. A heavy-traffic theorem for the $GI/G/1$ queue with a Pareto-type service time distribution. *J. Appl. Math. Stoch. Anal.*, **11**:247–254, 1998.
20. J. W. Cohen. Heavy-traffic theory for the heavy-tailed $M/G/1$ queue and ν-stable Lévy noise traffic. Report PNA-R9805, CWI, 1998.
21. J. W. Cohen. The ν-stable Lévy motion in heavy traffic analysis of queueing models with heavy-tailed distributions. Report PNA-R9808, CWI, 1998.
22. G. Doetsch. *Handbuch der Laplace Transformation. Vol. II.* Birkhäuser Verlag, Basel, 1950.
23. P. R. Jelenković and A. A. Lazar. Multiplexing on/off sources with subexponential on-periods. *Adv. Appl. Probab.*, 31, 1999.
24. O. Kella and W. Whitt. A storage model with a two-state random environment. *Oper. Res.*, **40**(S2):S257–S262, 1992.
25. J. F. C. Kingman. The heavy traffic approximation in the theory of queues. In W. L. Smith and W. E. Wilkinson, eds., *Proceedings of the Symposium on Congestion Theory*, pp. 137–159. The University of North Carolina Press, Chapel Hill, 1965.
26. W. E. Leland, M. S. Taqqu, W. Willinger, and D. V. Wilson. On the self- similar nature of Ethernet traffic (extended version). *IEEE/ACM Trans. Networking*, **2**:1–15, 1994.
27. A. De Meyer and J. L. Teugels. On the asymptotic behavior of the distributions of the busy period and service time in $M/G/1$. *J. Appl. Probab.*, **17**:802–813, 1980.
28. T. J. Ott. The stable $M/G/1$ queue in heavy traffic and its covariance function. *Adv. Appl. Probab.*, **9**:169–186, 1977.
29. A. G. Pakes. On the tails of waiting-time distributions. *J. Appl. Probab.*, **12**:555–564, 1975.
30. V. Paxson and S. Floyd. Wide area traffic: the failure of Poisson modeling. *IEEE/ACM Trans. Networking*, **3**:226–244, 1995.
31. N. U. Prabhu. *Stochastic Storage Processes*. Springer-Verlag, Berlin, 1980.
32. T. Rolski, S. Schlegel, and V. Schmidt. Asymptotics of Palm-stationary buffer content distributions in fluid flow queues. *Adv. Appl. Probab.* **31**:235–253, 1999.
33. G. Samorodnitsky and M. S. Taqqu. *Stable Non-Gaussian Random Processes*. Chapman and Hall, New York, 1994.
34. B. Sengupta. An approximation for the sojourn-time distribution for the $GI/G/1$ processor-sharing queue. *Stoch. Models*, **8**:35–57, 1992.
35. A. V. Skorokhod. Limit theorems for stochastic processes with independent increments. *Theory Probab. and Its Appl.*, **11**:138–171, 1957.

36. W. Whitt. Heavy traffic limit theorems for queues: a survey. In A. B. Clarke, ed., *Mathematical Methods in Queueing Theory*, pp. 307–350. Springer-Verlag, Berlin, 1974.
37. E. T. Whittaker and G. N. Watson. *A Course of Modern Analysis*. Cambridge University Press, Cambridge, 1946.
38. W. Willinger, M. S. Taqqu, W. E. Leland, and D. V. Wilson. Self- similarity in high-speed packet traffic: analysis and modeling of Ethernet traffic measurements. *Statist. Sci*, **10**:67–85, 1995.
39. S. F. Yashkov. Processor-sharing queues: some progress in analysis. *Queueing Syst.*, **2**:1–17, 1987.
40. S. F. Yashkov. On a heavy-traffic limit theorem for the $M/G/1$ processor-sharing queue. *Stoch. Models*, **9**:467–471, 1993.
41. A. P. Zwart. Sojourn times in a multiclass processor sharing queue. In P. Key and G. Smith, eds., *Proceedings of ITC-16*, pp. 335–344. North-Holland, Amsterdam, 1999.
42. A. P. Zwart and O. J. Boxma. Sojourn time asymptotics in the $M/G/1$ processor sharing queue. Report PNA-R9802, CWI, 1998; to appear in *Queueing Syst.*

7

FLUID QUEUES, ON/OFF PROCESSES, AND TELETRAFFIC MODELING WITH HIGHLY VARIABLE AND CORRELATED INPUTS

SIDNEY RESNICK AND GENNADY SAMORODNITSKY

Cornell University, School of Operations Research and Industrial Engineering, Ithaca, NY 14853

7.1 INTRODUCTION

Large teletraffic data sets exhibiting nonstandard features incompatible with classical assumptions of short-range dependence and exponentially decreasing tails can now be explored, for instance, at the ITA Web site `www.acm.org/sigcomm/ITA/`. These data sets exhibit the phenomena of *heavy-tailed marginal distributions* and *long-range dependence*. Tails can be so heavy that only infinite variance models are possible (e.g., see Willinger et al. [49]), and sometimes, as in file size data, even first moments are infinite [1]. See also Beran et al. [3], Crovella and Bestavros [12–14], Leland et al. [33], Resnick [38], Taqqu et al. [48], and Willinger et al. [49]. Other areas where heavy tails and long-range dependence are crucial properties are finance, insurance, and hydrology [4–7, 16, 17, 24–26, 35, 37].

New features in the teletraffic data discussed in recent studies suggest several issues for study and discussion.

- *Statistical.* How can statistical models be fit to such data? Finite variance black box time series modeling has traditionally been dominated by ARMA or Box–Jenkins models. These models can be adapted to heavy-tailed data and work very well on simulated data. However, for real nonsimulated data exhibiting

Self-Similar Network Traffic and Performance Evaluation, Edited by Kihong Park and Walter Willinger
ISBN 0-471-31974-0

dependencies, such ARMA models provide unacceptable fits and do not capture the correct dependence structure. For discussion see Davis and Resnick [15], Resnick [38, 39], Resnick et al. [42], and Resnick and van den Berg [43].

- *Probabilistic.* What probability models explain observed features in the data such as long-range dependence and heavy tails.
- *Consequences.* Do the new features revealed by current teletraffic data studies mean we have to give up Poisson derived models and exponentially bounded tails and the highly linear models of time series? Various bits of evidence emphasize the deficiencies of classical modeling. There are simulation studies [34] and the experimental queueing analysis of Erramilli, Narayan, and Willinger [18]. An analytic example [40] shows that for a simple $G/M/1$ queue, a stationary input with long-range dependence can induce heavy tails for the waiting time distribution and for the distribution of the number in the system.

Connections between long-range dependence and heavy tails need to be more systematically explored but it is clear that in certain circumstances, long-range dependent (LRD) inputs can cause heavy-tailed outputs and (as we discuss here) heavy tails can cause long-range dependence. We discuss three models where heavy tails induce long-range dependence:

1. A single channel on/off source feeding a single server working at constant rate $r > 0$. Transmission or *on* periods have heavy-tailed distributions.
2. A multisource system where a single server working at constant rate $r > 0$ is fed by $J > 1$ *on/off* sources. Transmission periods have heavy-tailed distributions.
3. An infinite source model feeding a single server working at constant rate $r > 0$. At Poisson time points, nodes or sources commence transmitting. Transmission times have heavy-tailed distributions.

In each of the three cases, our basic descriptor of system performance is the time for buffer content to reach a critical level. Such a measure of performance is path based and makes sense without regard to stability of the model, existence of moments of input variables, or properties of steady-state quantities.

7.2 A SINGLE CHANNEL ON/OFF COMMUNICATION MODEL

7.2.1 Basic Setup

We consider first communication between a single source and a single destination server. The source transmits for random *on* periods alternating with random *off* periods when the source is silent. During the *on* periods, transmission is at unit rate.

Let $\{X_{\text{on}}, X_n, n \geq 1\}$ be i.i.d. nonnegative random variables representing *on* periods. The common distribution is F_{on}. Similarly, $\{Y_{\text{off}}, Y_n, n \geq 1\}$ are i.i.d.

nonnegative random variables independent of $\{X_{\text{on}}, X_n, n \geq 1\}$ representing *off* periods and these have common distribution F_{off}. The means are

$$\mu_{\text{on}} = \int_0^\infty \bar{F}_{\text{on}}(s)\, ds, \quad \mu_{\text{off}} = \int_0^\infty \bar{F}_{\text{off}}(s)\, ds,$$

which are assumed finite and the sum of the means is $\mu := \mu_{\text{on}} + \mu_{\text{off}}$. Using these random variables we generate an alternating renewal sequence characterized as follows.

1. The interarrival distribution is $F_{\text{on}} * F_{\text{off}}$ and the mean interarrival time is $\mu = \mu_{\text{on}} + \mu_{\text{off}}$.
2. The renewal times are

$$\left\{0, \sum_{i=1}^{n}(X_i + Y_i), n \geq 1\right\}.$$

Because of the finiteness of the means, the renewal process has a stationary version:

$$\left\{D, D + \sum_{i=1}^{n}(X_i + Y_i), n \geq 1\right\}.$$

where D is a delay random variable satisfying

$$\begin{aligned} P[D > x] &= \int_x^\infty \frac{P[X_{\text{on}} + Y_{\text{off}} > s]}{\mu}\, ds \\ &= \int_x^\infty \frac{1 - F_{\text{on}} * F_{\text{off}}(s)}{\mu}\, ds. \end{aligned}$$

However, making the process stationary in this manner has the disadvantage that the initial delay period D does not decompose into an *on* and an *off* period the way subsequent inter-renewal periods do and the following procedure is preferable for generating the stationary alternating renewal process. Define independent random variables $B, X_{\text{on}}^{(0)}, Y_{\text{off}}^{(0)}$, which are assumed independent of $\{X_{\text{on}}, X_n, n \geq 1\}$ and $\{Y_{\text{off}}, Y_n, n \geq 1\}$, by

$$P[B = 1] = \frac{\mu_{\text{on}}}{\mu} = 1 - P[B = 0],$$

$$P[X_{\text{on}}^{(0)} > x] = \int_x^\infty \frac{1 - F_{\text{on}}(s)}{\mu_{\text{on}}}\, ds := 1 - F_{\text{on}}^{(0)}(x),$$

$$P[Y_{\text{off}}^{(0)} > x] = \int_x^\infty \frac{1 - F_{\text{off}}(s)}{\mu_{\text{off}}}\, ds := 1 - F_{\text{off}}^{(0)}(x).$$

The delay random variable $D^{(0)}$ is defined by

$$D^{(0)} = B(X_{\text{on}}^{(0)} + Y_{\text{off}}) + (1 - B)Y_{\text{off}}^{(0)}.$$

This delayed renewal sequence

$$\{S_n, n \geq 0\} := \left\{D^{(0)}, D^{(0)} + \sum_{i=1}^{n}(X_i + Y_i), n \geq 1\right\}$$

is a stationary renewal process.

7.2.2 High Variability Induces Long-Range Dependence

Consider the indicator process $\{Z_t\}$, which is 1 iff t is in an on period. Thus, for $t \geq D^{(0)}$,

$$Z_t = \begin{cases} 1, & \text{if } S_n \leq t < S_n + X_{n+1}, \text{ some } n \\ 0, & \text{if } S_n + X_{n+1} \leq t < S_{n+1}, \text{ some } n \end{cases}$$

and if $0 \leq t < D^{(0)}$ we define

$$Z_t = \begin{cases} 1, & \text{if } B = 1 \text{ and } 0 \leq t < X_{\text{on}}^{(0)}, \\ 0, & \text{otherwise.} \end{cases}$$

A standard renewal argument gives the following result [22].

Proposition 7.2.1. *$\{Z_t, t \geq 0\}$ is strictly stationary and*

$$P[Z_t = 1] = \frac{\mu_{\text{on}}}{\mu}.$$

Conditional on $Z_t = 1$, the subsequent sequence of on/off periods is the same as seen from time 0 in the stationary process with $B = 1$.

It is easiest to express long-range dependence in terms of slow decay of covariance functions so we consider the second-order properties of the stationary process $\{Z_t\}$ (See Heath et al. [22].) The basis for the next result is a renewal theory argument.

Theorem 7.2.2. *The covariance function*

$$\gamma(s) = \text{Cov}(Z_t, Z_{t+s})$$

of the stationary process $\{Z(t), t \geq 0\}$ *is*

$$\begin{aligned}\gamma(s) &= \frac{\mu_{\text{on}}}{\mu}\left[\frac{\mu_{\text{off}}}{\mu} - \int_0^s \bar{F}_{\text{off}}(s-u)F_{\text{on}}^{(0)} * U(du)\right] \\ &= \frac{\mu_{\text{on}}}{\mu}\left[\frac{\mu_{\text{off}}}{\mu} - F_{\text{on}}^{(0)} * U * (1 - F_{\text{off}})(s)\right] \\ &= \frac{\mu_{\text{on}}\mu_{\text{off}}}{\mu^2} - \frac{1}{\mu}\int_0^s z(s-\omega)U(dw),\end{aligned}$$

where

$$U = \sum_{n=0}^{\infty}(F_{\text{on}} * F_{\text{off}})^{n*}$$

and

$$\begin{aligned}z(t) &= \int_0^t \bar{F}_{\text{off}}(x)\bar{F}_{\text{on}}(t-x)\,dx \\ &= \mu_{\text{on}}F_{\text{on}}^{(0)} * (1 - F_{\text{off}}(t) \\ &= \mu_{\text{off}}F_{\text{off}}^{(0)} * (1 - F_{\text{on}})(t).\end{aligned}$$

How do we analyze the asymptotic behavior of $\gamma(\cdot)$ as a function of s? Note $\gamma(s)$ is of the form

$$\gamma(s) = \text{const}\left[\lim_{v\to\infty} z * U(v) - z * U(s)\right]$$

so we need rates of convergence in the key renewal theorem. This can be based on a theorem of Frenk [19] and is given in Heath et al. [22].

Theorem 7.2.3. *Assume that there is an* $n \geq 1$ *such that* $(F_{\text{on}} * F_{\text{off}})^{n*}$ *is non-singular. Suppose*

$$\bar{F}_{\text{on}}(t) = t^{-\alpha}L(t), \quad t \to \infty,$$

where $1 < \alpha < 2$ *and* L *is slowly varying at infinity and assume*

$$\bar{F}_{\text{off}}(t) = o(\bar{F}_{\text{on}}(t)), \quad t \to \infty.$$

Then

$$\gamma(t) \sim \frac{\mu_{\text{off}}^2}{(\alpha-1)\mu^3}t^{-(\alpha-1)}L(t), \quad t \to \infty.$$

So $\gamma(t)$ decreases like a constant times $t\bar{F}_{\text{on}}(t)$. Such a slow decay of $\gamma(t)$ at an algebraic rate is characteristic of long-range dependence. One way to think about this result is that, with heavy-tailed *on* periods, there is a significant probability that a very long on period can cover both the time points s and $t + s$, thereby inducing strong correlation between these two time points.

Taqqu et al. [48] use the long-range dependence of the on/off process and superimpose many such processes. This superposition is approximately a fractional Brownian motion, giving one explanation of the observed self-similarity of Ethernet traffic. See also Leland et al. [33]. Other limiting procedures leading to Lévy motion with heavy tails are possible and are also briefly discussed in Leland et al. [33]. See also Konstantopoulos and Lin [32].

7.2.3 Single Channel Fluid Queues with Constant Service Rates

Suppose work enters a communication system according to the on/off process. The server works off the load at constant rate r assuming there is load to work on.

Here are the formal model ingredients for the single source model.

1. *The Input Process*

$$A(t) := \int_0^t Z_v \, dv.$$

Since $A(t) \sim t\mu_{\text{on}}/\mu$, the long-term input rate is μ_{on}/μ.

2. *The Output Process*. There is a release rate function—the release rate from the system when contents are at level x is

$$r(x) := \begin{cases} r, & \text{if } x > 0, \\ 0, & \text{if } x = 0. \end{cases}$$

3. *The Stability Condition*. Input does not overwhelm the release rate. This necessitates

$$1 > r > \frac{\mu_{\text{on}}}{\mu}. \tag{7.1}$$

The restriction that $r < 1$ results from the normalization that work arrives at rate 1 during *on* periods and prevents the contents process from having 0 as an absorbing state.

4. The contents process $\{X(t)\}$ satisfies the storage equation

$$dX(t) = dA(t) - r(X(t))\, dt.$$

Note that during an *on* period, the net input rate is $1 - r$ since work is inputted at unit rate but the server works at rate r. During an *off* period, the release rate is r

(provided there is liquid to release). This means the paths of $X(\cdot)$ are sawtoothed shaped.

7.2.3.1 *Regeneration Times*

Recall that the stationary alternating renewal process is

$$\left\{ S_n = D^{(0)} + \sum_{i=1}^{n} (X_i + Y_i), n \geq 0 \right\}.$$

Since the contents process is stable, we can define regeneration times

$$\{C_n\} := \{S_n : X(S_n -) = 0\},$$

which are times when a dry period ends and input commences. So the standard limit theorems due to Smith for regenerative processes [45] guarantee limit distributions exist in discrete and continuous time:

$$P[X(S_n) > x] \to 1 - W(x), \quad P[X(t) > x] \to 1 - V(x).$$

There are also connections with standard random walk theory and Lindley's equation holds. If we compare $X(S_n)$ with $X(S_{n+1})$ we get

$$\begin{aligned} X(S_{n+1}) &= (X(S_n) + (1 - r)X_{n+1} - rY_{n+1})^+ \\ &= (X(S_n) + \xi_{n+1})^+, \end{aligned}$$

where

$$\xi_{n+1} = (1 - r)X_{n+1} - rY_{n+1}$$

and $\{\xi_j\}$ i.i.d. It is important to distinguish between the random walk with steps $\{\xi_n\}$ and the random walk with steps $\{X_n + Y_n\}$.

Assume

$$1 - F_{\text{on}}(x) = x^{-\alpha} L(x), \quad \alpha > 1, \ x \to \infty.$$

From standard random walk theory [11, 36] we get

$$\begin{aligned} 1 - W(x) &\sim \frac{\rho}{1 - \rho} \frac{(1 - r)^{\alpha - 1}}{(\alpha - 1)\mu_{\text{on}}} x^{-(\alpha - 1)} L(x) \\ &:= b x^{-(\alpha - 1)} L(x), \quad x \to \infty, \end{aligned}$$

so that the tail of W is heavy and comparable to the integral of the tail of F_{on}. The definition of ρ is

$$\rho = \frac{\mu_{\text{on}}}{\mu_{\text{off}}} \frac{1-r}{r} < 1.$$

As expected from the sawtooth shape of the paths, the tail of $V(x)$ is heavier because of a bigger multiplicative factor

$$1 - V(x) \sim \left(b + \frac{(1-r)^{\alpha-1}}{\mu(\alpha-1)} \right) x^{-(\alpha-1)} L(x).$$

See Boxma and Dumas [9] and Heath et al. [22] and the references therein and also the discussion in Chapter 10. The tails of both $W(x)$ and $V(x)$ are of the form const $\times\, x\bar{F}_{\text{on}}(x)$. So the equilibrium content of the system, in either continuous or discrete time, can get quite large with nontrivial probability. This point will be reinforced in the discussion of the time it takes for buffer content to reach a critical level.

7.2.4 Extremes, Level Crossings, and Buffer Overflow

The distributions $W(x)$ and $V(x)$ are standard queueing quantities and convey some performance information. Another performance measure, one that is less dependent on notions of stability and existence of moments, is the time until buffer overflow. We formulate the time to buffer overflow as the hitting time of a high level L:

$$\tau(L) := \inf\{t \geq 0 : X(t) \geq L\}.$$

Define $M(t) = \bigvee_{s=0}^{t} X(s) = \max\{X(s) : 0 \leq s \leq t\}$, and the reason for interest in the maximum content up to t is that as a process it is the inverse of the $\tau(\cdot)$ process since

$$\tau(L) = \inf\{s > 0 : X(s) \geq L\} = \inf\{s > 0 : M(s) \geq L\} := M^{\leftarrow}(L),$$

where $M^{\leftarrow}(\cdot)$ is the right continuous inverse of the monotone function $M(\cdot)$. If we understand the asymptotic behavior of $M(\cdot)$, then we will understand the asymptotic behavior of $\tau(\cdot)$. To understand the behavior of $M(\cdot)$ we first study the extremes of $\{X(S_n)\}$ and then fill in the behavior between the discrete points $\{S_n\}$. We study the maxima of the random walk generated by $\{\xi_n\}$ over cycles and then knit cycles together. Recall

$$\xi_{n+1} = (1-r)X_{n+1} - rY_{n+1}.$$

The first downgoing ladder epoch of $\{\sum_{i=0}^{n} \xi_i, n \geq 0\}$ is

$$\bar{N} = \inf\left\{n > 0 : \sum_{i=0}^{n} \xi_i \leq 0\right\}. \tag{7.2}$$

The tail behavior of the maximum in a (discrete) cycle is described next. See Asmussen [2] and Heath et al. [21].

Proposition 7.2.4. *For the stable queueing process $\{X(S_n)\}$ satisfying*

$$1 - F_{\text{on}}(x) = x^{-\alpha}L(x), \quad \alpha > 1, \ x \to \infty,$$

the maximum over a cycle has a distribution tail asymptotic to the tail of the on distribution and, in particular,

$$\begin{aligned} P\left[\bigvee_{n=0}^{\bar{N}} X(S_n) > x\right] &\sim P[\xi_1 > x]E(\bar{N}) \\ &\sim P[(1-r)X_1 > x]E(\bar{N}) \\ &\sim (1-r)^{\alpha}\bar{F}_{\text{on}}(x)E(\bar{N}). \end{aligned}$$

Note that this tail is lighter than the tail of W or V from the previous section. Interestingly, the *off* distribution influences this expression only through the factor $E(\bar{N})$.

From maxima over discrete cycles, it is relatively simple to derive tail behavior over a continuous time cycle: define $C_1 = S_{\bar{N}}$.

Corollary 7.2.5. *Assume the contents process $\{X(t)\}$ is stable and*

$$1 - F_{\text{on}}(x) = x^{-\alpha}L(x), \quad \alpha > 1, \ x \to \infty.$$

The distribution tail of the maximum of the contents process over one cycle is asymptotic to the tail of the on distribution;

$$P\left[\bigvee_{s=0}^{C_1} X(s) > x\right] \sim (1-r)^{\alpha}\bar{F}_{\text{on}}(x)E(\bar{N}).$$

Again note the minimal effect of F_{off}, which only affects the answer through the multiplicative factor $E(\bar{N})$.

Now the behavior up to an arbitrary time t is derived from the random number of cycles squeezed into $[0, t]$ and we have the following result from Heath et al. [21]. (Let $D_r(0, \infty)$ be the right continuous functions on $(0, \infty)$ with finite left limits; $D_l(0, \infty)$ are the left continuous functions with finite right limits.)

Theorem 7.2.6. *Assume* $\{X(t)\}$ *stable and*

$$1 - F_{\mathrm{on}}(x) = x^{-\alpha}L(x), \quad \alpha > 1, \ x \to \infty.$$

Define the quantile function

$$b(s) = \left(\frac{1}{1 - F_{\mathrm{on}}}\right)^{\leftarrow}(s).$$

Let $\{Y_\alpha(t), t > 0\}$ *be the extremal process [44] generated by*

$$\Phi_\alpha(x) = \exp\{-x^{-\alpha}\}, \quad x > 0$$

so that

$$P[Y_\alpha(t) \le x] = \Phi_\alpha^t(x).$$

Define the rescaled extremal process

$$S_\alpha(t) = \frac{1 - r}{\mu^{1/\alpha}} Y_\alpha(t).$$

Then in $D_r(0, \infty) \times D_l(0, \infty)$, *as* $u \to \infty$,

$$\left(\frac{M(u\cdot)}{b(u)}, \left(\frac{M(u\cdot)}{b(u)}\right)^{\leftarrow}\right) \Rightarrow (S_\alpha(\cdot), S_\alpha^{\leftarrow}(\cdot)).$$

In particular, we get for the first passage process, as $u \to \infty$,

$$(1 - F_{\mathrm{on}}(u))\tau(u\cdot) \Rightarrow Y_\alpha^{\leftarrow}\left(\frac{\mu^{1/\alpha}}{1 - r}\right).$$

and

$$\lim_{L \to \infty} P\left[\frac{(1 - r)^\alpha}{\mu}(1 - F_{\mathrm{on}}(L))\tau(L) \le x\right] = P[E(1) \le x] = 1 - e^{-x}, \quad x > 0,$$

where $E(1)$ *is a unit exponential random variable. Also, as* $x \to \infty$,

$$(1 - F_{\mathrm{on}}(x))E(\tau(x)) \to \frac{\mu}{(1 - r)^\alpha},$$

so equivalently

$$E(\tau(x)) \sim \frac{\mu}{(1-r)^{\alpha}} x^{\alpha}/L(x).$$

The approximation to $E(\tau(x))$ is only first order and not expected to be dazzlingly good but to test this we did two modest simulations with

$$F_{\text{on}} = \text{Pareto}, \ \alpha = 1.5, \ r = 0.53$$
$$F_{\text{off}} = \begin{cases} \text{the same Pareto or} \\ \text{constant } \textit{off} \text{ times equal to 3,} \end{cases}$$

The number of replications was 500 and the levels L were 2, 5, 10, 22, 46, 100, 215, 464. The simulation with the constant off times shows our approximation is surprisingly effective but, as expected, having variability in the off distribution makes the approximation less accurate. See Fig. 7.1.

As a final experiment, we decided to test the correctness of the intuition that a high level crossing by the content level was due to a single very long *on* period, rather than due to gradual buildup. We ran 1000 simulation runs of the system. Each simulation consisted of running the system until level $L = 64$ was crossed and

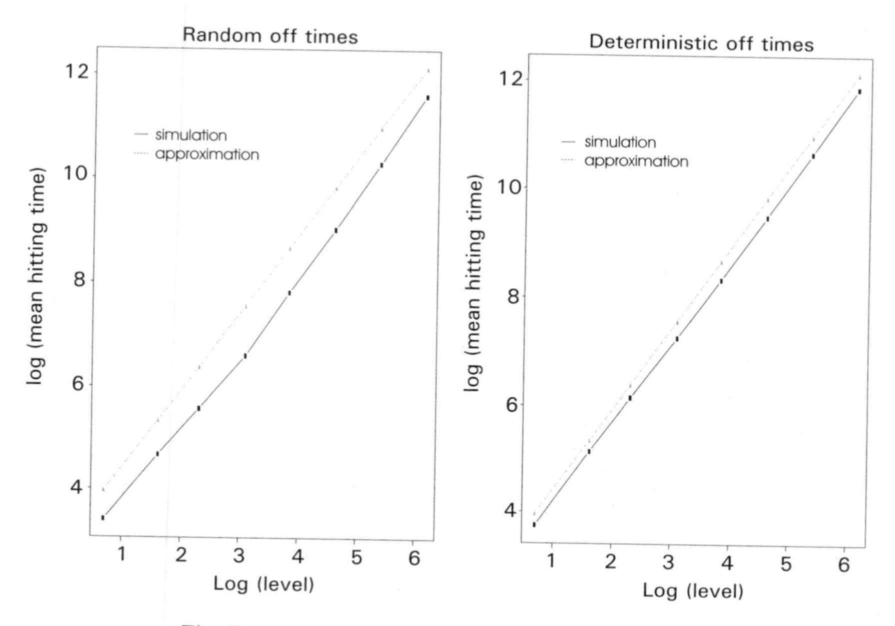

Fig. 7.1 Approximation versus simulated mean times.

keeping track of the length of the *on* period, which was in progress when level L was crossed. The histogram in Fig. 7.2 is for the ratios

$$\frac{(1-r)\times(\text{length of the on period resulting in level exceedence})}{L}\wedge 3$$

where $r = 0.53$ and $\alpha = 1.5$. The wedge symbol $\wedge$ means minimum. The *off* distribution was concentrated at 3 and the *on* distribution was Pareto. The histogram shows that 85% of the time, the contents process crosses level L due to a very long *on* period and not due to gradual buildup. This is due to the fact that for all but 150 out of 1000 of the runs, the ratio was ≥ 1.

7.2.5 Contrast with Exponential Tails

From results of Iglehart [27] we may contrast the heavy-tailed case with the case of exponentially decreasing tails. We make the following assumptions.

1. There exists $\gamma > 0$ such that $E(e^{\gamma\xi_1}) = 1$. This implies that the right tail of ξ_1 is bounded by $\exp\{-\gamma x\}$, $x > 0$. Note that F_{off} helps determine the growth rate γ.
2. For this value of γ, $E\xi_1 e^{\gamma\xi_1} := \mu_\gamma \in (0, \infty)$.
3. The random variable ξ_1 has a nonlattice distribution.

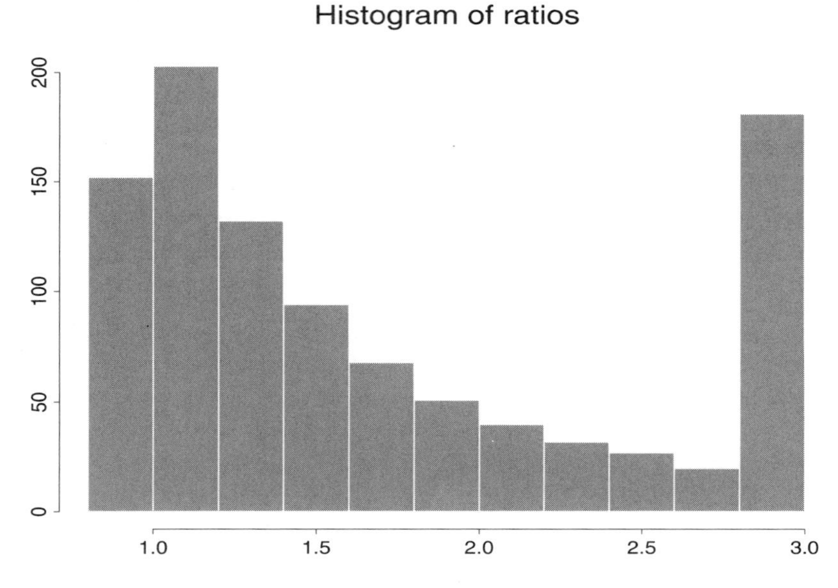

Fig. 7.2 Histogram giving effect of long *on* periods.

Define $\bar{N}$ as in Eq. (7.2) and

$$a(0) := \frac{(1 - E(\exp\{\gamma \sum_{i=1}^{\bar{N}} \xi_i\}))^2}{\gamma \mu_\gamma E(\bar{N})} E(\exp\{\gamma(1-r)X_1\}).$$

Then

$$\left(\frac{a(0)}{\mu E(\bar{N})} \frac{\tau(u)}{e^{\gamma u}} \right) \Rightarrow E(1)$$

as $u \to \infty$, where $E(1)$ is a unit exponential random variable. Also,

$$E\tau(s) \sim \frac{\mu E(\bar{N})}{a(0)} e^{\gamma s} \quad (s \to \infty).$$

So in this case, the time to hit a high level grows exponentially fast in the level (better) as opposed to algebraically fast (worse) in the level for the heavy-tailed case.

7.3 BUFFER OVERFLOW IN A MULTISOURCE MODEL

Suppose a single server is fed by J i.i.d. on/off sources. As before we assume the server works at constant rate r. For $i = 1, \ldots, J$ let $Z_i(t)$ be the indicator of the ith on/off source describing when this alternating renewal source is on and transmitting. Define

$$Z^{(J)}(t) = \sum_{i=1}^{J} Z_i(t), \quad t > 0$$

to be the total input rate at t; that is, the number of sources transmitting at t. The contents process corresponding to this input is

$$dX^{(J)}(t) = Z^{(J)}(t)\, dt - r(X^{(J)}(t)\, dt.$$

To guarantee stability, we assume the condition $J\mu_{\text{on}}/\mu < r$.

As in the single channel model, we focus on time to buffer overflow as a meaningful performance measure and thus consider

$$\tau(L) = \inf\{t \geq 0 : X^{(J)}(t) \geq L\}.$$

Because the superposition of renewal processes is no longer a renewal or regenerative process, this J-channel model is significantly more complex to analyze. We are only able to analyze the expected value of $\tau(L)$ and only in the case $r < 1$.

Theorem 7.3.1 [21]. *Let*

$$\bar{F}_{\text{on}}(x) = x^{-\alpha}L(x), \quad \alpha > 1, \ x \to \infty,$$

and assume that the off distribution tail is not too long. *If the service rate r satisfies the stability condition* and $r < 1$, *then we have for any initial conditions*

$$\lim_{L\to\infty} \bar{F}_{\text{on}}\left(\frac{L}{1 - r + (J-1)\mu_{\text{on}}/\mu}\right) E\tau(L) = \frac{1}{J}\mu.$$

The condition that the off distribution be *not too long* is needed for technical reasons and means there exist a distribution H^* and p with $1 < p < \alpha$ such that

$$\int_0^\infty x^p H^*(dx) < \infty, \quad \frac{\bar{F}_{\text{off}}(x+y)}{\bar{F}_{\text{off}}(y)} \le \bar{H}^*(x)$$

for every $x \ge 0$ and $y \ge 0$ such that $\bar{F}_{\text{off}}(y) > 0$. The class of distributions satisfying this condition includes Gamma distributions, distributions with compact support, and others.

For related finite buffer results, see Jelenković [29] and Zwart [50].

7.4 AN INFINITE SOURCE MODEL

The difficulty in analyzing the J-channel *on/off* model of the previous section suggests letting $J \to \infty$ in order to achieve additional tractability. So suppose now there are an infinite number of nodes in a network. Sources turn on and initiate *sessions* at Poisson time points $\{\Gamma_k, k \ge 1\}$. The rate of the Poisson process is λ. The lengths of sessions are i.i.d. random variables $\{X_n\}$, which have a common distribution F_{on} and assume $E(X_1) = \mu_{\text{on}}$. During a session, an active source transmits at rate 1 and whenever there is work, the single server works at rate r. An important quantity is

$$\begin{aligned} N(t) &= \text{number of sessions in progress at } t \\ &= \sum_{k=1}^{\infty} 1_{[\Gamma_k < t \le \Gamma_k + X_k]} \\ &= \text{the number of busy servers in the } M/G/\infty \text{ queue.} \end{aligned}$$

Traditional queueing theory recognizes $N(t)$ as the number of busy trunklines in an infinite line telephone model.

It is well known and easy to derive the following facts.

1. $N(t)$ is Poisson distributed. At equilibrium the mean is $\lambda\mu_{\text{on}}$.

2. At equilibrium, the stationary version of $N(\cdot)$ has covariance function at lag s equal to $\int_s^\infty \bar{F}_{\text{on}}(u)\,du$. So if $\bar{F}_{\text{on}}$ is heavy tailed, there will be slow decay of the covariance function and long-range dependence will be present.

Both facts follow readily from the fact that the point process with points $\{(\Gamma_k, X_k)\}$ is a two-dimensional Poisson process with mean measure $\lambda\, ds \times F(dy)$.

We define the contents process $X(\cdot)$ by

$$dX(t) = N(t)\,dt - r1_{[X(t)>0]}\,dt$$

so the input rate is random depending on the number of active sources at a given time. To ensure stability we assume

$$\text{long-term input rate} = \lambda\mu_{\text{on}} < r.$$

For additional discussion of this model, see Boxma [8], Brichet et al. [10], Konstantopoulos and Lin [32], and Resnick and Rootzen [46]. A discrete time analog of this model is discussed in Chapter 9.

7.4.1 Activity Periods

Following Jelenković and Lazar [30, 31], we define an activity period to be a busy period in the corresponding $M/G/\infty$ queue so if at time 0 the initial conditions are that $N(0) = 0$ and a node turns on, then the length of the initial activity period is $\inf\{t > 0 : N(t) = 0\}$. Note that an activity period could end with nonzero content; it just depends on there being no active sources. The length of an activity period is a random variable whose distribution is known by means of a Laplace transform formula as given in Takács [47] or Hall [20]. Based on this, we may derive another interesting quantity.

Let v be the number of sessions in an activity period and define

$$p_+(u) = P\left[\bigvee_{i=1}^{v} X_i > u\right]$$

to be the probability that at least one of the sessions forming an activity period is longer than u. Then based on the Tákacs formula [47] for the Laplace transform of the $M/G/\infty$ busy period, we may derive the following (see Resnick and Samorodnitsky [41]):

$$\begin{aligned} p_+(u) &= \frac{\bar{F}_{\text{on}}(u)}{\lambda\bar{F}_{\text{on}}(u)\displaystyle\int_0^u \exp\left(-\lambda\int_0^t \bar{F}_{\text{on}}(x)\,dx\right)dt + \exp\left(-\lambda\int_0^u \bar{F}_{\text{on}}(x)\,dx\right)} \\ &\sim e^{\lambda\mu_{\text{on}}}\bar{F}_{\text{on}}(u), \qquad u \to \infty. \end{aligned}$$

This allows us (see Resnick and Samorodnitsky [41] for the proof) to generalize a result of Jelenković and Lazar [28]. See also Chapter 10. Our result weakens assumptions on F_{on} and weakens the assumption $r < 1$.

Theorem 7.4.1. *Let I be the change in the buffer content during the activity period. Assume that the stability condition holds, and that*

$$\lambda\mu_{\text{on}} < r < 1 + \lambda\mu. \tag{7.3}$$

Assume the session length distribution F_{on} has a regularly varying (or only dominatedly varying) tail with index $\alpha > 1$. Then

$$\lim_{x\to\infty} \frac{P(I > x)}{\bar{F}_{\text{on}}\left(\dfrac{x}{1 - r + \lambda\mu_{\text{on}}}\right)} = e^{\lambda\mu_{\text{on}}}. \tag{7.4}$$

Plausibility Argument Heavy tails require a long session in order for $I > x$. Suppose a long session is in progress. During this long session, the input rate is $1 + \lambda\mu_{\text{on}} > r$ (since the overall rate is $\lambda\mu_{\text{on}}$ and the long session provides added input at rate 1) and therefore content increases by amount

$$(1 + \lambda\mu_{\text{on}} - r) \times (\text{length of long session}).$$

Thus, if we have a long session whose length exceeds

$$x/(1 - r + \lambda\mu_{\text{on}}),$$

the content will increase by at least x. So

$$P[I > x] \approx p_+\left(\frac{x}{1 - r + \lambda\mu_{\text{on}}}\right).$$

Using the formula for $p_+(u)$ yields the result.

7.4.2 Extremes, First Passages, and Buffer Overflow

Suppose the added restriction that

$$\lambda\mu_{\text{on}} < r < 1.$$

Although the assumption $r < 1$ is a serious restriction, assuming it clarifies when the contents process increases or decreases. In fact,

$$X(t) \uparrow \text{ iff } N(t) \geq 1 \quad \text{and} \quad X(t) \downarrow \text{ iff } N(t) = 0.$$

This allows us to mimic the single channel model analysis and makes random walk theory applicable. Observe that $X(\cdot)$ regenerates whenever $X(t) = N(t) = 0$ and assuming $X(0) = 0$ we set

$$T_1 = \inf\{t > 0 : X(t) = N(t) = 0\}$$

for the initial cycle length. Define as before

$$M(t) = \bigvee_{s=0}^{t} X(s),$$

and think of $M(t)$ as the maximum over a random number of cycles. Let

$$A_1, E_1, A_2, E_2, \ldots$$

be the alternating sequence of activity periods and idle periods. During A-intervals, $X(t)$ is increasing and during the idle periods $X(t)$ decreases. Let $I_1, I_2, \ldots$ be the change in X over the successive activity periods. So $I_1 = X(A_1) - X(0)$, and so on. Then we set

$$\bar{N} = \inf\left\{n \geq 1 : \sum_{i=1}^{n} (I_i - rE_i) \leq 0\right\}$$

and get by mimicking the single channel analysis in Proposition 7.2.4 and Corollary 7.2.5 that

$$\begin{aligned} P\left[\bigvee_{s=0}^{T_1} X(s) > x\right] &\sim E(\bar{N}) P[I_1 > x] \\ &\sim E(\bar{N}) e^{\lambda\mu_{\text{on}}} (1 + \lambda\mu_{\text{on}} - r)^{\alpha} \bar{F}_{\text{on}}(x), \end{aligned}$$

and as $n \to \infty$, $M(nt)/b(n)$ converges weakly to an extremal process (where, as before in Theorem 7.2.6, $b(\cdot)$ is the quantile function associated with F_{on}), which necessitates

$$\bar{F}_{\text{on}}(L)\tau(L) \Rightarrow \frac{E}{\lambda E(\bar{N})(1 + \lambda\mu_{\text{on}} - r)^{\alpha}},$$

where E is a unit exponential random variable. The mean of $\tau(L)$ also also converges to the corresponding limit.

7.4.3 The General Case: What Determines Buffer Overflow?

We now remove the restriction that $r < 1$. We continue to suppose sources commence transmission at Poisson time points. Detailed analysis of this general case is difficult because of inability to readily identify useful regeneration times. The question requiring a solution is: How is performance loss related to the mean drift parameters of the process and the heaviness of the tail?

Again, a sensible performance measure is

$$\tau(L) = \inf\{t > 0: X(t) \geq L\}$$

and we continue to think of L as a critical level or the capacity of a buffer.

For the general analysis, we assume

- Stability: $r > \lambda\mu_{\text{on}}$.
- Heavy-tailed (or just dominatedly varying) transmission times:

$$1 - F_{\text{on}}(x) = x^{-\alpha}L(x), \quad x \to \infty, \ \alpha > 1.$$

Key Parameter We define the key parameter k, which is the minimum number of simultaneously running sessions that causes the mean drift to flip from negative to positive:

$$k := \inf\{j \geq 1: \lambda\mu_{\text{on}} + j > r\}.$$

It is only when k sessions are running simultaneously that content increases significantly. The philosophy is that of large deviations, namely, that an unlikely event (buffer overflow) has to happen in the most likely way possible (simultaneous transmission by k active sources). We can state the following result about expected time to buffer overflow. Note that the complexity allows only analysis of expected times and that only the order is obtained. For functions f, g we write

$$f \asymp g$$

to mean

$$0 < \liminf_{x \to \infty} \frac{f(x)}{g(x)} \leq \limsup_{x \to \infty} \frac{f(x)}{g(x)} < \infty.$$

Theorem 7.4.2 [23]. *Suppose the session length distribution F_{on} is heavy tailed,*

$$1 - F_{\text{on}}(x) = x^{-\alpha}L(x), \quad \alpha > 1,$$

and

$$r - \lambda\mu_{\text{on}} > 0 \textit{ is noninteger}.$$

Then

$$E\tau(\gamma) \asymp \gamma^{1-k}\left(\frac{1}{1 - F_{\text{on}}(\gamma)}\right)^k.$$

If

$$1 - F_{\text{on}}(x) \asymp x^{-\alpha}, \quad \alpha > 1,$$

then

$$E(\tau(\gamma)) \asymp \gamma^{1-k+\alpha k} = \gamma^{1+k(\alpha-1)}.$$

The obvious conclusion from this result is that paying for an increase in r, which does not change k, is foolish since it will not affect the asymptotic growth rate of $E(\tau(\gamma))$. One should increase r enough so that

$$k \longmapsto k + 1$$

and then $E(\tau(\gamma))$ improves from

$$\gamma^{1+k(\alpha-1)} \longmapsto \gamma^{1+(k+1)(\alpha-1)} = (\gamma^{1+k(\alpha-1)})(\gamma^{\alpha-1}).$$

7.5 CONCLUSION

Fluid queues have frequently been used as models for telecommunication networks. We focused on models where a server works off the load at constant rate r. We analyzed a fluid queue being fed by (1) a single on/off renewal process, (2) k such on/off models, and (3) an infinite number of sources. In models (1) and (2) sources turn on at renewal time points and in (3) sources turn on at Poisson times. Transmissions last for i.i.d. periods governed by a long-tailed distribution of session lengths. The heavy tails induce long-range dependence in the system and result in performance deterioration. We studied the time it takes until such fluid queues with finite but large holding capacity reach overflow. The expected time until overflow, as a function of capacity L, increases only polynomially fast, and so overflows happen much more often than in the "classical" light-tailed case, where the expected overflow time increases as an exponential function of L.

Tails of many teletraffic quantities are heavy and there is suspicion that they are getting heavier as WWW users become more demanding. However, this hypothesis is yet to be verified statistically and the data currently collected may be inadequate to prove the hypothesis. Preliminary month-by-month analysis of the Boston University [12, 13] data is inconclusive.

The model with an infinite number of sources is receiving continuing attention. A detailed study of the steady-state contents distribution is under way and this will hopefully lead to more detailed information about the extreme value behavior of the system. Also, by rescaling time by T (with $T \to \infty$) but keeping the Poisson rate fixed, Lévy stable motion is an approximation to the cumulative input [32]. However, if the input rate $\lambda = \lambda(T)$ is allowed to vary with T, then Gaussian limits occur if $\lambda(T)$ grows faster than a critical rate. This provides a different approximation to cumulative inputs. The usefulness of both approximations awaits detailed investigation.

ACKNOWLEDGMENT

The research of S. Resnick and G. Samorodnitsky was partially supported by NSF Grant DMS-97-04982 and NSA Grant MDA904-98-1-0041 at Cornell University.

REFERENCES

1. M. Arlitt and C. Williamson. Web server workload characterization: the search for invariants. Master's thesis, University of Saskatchewan, 1996.
2. S. Asmussen. Subexponential asymptotics for stochastic processes: extremal behavior, stationary distributions and first passage probabilities. *Ann. Appl. Probab.*, **8**:354–374, 1998.
3. J. Beran, R. Sherman, W. Willinger, and M. S. Taqqu. Long range dependence in variable-bit-rate video. *IEEE Trans. Commun.*, **43**:1566–1579, 1995.
4. D. C. Boes. Schemes exhibiting Hurst behavior. In J. Srivastava, ed., *Essays in Honor of Franklin Graybill*, pp. 21–42. North Holland, Amsterdam, 1988.
5. D. C. Boes and J. D. Salas-La Cruz. On the expected range and expected adjusted range of partial sums of exchangeable random variables. *J. Appl. Probab.*, **10**:671–677, 1973.
6. T. Bollerslev. Generalized autoregressive conditional heteroskedasticity. *J. Econometrics*, **31**:307–327, 1986.
7. T. Bollerslev, R. Y. Chow, and K. F. Kroner. ARCH modeling in finance: a review of the theory and empirical evidence. *J. Econometrics*, **52**:5–59, 1992.
8. O. Boxma. Fluid queues and regular variation. *Perf. Eval.*, **27**, **28**:699–712, 1996.
9. O. Boxma and V. Dumas. Fluid queues with long-tailed activity period distributions. *Comput. Commun.*, **21**:1509–1529, 1998. Special issue on Stochastic Analysis and Optimisation of Communication Systems.
10. F. Brichet, J. Roberts, A. Simonian, and D. Veitch. Heavy traffic analysis of a storage model with long range dependent on/off sources. *Queueing Syst.*, **23**:197–215, 1996.
11. J. W. Cohen. Some results on regular variation for distributions in queueing and fluctuation theory. *J. Appl. Probab.*, **10**:343–353, 1973.
12. M. Crovella and A. Bestavros. Explaining World Wide Web traffic self-similarity. Preprint available as TR-95-015 from {`crovella,best`}cs.bu.edu, 1995.
13. M. Crovella and A. Bestavros. Self-similarity in World Wide Web traffic: evidence and possible causes. In *Proceedings of the 1996 ACM SIGMET RICS International Conference on Measurement and Modeling of Computer Systems*, **24**:160–169, 1996.

14. M. Crovella and A. Bestavros. Self-similarity in World Wide Web traffic: evidence and possible causes. *IEEE/ACM Trans. Networking*, **5**(6):835–846, 1997.
15. R. Davis and S. Resnick. Limit theory for bilinear processes with heavy tailed noise. *Ann. Appl. Probab.*, **6**:1191–1210, 1996.
16. L. de Haan, D. W. Jansen, K. Koedijk, and Casper G. de Vries. Safety first portfolio selection, extreme value theory and long run asset risks. In J. Galambos, ed., *Extreme Value Theory and Applications*, pp. 471–487. Kluwer, Dordrecht, 1994.
17. C. G. de Vries. On the relation between GARCH and stable processes. *J. Econometrics*, **48**:313–324, 1991.
18. A. Erramilli, O. Narayan, and W. Willinger. Experimental queueing analysis with long-range dependent packet traffic. *IEEE/ACM Trans. Network Comput.*, **4**:209–223, 1996.
19. J. B. G. Frenk. *On Banach Algebras, Renewal Measures and Regenerative Processes*, Vol. 38 of *CWI Tracts*. Stichting Mathematisch Centrum, Centrum voor Wiskunde en Informatica, Amsterdam, 1987.
20. P. Hall. *Introduction to the Theory of Coverage Processes*. Wiley, New York, 1988.
21. D. Heath, S. Resnick, and G. Samorodnitsky. Patterns of buffer overflow in a class of queues with long memory in the input stream. *Ann. Appl. Probab.*, **7**(4):1021–1057, 1997.
22. D. Heath, S. Resnick, and G. Samorodnitsky. Heavy tails and long range dependence in on/off processes and associated fluid models. *Math. Oper. Res.*, **23**(1):145–165, 1998.
23. D. Heath, S. Resnick, and G. Samorodnitsky. How system performance is affected by the interplay of averages in a fluid queue with long range dependence induced by heavy tails. *Ann. Appl. Probab.*, **9**:352–375, 1999.
24. M. C. A. B. Hols and C. G. de Vries. The limiting distribution of extremal exchange rate returns. *J. Appl. Econometrics*, **6**:287–302, 1991.
25. H. E. Hurst. Long-term storage capacity of reservoirs. *Trans. Am. Soc. Civ. Eng.*, **116**:770–808, 1951.
26. H. E. Hirst. Methods of using long-term storage in reservoirs. *Proc. Inst. Civ. Eng., Part I*, pp. 519–577, 1955.
27. D. L. Iglehart. Extreme values in the $GI/G/1$ queue. *Ann. Math. Statist.*, **43**:627–635, 1972.
28. P. R. Jelenković and A. A. Lazar. Asymptotic results for multiplexing subexponential on–off processes. *Adv. Appl. Probab.*, **31**:394–421, 1999.
29. P. R. Jelenković. Subexponential loss rates in a $GI/GI/1$ queue with applications. *Queueing Syst.*, **33**:91–123, 1999.
30. P. R. Jelenković and A. A. Lazar. A network multiplexer with multiple time scale and subexponential arrivals. In *Stochastic Networks*, pp. 215–235. Springer, New York, 1996.
31. P. R. Jelenković and A. A. Lazar. Subexponential asymptotics of a Markov-modulated random walk with queueing applications. *J. Appl. Probab.*, **35**(2):325–347, 1998.
32. T. Konstantopoulos and S.-J. Lin. Macroscopic models for long-range dependent network traffic. *Queueing Syst. Theory Appl.*, **28**(1–3):215–243, 1998.
33. W. E. Leland, M. S. Taqqu, W. Willinger, and D. V. Wilson. On the self-similar nature of Ethernet traffic (extended version). *IEEE/ACM Trans. Networking*, **2**:1–15, 1994.
34. B. Melamed, M. Livny, and A. K. Tsiolis. The impact of autocorrelations on queueing systems. *Manage. Sci.*, **39**:322–339, 1993.

35. A. McNeil. Estimating the tails of loss severity distributions using extreme value theory. *Astin Bull.*, **27**:117–137, 1997.

36. A. G. Pakes. On the tails of waiting-time distributions. *J. Appl. Probab.*, **12**:555–564, 1975.

37. S. Resnick. Discussion of the Danish data on large fire insurance losses. *Astin Bull.*, **27**:139–151, 1997.

38. S. Resnick. Heavy tail modeling and teletraffic data. *Ann. Statist.*, **25**:1805–1869, 1997.

39. S. Resnick. Why non-linearities can ruin the heavy tailed modeler's day. In M. S. Taqqu, R. Adler, and R. Feldman, eds., *A Practical Guide to Heavy Tails: Statistical Techniques for Analysing Heavy Tailed Distributions*. Birkhäuser, Boston, 1998.

40. S. Resnick and G. Samorodnitsky. Performance decay in a single server exponential queueing model with long range dependence. *Oper. Res.*, **45**:235–243, 1997.

41. S. Resnick and G. Samorodnitsky. Activity periods of an infinite server queue and performance of certain heavy tailed fluid queues. Available as TR1201.ps.Z at `http://www.orie.cornell.edu/trlist/trlist.html`, 1998. To appear: *QUESTA*

42. S. Resnick, G. Samorodnitsky, and F. Xue. How misleading can sample acf's of stable ma's be? (very!). *Ann. Appl. Probab.*, 1998. Technical report 1209; available at `www.orie.cornell.edu/trlist/trlist.html`. To appear: *Ann. Appl. Prob.*

43. S. Resnick and E. van den Berg. Sample correlation behavior for the heavy tailed general bilinear process. Technical report 1210; available at `www.orie.cornell.edu/trlist/trlist.html`, 1998. To appear: *Stochastic Models*.

44. S. I. Resnick. *Extreme Values, Regular Variation and Point Processes*. Springer-Verlag, New York, 1987.

45. S. I. Resnick. *Adventures in Stochastic Processes*. Birkhäuser, Boston, 1992.

46. S. I. Resnick and H. Rootzén. Self-similar communication models and very heavy tails. Available as TR1228.ps.Z at `www.orie.cornell.edu/trlist/trlist.html`, 1998. To appear: *Ann. Appl. Prob.*

47. L. Takács. *An Introduction to Queueing Theory*. Oxford University Press, New York, 1962.

48. M. Taqqu, W. Willinger, and R. Sherman. Proof of a fundamental result in self-similar traffic modeling. *Comput. Commun. Rev.*, **27**:5–23, 1997.

49. W. Willinger, M. S. Taqqu, M. Leland, and D. Wilson. Self-similarity through high variability: statistical analysis of Ethernet LAN traffic at the source level. *Comput. Commun. Rev.*, **25**:100–113, 1995. *Proceedings of the ACM SIGCOMM'95*, Cambridge, MA.

50. A. P. Zwart. A fluid queue with a finite buffer and subexponential input. COSOR Memorandum 98-25, Eindhoven University at Technology; available at `http://www.win.tue.nl/math/bs/cosor98.html`, 1998.

8

BOUNDS ON THE BUFFER OCCUPANCY PROBABILITY WITH SELF-SIMILAR INPUT TRAFFIC

N. LIKHANOV
Institute for Problems of Information Transmission, Russian Academy of Science, Moscow, Russia

8.1 INTRODUCTION

High-quality traffic measurements indicate that actual traffic behavior over high-speed networks shows self-similar features. These include an analysis of hundreds of millions of observed packets on several Ethernet LANs [7, 8], and an analysis of a few million observed frame data by variable bit rate (VBR) video services [1]. In these studies, packet traffic appears to be statistically self-similar [2, 11]. Self-similar traffic is characterized by "burstiness" across an extremely wide range of time scales [7]. This behavior of aggregate Ethernet traffic is very different from conventional traffic models (e.g., Poisson, batch Poisson, Markov modulated Poisson process [4]).

A lot of studies have been made for the design, control, and performance of high-speed and cell-relay networks, using traditional traffic models. It is likely that many of those results need major revision when self-similar traffic models are considered [18].

Self-similarity manifests itself in a variety of different ways: a spectral density that diverges at the origin, a nonsummable autocorrelation function (indicating long-range dependence), an index of dispersion of counts (IDCs) that increases monotonically with the sample time T, and so on [7]. A key parameter characterizing self-similar processes is the so-called Hurst parameter, H, which is designed to capture the degree of self-similarity.

Self-Similar Network Traffic and Performance Evaluation, Edited by Kihong Park and Walter Willinger
ISBN 0-471-31974-0

Self-similar process models can be derived in different ways. One way is to construct the self-similar process as a sum of independent sources with a special form of the autocorrelation function. If we put the peak rate of each source going to zero as the number of sources goes to infinity, models like those of Mandelbrot [11] and Taqqu and Lévy [15] will be obtained. Queueing analysis for these kinds of processes is given in Chapters 4 and 5 in this volume. Another approach is to consider on/off sources with constant peak rate, while the number of sources goes to infinity. In this way, we obtain self-similar processes with sessions arrived as Poisson r.v.'s [9]. Originally this process was proposed by Cox [2] and queueing analysis was done recently by many authors [3, 6, 9, 10, 13, 17]. The main results of these papers are presented in this volume. In Chapter 9 we can find a complete overview of this topic. Chapters 7 and 10, present some particular results for the above model as well as results for the model with finite number of on/off sources. Models close to on/off processes arrived as Poisson r.v.'s are considered in Chapter 11. In Chapter 6 a queueing system with instantaneous arrivals is given.

In this chapter we will find the class of all self-similar processes with independent sessions arrived as Poission r.v.'s. For the particular case of the Cox model, we will find asymptotic bounds for buffer overflow probability. Compare with Chapter 9, Section 10.4 of Chapter 10, and Section 7.4 of Chapter 7, where asymptotic bounds are presented for a wide class of processes beyond the self-similar one we will focus on the self-similar case (Pareto distribution of the active period). For this case we present some new bounds, which are more accurate compared to the best-known current bounds [3, 10].

This chapter is organized in the following way. First, we give the definition of second-order self-similar traffic and some well-known, but useful, relations between variance, autocorrelation, and spectral density functions. This is followed by a construction of a class of second-order self-similar processes. Finally, asymptotic queueing behavior for a particular form of the processes from our class is analyzed.

8.2 SECOND-ORDER SELF-SIMILAR PROCESSES

We consider a discrete-time stationary stochastic process

$$X = (\ldots, X_{-1}, X_0, X_1, \ldots)$$

with a constant mean $\mu = \mathbf{E}X_i$ and finite variance $\sigma^2 = \operatorname{Var} X_i \triangleq \mathbf{E}\{(X_i - \mu)^2\}$. For convenience, we also consider the process $x = (\ldots, x_{-1}, x_0, x_1, \ldots)$, which is equal to $x_i \triangleq X_i - \mu$, so the process x is the same as process X, but with zero mean. $\mathbf{E}x_i = 0$, $\operatorname{Var} x_i = Ex_i^2 = \sigma^2$. The autocorrelation function of processes x and X is

$$r(k) = \mathbf{E}\{(X_i - \mu)(X_{i+k} - \mu)\}/\sigma^2 = \mathbf{E}\{x_i x_{i+k}\}/\sigma^2, \quad k = 0, \pm 1, \pm 2, \ldots.$$

Since the process X is a stationary one, $r(k)$, $\operatorname{Var} X_i$, and $\mathbf{E}X_i$ do not depend on i.

Denote

$$X^{(m)} = (\dots, X_{-1}^{(m)}, X_0^{(m)}, X_1^{(m)}, \dots),$$
$$x^{(m)} = (\dots, x_{-1}^{(m)}, x_0^{(m)}, x_1^{(m)}, \dots), \quad m = 1, 2, 3, \dots,$$

where

$$X_i^{(m)} \triangleq \frac{1}{m}(X_{im+1} + \dots + X_{im+m}),$$
$$x_i^{(m)} \triangleq (x_{im+1} + \dots + x_{im+m}).$$

We have

$$\mathbf{E}X_i^{(m)} = \mu, \quad \text{Var } X_i^{(m)} = \frac{1}{m^2} \text{Var } x_i^{(m)}$$

and autocorrelation function $r^{(m)}(k)$ of the processes $X_i^{(m)}, x_i^{(m)}$:

$$r^{(m)} = \mathbf{E}\{(X_i^{(m)} - \mu)(X_{i+k}^{(m)} - \mu)\}/\text{Var } X_i^{(m)} = \mathbf{E}\{x_i^{(m)} x_{i+k}^{(m)}\}/\text{Var } x_i^{(m)},$$
$$k = 0, \pm 1, \pm 2, \dots.$$

First we recall some well-known relations between the autocorrelation function and the variance of the sum [2, 16] (see also Chapter 2). We have

$$\begin{aligned} m^2 \text{ Var } X_0^{(m)} &= \text{Var } x_0^{(m)} = \mathbf{E}(x_1 + x_2 + \dots + x_m)(x_1 + x_2 + \dots + x_m) \\ &= \sigma^2 \sum_{i=1}^{m} \sum_{j=1}^{m} r(i-j) = \sigma^2 \left(m + 2 \sum_{i=1}^{m-1} (m-i) r(i) \right), \end{aligned} \tag{8.1}$$

where we use $r(-k) = r(k)$, $k = 0, 1, 2, \dots$. From the above equations we have

$$\text{Var } x_0^{(m+1)} - \text{Var } x_0^{(m)} = \sigma^2 + 2\sigma^2 \sum_{i=1}^{m} r(i).$$

Next,

$$(\text{Var } x_0^{(m+1)} - \text{Var } x_0^{(m)}) - (\text{Var } x_0^{(m)} - \text{Var } x_0^{(m-1)}) = 2\sigma^2 r(m).$$

If we define $\text{Var } x_i^{(0)} = 0$, then

$$r(m) = \frac{1}{2\sigma^2}[\text{Var } x_0^{(m+1)} - 2 \text{ Var } x_0^{(m)} + \text{Var } x_0^{(m-1)}]. \tag{8.2}$$

The same equation for $r^{(k)}(m)$ is

$$r^{(k)}(m) = \frac{1}{2\,\mathrm{Var}\, x_0^{(k)}}[\mathrm{Var}\, x_0^{(k(m+1))} - 2\,\mathrm{Var}\, x_0^{(km)} + \mathrm{Var}\, x_0^{(k(m-1))}]. \tag{8.3}$$

One more important statistical characteristic of a stationary process is its spectral density function $f(\omega)$. For processes with a continuous spectrum we have

$$\begin{aligned} f(\omega) &\triangleq \frac{\sigma^2}{2\pi}\sum_{k=-\infty}^{\infty} r(k)e^{-\mathrm{i}k\omega}, \\ r(k) &= \frac{1}{\sigma^2}\int_{-\pi}^{\pi} f(\omega)e^{\mathrm{i}k\omega}\,d\omega. \end{aligned} \tag{8.4}$$

Let us express $\mathrm{Var}\, x_0^{(m)}$ via the spectral density function $f(\omega)$. Substituting Eq. (8.4) into (8.1) we have

$$\begin{aligned} \mathrm{Var}\, x_0^{(m)} &= \sum_{k=1}^{m}\sum_{j=1}^{m}\int_{-\pi}^{\pi} f(\omega)e^{\mathrm{i}\omega(k-j)}\,d\omega = \int_{-\pi}^{\pi} f(\omega)\sum_{k=1}^{m}\sum_{j=1}^{m} e^{\mathrm{i}\omega(k-j)}\,d\omega \\ &= \int_{-\pi}^{\pi}\left|\frac{1-e^{\mathrm{i}\omega m}}{1-e^{\mathrm{i}\omega}}\right|^2 f(\omega)\,d\omega = \int_{-\pi}^{\pi}\frac{1-\cos\omega m}{1-\cos\omega} f(\omega)\,d\omega. \end{aligned} \tag{8.5}$$

Now we can give the definition of a second-order self-similar process.

Definition 8.2.1. A stationary process $X = (\ldots, X_{-1}, X_0, X_1, \ldots)$ with finite mean $\mu = \mathbf{E}X_i < \infty$ and variance $\sigma^2 = \mathrm{Var}\, X_i < \infty$ is called exactly second-order self-similar with parameter $0 < \beta < 1$, if

$$r(k) = \tfrac{1}{2}[(k+1)^{2-\beta} - 2k^{2-\beta} + (k-1)^{2-\beta}] \tag{8.6}$$

for all $k = 1, 2, \ldots$.

Parameter β and the *Hurst* parameter H are related as $H = 1 - \beta/2$, $\frac{1}{2} < H < 1$.

Let us discuss the given definition. Substituting Eq. (8.6) into (8.1), it is easy to see that for a self-similar process

$$\mathrm{Var}\, x_0^{(m)} = \sigma^2 m^{2-\beta}, \quad m = 1, 2, \ldots. \tag{8.7}$$

From Eq. (8.2) we can easily conclude that if $\mathrm{Var}\, x_0^{(m)} = \sigma^2 m^{2-\beta}$, then the autocorrelation function $r(k)$ satisfies Eq. (8.6). This means that Eq. (8.7) is equivalent to Eq. (8.6) and can be used in the above definition instead of Eq. (8.6).

Substituting Eq. (8.7) into (8.3) we get

$$r^{(k)}(m) = r(m), \quad m = \pm 1, \pm 2, \ldots, \quad k = 1, 2, \ldots. \tag{8.8}$$

Equation (8.8) can clarify the sense of the definition of the self-similar process: the original process X and its aggregated process $X^{(m)}$ have the same correlation structure.

The spectral density function of the self-similar process was found by Sinai [14]:

$$f(\omega) = \frac{1}{2}\int_{-\pi}^{\pi} (1 - \cos(\omega n)) \sum_{k=-\infty}^{\infty} \frac{1}{|\omega/2\pi + k|^{2-\beta}}\, d\omega \tag{8.9}$$

It is easy to see from Eq. (8.5) that, if process X has spectral density function (8.9), its variance will be equal to (8.7). As $\omega \to 0$, from Eq. (8.9) we obtain $f(\omega) \sim \text{const} \cdot \omega^{1-\beta}$. We use $f(x) \sim g(x)$ in the sense that $\lim_{x\to\infty} (f(x)/g(x)) = 1$.

Now we conclude that an exactly second-order self-similar process can be defined via its autocorrelation function (8.6), variance of the sum (8.7), or spectral density function (8.9), and all these definitions are equivalent. Meanwhile, for the definition of an asymptotically second-order self-similar process, it is important to define the kind of characteristic to be used. We will use the following definition.

Definition 8.2.2. A stationary process $X = (\ldots, X_{-1}, X_0, X_1, \ldots)$ with finite mean $\mu = \mathbf{E}X_i < \infty$ and variance $\sigma^2 = \text{Var}\, X_i < \infty$ is called asymptotically second-order self-similar with parameter $0 < \beta < 1$, if

$$\lim_{m\to\infty} r^{(m)}(k) = \tfrac{1}{2}[(k+1)^{2-\beta} - 2k^{2-\beta} + (k-1)^{2-\beta}] \tag{8.10}$$

for all $k = 1, 2, \ldots$.

The sense of this definition is that, for sufficiently large m, the processes $X^{(m)}$ will have the same autocorrelation function equal to (8.6).

As we can see from Eqs. (8.1) and (8.5), $\text{Var}\, X_0^{(m)}$ is a double integral of the autocorrelation function or an integral of the spectral density function. These relations can clarify the behavior of $r(k)$, $\text{Var}\, X_0^{(m)}$, and $f(\omega)$ obtained from the real traffic measurements. Usually self-similarity of the data traffic is established by analyzing the traffic variance $\text{Var}\, X_0^{(m)}$ [5, 12], but the behavior of the variance $\text{Var}\, X_0^{(m)}$ can be close to a self-similar one, while the behavior of the autocorrelation can be quite different from the theoretical one since the autocorrelation function is a second derivative of the variance $\text{Var}\, X_0^{(m)}$. It is also important that, as can be seen from Eq. (8.5), even small harmonics at low frequency can have a dramatic influence on the behavior of $\text{Var}\, X_0^{(m)}$ in a sufficient region of m values.

8.3 MODEL OF SELF-SIMILAR TRAFFIC

Consider a Poission process on the time axis with intensity λ. Let

$$\theta = (\ldots, \theta_{-1}, \theta_0, \theta_1, \ldots),$$

where θ_t is the number of Poisson points in the interval $[t, t+1)$, $t = 0, \pm 1, \pm 2, \ldots$. The random variables θ_t are independent and identically distributed with $\Pr\{\theta_t = k\} = (\lambda^k / k!)e^{-\lambda}$.

Suppose that θ_t is the number of new active sources arriving to the system at moment t. For each source we assign a random variable $\tau_{t,i}$—the length of the active period, $\tau_{t,i} \in \{1, 2, \ldots\}$, $t = 0, \pm 1, \pm 2, \ldots, i = 1, 2, \ldots$—and a random process $\psi_{t,i}(n)$—the rate of cell generation during the active period at the moment n from the beginning of the period, $\psi_{t,i}(n) \in R_+$, $n = 0, 1, 2, \ldots$. The random couples $(\tau_{t,i}, \psi_{t,i}(n))$ are independent and identically distributed and also independent of the process θ. We define our process $Y = (\ldots, Y_{-1}, Y_0, Y_1, \ldots)$ as

$$Y_t \triangleq \sum_{k=-\infty}^{t} \sum_{j=1}^{\theta_k} \psi_{k,j}(t-k) I(\{\tau_{k,j} > t-k\}), \tag{8.11}$$

where $I(A)$ is an indicator function of the event,

$$I(\{\tau_{k,j} > t-k\}) = \begin{cases} 1, & \text{if } \tau_{k,j} > t-k, \\ 0, & \text{otherwise.} \end{cases}$$

Here and later we assume that the sum $\sum_{i=j}^{k}$ equals zero (independent of its argument) if $k < j$.

The random variable Y_t is the total cell generation rate from all active sources at time t. If $\mathbf{E}Y_t < \infty$, the process Y is stationary and ergodic.

Process Y can also be obtained as a sum of the processes of M independent sources, where each source can be in only the active or passive state. In the passive state, a source does not generate cells, while in the active state it generates cells with rate $\psi_{t,i}(n)$. Suppose that the length of the passive period u is an arbitrary random variable with mean $\bar{u}$, and the length of active periods is $\tau_{t,i}$. If we now consider $M \to \infty$, so that $M/(\bar{u} + \mathbf{E}\tau_{t,i}) \to \lambda$ and $\Pr\{u < M\} \to 0$, then the obtained process will be statistically the same as the process Y.

Let us calculate the mean, variance, and autocorrelation function of the process Y.

From Eq. (8.11), considering that the variance of the sum of independent random variables is equal to the sum of variance, we have

$$\sigma^2 = \text{Var } Y_t = \sum_{k=-\infty}^{t} \text{Var}\left[\sum_{j=1}^{\theta_k} \psi_{k,j}(t-k) I(\{\tau_{k,j} > t-k\})\right].$$

Replacing k with $i = t - k + 1$ and taking into account that $(\tau_{t,i}, \psi_{t,i}(n))$ are identically distributed, we can write

$$\text{Var } Y_t = \sum_{i=1}^{\infty} \text{Var}\left[\sum_{j=1}^{\theta_0} \psi_{0,j}(i-1) I(\{\tau_{0,j} > i-1\})\right]. \tag{8.12}$$

Denote

$$\kappa_{i,j} \triangleq \psi_{0,j}(i-1) I(\{\tau_{0,j} > i-1\}), \quad i = 1, 2, \ldots, \quad j = 1, 2, \ldots.$$

We have

$$S \triangleq \mathbf{E}\left[\sum_{j=1}^{\theta_0} \kappa_{i,j}\right] = \lambda \mathbf{E}\kappa_{i,1}.$$

Let us consider

$$\begin{aligned}
\operatorname{Var}&\left[\sum_{j=1}^{\theta_0} \psi_{0,j}(i-1) I(\{\tau_{0,j} > i-1\})\right] = \operatorname{Var}\left[\sum_{j=1}^{\theta_0} \kappa_{i,j}\right] \\
&= \mathbf{E}\left[\sum_{j=1}^{\theta_0} \kappa_{i,j} - S\right]^2 = \sum_{m=0}^{\infty} \Pr\{\theta_0 = m\} \mathbf{E}\left[\left(\sum_{j=1}^{\theta_0} \kappa_{i,j} - S\right)^2 |\theta_0 = m\right] \\
&= \sum_{m=0}^{\infty} \Pr\{\theta_0 = m\} \mathbf{E}\left(\sum_{j=1}^{m} \kappa_{i,j} - S\right)^2 \\
&= \sum_{m=0}^{\infty} \Pr\{\theta_0 = m\} \left[\mathbf{E}\left(\sum_{j=1}^{m} \kappa_{i,j}\right)^2 - 2Sm\mathbf{E}\kappa_{i,1} + S^2\right] \\
&= \sum_{m=0}^{\infty} \Pr\{\theta_0 = m\}[m \operatorname{Var} \kappa_{i,1} + m^2(\mathbf{E}\kappa_{i,1})^2 - 2Sm\mathbf{E}\kappa_{i,1} + S^2] \\
&= \lambda \operatorname{Var} \kappa_{i,1} + \lambda(\mathbf{E}\kappa_{i,1})^2 = \lambda \mathbf{E}\kappa_{i,1}^2,
\end{aligned} \tag{8.13}$$

where $\Pr\{\theta_0 = m\}$ is a Poisson distribution.

From Eqs. (8.12) and (8.13) we have

$$\sigma^2 = \operatorname{Var} Y_t = \lambda \sum_{i=1}^{\infty} \mathbf{E}\kappa_{i,1}^2 = \lambda \sum_{i=1}^{\infty} \mathbf{E}[\psi_{0,1}^2(i-1) I(\{\tau_{0,1} > i-1\})].$$

Now we will calculate the autocorrelation function $r(k)$. For any given t and k define the following random variables:

$$\begin{aligned}
\eta_{1,i} &\triangleq \sum_{j=1}^{\theta_i} \psi_{i,j}(t-1) I(\{\tau_{i,j} > t+k-i\}), \\
\eta_{2,i} &\triangleq \sum_{j=1}^{\theta_i} \psi_{i,j}(t+k-i) I(\{\tau_{i,j} > t+k-i\}), \\
\zeta &\triangleq \sum_{i=-\infty}^{t} \sum_{j=1}^{\theta_i} \psi_{i,j}(t-i) I(\{t+k-i \geq \tau_{i,j} > t-i\}), \\
\eta_1 &\triangleq \sum_{i=-\infty}^{t} \eta_{1,i}, \quad \eta_2 \triangleq \sum_{i=-\infty}^{t} \eta_{2,i}, \quad \chi \triangleq \sum_{i=t+1}^{t+k} \eta_{2,i}.
\end{aligned} \tag{8.14}$$

We have

$$Y_t = \zeta + \eta_1, \quad Y_{t+k} = \chi + \eta_2. \tag{8.15}$$

Since χ, ζ are independent random variables and also independent from η_1, η_2,

$$r(k) = \frac{1}{\sigma^2}\mathbf{E}[(Y_t - \mathbf{E}Y_t)(Y_{t+k} - \mathbf{E}Y_{t+k})] = \frac{1}{\sigma^2}\mathbf{E}[(\eta_1 - \mathbf{E}\eta_1)(\eta_2 - \mathbf{E}\eta_2)].$$

Using independence of $\eta_{1,j}, \eta_{2,i}$ for $i \neq j$ we have

$$\begin{aligned}
\sigma^2 r(k) &= \mathbf{E}[(\eta_1 - \mathbf{E}\eta_1)(\eta_2 - \mathbf{E}\eta_2)] \\
&= \mathbf{E}\left[\left(\sum_{i=-\infty}^{t}(\eta_{1,i} - \mathbf{E}\eta_{1,i})\right)\left(\sum_{i=-\infty}^{t}(\eta_{2,i} - \mathbf{E}\eta_{2,i})\right)\right] \\
&= \sum_{i=-\infty}^{t}\mathbf{E}[(\eta_{1,i} - \mathbf{E}\eta_{1,i})(\eta_{2,i} - \mathbf{E}\eta_{2,i})].
\end{aligned} \tag{8.16}$$

Consider the term $\mathbf{E}[(\eta_{1,i} - \mathbf{E}\eta_{1,i})(\eta_{2,i} - \mathbf{E}\eta_{2,i})]$. For any given i define random variables

$$\begin{aligned}
\vartheta_{1,j} &\triangleq \psi_{i,j}(t-i)I(\{\tau_{i,j} > t+k-i\}), \\
\vartheta_{2,j} &\triangleq \psi_{i,j}(t+k-i)I(\{\tau_{i,j} > t+k-i\}).
\end{aligned}$$

We have

$$\begin{aligned}
&\mathbf{E}[(\eta_{1,i} - \mathbf{E}\eta_{1,i})(\eta_{2,i} - \mathbf{E}\eta_{2,i})] \\
&\quad = \sum_{m=0}^{\infty}\Pr\{\theta_0 = m\}\mathbf{E}[(\eta_{1,i} - \mathbf{E}\eta_{1,i})(\eta_{2,i} - \mathbf{E}\eta_{2,i})|\theta_0 = m] \\
&\quad = \sum_{m=0}^{\infty}\Pr\{\theta_0 = m\}\mathbf{E}\left[\left(\sum_{j=1}^{m}\vartheta_{1,j} - \mathbf{E}\eta_{1,i}\right)\left(\sum_{j=1}^{m}\vartheta_{2,j} - \mathbf{E}\eta_{2,i}\right)\right] \\
&\quad = \sum_{m=0}^{\infty}\Pr\{\theta_0 = m\}\left[\mathbf{E}\left[\left(\sum_{j=1}^{m}\vartheta_{1,j}\right)\left(\sum_{j=1}^{m}\vartheta_{2,j}\right)\right] - \mathbf{E}\eta_{1,i}\mathbf{E}\eta_{2,i}\right].
\end{aligned} \tag{8.17}$$

Using the fact that couples $(\vartheta_{1,j}, \vartheta_{2,j})$ are independent and identically distributed, we obtain

$$\begin{aligned}
\mathbf{E}\left[\left(\sum_{j=1}^{m}\vartheta_{1,j}\right)\left(\sum_{j=1}^{m}\vartheta_{2,j}\right)\right] &= (m-1)m\mathbf{E}\vartheta_{1,1}\mathbf{E}\vartheta_{2,1} + m\mathbf{E}[\vartheta_{1,1}\vartheta_{2,1}], \\
\mathbf{E}\eta_{1,i} = \lambda\mathbf{E}\vartheta_{1,1}, &\quad \mathbf{E}\eta_{2,i} = \lambda\mathbf{E}\vartheta_{2,1}.
\end{aligned} \tag{8.18}$$

Substituting Eq. (8.18) into (8.17) and taking into account that θ_0 has a Poisson distribution, we get

$$\begin{aligned}\mathbf{E}[(\eta_{1,i} - \mathbf{E}\eta_{1,i})(\eta_{2,i} - \mathbf{E}\eta_{2,i})] \\ &= \sum_{m=0}^{\infty} \Pr\{\theta_0 = m\}[((m-1)m - \lambda^2)\mathbf{E}\vartheta_{1,1}\mathbf{E}\vartheta_{2,1} + m\mathbf{E}[\vartheta_{1,1}\vartheta_{2,1}]] = \lambda\mathbf{E}[\kappa_{1,1}\kappa_{2,1}] \\ &= \lambda\mathbf{E}[\psi_{i,1}(t-i)\psi_{i,1}(t+k-i)I(\{\tau_{i,1} > t+k-i\})].\end{aligned}$$

Finally, using Eq. (8.16), we have

$$\begin{aligned}\sigma^2 r(k) &= \lambda \sum_{i=-\infty}^{t} \mathbf{E}[\psi_{0,1}(t-i)\psi_{0,1}(t+k-i)I(\{\tau_{0,1} > t+k-i\})] \\ &= \lambda \sum_{i=0}^{\infty} \mathbf{E}[\psi_{0,1}(i)\psi_{0,1}(i+k)I(\{\tau_{0,1} > i+k\})]. \end{aligned} \tag{8.19}$$

The mean value of Y_t will be equal to

$$\mathbf{E}Y_t = \lambda \sum_{i=0}^{\infty} \mathbf{E}[\psi_{0,1}(i)I(\{\tau_{0,1} > i\})]. \tag{8.20}$$

A self-similar process Y can be obtained from the following theorem.

Theorem 8.3.1. *Process $Y = (\ldots, Y_{-1}, Y_0, Y_1, \ldots)$ defined by Eq. (8.11) with finite mean $\mu = \mathbf{E}Y_t < \infty$ and variance $\sigma^2 = \mathrm{Var}\, Y_t < \infty$ is exactly second-order self-similar with parameter $0 < \beta < 1$, if*

$$\begin{aligned}\sum_{i=0}^{\infty} \mathbf{E}[\psi_{0,1}(i)\psi_{0,1}(i+k)I(\{\tau_{0,1} > i+k\})] \\ = \frac{\sigma^2}{2\lambda}[(k+1)^{2-\beta} - 2k^{2-\beta} + (k-1)^{2-\beta}]\end{aligned} \tag{8.21}$$

for all $k = 1, 2, \ldots$.

Proof. This theorem directly follows from the definition of a self-similar process (8.6) and the expression for the autocorrelation function (8.19). ■

Corollary 8.3.2. *If a random process $\psi_{k,j}(i)$ is a constant one, $\psi_{k,j}(i) \equiv \psi_{k,j}$ for all $i = 0, 1, 2, \ldots$, then a process $Y = (\ldots, Y_{-1}, Y_0, Y_1, \ldots)$ defined by Eq. (8.11), with finite mean $\mu = \mathbf{E}Y_t < \infty$ and variance $\sigma^2 = \mathrm{Var}\, Y_t < \infty$, will be exactly second-order self-similar with parameter $0 < \beta < 1$, if*

$$\begin{aligned}\Pr\{\tau_{0,1} > k\}\mathbf{E}[\psi_{0,1}^2 | \tau_{0,1} > k] &= -\frac{\sigma^2}{2\lambda}\Delta^3((k-1)^{2-\beta}) \\ &\triangleq -\frac{\sigma^2}{2\lambda}[(k+2)^{2-\beta} - 3(k+1)^{2-\beta} + 3k^{2-\beta} - (k-1)^{2-\beta}]\end{aligned} \tag{8.22}$$

for all $k = 0, 1, 2, \ldots,$ *where, for convenience, we will use that* $(-1)^{2-\beta} \triangleq 1$, $0^{2-\beta} \triangleq 0$.

Proof. Since $\psi_{0,1}(i)$ does not depend on i, we can write

$$\mathbf{E}[\psi_{0,1}(i)\psi_{0,1}(i+k)I(\{\tau_{0,1} > i+k\})] = \Pr\{\tau_{0,1} > k\}\mathbf{E}[\psi_{0,1}^2 | \tau_{0,1} > k].$$

Substituting this into Eq. (8.21) and subtracting Eqs. (8.21) for k and $k+1$, we obtain Eq. (8.22). ■

Equation (8.22) can also be written in the following form:

$$\begin{aligned}\Pr\{\tau_{0,1} = k\}\mathbf{E}[\psi_{0,1}^2 | \tau_{0,1} = k] &= \frac{\sigma^2}{2\lambda}\Delta^4((k-2)^{2-\beta}) \\ &\triangleq \frac{\sigma^2}{2\lambda}[(k+2)^{2-\beta} - 4(k+1)^{2-\beta} + 6k^{2-\beta} - 4(k-1)^{2-\beta} + (k-2)^{2-\beta}],\end{aligned} \tag{8.23}$$

$k = 1, 2, \ldots.$

Corollary 8.3.3. *If a random process* $\psi_{k,j}(i)$ *is independent of* $\tau_{k,j}$ *and stationary, then a process* $Y = (\ldots, Y_{-1}, Y_0, Y_1, \ldots)$ *defined by Eq. (8.11), with finite mean* $\mu = \mathbf{E}Y_t < \infty$ *and variance* $\sigma^2 = \mathrm{Var}\ Y_t < \infty$, *will be exactly second-order self-similar with parameter* $0 < \beta < 1$, *if*

$$(\sigma_\psi^2 r_\psi(k) + \mu_\psi^2)\ \Pr\{\tau_{0,1} > k\} = -\frac{\sigma^2}{2\lambda}\Delta^3((k-1)^{2-\beta}) \tag{8.24}$$

for all $k = 0, 1, 2, \ldots,$ *where* $\sigma_\psi^2 r_\psi(k) \triangleq \mathbf{E}[(\psi_{0,1}(i) - \mu_\psi)(\psi_{0,1}(i+k) - \mu_\psi)]$, $\mu_\psi = \mathbf{E}\psi_{0,1}(i)$.

Proof. Since $\psi_{0,1}(i)$ is a stationary process and does not depend on $\tau_{0,1}$, we have

$$\begin{aligned}\mathbf{E}[\psi_{0,1}(i)\psi_{0,1}(i+k)I(\{\tau_{0,1} > i+k\})] &= \mathbf{E}[\psi_{0,1}(i)\psi_{0,1}(i+k)]\ \Pr\{\tau_{0,1} > i+k) \\ &= (\sigma_\psi^2 r_\psi(k) + \mu_\psi^2)\ \Pr\{\tau_{0,1} > i+k\}.\end{aligned}$$

Substituting this into Eq. (8.21) and subtracting Eq. (8.21) for k and $k+1$, we obtain Eq. (8.24). ■

Equation (8.24) can also be rewritten in the following form:

$$(\sigma_\psi^2 r_\psi(k) + \mu_\psi^2)\ \Pr\{\tau_{0,1} = k\} = \frac{\sigma^2}{2\lambda}\Delta^4((k-2)^{2-\beta}), \tag{8.25}$$

$k = 1, 2, \ldots.$

In the case where the random process $\psi_{k,j}(i) \equiv 1$, equations similar to (8.25) were found by Cox [2].

Now we give some results for an asymptotically self-similar process Y.

Theorem 8.3.4. *Process* $Y = (\ldots, Y_{-1}, Y_0, Y_1, \ldots)$ *defined by Eq. (8.11), with finite mean* $\mu = \mathbf{E}Y_t < \infty$ *and variance* $\sigma^2 = \operatorname{Var} Y_t < \infty$ *is asymptotically second-order self-similar with parameter* $0 < \beta < 1$, *if*

$$r(k) = \sum_{i=0}^{\infty} \mathbf{E}[\psi_{0,1}(i)\psi_{0,1}(i+k)I(\{\tau_{0,1} > i+k\})] \sim \text{const} \cdot k^{-\beta} \tag{8.26}$$

as $k \to \infty$.

Proof. Since $r(k) \sim \text{const} \cdot k^{-\beta}$, as $k \to \infty$, from Eq. (8.1) we have

$$\begin{aligned} m^2 \operatorname{Var} Y_0^{(m)} &= \sigma^2\left(m + 2\sum_{i=1}^{m-1}(m-i)r(i)\right) \sim 2c_1\sigma^2\left(\frac{1}{1-\beta} - \frac{1}{2-\beta}\right)m^{2-\beta} \\ &= c_2 m^{2-\beta}, \quad \text{as } m \to \infty, \end{aligned} \tag{8.27}$$

where c_1 and c_2 are some positive constants and $0 < \beta < 1$.

Substituting Eq. (8.27) into (8.3) we get

$$\begin{aligned} r^{(m)}(k) &\sim \frac{1}{2c_2 m^{2-\beta}}[c_2(m(k+1))^{2-\beta} - 2c_2(mk)^{2-\beta} + c_2(m(k-1))^{2-\beta}] \\ &= \tfrac{1}{2}[(k+1)^{2-\beta} - 2k^{2-\beta} + (k-1)^{2-\beta}], \quad \text{as } m \to \infty. \end{aligned}$$

It means, according to the definition (8.10), that process Y is asymptotically second-order self-similar with parameter $0 < \beta < 1$. ■

Corollary 8.3.5. *If a random process* $\psi_{k,j}(i)$ *is a constant one,* $\psi_{k,j}(i) \equiv \psi_{k,j}$ *for all* $i = 0, 1, 2, \ldots,$ *then a process* $Y = (\ldots, Y_{-1}, Y_0, Y_1, \ldots)$ *defined by Eq. (8.11), with finite mean* $\mu = \mathbf{E}Y_t < \infty$ *and variance* $\sigma^2 = \operatorname{Var} Y_t < \infty$, *will be asymptotically second-order self-similar with parameter* $0 < \beta < 1$, *if*

$$\Pr\{\tau_{0,1} > k\}\mathbf{E}[\psi_{0,1}^2 | \tau_{0,1}] > k \sim \text{const} \cdot k^{-1-\beta}, \quad \text{as } k \to \infty, \tag{8.28}$$

or

$$\Pr\{\tau_{0,1} = k\}\mathbf{E}[\psi_{0,1}^2 | \tau_{0,1} = k] \sim \text{const} \cdot k^{-2-\beta}, \quad \text{as } k \to \infty. \tag{8.29}$$

Proof. Since $\psi_{0,1}(i)$ does not depend on i, from Eq. (8.19) we have

$$r(k) = \frac{\lambda}{\sigma^2}\sum_{i=0}^{\infty} \Pr\{\tau_{0,1} > k\}\mathbf{E}[\psi_{0,1}^2 | \tau_{0,1} > k].$$

Substituting Eq. (8.28) in the above equation, we obtain

$$r(k) \sim \text{const} \cdot \sum_{i=0}^{\infty}(i+k)^{-1-\beta} \sim \text{const} \cdot k^{-\beta}, \quad \text{as } k \to \infty.$$

Then from Theorem 8.3.4 it immediately follows that Y is an asymptotically second-order self-similar process.

Statement (8.29) can be proved in the same way. ■

8.4 ASYMPTOTICAL BOUNDS FOR BUFFER OVERFLOW PROBABILITY

In this section we consider the process Y defined by Eq. (8.11) as the input traffic of a single server queueing system with constant server rate equal to C and infinite buffer size. Suppose process Y has finite mean $\mu = \mathbf{E}Y_t < C < \infty$ and finite variance $\sigma^2 = \text{Var } Y_t < \infty$. We will consider the particular form of the process Y. Namely, we consider the case when the random process $\psi_{i,j}(t) \equiv 1$. Let

$$\Pr\{\tau_{0,1} = i\} \sim c_0 i^{-2-\beta} \tag{8.30}$$

as $i \to \infty$, with $0 < \beta < 1$. Then, according to Eq. (8.29) with $\psi_{0,1} \equiv 1$, process Y will be asymptotically second-order self-similar with Hurst parameter $H = 1 - \beta/2$.

Now we are interested in the queue length behavior. Let n_t be the length of the queue at the moment t. Then we have

$$n_t = \max(0, n_{t-1} + Y_t - C). \tag{8.31}$$

We will estimate the probability $\Pr\{n_t > z\}$, that is, the stationary probability to find, at moment t, the length of the queue bigger than z, for large value of z.

For any given z, let us split the process Y_t into two processes $Y_t^{(1)}$ and $Y_t^{(2)}$, that is,

$$Y_t = Y_t^{(1)} + Y_t^{(2)},$$

according to the following rule:

$$Y_t^{(1)} \triangleq \sum_{k=-\infty}^{t} \sum_{j=1}^{\theta_k} I(\{\tau_{k,j} > t-k\}) I(\{\tau_{k,j} \leq \epsilon z\}), \tag{8.32}$$

$$Y_t^{(2)} \triangleq \sum_{k=-\infty}^{t} \sum_{j=1}^{\theta_k} I(\{\tau_{k,j} > t-k\}) I(\{\tau_{k,j} > \epsilon z\}), \tag{8.33}$$

where ϵ is some positive constant, and θ_k was defined at the beginning of Section 8.3. As $z \to \infty$, $\mathbf{E}Y_t^{(2)} \to 0$; nevertheless, the process $Y_t^{(2)}$ will produce the main contribution to the probability $\Pr\{n_t > z\}$ for large values of z, while process $Y_t^{(1)}$ will give an average load to the queueing system.

First, we derive an upper bound for the probability $\Pr\{n_t > z\}$.

Define

$$S_k \triangleq \sum_{i=0}^{k-1} Y_{t-i}, \quad S_k^{(j)} \triangleq \sum_{i=0}^{k-1} Y_{t-i}^{(j)}, \quad j = 1, 2. \tag{8.34}$$

We have

$$\Pr\{n_t > z\} = \Pr\left\{\max_{k\geq 1}\{S_k - kC\} > z\right\}.$$

For any $0 < \delta_1 < C - \mu$, $0 < \delta_2 < 1$, the following inequality is true:

$$\Pr\left\{\max_{k\geq 1}\{S_k - kC\} > z\right\} \leq \Pr\left\{\max_{k\geq 1}\{S_k^{(1)} - (\mu+\delta_1)k\} > \delta_2 z\right\} + \Pr\left\{\max_{k\geq 1}\{S_k^{(2)} - (C-\mu-\delta_1)k\} > z(1-\delta_2)\right\}. \tag{8.35}$$

Let us first estimate probability $\Pr\{\max_{k\geq 1}\{S_k^{(1)} - (\mu+\delta_1)k\} > \delta_2 z\}$. We present the result in the form of a lemma.

Lemma 8.4.1. *Let $Y_t^{(1)}$ be a process defined by Eq. (8.32) with*

$$\Pr\{\tau_{0,1} = i\} \sim c_0 i^{-2-\beta}, \quad \text{as } i \to \infty,\ 0 < \beta < 1.$$

Then for any given, $\delta_1 > 0$, $\delta_2 > 0$, $c_1 > 1$, $0 < \epsilon < \min\{\delta_2/(\mu+\delta_1), \beta/c_1\}$, and sufficiently large z,

$$\Pr\left\{\max_{k\geq 1}\{S_k^{(1)} - (\mu+\delta_1)k\} > \delta_2 z\right\} \leq z^{1-c_1\tilde{\delta}_2},$$

where $\tilde{\delta}_2 \triangleq \delta_2 - (\mu+\delta_1)\epsilon$, $\mu = \mathbf{E}Y_t$.

Proof. Let us estimate the probability

$$\Pr\{S_k^{(1)} - (\mu + \delta_1)k > \delta_2 z\}.$$

First, we have

$$S_k^{(1)} \leq \sum_{j=t-\tilde{k}}^{t} q_j, \tag{8.36}$$

where

$$q_j \triangleq \sum_{i=1}^{\theta_j} \tau_{j,i} I(\tau_{j,i} \leq \epsilon z), \quad \tilde{k} \triangleq k + \epsilon z.$$

Random variables q_j are independent and identically distributed. Denote via $\phi(s)$ the logarithm of the moment-generating function of the sum (8.36). Clearly,

$$\phi(s) \triangleq \ln\left[\sum_x e^{sx} \Pr\left\{\sum_{j=t-\tilde{k}}^{t} q_j = x\right\}\right] = \tilde{k} \sum_{j=1}^{\epsilon z} \lambda_j (e^{sj} - 1), \tag{8.37}$$

where the sum on x is taken over all possible values of the sum (8.36),

$$\lambda_j \triangleq \lambda \Pr\{\tau_{0,1} = j\}.$$

Since $\phi(s) < \infty$ for all $\infty > s > 0$, it can be shown that

$$\Pr\left\{\sum_{j=t-\tilde{k}}^{t} q_j > y\right\} \leq e^{-sy+\phi(s)}$$

for any given $s > 0$, $y > 0$. Thus

$$\Pr\{S_k^{(1)} > \delta_2 z + k(\mu + \delta_1)\} \leq P_u \triangleq \exp\{-s(\delta_2 z + k(\mu + \delta_1)) + \phi(s)\}.$$

Substituting expression (8.37) and using definitions of $\tilde{k}$ and $\tilde{\delta}_2$, we have

$$\begin{aligned} P_u &= \exp\left\{\tilde{k} \sum_{j=1}^{\epsilon z} \lambda_j (e^{sj} - 1) - s(\tilde{\delta}_2 z + \tilde{k}(\mu + \delta_1))\right\} \\ &= \exp\left\{\tilde{k} \sum_{j=1}^{\epsilon z} \lambda_j (e^{sj} - 1 - sj) - s\tilde{\delta}_2 z - s\tilde{k}\delta_1 + \tilde{k}s\left(\sum_{j=1}^{\epsilon z} \lambda_j j - \mu\right)\right\}. \end{aligned}$$

As we can see from Eq. (8.20) $\mu = \sum_{j=1}^{\infty} \lambda_j j$, so that $\sum_{j=1}^{\epsilon z} \lambda_j j - \mu < 0$ and, hence,

$$P_u \le \exp\left\{\tilde{k} \sum_{j=1}^{\epsilon z} \lambda_j (e^{sj} - 1 - sj) - s\tilde{\delta}_2 z - s\tilde{k}\delta_1\right\}. \tag{8.38}$$

Since $s_j > 0$ we can write

$$\sum_{j=1}^{\epsilon z} \lambda_j (e^{sj} - 1 - sj) \le \sum_{j=1}^{\epsilon z} \lambda_j (sj)^2 e^{sj} \le \sum\nolimits_1 + \sum\nolimits_2,$$

where

$$\sum\nolimits_1 \triangleq \sum_{j=1}^{\epsilon z} \lambda_j (sj)^2, \quad \sum\nolimits_2 \triangleq \sum_{j=1}^{\epsilon z} \lambda_j (sj)^3 \epsilon^{sj}.$$

In further consideration we put $s = c_1(\ln z)/z$. As we can see from Eq. (8.30), there is a constant $c_2 < \infty$ such that $\lambda_j \le c_2 j^{-2-\beta}$ for any $j > 0$. Using this fact we have, for sufficiently large z,

$$\sum\nolimits_1 \le s^2 c_2 \sum_{j=1}^{\epsilon z} j^{-\beta} \le s^2 c_2 \int_{j=1}^{\epsilon z} x^{-\beta}\, dx \le \text{const} \cdot z^{-1-\beta} \ln^2 z.$$

We used here that $j^{-\beta}$ is monotonically decreasing function, when we replaced the sum by an integral.

In the same way, for $\sum_2$ we have

$$\sum\nolimits_2 \le s^3 c_2 \sum_{j=1}^{\epsilon z} j^{1-\beta} e^{sj} \le s^3 c_2 (\epsilon z)^{1-\beta} \sum_{j=1}^{\epsilon z} e^{sj} \le s^2 c_2 (\epsilon z)^{1-\beta} e^{s\epsilon z}.$$

Substituting $s = c_1(\ln z)/z$, we obtain

$$\sum\nolimits_2 \le \text{const} \cdot z^{-1-\beta+\epsilon c_1} \ln^2 z.$$

Since, for sufficiently large z, the bound on $\sum_2$ will dominate compared to the bound on $\sum_1$, we get for $s = c_1(\ln z)/z$,

$$\begin{aligned} &\tilde{k} \sum_{j=1}^{\epsilon z} \lambda_j (e^{sj} - 1 - sj) - s\tilde{\delta}_2 z - s\tilde{k}\delta_1 \\ &\quad \le k(z^{-1-\beta+\epsilon c_1} \ln^2 z - \delta_1 c_1 z^{-1} \ln z) - c_1 \tilde{\delta}_2 \ln z - \delta_1 c_1 \epsilon \ln z \\ &\quad \le -kz^{-1} - \tilde{\delta}_2 c_1 \ln z, \end{aligned}$$

where we used that $\epsilon c_1 < \beta$. Substituting this in Eq. (8.38) we have for sufficiently large z

$$P_u \leq e^{-k/z} z^{-\tilde{\delta}_2 c_1}.$$

Finally, we get

$$\Pr\left\{\max_{k\geq 1}\{S_k^{(1)} - (\mu+\delta_1)k\} > \delta_2 z\right\} \leq \sum_{k=1}^{\infty} \Pr\{S_k^{(1)} - (\mu+\delta_1)k > \delta_2 z\}$$
$$\leq \sum_{k=1}^{\infty} P_u \leq z^{-\tilde{\delta}_2 c_1} \sum_{k=1}^{\infty} e^{-k/z} \leq z^{1-\tilde{\delta}_2 c_1}. \qquad \blacksquare$$

Now we consider the process $Y_t^{(2)}$. We have

$$\Pr\left\{\max_{k\geq 1}\{S_k^{(2)} - (C-\mu-\delta_1)k\} > z(1-\delta_2)\right\} = \Pr\{n_t^{(2)} > z(1-\delta_2)\},$$

where

$$n_t^{(2)} = \max\{0, n_{t-1}^{(2)} + Y_t^{(2)} - (C-\mu-\delta_1)\}.$$

Define

$$y_t \triangleq \begin{cases} 1, & \text{if } Y_t^{(2)} = m, \\ 0, & \text{if } Y_t^{(2)} < m, \\ 1 + \sum_{j=1}^{\theta_t} \tau_{1,j} I(\tau_{t,j} > \epsilon z), & \text{if } Y_t^{(2)} > m, \end{cases}$$

$$\bar{n}_t^{(2)} \triangleq \max(0, \bar{n}_{t-1}^{(2)} + y_t - a),$$

where $m = 1 + \lfloor \tilde{C} \rfloor$, $\tilde{C} = C - \mu - \delta_1$, $\lfloor \tilde{C} \rfloor$ is the integer part of $\tilde{C}$, and $a = \tilde{C} - \lfloor \tilde{C} \rfloor$. We will consider only the case when $\tilde{C} \neq \lfloor \tilde{C} \rfloor$, $a > 0$. For any given realization of the process Y, we have $\bar{n}_t^{(2)} \geq n_t^{(2)}$. It means that $\Pr\{n_t^{(2)} > z(1-\delta_2)\} \leq \Pr\{\bar{n}_t^{(2)} > z(1-\delta_2)\}$. For given moment t, define the moments

$$i_1 = \max(i\colon i \leq t, \bar{n}_{i-1}^{(2)} = 0\}, \quad i_2 = \min\{i\colon i \geq t, \bar{n}_t^{(2)} = 0\},$$

and the events

$$B_0 \triangleq \{\bar{n}_t^{(2)} > 0, \theta_{i_1}^{(2)} = 1, \theta_i^{(2)} = 0 \text{ for any } i_1 < i \leq i_2\},$$
$$B_1 \triangleq \{\bar{n}_t^{(2)} > 0, \theta_{i_1}^{(2)} = 1, \theta_i^{(2)} = 0 \text{ for any } i_1 < i \leq t\},$$
$$B_2 \triangleq \{\bar{n}_t^{(2)} > 0\} \backslash B_1.$$

To analyze probability $\Pr\{\bar{n}_t^{(2)} > \tilde{z}\}$, $\tilde{z} \triangleq z(1-\delta_2)$, we can write

$$\Pr\{\bar{n}_t^{(2)} > \tilde{z}\} = \Pr\{B_1 \cup B_2, \bar{n}_t^{(2)} > \tilde{z}\} \le \Pr\{B_2\} + \Pr\{B_1 \backslash B_0\} + \Pr\{\bar{n}_t^{(2)} > \tilde{z}, B_0\}.$$

First, let us estimate probability $\Pr\{B_2\}$. If $\lambda \mathbf{E}[\tau_{0,1} I(\tau_{0,1} > \epsilon z)] < 1$, it can be shown that

$$\begin{aligned} &\Pr\{B_1\} \le \Pr\{Y_t^{(2)} = m\}/(1-a), \\ &\Pr\{B_2\} \le (\Pr\{B_1\} + \Pr\{B_2\})\lambda \mathbf{E}[\tau_{0,1} I(\tau_{0,1} > \epsilon z)]/(1-a). \end{aligned} \tag{8.39}$$

By induction on k, we find

$$\Pr\{Y_t^{(2)} = k\} \le (\lambda \mathbf{E}[\tau_{0,1} I(\tau_{0,1} > \epsilon z)])^k \sim \left(\frac{\lambda c_0}{\beta}\right)^k (\epsilon z)^{-k\beta}$$

as $z \to \infty$. Applying this to Eq. (8.39) we obtain

$$\Pr\{B_1\} \le \text{const} \cdot z^{-\beta m} + o(z^{-\beta m}), \quad \Pr\{B_2\} \le \text{const} \cdot z^{-\beta(m+1)} + o(z^{-\beta(m+1)})$$

Next, we estimate probability

$$\begin{aligned} \Pr\{B_1 \backslash B_0\} &\le \Pr\{Y_1 = m-1\} \sum_{k=\epsilon z}^{\infty} \frac{\lambda}{a} k \Pr\{\tau_{0,1} = k) \\ &\times \Pr\left\{\sum_{j=2}^{k/a} \theta_j^{(2)} \ge 1\right\} \sim \frac{\lambda c_0 \Pr\{Y_1 = m-1\}}{a} \int_{\epsilon z}^{\infty} x^{-1-\beta}(1-e^{-vx})\, dx, \\ &\text{as } z \to \infty, \end{aligned}$$

where

$$v = \lambda \Pr\{\tau_{0,1} \ge \epsilon z\} \sim \frac{\lambda c_0}{(1+\beta)} (\epsilon z)^{-\beta-1}, \quad \text{as } z \to \infty.$$

Calculating this integral we get

$$\begin{aligned} \int_{\epsilon z}^{\infty} x^{-1-\beta}(1-e^{-vx})\, dx &= \int_{\epsilon z}^{(\epsilon z)^{1+\beta}} x^{-1-\beta}(1-e^{-vx})\, dx \\ &+ \int_{(\epsilon z)^{1+\beta}}^{\infty} x^{-1-\beta}(1-e^{-vx})\, dx \le \text{const} \cdot z^{-\beta-\beta^2}, \\ &\text{as } z \to \infty \end{aligned}$$

and, consequently,

$$\Pr\{B_1 \backslash B_0\} \le \text{const} \cdot z^{-\beta m - \beta^2} + o(z^{-\beta m - \beta^2}), \quad \text{as } z \to \infty.$$

Finally, we need to estimate the probability

$$\Pr\{\bar{n}_t^{(2)} > \tilde{z}, B_0\} \le \lambda^m \sum_{i=\tilde{z}/(1-a)}^{\infty} (\mathbf{E}[\tau_{0,1} I(\tau_{0,1} > l)])^{m-1} \Pr\{\tau_{0,1} = l\}$$
$$\times \frac{l(1-a) - \tilde{z}}{a(1-a)} \sim \left(\frac{\lambda c_0}{\beta}\right)^m \frac{\beta}{a(1-a)} \int_{\tilde{z}/(1-a)}^{\infty} x^{-2-\beta m}(x(1-a) - \tilde{z})\, dx$$
$$= \left(\frac{\lambda c_0}{\beta}\right)^m \frac{(1-a)^{\beta m}}{am(1+\beta m)} \tilde{z}^{-\beta m}, \quad \text{as } z \to \infty.$$

From bounds for probabilities $\Pr\{B_2\}$, $\Pr\{B_1 \backslash B_0\}$, and $\Pr\{\bar{n}_t^{(2)} > \tilde{z}, B_0\}$ it follows that probability $\Pr\{\bar{n}_t^{(2)} > \tilde{z}, B_0\}$ will produce the main contribution to $\Pr\{\bar{n}_t^{(2)} > \tilde{z}\}$. Taking this into account, using Eq. (8.35), and applying Lemma 8.4.1 with arbitrarily small $\delta_1 > 0$ and $\delta_2 > 0$, and $c_1 = 2(2+\beta m)/\delta_2$ and $\epsilon = \frac{1}{2}\min[\delta_2/2(\mu+\delta_1), \beta/c_1]$, we can get the upper bound for the probability $\Pr\{n_t > z\}$. We present this upper bound as well as a lower one in the form of the following theorem.

Theorem 8.4.2. *Let Y be an asymptotically second-order self-similar process defined by Eq. (8.11) with Poisson intensity λ, Hurst parameter $H = 1 - \beta/2$, $0 < \beta < 1$, probability $\Pr\{\tau_{0,1} = i\} \sim c_0 i^{-2-\beta}$, as $i \to \infty$, mean $\mu = \mathbf{E}Y_t$, and $\psi_{i,j}(t) \equiv 1$. Then for queue length n_t defined by Eq. (8.31), we have*

$$\Pr\{n_t > z\} \le (1+\epsilon_0)\left(\frac{\lambda c_0}{\beta}\right)^m \frac{(1-a)^{\beta m}}{am(1+\beta m)} z^{-\beta m} + o(z^{-\beta m}),$$
$$\Pr\{n_t > z\} \ge (1-\epsilon_0)\frac{(1-a)^{\beta m}}{(1+\beta)^{m-1}\beta m(1+\beta m)} \left(\frac{\beta m}{2+\beta m}\right)^{\beta m(m-1)/2} \left(\frac{2}{2+\beta m}\right)^{m-1}$$
$$\times (\lambda c_0)^m z^{-\beta m} + o(z^{-\beta m}), \quad \text{as } z \to \infty,$$

where $\epsilon_0 > 0$ is any arbitrary small constant, $m = 1 + \lfloor C - \mu \rfloor$, $a = C - \mu - \lfloor C - \mu \rfloor$, C is a server rate, $C > \mu$, and $C - \mu \ne \lfloor C - \mu \rfloor$.

Proof. We still need to prove only the lower bound for the probability $\Pr\{n_t > z\}$.

First, we can see that

$$\Pr\{n_t > z\} = \Pr\left\{\max_{k \ge 1}\{S_k - kC\} > z\right\} \ge \Pr\left\{\min_{k \ge 1}\{S_k^{(1)} - k(\mu - \delta_1)\} > -z\delta_2\right\}$$
$$\times \Pr\left\{\max_{k \ge 1}\{S_k^{(2)} - k(C + \delta_1 - \mu)\} > z(1+\delta_2)\right\}, \tag{8.40}$$

where S_k was defined by Eq. (8.34), with $\delta_1 > 0$ and $\delta_2 > 0$ being any positive constants. Let us estimate the first probability on the right-hand side of the inequality (8.40). For any integer number $J > 0$, if $k > J$, we have

$$S_k^{(1)} \geq \sum_{i=1}^{\eta_{k_1}} \xi_i, \quad k_1 = k - J,$$

where ξ_i is a random variable with distribution

$$\Pr\{\xi = j\} = \Pr\{\tau_{0,1} = j\}, \quad j \leq J,$$

η_k is a Poisson random variable with mean λk. If J does not depend on z, then Var $\xi_i \leq$ const. Let us take J such that

$$\lambda \mathbf{E}\xi_i > \mu - \delta_1$$

for $z > J$ and $z > \delta_2/J(\mu - \delta_1)$. By the strong law of large numbers

$$\lim_{z\to\infty} \Pr\left\{\min_{k\geq 1}\{S_k^{(1)} - k(\mu - \delta_1)\} > -z\delta_2\right\} = 1. \tag{8.41}$$

Now we will find the second probability in the inequality (8.40). We have

$$\Pr\left\{\max_{k\geq 1}\{S_k^{(2)} - k(C + \delta_1 - \mu)\} > z(1 + \delta_2)\right\} = \Pr\{\underline{n}_t^{(2)} > z(1 + \delta_2)\},$$

where

$$\underline{n}_t^{(2)} \triangleq \max(0, \underline{n}_{t-1}^{(2)} + Y_t^{(2)} - (C + \delta_1 - \mu)).$$

Next,

$$\Pr\{n_t^{(2)} > z(1 + \delta_2)\} \geq \sum_{l=z(1+\delta_2)/(1-a_1)}^{\infty} \left[\prod_{i=1}^{m_1-1} \lambda \Pr\{\tau_{0,1} > l_i\}(\gamma - 1)l_{i-1}\right]$$
$$\times \lambda \Pr\{\tau_{0,1} = l\}(l(1 - a_1) - z(1 + \delta_2))/(1 - a_1),$$

where $l_i = \gamma^i l$, $\gamma > 1$, $m_1 = \lfloor C + \delta_1 - \mu \rfloor + 1$, and $a_1 = C + \delta_1 - \mu - m_1 + 1$. Putting $\gamma = 1 + 2/\beta m_1$ we get

$$\Pr\{n_t^{(2)} > z(1 + \delta_2)\} \geq \frac{(1 - a_1)^{\beta m_1}}{(1 + \beta)^{m_1-1}\beta m_1(1 + \beta m_1)} \left(\frac{\beta m_1}{2 + \beta m_1}\right)^{\beta m_1(m_1-1)/2}$$
$$\times \left(\frac{2}{2 + \beta m_1}\right)^{m_1-1} (\lambda c_0)^{m_1}(z(1 + \delta_2))^{-\beta m_1} + o(z^{-\beta m_1}). \tag{8.42}$$

Taking δ_1 and δ_2 arbitrarily small from Eqs. (8.40), (8.41), and (8.42), we obtain the lower bound for $\Pr\{n_t > z\}$. ■

Lower and upper bounds given by Theorem 8.4.2 have the same exponent $-2(1-H)\lfloor C+1-\mu \rfloor$, which depends only on the Hurst parameter H, server rate C expressed in units of single source rate, and average load of input traffic μ. If $\lfloor C+1-\mu \rfloor < 1/2(1-H)$, random variable n_t will have stationary distribution with infinite mean $\mathbf{E}n_t = \infty$, while for large values of $C-\mu$ the tail of the n_t distribution will decrease sufficiently quickly. The variance of the input process Y does not appear in bounds; meanwhile, for exactly second-order self-similar processes, parameters c_0 and λ will be defined by the mean and variance of the process Y.

The best logarithmic bounds, obtained previously, for the probability $\Pr\{n_t > z\}$ are presented in the papers by Duffield [3] and Liu et al. [10] (Pareto case). In our notation, exponents of these bounds are $-\beta\lfloor C+1-\mu \rfloor$ for the lower bound and $\min[-\beta, 1-\beta(C-\mu]$ for the upper bound. As we can see, Theorem 8.4.2 corresponds to the previous results. Moreover, from Theorem 8.4.2 it follows that the lower bound obtained in Liu et al. [10] is indeed logarithmically exact.

8.5 CONCLUSION

A wide class of discrete-time, second-order self-similar processes has been presented in this chapter. Theorem 8.3.1 gives all self-similar traffic models with independent sources and Poisson arrivals. This class of processes is simple for simulation, by nature close to real network traffic, and suitable for analytical study. In Section 8.4 lower and upper bounds for the probability distribution of queue length with self-similar input traffic were presented. In order to get these bounds, we split input traffic into two processes. The first process represents typical behavior of self-similar traffic, while the second one represents large deviation cases. To analyze queue behavior under the first process, we used the law of large numbers and results similar to the Chernov bound. The second process can be analyzed on the basis of probability computations for rare events. As we discovered, asymptotic behavior of queue length probability is strongly dependent on the ratio of server rate to the average rate of the source. Thus, to achieve good performance of a queueing system driven with self-similar traffic, the average rate of heavy-tail sources should be bounded by a sufficiently small value.

When we get bounds for a probability distribution, we use one simple particular form of the self-similar process from our class. Nevertheless, methods developed in Section 8.4 can be applied to the more complicated processes presented in Section 8.3.

REFERENCES

1. J. Beran, R. Sherman, M. S. Taqqu, and W. Willinger. Long-range dependence in variable bit-rate video traffic. *IEEE Trans. Commun.*, COM-43:1566–1579, 1995.
2. D. R. Cox. Long-range dependence: a review. In H. A. David and H. T. David, eds., *Statistics: An Appraisal*, pp. 55–74. The Iowa State University Press, Ames, 1984.
3. N. G. Duffield. On the relevance of long-tailed durations for the statistical multiplexing of large aggregations. *Proceedings of the 34th Annual Allerton Conference on Communication, Control and Computing*, Oct. 2–4, 1996.
4. V. S. Frost and B. Melamed. Traffic modeling for telecommunications networks. *IEEE Commun. Mag.*, **32**(3):70–81, March 1994.
5. M. W. Garrett and W. Willinger. Analysis, modeling and generation of self-similar VBR video traffic. In *Proc. ACM SIGCOMM'94*, London, pp. 269–280, 1994.
6. P. R. Jelenković and A. A. Lazar. Multiplexing on–off sources with subexponential on periods: Part II. In *Proceedings of the 15th International Teletraffic Congress*, Washington, DC, pp. 965–974, June 1997.
7. W. E. Leland, M. S. Taqqu, W. Willinger, and D. V. Wilson. On the self-similar nature of Ethernet traffic (extended version). *IEEE/ACM Trans. Networking*, **2**(1):1–15, February 1994.
8. W. E. Leland and D. V. Wilson. High time resolution measurement and analysis of LAN traffic: implications for LAN interconnection. In *Proc. IEEE INFOCOM'91*, Bal Harbour, FL, pp. 1360–1366, 1991.
9. N. Likhanov, B. Tsybakov, and N. D. Georganas. Analysis of an ATM buffer with self-similar ("fractal") input traffic. In *Proc. IEEE INFOCOM'95*, Boston, pp. 985–992, April 1995.
10. Z. Liu, Ph. Nain, D. Towsley, and Z.-L. Zhang. Asymptotic behavior of a multiplexer fed by a long-range dependent process. *Appl. Probab.*, **36**(1):105–118, 1998.
11. B. B. Mandelbrot. Self-similar error clusters in communication systems and the concept of conditional stationarity. *IEEE Trans. Commun. Technol.*, **COM-13**:71–90, 1965.
12. I. Norros. On the use of fractional Brownian motion in the theory of connectionless networks. Technical Report TD(94)033, COST242, 1994.
13. M. Parulekar and A. Makowski. Tail probabilities for a multiplexer driven by $M/G/\infty$ input processes (I): preliminary asymptotics. *Queueing Syst. Theory Applic.*, **27**:271–296, 1998.
14. Ya. G. Sinai. Automodel probability distributions. *Teor. Veroytn. Ee Primen.*, **21**(1):63–80, 1976.
15. M. S. Taqqu and J. B. Lévy. Using renewal process to generate long-range dependence and high variability. In E. Eberlein and M. S. Taqqu, eds., *Dependence in Probability and Statistics*, Vol. 11, pp. 73–89. Birkhauser, Boston, 1986.
16. B. Tsybakov and N. D. Georganas. Self-similar processes in communications networks. *IEEE Trans. Inf. Theory*, **44**(5):1713–1725, September 1998.
17. B. Tsybakov and N. D. Georganas. Self-similar traffic and upper bounds to buffer overflow in an ATM queue. *Perf. Eval.*, **32**(1):57–80, February 1998.
18. D. Veitch. Novel models of broadband traffic. In *Proc. IEEE Globecom'93*, Houston, December 1993.

9

BUFFER ASYMPTOTICS FOR $M/G/\infty$ INPUT PROCESSES

ARMAND M. MAKOWSKI AND MINOTHI PARULEKAR

Institute for Systems Research, and Department of Electrical and Computer Engineering, University of Maryland, College Park, MD 20742

9.1 INTRODUCTION

Several recent measurement studies have concluded that classical Poisson-like traffic models do not account for time dependencies observed at multiple time scales in a wide range of networking applications, for example, Ethernet LANs [13, 20, 33], variable bit rate (VBR) traffic [3, 15], Web traffic [6], and WAN traffic [31]. As the resulting temporal correlations are expected to have a significant impact on buffer engineering practices, this "failure of Poisson modeling" has generated an increased interest in a number of alternative traffic models that capture observed (long-range) dependencies [14, 24]. Proposed models include fractional Brownian motion [25] and its discrete-time analog, fractional Gaussian noise [1]. Already both have exposed clearly the limitations of traditional traffic models in predicting storage requirements and devising congestion controls. A discussion of these issues in the case of fractional Brownian motion is summarized in Chapter 4.

In this chapter we focus instead on the class of $M/G/\infty$ input processes as potential traffic models. An $M/G/\infty$ input process is understood as the busy server process of a discrete-time infinite server system fed by a discrete-time Poisson process of rate λ (customers/slot) and with generic service time σ distributed according to G. As argued in Parulekar [27] and Parulekar and Makowski [29], these $M/G/\infty$ input processes constitute a viable alternative to existing traffic models; reasons range from flexibility to tractability.

Self-Similar Network Traffic and Performance Evaluation, Edited by Kihong Park and Walter Willinger
ISBN 0-471-31974-0

First, the relevance of the $M/G/\infty$ input model to network traffic modeling is perhaps best explained through its connection to an attractive model for aggregate packet streams proposed by Likhanov et al. [21]. They show that the combined traffic generated by several independent, identically distributed (i.i.d.) on/off sources with Pareto distributed activity periods behaves in the limit, as the number of sources increases, like the $M/G/\infty$ input stream with a Pareto distributed σ. This provides a rationale for the view that $M/G/\infty$ input processes could provide a natural alternative to existing traffic models, at least for certain multiplexed applications.

Second, the class of $M/G/\infty$ input processes is stable under multiplexing; that is, the superposition of several $M/G/\infty$ processes can be represented by an $M/G/\infty$ input process.

Third, the $M/G/\infty$ model displays great flexibility in capturing positive dependencies over a wide range of time scales; this is achieved very simply through the tail behavior of σ (Proposition 9.4.1). The degree of positive correlation can further be characterized by the sum of the autocovariances, or index of dispersion of counts (IDCs), with the process being short-range dependent (SRD) (i.e., IDC finite) if and only if $\mathbf{E}[\sigma^2]$ is finite (Proposition 9.5.1).

Insights into how temporal correlations of $M/G/\infty$ input processes will affect queueing performance can be gained by analyzing the behavior of a multiplexer fed by an $M/G/\infty$ input process. For simplicity, we model the multiplexer as a discrete-time single server system consisting of an infinite size buffer and a server with a constant release rate c (cells/slot). The number of customers in the input buffer at time t is denoted by q_t. Our performance index is the steady-state buffer tail probability $\mathbf{P}[q_\infty > b]$, as this quantity is indicative of the buffer overflow probability in a corresponding finite buffer system with b positions.

Computing these tail probabilities, either analytically or numerically, represents a challenging problem in the absence of any underlying Markov property for $M/G/\infty$ inputs. Instead, we focus on the simpler task of determining the asymptotic tail behavior of the queue-length distribution for large buffer size. More precisely, we seek results of the form

$$\lim_{b\to\infty} \frac{1}{h(b)} \ln \mathbf{P}[q_\infty > b] = -\gamma \tag{9.1}$$

for some positive constant γ and mapping $h\colon \mathbb{R}_+ \to \mathbb{R}_+$; these quantities are characterized by λ, G, and c and should be easily computable. Limits such as Eq. (9.1) suggest approximations of the form

$$\mathbf{P}[q_\infty > b] \sim e^{-h(b)\gamma} \quad (b \to \infty). \tag{9.2}$$

Needless to say, such estimates should be approached with care [5]. Nonetheless, Eq. (9.1) already provides some *qualitative* insights into the queueing behavior at the multiplexer and could, in principle, be used to produce guidelines for sizing up its buffers.

In this chapter we provide an overview of some recent work on this issue. Drastically different behaviors emerge depending on whether $v_t^* = O(t)$ or $v_t^* = o(t)$ (with $t \to \infty$), where $v_t^* = -\ln \mathbf{P}[\hat{\sigma} > t]$, $t = 1, 2, \ldots$, and $\hat{\sigma}$ is the forward recurrence time (9.9) associated with σ. The case $v_t^* = O(t)$ is associated with the service time σ having exponential tails, while the case $v_t^* = o(t)$ corresponds to heavy or subexponential tails for σ.

Our focus here is primarily (but not exclusively) on large deviations techniques in order to obtain Eq. (9.1). This approach has already been adopted by a number of authors [10, 16, 19]. Applying results by Duffield and O'Connell [10] (and some recent extensions thereof [11, 30]), we are able to compute $h(b)$ and γ under reasonably general conditions. In fact, for a large class of distributions, we can select $h(b) = v^*_{\lceil b \rceil}$, and the asymptotics (9.1) and (9.2) then take the compact form

$$\mathbf{P}[q_\infty > b] \sim \mathbf{P}[\hat{\sigma} > b]^\gamma \quad (b \to \infty). \tag{9.3}$$

Hence, in many cases, including Weibull, lognormal, and Pareto service times, q_∞ and $\hat{\sigma}$ (thus σ) belong to the same distributional class as characterized by tail behavior.

In many cases of interest, in lieu of Eq. (9.1), these large deviations techniques yield only the weaker asymptotic bounds

$$-\gamma_* \leq \liminf_{b \to \infty} \frac{1}{h(b)} \ln \mathbf{P}[q_\infty > b] \tag{9.4}$$

and

$$\limsup_{b \to \infty} \frac{1}{h(b)} \ln \mathbf{P}[q_\infty > b] \leq -y^* \tag{9.5}$$

with $\gamma_* \neq \gamma^*$. This situation typically occurs when σ is heavy tailed (more generally, subexponential) with either finite or infinite $\mathbf{E}[\sigma^2]$, in which case large deviations excursions are only one of several causes for buffer exceedances [19]. While Eqs. (9.4) and (9.5) are still useful in providing *bounds* on decay rates, they will not be tight in the heavy-tail case and other approaches are needed. Of particular relevance are the approaches of Liu et al. [22] (summarized in Section 9.11) and of Likhanov (discussed in Chapter 8). Liu et al. [22] derive bounds through direct arguments that rely on the asymptotics of Pakes [26] for the $GI/GI/1$ queue under subexponential assumptions [12]. While Likhanov presents lower and upper bounds only when σ is Pareto, these bounds are asymptotically tight. Results for the continuous-time model can be found in Jelenković and Lazar [17], and in Chapters 10 and 7.

Comparison of Eq. (9.3) with results from Norros [25] and Parulekar and Makowski [28] points already to the complex and subtle impact of (long-range) dependencies on the tail probability $\mathbf{P}[q_\infty > b]$. Indeed, in Norros [25] the input stream to the multiplexer was modeled as a fractional Gaussian noise process (or rather its continuous-time analog) exhibiting long-range dependence (in fact, self-similarity), and the buffer asymptotics displayed Weibull-like characteristics. On the other hand, by the results described above, an $M/G/\infty$ input process with a Weibull service time also yields Weibull-like buffer asymptotics although the input process is now short-range dependent. Hence, the *same* asymptotic buffer behavior can be induced by two vastly different input streams, one long-range dependent and the other short-range dependent! To make matters worse, if the pmf G were Pareto instead of Weibull, the input process would now be long-range dependent, in fact asymptotically self-similar [28], but the buffer distribution would now exhibit Pareto-like asymptotics, in sharp contrast with the results of Norros [25].

To reiterate the main conclusion of Parulekar and Makowski [28], the value of the Hurst parameter as the sole indicator of long-range dependence (via asymptotic self-similarity) is at best questionable as it does not characterize buffer asymptotics by itself. Furthermore, buffer sizing cannot be determined adequately by appealing solely to the short- versus long-range dependence characterization of the input model used, be it of the $M/G/\infty$ type or otherwise. Of course, this is not too surprising since long-range dependence (and its close cousin, second-order self-similarity) is determined by second-order properties of the input process, while asymptotics of the form (9.1) invoke much finer probabilistic properties, which are embedded here in the sequence $\{v_t^*,\ t = 1, 2, \ldots\}$. The finiteness of $\mathbf{E}[\sigma^2]$ (which characterizes the SRD nature of the $M/G/\infty$ input process) is obviously a poor marker for predicting the behavior of this sequence.

To close, we note that the diverse queueing behavior demonstrated here is tied to the tail behavior of σ, which determines the correlation structure of $M/G/\infty$ inputs. This clearly illustrates the tremendous impact that the correlation structure of an input stream can have on the corresponding queueing performance given that the $M/G/\infty$ inputs all have *Poisson* marginals! One more data point for the need of a cautious approach in modeling network traffic when time dependencies are either observed or suspected.

9.2 THE $M/G/\infty$ INPUT PROCESS

We summarize various facts concerning the busy server process of a discrete-time $M/G/\infty$ system; details are available in Parulekar [27].

9.2.1 The Model

Consider a system with infinitely many servers. During time slot $[t, t+1)$, β_{t+1} new customers enter the system. Customer i, $i = 1, \ldots, \beta_{t+1}$, is presented to its own server and begins service by the start of slot $[t+1, t+2)$; its service time has duration $\sigma_{t+1,i}$ (expressed in number of slots). Let b_t denote the number of busy servers or, equivalently, of customers still present in the system, at the beginning of slot $[t, t+1)$. We assume that b servers are initially present in the system at $t = 0$ (i.e., at the beginning of slot $[0, 1)$) with customer i, $i = 1, \ldots, b$, requiring an amount of work of duration $\sigma_{0,i}$ from its own server. The busy server process $\{b_t,\ t = 0, 1, \ldots\}$ is what we refer to as the $M/G/\infty$ input process.

The following assumptions are enforced on the **R**-valued random variables (rvs) b, $\{\beta_{t+1}, t = 0, 1, \ldots\}$ and $\{\sigma_{t,i},\ t = 0, 1, \ldots;\ i = 1, 2, \ldots\}$: (1) The rvs are mutually independent; (2) The rvs $\{\beta_{t+1},\ t = 0, 1, \ldots\}$ are i.i.d. Poisson rvs with parameter $\lambda > 0$; (3) The rvs $\{\sigma_{t,i},\ t = 1, 2, \ldots;\ i = 1, 2, \ldots\}$ are i.i.d. with common pmf G on $\{1, 2, \ldots\}$. We denote by σ a generic **R**-valued rv distributed according to the pmf G. Throughout we assume this pmf G to have a finite first moment, or equivalently, $\mathbf{E}[\sigma] < \infty$. At this point, no additional assumptions are made on the rvs $\{\sigma_{0,i},\ i = 1, 2, \ldots\}$.

For each $t = 0, 1, \ldots,$ we note the decomposition

$$b_t = b_t^{(0)} + b_t^{(a)}, \tag{9.6}$$

where the rvs $b_t^{(0)}$ and $b_t^{(a)}$ describe the contributions to the number of customers in the system at the beginning of slot $[t, t+1)$ from those initially present (at $t = 0$) and from the new arrivals in the interval $[0,\ t]$, respectively. Under the enforced operational assumptions, we readily check that

$$b_t^{(a)} = \sum_{s=1}^{t} \sum_{i=1}^{\beta_s} \mathbf{1}[\sigma_{s,i} > t - s] \quad \text{and} \quad b_t^{(0)} = \sum_{i=1}^{b} \mathbf{1}[\sigma_{0,i} > t]. \tag{9.7}$$

The rv $b_t^{(a)}$ can also be interpreted as the number of busy servers in the system at the beginning of slot $[t, t+1)$ given that the system was initially empty (i.e., $b = 0$).

9.2.2 The Stationary Version

Although the busy server process $\{b_t,\ t = 0, 1, \ldots\}$ is in general *not* a (strictly) stationary process, it does admit a stationary and ergodic version $\{b_t^*,\ t = 0, 1, \ldots\}$. This stationary version satisfies the decomposition (9.6) with the portion in (9.7) due to the initial condition replaced by

$$b_t^{*(0)} = \sum_{n=1}^{b} \mathbf{1}[\hat{\sigma}_n > t], \quad t = 0, 1, \ldots, \tag{9.8}$$

where (1) the rvs b and $\{\hat{\sigma}_n,\ n = 1, 2, \ldots\}$ are independent of the rvs $\{\beta_{t+1},\ t = 0, 1, \ldots\}$ and $\{\sigma_{t,i},\ t = 1, 2, \ldots;\ i = 1, 2, \ldots\}$; (2) the rvs $\{\hat{\sigma}_n,\ n = 1, 2, \ldots\}$ are independent of the rv b, which is Poisson distributed with parameter $\lambda\mathbf{E}[\sigma]$; and (3) the rvs $\{\hat{\sigma}_n,\ n = 1, 2, \ldots\}$ are i.d.d. rvs distributed according to the forward recurrence time $\hat{\sigma}$ associated with σ; the corresponding *equilibrium* pmf $\hat{G}$ of $\hat{\sigma}$ is given by

$$\mathbf{P}[\hat{\sigma} = r] = \frac{\mathbf{P}[\sigma \geq r]}{\mathbf{E}[\sigma]}, \quad r = 1, 2, \ldots. \tag{9.9}$$

The following properties of $\{b_t^*,\ t = 0, 1, \ldots\}$ follow readily from this representation [7, 18 (Theorem 3.11, p. 79), 27].

Proposition 9.2.1. *The stationary and ergodic version* $\{b_t^*,\ t = 0, 1, \ldots\}$ *of the busy server process has the following properties:*

(i) *For each* $t = 0, 1, \ldots$, *the rv* b_t^* *is a Poisson rv with parameter* $\lambda \mathbf{E}[\sigma]$.

(ii) *The process is reversible in that*

$$(b_0^*, b_1^*, \ldots, b_t^*) =_{\text{st}} (b_t^*, b_{t-1}^*, \ldots, b_0^*), \quad t = 0, 1, \ldots . \tag{9.10}$$

9.3 THE BUFFER SIZING PROBLEM

As we shall see shortly in Sections 9.4 and 9.5, $M/G/\infty$ input processes display an extremely rich correlation structure. We expect these temporal correlations to have a significant impact on queueing performance when such processes are offered to a multiplexer. To gain some insights into this basic issue we map a multiplexer into a discrete-time single server queue with infinite capacity and constant release rate of c cells/slot under the first-come-first-served discipline. The cell stream is modeled by an $M/G/\infty$ input process as defined above, with b_{t+1} representing the number of new cells that arrive at the start of time slot $[t, t+1)$. Let q_t^b denote the number of cells remaining in the buffer by the end of slot $[t-1, t)$, so that $q_t^b + b_{t+1}$ cells are ready for transmission during slot $[t, t+1)$. If the multiplexer output link can transmit c cells/slot, then the buffer content sequence $\{q_t^b,\ t = 0, 1, \ldots\}$ evolves according to the Lindley recursion

$$q_0^b = q; \quad q_{t+1}^b = [q_t^b + b_{t+1} - c]^+, \quad t = 0, 1, \ldots, \tag{9.11}$$

for some initial condition q.

Conditions under which the queueing system (9.11) admits a steady-state regime are well known and are given next.

Proposition 9.3.1. *If* $\lambda \mathbf{E}[\sigma] < c$, *then there exists an* $\mathbb{R}_+$*-valued rv* q_∞^b *such that* $q_t^b \Rightarrow_t q_\infty^b$ *for any choice of the initial conditions* q, b, *and* $\{\sigma_{0,i},\ i = 1, 2, \ldots\}$. *The system is then said to be stable.*

This characterization of stability follows by extending Loynes's result [23] to Lindley recursions driven by sequences that *couple* with their stationary and ergodic versions [2]. Here, for *any* choice of the initial conditions b and $\{\sigma_{0,i},\ i = 1, 2, \ldots\}$, the sequence $\{b_{t+1},\ t = 0, 1, \ldots\}$ indeed couples in finite time with the stationary and ergodic version introduced in Proposition 9.2.1.

Stationary $M/G/\infty$ processes being time-reversible, we have the representation

$$q_\infty^b =_{\text{st}} \sup(S_t^b - ct,\ t = 0, 1, \ldots) \tag{9.12}$$

for the steady-state buffer content q_∞^b with

$$S_0^b = 0; \quad S_t^b = b_1^* + \cdots + b_t^*, \quad t = 1, 2, \ldots. \tag{9.13}$$

Hereafter, by an $M/G/\infty$ input process we mean its *stationary* version $\{b_t^*,\ t = 0, 1, \ldots\}$, which is fully characterized by the pair (λ, G). Moreover, from now on, we always assume the *stability* condition

$$r_{\text{in}} \equiv \lambda \mathbf{E}[\sigma] < c. \tag{9.14}$$

9.4 SECOND-ORDER CORRELATIONS

Before discussing the asymptotics associated with buffer overflow induced by $M/G/\infty$ input processes, we make a slight detour to explore the correlation structure of such input processes.

9.4.1 Correlation Properties

In view of Proposition 9.2.1, the stationary version $\{b_t^*,\ t = 0, 1, \ldots\}$ has a well-defined (auto)covariance function $\Gamma\colon \mathbf{R} \to \mathbb{R}$, say,

$$\Gamma(h) \equiv \text{Cov}[b_t^*, b_{t+h}^*], \quad t, h = 0, 1, \ldots. \tag{9.15}$$

Proposition 9.4.1. *We have*

$$\Gamma(h) = \lambda \mathbf{E}[(\sigma - h)^+] = \lambda \mathbf{E}[\sigma]\mathbf{P}[\hat{\sigma} > h], \quad h = 0, 1, \ldots. \tag{9.16}$$

The first equality in Eq. (9.16) is established in Cox and Isham [7] and the second equality follows readily from the definition (9.9). From Eq. (9.16) we find the autocorrelation function $\gamma\colon \mathbf{R} \to \mathbb{R}$ of the $M/G/\infty$ process (λ, σ) to be given by

$$\gamma(h) \equiv \frac{\Gamma(h)}{\Gamma(0)} = \mathbf{P}[\hat{\sigma} > h], \quad h = 0, 1, \ldots. \tag{9.17}$$

Note that $\gamma(0) = 1$ as we recall that $\mathbf{P}[\sigma > 0] = 1$.

9.4.2 Inverting γ

Proposition 9.4.1 shows that the correlation structure of the stationary $M/G/\infty$ input process (λ, σ) is completely determined by the pmf of $\hat{\sigma}$ (thus of σ). It turns out that the inverse is true as well. Indeed, Eqs. (9.9) and (9.17) together imply

$$\begin{aligned}\gamma(h) - \gamma(h+1) &= \mathbf{P}[\hat{\sigma} > h] - \mathbf{P}[\hat{\sigma} > h+1] \\ &= \frac{1}{\mathbf{E}[\sigma]}\mathbf{P}[\sigma > h], \quad h = 0, 1, \ldots,\end{aligned} \tag{9.18}$$

so that the mapping $h \to \gamma(h)$ is necessarily decreasing and integer-convex. Taking into account the facts $\gamma(0) = 1$ and $\mathbf{P}[\sigma > 0] = 1$, we conclude from Eq. (9.18) (with $h = 0$) that

$$\mathbf{E}[\sigma]^{-1} = 1 - \gamma(1) \tag{9.19}$$

with $\gamma(1) < 1$ necessarily by the finiteness of $\mathbf{E}[\sigma]$. Combining Eqs. (9.18) and (9.19) we find that

$$\mathbf{P}[\sigma > h] = \frac{\gamma(h) - \gamma(h+1)}{1 - \gamma(1)}, \quad h = 0, 1, \ldots. \tag{9.20}$$

Note also from Eq. (9.20) that

$$\mathbf{E}[\sigma] = \sum_{h=0}^{\infty} \mathbf{P}[\sigma > h] = \frac{1 - \lim_{h\to\infty} \gamma(h)}{1 - \gamma(1)} \tag{9.21}$$

and Eq. (9.19) imposes $\lim_{h\to\infty} \gamma(h) = 0$. A moment of reflection readily leads to the following invertibility result.

Proposition 9.4.2. *An $\mathbb{R}_+$-valued sequence $\{\gamma(h),\ h = 0, 1, \ldots\}$ is the autocorrelation function of the $M/G/\infty$ process (λ, σ) with integrable σ if and only if the corresponding mapping $h \to \gamma(h)$ is decreasing and integer-convex with $\gamma(0) = 1 > \gamma(1)$ and $\lim_{h\to\infty} \gamma(h) = 0$, in which case the pmf G of σ is given by Eq. (9.20).*

9.5 LONG-RANGE DEPENDENCE

The existence of positive correlations in the sequence $\{b_t^*,\ t = 0, 1, \ldots\}$ is clearly apparent from Eq. (9.16). The strength of such positive correlations can be formalized in several ways, which we now describe; additional material is available in Cox [8] and we refer the reader to Tsybakov and Georganas [32] for a discussion of alternative definitions.

The sequence $\{b_t^*,\ t = 0, 1, \ldots\}$ is said to be *short-range dependent* (SRD) if

$$\sum_{h=0}^{\infty} \Gamma(h) < \infty. \tag{9.22}$$

Otherwise, the sequence $\{b_t^*,\ t = 0, 1, \ldots\}$ is said to be *long-range dependent* (LRD). Easy calculations using Eq. (9.16) readily lead to the following simple characterization.

Proposition 9.5.1. *We have*

$$\sum_{h=0}^{\infty} \Gamma(h) = \frac{\lambda}{2} \mathbf{E}[\sigma(\sigma + 1)] \tag{9.23}$$

so that the process is SRD if and only if $\mathbf{E}[\sigma^2]$ *is finite.*

Interesting subclasses of LRD processes can further be identified through the notion of second-order *self-similarity*. To do so, we introduce the rvs

$$b_t^{(m)} \equiv \frac{1}{m} \sum_{k=0}^{m-1} b_{mt+k}^*, \quad m = 1, 2, \ldots; t = 0, 1, \ldots. \tag{9.24}$$

For each $m = 1, \ldots,$ the rvs $\{b_t^{(m)},\ t = 0, 1, \ldots\}$ form a (wide-sense) stationary sequence with correlation structure defined by

$$\Gamma^{(m)}(h) \equiv \mathrm{Cov}[b_t^{(m)}, b_{t+h}^{(m)}] \quad \text{and} \quad \gamma^{(m)}(h) \equiv \frac{\Gamma^{(m)}(h)}{\Gamma^{(m)}(0)}, \quad h = 0, 1, \ldots. \tag{9.25}$$

For each $H > 0$ consider the mapping $\gamma_H : \mathbf{R} \to \mathbb{R}_+$ given by

$$\gamma_H(h) \equiv \tfrac{1}{2}(|h+1|^{2H} - 2|h|^{2H} + |h-1|^{2H}), \quad h = 0, 1, \ldots. \tag{9.26}$$

We say that the sequence $\{b_t^*,\ t = 0, 1, \ldots\}$ is *exactly (second-order) self-similar* if

$$\mathrm{Var}[b_t^{(m)}] = \delta^2 m^{-\beta}, \quad m = 1, 2, \ldots \tag{9.27}$$

for some constants $\delta^2 > 0$ and $0 < \beta < 1$, a requirement equivalent to

$$\Gamma(h) = \delta^2 \gamma_H(h), \quad h = 0, 1, \ldots, \tag{9.28}$$

where $H \equiv 1 - \beta/2$ is known as the Hurst parameter of the process. The parameter H being in the range (0.5, 1), the mapping γ_H is strictly decreasing and integer-convex, with $\gamma_H(0) = 1$, and behaves asymptotically as

$$\gamma_H(h) \sim H(2H-1)h^{2H-2} \quad (h \to \infty). \tag{9.29}$$

By Proposition 9.4.2 we can interpret γ_H as the autocorrelation function of the $M/G/\infty$ input process (λ, σ_H) with

$$\mathbf{P}[\sigma_H > r] = \frac{|r+2|^{2H} - 3|r+1|^{2H} + 3|r|^{2H} - |r-1|^{2H}}{4(1-2^{2H-2})}, \qquad r = 1, 2, \ldots,$$

so that the $M/G/\infty$ input process (λ, σ_H) is exactly second-order self-similar with Hurst parameter H.

In applications, the notion of exact self-similarity is often too restrictive and is weakened as follows. The sequence $\{b_t^*,\ t = 0, 1, \ldots\}$ is said to be *asymptotically (second-order) self-similar* if

$$\lim_{m\to\infty} \gamma^{(m)}(h) = \gamma_H(h), \qquad h = 1, 2, \ldots. \tag{9.30}$$

This will happen for the $M/G/\infty$ input process (λ, σ) if

$$\mathbf{P}[\sigma > r] \sim r^{-\alpha} L(r), \tag{9.31}$$

with $1 < \alpha < 2$, for some slowly varying function $L\colon \mathbb{R}_+ \to \mathbb{R}_+$, in which case $H = (3-\alpha)/2$.

9.6 GENERAL BUFFER ASYMPTOTICS

Several authors [10, 16, 19] have derived asymptotics such as (9.1) by means of large deviations estimates associated with the sequence $\{t^{-1}(S_t^b - ct),\ t = 0, 1, \ldots\}$. These results (and their necessary extensions) are summarized below as they apply to the present context.

9.6.1 A General Setup

With a given $\mathbb{R}$-valued sequence $\{\xi_{t+1},\ t = 0, 1, \ldots\}$, we associate the $\mathbb{R}_+$-valued rv q_∞ given by

$$q_\infty \equiv \sup(S_t,\ t = 0, 1, \ldots), \tag{9.32}$$

where

$$S_0 = 0; \quad S_t = \xi_1 + \cdots + \xi_t, \quad t = 1, 2, \ldots. \tag{9.33}$$

If the sequence $\{\xi_{t+1},\ t = 0, 1, \ldots\}$ is assumed stationary and ergodic with $\mathbf{E}[\xi_1] < 0$, then q_∞ is a.s. finite. We are interested in characterizing the asymptotic behavior of the tail probability $\mathbf{P}[q_\infty > b]$ for large b.

To fix the terminology, a scaling sequence is any monotone increasing $\mathbb{R}$-valued sequence $\{v_t,\ t=0,1,\ldots\}$ such that $\lim_{t\to\infty} v_t = \infty$. The sequence $\{t^{-1}S_t,\ t=1,2,\ldots\}$ is said to satisfy the *large deviations principle under scaling* v_t if there exists a lower-semicontinuous function $I\colon \mathbb{R} \to [0,\infty]$ such that for every open set G,

$$-\inf_{x\in G} I(x) \le \liminf_{t\to\infty} \frac{1}{v_t} \ln \mathbf{P}[t^{-1}S_t \in G] \tag{9.34}$$

and for every closed set F,

$$\limsup_{t\to\infty} \frac{1}{v_t} \ln \mathbf{P}[t^{-1}S_t \in F] \le -\inf_{x\in F} I(x). \tag{9.35}$$

We refer to Eqs. (9.34) and (9.35) as the large deviations lower and upper bounds, respectively. The *rate function* I is said to be good if for each $r>0$, the level set $\{x\in\mathbb{R}\colon I(x)\le r\}$ is a compact subset of $\mathbb{R}$. Additional information on large deviations can be found in Dembo and Zeitouni [9].

We consider a given scaling sequence $\{v_t,\ t=0,1,\ldots\}$ with the property that there exist functions $g,\ h\colon \mathbb{R}_+ \to \mathbb{R}_+$ such that h is monotone increasing with $\lim_{b\to\infty} h(b) = \infty$, and the limits

$$\lim_{b\to\infty} \frac{v_{\lfloor b/y\rfloor}}{h(b)} = \lim_{b\to\infty} \frac{v_{\lceil b/y\rceil}}{h(b)} = g(y), \quad y>0 \tag{9.36}$$

exist with $\lceil x\rceil$ (resp. $\lfloor x\rfloor$) denoting the ceiling (resp. floor) of x.

9.6.2 A Lower Bound

The following theorem is essentially due to Duffield and O'Connell [10]; a proof is included here for the sake of completeness.

Proposition 9.6.1. *If the process $\{t^{-1}S_t,\ t=1,2,\ldots\}$ satisfies the large deviations lower bound (9.34) with good rate function $I\colon \mathbb{R}\to[0,\infty]$ (under scaling v_t), then we have Eq. (9.4) with*

$$\gamma_* = \inf_{y>0}\left(g(y)\inf_{x>y} I(x)\right). \tag{9.37}$$

Proof. Fix $b>0$ and $y>0$. From the definition of q_∞, we find

$$\mathbf{P}[q_\infty > b] \ge \mathbf{P}\left[\frac{S_{\lceil b/y\rceil}}{\lceil b/y\rceil} > \frac{b}{\lceil b/y\rceil}\right] \ge \mathbf{P}\left[\frac{S_{\lceil b/y\rceil}}{\lceil b/y\rceil} > y\right] \tag{9.38}$$

so that

$$\frac{1}{h(b)} \ln \mathbf{P}[q_\infty > b] \geq \frac{v_{\lceil b/y \rceil}}{h(b)} \cdot \frac{1}{v_{\lceil b/y \rceil}} \ln \mathbf{P}\left[\frac{S_{\lceil b/y \rceil}}{\lceil b/y \rceil} > y\right]. \tag{9.39}$$

Letting b go to infinity in this last inequality, we get

$$\liminf_{b \to \infty} \frac{1}{h(b)} \ln \mathbf{P}[q_\infty > b] \geq -g(y) \inf_{x > y} I(x) \tag{9.40}$$

upon invoking Eq. (9.36) and the lower bound (9.34) (with $G = (-\infty, y)$). This is essentially Theorem 2.1 of Duffield and O'Connell [10] and shows the local nature of the lower bound. As the best lower bound is the largest, we can immediately sharpen (9.40) into the lower bound (9.4) with γ_* given by (9.37). ■

The existence of (9.34) (and for that matter, of (9.35)) is typically validated through the Gärtner–Ellis theorem [9, Theorem 2.3.6, p. 45]. In that context, for each $t = 1, 2, \ldots$, we define

$$\Lambda_t(\theta) \equiv \frac{1}{v_t} \ln \mathbf{E}\left[\exp\left(\frac{\theta v_t}{t} S_t\right)\right], \quad \theta \in \mathbb{R}, \tag{9.41}$$

and it is required that for each θ in $\mathbb{R}$, the limit

$$\Lambda(\theta) \equiv \lim_{t \to \infty} \Lambda_t(\theta) \tag{9.42}$$

exists (possibly as an extended real number). Under broad conditions, the process $\{t^{-1}S_t,\ t = 1, 2, \ldots\}$ then satisfies the large deviations principle under scaling v_t with good rate function $\Lambda^*\colon \mathbb{R} \to [0, \infty]$, where Λ^* is the Legendre–Fenchel transform of the mapping $\Lambda\colon \mathbb{R} \to (-\infty, \infty]$ defined through Eq. (9.42), namely,

$$\Lambda^*(z) \equiv \sup_{\theta \in \mathbb{R}} (\theta z - \Lambda(\theta)), \quad z \in \mathbb{R}. \tag{9.43}$$

Expression (9.37) simplifies when the large deviations principle for the process $\{t^{-1}S_t,\ t = 1, 2, \ldots\}$ holds with a good rate function $I\colon \mathbb{R} \to [0, \infty]$, which is *convex*. Indeed, the relation

$$\inf_{x \in \mathbb{R}} I(x) = I(\mathbf{E}[\xi_1]) \tag{9.44}$$

follows readily from the goodness of I and the fact that $\lim_{t \to \infty} t^{-1}S_t = \mathbf{E}[\xi_1] < 0$ a.s. under the ergodic assumption. However, by convexity we have I increasing (resp. decreasing) on $(\mathbf{E}[\xi_1], \infty)$ (resp. on $(-\infty, \mathbf{E}[\xi_1])$, and the conclusion $\inf_{x>y} I(x) = I(y+)$ holds for all $y > 0$. The interior of the effective domain of I

is an interval of the form (y_*, y^*) with $-\infty \le y_* \le \mathbf{E}[\xi_1] \le y^* \le \infty$, and Eq. (9.37) becomes

$$\gamma_\infty = \inf_{y>0} g(y)I(y+) = \inf_{0<y<y^*} g(y)I(y). \tag{9.45}$$

The nondegeneracy condition $y^* > 0$ holds in most applications.

9.6.3 An Upper Bound

In Duffield and O'Connell [10], the companion upper bound (9.5) was derived under a set of conditions that, unfortunately, do not cover some instances of the $M/G/\infty$ process considered here. Upon refining the arguments of Duffield and O'Connell [10], we have established the following asymptotic upper bound; details are available in Section 9.14 and in Parulekar [27]. An alternative approach was given by Duffield [11] but more *explicit* expressions are given here for the upper bound.

Proposition 9.6.2. *Assume the following conditions:*

(i) *For each θ in $\mathbb{R}$, the limit (9.42) exists (possibly as an extended real number) with*

$$\inf_{\theta>0} \Lambda(\theta) \in \mathbb{R}. \tag{9.46}$$

(ii) *For some finite $K \ge 0$, we have*

$$\lim_{t\to\infty} \frac{\ln t}{v_t} = K. \tag{9.47}$$

If $K = 0$, we further assume that the sequence $\{\ln t/v_t,\ t = 1, 2, \ldots\}$ is eventually decreasing.

Then, for each $y > 0$ we have

$$\limsup_{b\to\infty} \frac{1}{h(b)} \ln \mathbf{P}[q_\infty > b] \le -\min(\alpha(y), \beta(y)) \tag{9.48}$$

with the notation

$$\alpha(y) = -Kg(y) + \sup_{\theta>0} \liminf_{n\to\infty} \left(\inf_{x>y} \left(\frac{v_n}{h(nx)} (\theta x - \Lambda(\theta)) \right) \right) \tag{9.49}$$

and

$$\beta(y) = (\Lambda^*(0) - K) g(y). \tag{9.50}$$

The case $K = 0$ is equivalent to Hypothesis 2.2(iv) in Duffield and O'Connell [10]. Moreover, the upper bound is trivial in cases where $\Lambda^*(0) \le K$. As the least upper bound is the sharpest, under the assumptions of Proposition 9.6.2 we immediately get Eq. (9.5) with

$$\gamma^* = \sup_{y>0}(\min(\alpha(y), \beta(y))). \tag{9.51}$$

9.7 EVALUATION OF $\Lambda(\theta)$ ($\theta \in \mathbb{R}$)

An important step in applying the results of the previous section consists in finding a scaling sequence $\{v_t,\ t = 0, 1, \ldots\}$ such that for each θ in $\mathbb{R}$, the limit (9.42) exists (possibly as an extended real number). In Parulekar and Makowski [28–30], we show that the selection of this scaling is governed by the behavior of the sequence $\{v_t^*,\ t = 0, 1, \ldots\}$ given by

$$v_t^* = -\ln \mathbf{P}[\hat{\sigma} > t], \quad t = 0, 1, \ldots. \tag{9.52}$$

This is done under the assumption that the limit

$$\lim_{t\to\infty} \frac{v_t^*}{t} = R \tag{9.53}$$

exists (possibly infinite); this is a very mild assumption, which holds in all known cases. The choice of the appropriate scaling v_t turns out to depend on whether $R = \infty$ (Case I), $0 < R < \infty$ (Case II), or $R = 0$ (Case III).

To state the results more conveniently, we set

$$\Lambda_{b,t}(\theta) \equiv \frac{1}{v_t} \ln \mathbf{E}\left[\exp\left(\frac{v_t}{t}\theta S_t^b\right)\right], \quad \theta \in \mathbb{R} \tag{9.54}$$

for each $t = 1, 2, \ldots$. Obviously, if the limit

$$\Lambda_b(\theta) \equiv \lim_{t\to\infty} \Lambda_{b,t}(\theta), \quad \theta \in \mathbb{R} \tag{9.55}$$

exists (possibly infinite), so does (9.42) with

$$\Lambda(\theta) = \Lambda_b(\theta) - c\theta, \quad \theta \in \mathbb{R}, \tag{9.56}$$

and it suffices to concentrate on finding the limit (9.55). The main facts along these lines are developed in the next two theorems; proofs are available in Parulekar [27] and Parulekar and Makowski [30]. Cases I and II are covered first.

Theorem 9.7.1. *Assume $R > 0$, possibly infinite, and take the linear scaling*

$$v_t = t, \quad t = 1, 2, \ldots. \tag{9.57}$$

Then, for each θ in $\mathbb{R}$, the limit $\Lambda_b(\theta) \equiv \lim_{t\to\infty} \Lambda_{b,t}(\theta)$ exists and is given by

$$\Lambda_b(\theta) = \lambda \mathbf{E}[e^{\theta\sigma} - 1], \tag{9.58}$$

with $\mathbf{E}[e^{\theta\sigma}]$ finite (resp. infinite) if $\theta < R$ (resp. $R < \theta$).

We now turn to Case III.

Theorem 9.7.2. *Assume $R = 0$ with $\{v_t^*/t, \ t = 1, 2, \ldots\}$ eventually monotone decreasing. Then, with the scaling*

$$v_t = v_t^*, \quad t = 1, 2, \ldots, \tag{9.59}$$

it holds for each $\theta \neq 1$ in $\mathbb{R}$ that the limit $\Lambda_b(\theta) \equiv \lim_{t\to\infty} \Lambda_{b,t}(\theta)$ exists and is given by

$$\Lambda_b(\theta) = \begin{cases} \lambda \mathbf{E}[\sigma]\theta, & \text{if } \theta < 1, \\ \infty, & \text{if } \theta > 1, \end{cases} \tag{9.60}$$

provided there exists a mapping $\Gamma : \mathbb{R} \to \mathbb{R}$ such that (i) $\Gamma(t) < t$ for large $t = 1, 2, \ldots$, (ii) $\lim_{t\to\infty} v_t^*(\Gamma(t)/t) = \infty$, *and* (iii) $\lim_{t\to\infty} (v_t^*/t)(\Gamma(t)/v_{\Gamma(t)}^*) = 0$.

The assumptions of Theorem 9.7.2 are satisfied in all cases known to the authors and are easy to check for broad classes of distributions. If $v_t^* \sim t^\beta$ $(0 < \beta < 1)$, we can take $\Gamma(t) = t^\gamma$ with $1 - \beta < \gamma < 1$. If $v_t^* \sim (\ln t)^\beta$ $(\beta > 0)$, then the choice $\Gamma(t) = t(\ln t)^{-\gamma}$ with $0 < \gamma < \beta$ will do.

9.8 EXPONENTIAL TAILS

By exponential tails we refer to the situation where

$$\mathbf{E}[e^{\theta\sigma}] < \infty \quad \text{for some } \theta > 0. \tag{9.61}$$

It is easy to see [30, Section 6] that for each θ in $\mathbb{R}$, the quantities $\mathbf{E}[e^{\theta\sigma}]$ and $\mathbf{E}[e^{\theta\hat{\sigma}}]$ are simultaneously finite (resp. infinite). Moreover, with $R > 0$, possibly infinite, these quantities are finite (resp. infinite) whenever $\theta < R$ (resp. $\theta > R$). Thus, under limit (9.53), exponential tails correspond to Cases I and II.

9.8.1 The Asymptotics Under Exponential Tails

When $R > 0$, Theorem 9.7.1 suggests $v_t = t$, so that $h(b) = b$, $g(y) = y^{-1}$, and $K = 0$. Therefore,

$$\Lambda(\theta) = \lambda \mathbf{E}[e^{\theta\sigma} - 1] - c\theta, \quad \theta \in \mathbb{R} \tag{9.62}$$

and the Gärtner–Ellis theorem [9, Theorem 2.3.6, p. 45] guarantees the full large deviations principle (under scaling t) for $\{t^{-1}(S_t^b - ct),\ t = 1, 2, \ldots\}$ with rate function Λ^*. From Eq. (9.45) we readily conclude

$$\gamma_* = \inf_{y>0} \frac{\Lambda^*(y)}{y}, \tag{9.63}$$

while easy calculations show that

$$\alpha(y) = \sup_{\theta>0} \left(\inf_{x>y} \left(\frac{\theta x - \Lambda(\theta)}{x} \right) \right) \quad \text{and} \quad \beta(y) = \frac{\Lambda^*(0)}{y} \tag{9.64}$$

for all $y > 0$. This leads easily to

$$\gamma^* = \sup\{\theta > 0 \colon \Lambda(\theta) < 0\}. \tag{9.65}$$

Combining these facts with Propositions 9.6.1 and 9.6.2, we finally obtain the following asymptotics.

Proposition 9.8.1. *Assume $R > 0$. Then, the asymptotics*

$$\lim_{b\to\infty} \frac{1}{b} \ln \mathbf{P}[q_\infty^b > b] = -\gamma \tag{9.66}$$

holds with $\gamma_* = \gamma^* \equiv \gamma$.

Proposition 9.8.1 covers the situations where the tail of G decays *at least* exponentially fast, for example, the Rayleigh, Gamma, and geometric cases. This result is of course not new; it has been obtained earlier by several authors [10, 16, 19] and paves the way to the notion of *effective bandwidth*. However, the arguments leading to it, which are detailed in Parulekar [27], already show that the upper bound of Proposition 9.6.2 is good enough to recover this "classical" case.

9.8.2 Comparison with Instantaneous Inputs

Another input process closely related to the $M/G/\infty$ input process (λ, σ) is the input process according to which the work associated with a session is offered *instantaneously* to the buffer, rather than *gradually* as was the case for $M/G/\infty$ input processes.

Such an instantaneous input process, say, $\{a_{t+1},\ t = 0, 1, \ldots\}$, is composed of i.i.d. $\mathbb{R}$-valued rvs given by

$$a_{t+1} := \sum_{i=1}^{\beta_{t+1}} \sigma_{t+1,i}, \quad t = 0, 1, \ldots, \tag{9.67}$$

where the two families of i.i.d. rvs $\{\beta_{t+1},\ t = 0, 1, \ldots\}$ and $\{\sigma_{t+1,i},\ t = 0, 1, \ldots, i = 1, 2, \ldots\}$ are as in Section 9.2.1. Let a denote a generic rv of the i.i.d. sequence $\{a_{t+1},\ t = 0, 1, \ldots\}$. When offered to the multiplexer described by Eq. (9.11), the instantaneous arrival process generates a sequence of buffer contents $\{q_t^a,\ t = 0, 1, \ldots\}$ via the recursion

$$q_0^a = q; \quad q_{t+1}^a = [q_t^a + a_{t+1} - c]^+, \quad t = 0, 1, \ldots \tag{9.68}$$

for some initial condition q. This Lindley recursion is simply the one associated with a $D/GI/1$ queue. Hence, the system will be stable if $\mathbf{E}[a] < c$, a condition equivalent to (9.14), in which case $q_t^a \Rightarrow_t q_\infty^a$ for some $\mathbb{R}_+$-valued rv q_∞^a. Here as well, we are interested in the tail probabilities $\mathbf{P}[q_\infty^a > b]$ for b large, and in comparing these asymptotics with those of $\mathbf{P}[q_\infty > b]$.

To apply Propositions 9.6.1 and 9.6.2, it is natural to consider the partial sum sequence $\{S_t^a,\ t = 0, 1, \ldots\}$ associated with the instantaneous input process $\{a_{t+1},\ t = 0, 1, \ldots\}$, namely,

$$S_0^a = 0; \quad S_t^a = a_1 + \cdots + a_t, \quad t = 1, 2, \ldots, \tag{9.69}$$

so that

$$q_\infty^a =_{\text{st}} \sup(S_t^a - ct,\ t = 0, 1, \ldots). \tag{9.70}$$

Under the enforced independence assumptions, for each $t = 1, 2, \ldots$ we have

$$\mathbf{E}[e^{\theta S_t^a}] = \prod_{s=1}^{t} \mathbf{E}[\mathbf{E}[e^{\theta\sigma}]^{\beta_s}] = e^{-\lambda t(1-\mathbf{E}[e^{\theta\sigma}])}, \quad \theta \in \mathbb{R}, \tag{9.71}$$

so that

$$\frac{1}{t}\lim_{t\to\infty} \ln \mathbf{E}[e^{\theta S_t^a}] = \lambda \mathbf{E}[e^{\theta\sigma} - 1], \quad \theta \in \mathbb{R}. \tag{9.72}$$

Consequently, by the Gärtner–Ellis theorem [9, Theorem 2.3.6, p. 45], the sequence $\{t^{-1}(S_t^a - ct),\ t = 1, 2, \ldots\}$ does satisfy the full large deviations principle. Its rate functional coincides with that of $\{t^{-1}(S_t^b - ct),\ t = 1, 2, \ldots\}$, thereby implying

$$\lim_{b\to\infty} \frac{1}{b} \ln \mathbf{P}[q_\infty^a > b] = \lim_{b\to\infty} \frac{1}{b} \ln \mathbf{P}[q_\infty^b > b] = -\gamma \tag{9.73}$$

by another combined use of Propositions 9.6.1 and 9.6.2. This is not too surprising as we note that the processes $\{t^{-1}(S_t^a - ct),\ t = 1, 2, \ldots\}$ and $\{t^{-1}(S_t^b - ct),\ t = 1, 2, \ldots\}$ are *exponentially equivalent* (under linear scaling) [9, p. 114]; this is an easy consequence of the relations

$$\sum_{s=1}^{t} b_s^{*(0)} = \sum_{n=1}^{b} \min(t, \hat{\sigma}_n - 1) \tag{9.74}$$

and

$$\sum_{r=1}^{t} b_r^{(a)} = \sum_{s=1}^{t} \sum_{i=1}^{\beta_s} \min(t - s + 1, \sigma_{s,i}) \tag{9.75}$$

derived elsewhere [30, Section 5].

9.9 NONEXPONENTIAL TAILS

This case is characterized by

$$\mathbf{E}[e^{\theta\sigma}] = \infty, \quad \theta > 0 \tag{9.76}$$

or equivalently, under (9.53), by $R = 0$. We are thus in Case III, which is covered by Theorem 9.7.2, so that the appropriate scaling is given by $v_t = v_t^*$. Under the conclusion of Theorem 9.7.2, we have

$$\Lambda(\theta) = \begin{cases} (\lambda\mathbf{E}[\sigma] - c)\theta, & \text{if } \theta < 1, \\ \infty, & \text{if } \theta > 1. \end{cases} \tag{9.77}$$

We now consider the evaluation of the constants γ_* and γ^*. The boundary case $\theta = 1$ in Eq. (9.60) (thus in Eq. (9.77)) appears to depend crucially on the pmf G. However, it can be shown [27] that $\lim_{t\to\infty} \Lambda_{b,t}(1)$ does exist for the examples discussed here. In fact, the *existence* of this limit suffices for our purpose, in that its *specific* value is of no consequence in evaluating the various quantities of interest. Indeed, from Eqs. (9.43) and (9.56) we then get

$$\begin{aligned} A^*(z) &= \sup_{\theta \le 1}((c+z)\theta - \Lambda_b(\theta)) \\ &= \sup_{\theta < 1}((c+z)\theta - \lambda\mathbf{E}[\sigma]\theta), \quad z \in \mathbb{R}, \end{aligned}$$

with the last step exploiting the fact $\lim_{\theta\uparrow 1} \Lambda_b(\theta) \le \Lambda_b(1)$ (together with the existence of $\Lambda_b(1)$). Finally, the stability condition (9.14) yields

$$\Lambda^*(z) = z + c - r_{\text{in}}, \quad z \ge 0. \tag{9.78}$$

9.9.1 Evaluation of γ^*

We work under the assumption that the scaling sequence $\{v_t^*,\ t = 1, 2, \ldots\}$ satisfies Condition (ii) of Proposition 9.6.2. Fix $y > 0$. Setting $z = 0$ in Eq. (9.78) we obtain

$$\beta(y) = (c - r_{\text{in}} - K)g(y). \tag{9.79}$$

Next, going back to the expression (9.49) we note that

$$\begin{aligned}
&\sup_{\theta>0} \liminf_{t\to\infty}\left(\inf_{x>y}\left(\frac{v_t}{h(tx)}(\theta x - \Lambda(\theta))\right)\right) \\
&\quad = \sup_{0<\theta<1} \liminf_{t\to\infty}\left(\inf_{x>y}\left(\frac{v_t}{h(tx)}(x + c - r_{\text{in}})\theta\right)\right) \\
&\quad = \theta \liminf_{t\to\infty}\left(\inf_{x>y}\left(\frac{v_t}{h(tx)}(x + c - r_{\text{in}}\right)\right)
\end{aligned} \tag{9.80}$$

so that

$$\alpha(y) = -Kg(y) + \liminf_{t\to\infty}\left(\inf_{x>y}\left(\frac{v_t}{h(tx)}(x + c - r_{\text{in}}\right)\right) \tag{9.81}$$

with closed forms available for specific choices of the pmf G (Section 9.10).

9.9.2 Evaluation of γ^*

If we *assume* that $\{t^{-1}(S_t^b - ct),\ t = 1, 2, \ldots\}$ satisfies a large deviations principle with good rate function Λ^*, then blind substitution into Eq. (9.78) yields

$$\gamma_* = \inf_{y>0} g(y)(y + c - r_{\text{in}}). \tag{9.82}$$

Unfortunately, the Gärtner–Ellis theorem [9, Theorem 2.3.6, p. 45] here gives a *trivial* lower bound (9.34) for $\{t^{-1}(S_t^b - ct),\ t = 1, 2, \ldots\}$. This fact, pointed out by Duffield [11], precludes the use of Proposition 9.6.1 to conclude the lower bound (9.4). Nevertheless, as discussed in Section 9.11, such a lower bound does hold in many cases of interest, with case-specific arguments needed to establish it.

9.10 EXAMPLES

The examples considered here are constructed by taking the $\{1, 2, \ldots\}$-valued rv σ to be of the form $\sigma =_{\text{st}} [X]$, where X is an integrable $\mathbb{R}_+$-valued rv with $\mathbf{P}[X = 0] = 0$.

9.10.1 Integrated Tails and Forward Recurrence Times

Recall that with any $\mathbb{R}_+$-valued rv X with $0 < \mathbf{E}[X] < \infty$, we can associate the $\mathbb{R}_+$-valued rv X^* whose distribution is the *integrated tail* distribution of X, namely,

$$\mathbf{P}[X^* \leq x] = \frac{1}{\mathbf{E}[X]} \int_0^x \mathbf{P}[X > u]\, du, \quad x \geq 0. \tag{9.83}$$

For any $\{1, 2, \ldots\}$-valued rv σ, the difference between the $\{1, 2, \ldots\}$-valued rv $\hat{\sigma}$ and the $\mathbb{R}_+$-valued rv σ^* is emphasized through the relation

$$\mathbf{P}[\sigma^* > x] = \mathbf{P}[\hat{\sigma} > r] - (x - r)\mathbf{P}[\hat{\sigma} = r + 1], \quad r \le x < r + 1, \tag{9.84}$$

with $r = 0, 1, \ldots$. Moreover, if $\sigma =_{\text{st}} [X]$ for some $\mathbb{R}_+$-valued rv with $\mathbf{P}[X = 0] = 0$ and $0 < \mathbf{E}[X] < \infty$, simple algebra shows that

$$\frac{\mathbf{E}[X]}{\mathbf{E}[\sigma]}\mathbf{P}[X^* > r] \le \mathbf{P}[\hat{\sigma} > r] \le \frac{\mathbf{E}[X]}{\mathbf{E}[\sigma]}\mathbf{P}[X^* > r - 1] \tag{9.85}$$

for all $r = 1, 2, \ldots$. Consequently,

$$\mathbf{P}[\hat{\sigma} > r] \sim \frac{\mathbf{E}[X]}{\mathbf{E}[\sigma]}\mathbf{P}[X^* > r] \tag{9.86}$$

provided

$$\lim_{r\to\infty} \frac{\mathbf{P}[X^* > r - 1]}{\mathbf{P}[X^* > r]} = 1. \tag{9.87}$$

This will happen if the rv X^* belongs to the class $\mathcal{L}$ of long-tailed distributions on $\mathbb{R}_+$ [12]. This will be so for all examples considered here, in which case Eq. (9.86) implies

$$-v_t^* = \ln \mathbf{P}[\hat{\sigma} > t] \sim \ln \mathbf{P}[X^* > t]. \tag{9.88}$$

9.10.2 Specific Cases

9.10.2.1 Weibull The rv X is a Weibull rv with parameters $(a, v)(a > 0, 0 < v < 1)$; that is,

$$\mathbf{P}[X > x] = e^{-ax^v}, \quad x \ge 0, \tag{9.89}$$

and it is well known that

$$\mathbf{P}[X^* > x] \sim C(a, v)x^{1-v}e^{-ax^v} \tag{9.90}$$

for some constant $C(a, v)$ determined solely by the parameters (a, v). Consequently, $v_t^* \sim at^v$, so that $h(b) = b^v$, $g(y) = ay^{-v}$, and $K = 0$. Elementary calculus shows that

$$\gamma_{\text{W}}^* = \frac{a}{v}\left(\frac{v(c - r_{\text{in}})}{1 - v}\right)^{1-v} = ae^{H(v)+(1-v)\ln(c-r_{\text{in}})}, \tag{9.91}$$

where $H(v)$ denotes the natural entropy of the pmf $(v, 1-v)$—that is, $H(v) = -v \ln v - (1-v)\ln(1-v)$, and Proposition 9.6.2 yields

$$\limsup_{b\to\infty} \frac{1}{b^v} \ln \mathbf{P}[q_\infty^b > b] \le -\gamma_{\mathrm{W}}^*. \tag{9.92}$$

9.10.2.2 Lognormal The rv X is a lognormal rv with parameters $(\mu, \delta)(\delta > 0)$ if $X =_{\mathrm{st}} \exp(Y)$ for some Gaussian rv Y with mean μ and variance δ^2. We have

$$\mathbf{P}[X^* > x] \sim D(\delta, \mu)\frac{xe^{-(\log x-\mu)^2/2\delta^2}}{(\ln x - \mu)^2} \tag{9.93}$$

for some constant $D(\delta, \mu)$ determined solely by the parameters (δ, μ), so that $v_t^* \sim (1/2\delta^2)(\ln t)^2$. Consequently, take $h(b) = (\ln b)^2$, $g(y) = 1/2\delta^2$, and $K = 0$. Easy calculations show that

$$\alpha_{\mathrm{L}}(y) = \frac{1}{2\delta^2}(c - r_{\mathrm{in}} + y) \quad \text{and} \quad \beta_{\mathrm{L}}(y) = \frac{1}{2\delta^2}(c - r_{\mathrm{in}}) \tag{9.94}$$

for all $y > 0$. Therefore,

$$\gamma_{\mathrm{L}}^* = \frac{c - r_{\mathrm{in}}}{2\delta^2} \tag{9.95}$$

and by Proposition 9.6.2 we find

$$\limsup_{b\to\infty} \frac{1}{(\ln b)^2} \ln \mathbf{P}[q_\infty^b > b] \le -\gamma_{\mathrm{L}}^*. \tag{9.96}$$

In both cases considered thus far $\mathbf{E}[\sigma^2]$ is finite, and the process $\{b_t^*,\ t = 0, 1, \ldots\}$ is therefore short-range dependent by Proposition 9.5.1. Moreover, by using Eq. (9.82) we note the equalities

$$\gamma_{\mathrm{W}}^* = \gamma_{\mathrm{W},*} \quad \text{and} \quad \gamma_{\mathrm{L}}^* = \gamma_{\mathrm{L},*}. \tag{9.97}$$

Hence, were we to apply Proposition 9.6.1 *blindly* (without further justification), we could then replace Eqs. (9.92) and (9.96) by their appropriate version of the stronger limiting equality (9.1). The equalities (9.97) thus hold out the possibility that in the Weibull and lognormal cases, $\{t^{-1}(S_t^b - ct),\ t = 1, 2, \ldots\}$ satisfies the large deviations lower bound (9.34) with rate functional Λ^*.

9.10.2.3 Generalized Pareto The rv X is a generalized Pareto or regularly varying rv with parameter $\alpha(1 < \alpha)$ if

$$\mathbf{P}[X > x] = x^{-\alpha}L(x), \quad x > 0 \tag{9.98}$$

for some slowly varying function $L\colon \mathbb{R}_+ \to \mathbb{R}_+$, in which case

$$\mathbf{P}[X^* > x] = x^{-\alpha+1}L(x), \quad x > 0. \tag{9.99}$$

Hence, $v_t^* \sim (\alpha - 1)\ln t$ so that $h(b) = \ln b$, $g(y) = \alpha - 1$, and $K = (\alpha - 1)^{-1}$. We find

$$\gamma_{\mathrm{P},*} = (\alpha - 1)(c - r_{\mathrm{in}}) \tag{9.100}$$

together with $\alpha_{\mathrm{P}}(y) = -1 + (c - r_{\mathrm{in}} + y)(\alpha - 1)$ and $\beta_{\mathrm{P}}(y) = -1 + \gamma_{\mathrm{P},*}$ for all $y > 0$. Invoking Proposition 9.6.2 we find

$$\limsup_{b\to\infty} \frac{1}{\ln b} \ln \mathrm{P}[q_\infty^b > b] \le -\gamma_{\mathrm{P}}^*, \tag{9.101}$$

where

$$\gamma_{\mathrm{P}}^* = (\alpha - 1)(c - r_{\mathrm{in}}) - 1 = \gamma_{\mathrm{P},*} - 1. \tag{9.102}$$

Here, the process $\{b_t^*,\ t = 0, 1, \ldots\}$ is long-range dependent since $\mathbf{E}[\sigma^2] = \infty$. The condition $(\alpha - 1)(c - r_{\mathrm{in}}) > 1$ is required to make the bound (9.101) nontrivial, and we have $\gamma_{\mathrm{P}}^* < \gamma_{*,\mathrm{P}}$. Although this inequality is not sufficient by itself to reach any negative conclusion concerning the existence of the lower bound (9.34) for $\{t^{-1}(S_t^b - ct),\ t = 1, 2, \ldots\}$ in the Pareto case, it strongly suggests that in the long-range dependent case, the investigation of the buffer asymptotics will require that we look beyond large deviations techniques. Going back to the heuristics given in Kesidis et al. [19], we attribute this to the fact that now buffer exceedances cannot be explained entirely by large deviations excursions in the arrival stream, as there is a need to take into consideration the effect of a single customer with a large workload—the tail of the distribution has become too heavy to neglect such a customer! Hence, any argument based on large deviations techniques *alone* is bound to fall short. However, we conjectured [29] that Eq. (9.1) still holds with scaling $h(b) = \ln b$ as specified through Eq. (9.36) but of course with a different value for γ. Progress on this question has recently been made and is summarized in Sections 9.11 and 9.12.

9.11 SUBEXPONENTIAL TAILS

At this point, we have seen through examples that in the nonexponential case a nontrivial upper bound (9.4) always holds (perhaps not with the best constant γ^*), whereas the lower bound (9.5) is in doubt, at least if one insists on going through Proposition 9.6.1. The suspicion that this shortcoming is linked to the methodology based on large deviations was confirmed by Liu et al. [22], where the issue of devising asymptotics for $\mathbf{P}[q_\infty^b > b]$ was revisited by means of basic principles; both lower and upper bound asymptotics were proposed and in some cases the latter are tighter than the ones given here. This section is devoted to a discussion of the results of Liu et al. [22], together with a comparison with the bounds obtained here.

The approach of Liu et al. [22] is most informative when applied to the subset of distributions with nonexponential tails known as *subexponential* distributions [4, 12]: An $\mathbb{R}_+$-valued rv X is said to be *subexponential*, written $X \in \mathcal{S}$, if

$$\lim_{x \to \infty} \frac{\mathbf{P}[X + X' > x]}{\mathbf{P}[X > x]} = 2, \tag{9.103}$$

where X' is an independent copy of X. The terminology is substantiated [12] by the fact that, under Eq. (9.103), we have

$$\lim_{x \to \infty} e^{\delta x} \mathbf{P}[X > x] = \infty, \quad \delta > 0. \tag{9.104}$$

9.11.1 Instantaneous Inputs with Subexponential Tails

The clue that the bounds obtained thus far could indeed be improved is already apparent from the following well-known result of Pakes [26] when applied to the model with instantaneous inputs [Section 9.8.2].

Proposition 9.11.1. *If $a^* \in \mathcal{S}$, then it holds that $q_\infty^a \in \mathcal{S}$ with*

$$\mathbf{P}[q_\infty^a > b] \sim \frac{r_{\text{in}}}{c - r_{\text{in}}} \mathbf{P}[a^* > b]. \tag{9.105}$$

If $\sigma \in \mathcal{S}$, then $a \in \mathcal{S}$ with $\mathbf{P}[a > t] \sim \mathbf{E}[\beta]\mathbf{P}[\sigma > t]$ [12], and standard arguments now yield

$$\int_b^\infty \mathbf{P}[a > t]\, dt \sim \mathbf{E}[\beta] \int_b^\infty \mathbf{P}[\sigma > t]\, dt, \tag{9.106}$$

so that $\mathbf{P}[a^* > b] \sim \mathbf{P}[\sigma^* > b]$ and $a^* \in \mathcal{S}$ whenever $\sigma^* \in \mathcal{S}$. Combining these comments, we immediately get the following proposition.

Proposition 9.11.2. *If $\sigma \in \mathcal{S}$ and $\sigma^* \in \mathcal{S}$,* then $q_\infty^a \in \mathcal{S}$ with

$$\mathbf{P}[q_\infty^a > b] \sim \frac{r_{\text{in}}}{c - r_{\text{in}}} \mathbf{P}[\sigma^* > b]. \tag{9.107}$$

9.11.2 Improved Upper Bounds

The expressions (9.74) and (9.75) readily lead to the bound

$$q_\infty^b \leq \sum_{n=1}^{b} \hat{\sigma}_n + q_\infty^b. \tag{9.108}$$

This observation, when coupled with the asymptotics (9.107), forms the basis for showing the following asymptotics.

Proposition 9.11.3. *If $\sigma \in \mathcal{S}$ and $\sigma^* \in \mathcal{S}$,* then

$$\limsup_{b\to\infty} \frac{\mathbf{P}[q_\infty^b > b]}{\mathbf{P}[\sigma^* > b]} \leq r_{\text{in}} + \frac{r_{\text{in}}}{c - r_{\text{in}}}; \tag{9.109}$$

hence,

$$\limsup_{b\to\infty} \frac{\ln \mathbf{P}[q_\infty^b > b]}{-\ln \mathbf{P}[\sigma^* > b]} \leq -1. \tag{9.110}$$

The derivation of Eq. (9.109) relies on well-known properties of subexponential rvs [12]; details are available in Liu et al. [22]. The release rate c does *not* appear in Eq. (9.110), thereby suggesting that Eq. (9.110) will not always improve on upper asymptotics obtained previously. This is illustrated in the special cases treated in Section 9.10.2, which we now revisit in light of Proposition 9.11.3; the asymptotics are easily validated through Eqs. (9.84) and (9.88).

Weibull Since $\ln \mathbf{P}[\sigma^* > b] \sim -ab^v$, we have

$$\limsup_{b\to\infty} \frac{1}{b^v} \ln \mathbf{P}[q_\infty^b > b] \leq -a, \tag{9.111}$$

an improvement on Eq. (9.92) only if $\gamma_W^* < a$; that is, if $(1-v)\ln(c - r_{\text{in}}) + H(v) < 0$. This is never the case for all v in $(0, 1)$ when $c - r_{\text{in}} \geq 1$ and true only for small v when $c - r_{\text{in}} < 1$.

Lognormal This time, $\ln \mathbf{P}[\sigma^* > b] \sim -(1/2\delta^2)(\ln b)^2$ so that

$$\limsup_{b\to\infty} \frac{1}{(\ln b)^2} \ln \mathbf{P}[q_\infty^b > b] \leq -\frac{1}{2\delta^2}. \tag{9.112}$$

This asymptotic upper bound is tighter than (9.96) only if $c - r_{\text{in}} < 1$.

Generalized Pareto Noting $\ln \mathbf{P}[\sigma^* > b] \sim (1-\alpha)\ln b$, we get

$$\limsup_{b\to\infty} \frac{1}{\ln b} \ln \mathbf{P}[q_\infty^b > b] \le -(\alpha - 1), \tag{9.113}$$

and we do better than (9.101) only if $(\alpha - 1)(c - r_{\text{in}}) < \alpha$.

9.12 GENERAL LOWER BOUNDS

The following lower bound holds in great generality and is essentially Proposition 3.1 in Liu et al. [22] couched in the notation used here.

Proposition 9.12.1. *For any* $\{1, 2, \ldots\}$*-valued rv* σ*, it holds that*

$$-\Gamma_* \le \liminf_{b\to\infty} \frac{1}{v^*_{\lfloor b\rfloor}} \ln \mathbf{P}[q_\infty^b > b] \tag{9.114}$$

with

$$\Gamma_* = \inf_{y>0} \left((\lfloor c - r_{\text{in}} + y\rfloor + 1) \limsup_{b\to\infty} \frac{v^*_{\lfloor b\rfloor}}{v^*_{\lfloor by\rfloor}} \right). \tag{9.115}$$

We now apply this result to the special cases of Section 9.10.2. Details of the calculations, which are available in Liu et al. [22, Section 3], are omitted in the interest of brevity.

Weibull Here, we have

$$\Gamma_{W,*} = \inf_{y>0} ((1 + \lfloor c - r_{\text{in}} + y\rfloor) y^{-\nu}) \tag{9.116}$$

and Proposition 9.12.1 gives

$$-a\Gamma_{W,*} \le \liminf_{b\to\infty} \frac{1}{b^\nu} \ln \mathbf{P}[q_\infty^b > b]. \tag{9.117}$$

Explicit expressions for $\Gamma_{W,*}$ are given in Liu et al. [22, Section 3].

Lognormal We have

$$\Gamma_{L,*} = \inf_{y>0} ((1 + \lfloor c - r_{\text{in}} + y\rfloor)) = (1 + \lfloor c - r_{\text{in}}\rfloor) \tag{9.118}$$

and Eq. (9.114) gives

$$-\frac{1}{2\delta^2}(1+\lfloor c-r_{\text{in}}\rfloor) \leq \liminf_{b\to\infty} \frac{1}{(\ln b)^2} \ln \mathbf{P}[q_\infty^b > b]. \tag{9.119}$$

Generalized Pareto Here as well we find

$$\Gamma_{P,*} = \inf_{y>0} ((1+\lfloor c-r_{\text{in}}+y\rfloor)) = (1+\lfloor c-r_{\text{in}}\rfloor) \tag{9.120}$$

and Eq. (9.114) takes the form

$$-(1+\lfloor c-r_{\text{in}}\rfloor)(\alpha-1) \leq \liminf_{b\to\infty} \frac{1}{\ln b} \ln \mathbf{P}[q_\infty^b > b]. \tag{9.121}$$

We now compare these expressions with those obtained in Sections 9.10.2 and 9.11.2. Under the condition

$$c - r_{\text{in}} < 1, \tag{9.122}$$

we have

$$\lim_{b\to\infty} \frac{1}{(\ln b)^2} \ln \mathbf{P}[q_\infty^b > b] = -\frac{1}{2\delta^2} \tag{9.123}$$

in the lognormal case, and

$$\lim_{b\to\infty} \frac{1}{\ln b} \ln \mathbf{P}[q_\infty^b > b] = -(\alpha-1) \tag{9.124}$$

in the regularly varying case, while no such conclusion can be drawn in the Weibull case. We note that the bounds given by Likhanov in Chapter 8 in this volume hold under the weaker condition that $c - r_{\text{in}}$ be noninteger, and automatically imply Eq. (9.124) (under (9.122)).

In Case II with $0 < R < \infty$, we have $v_t^* \sim Rt$ so that

$$\Gamma_* = \inf_{y>0} \frac{\lfloor c-r_{\text{in}}+y\rfloor+1}{y} = 1 \tag{9.125}$$

and Proposition 9.12.1 yields

$$-R \leq \liminf_{b\to\infty} \frac{1}{b} \ln \mathbf{P}[q_\infty^b > b]. \tag{9.126}$$

Interestingly enough, this lower bound is not as good as the one obtained in Proposition 9.8.1 by applying the general buffer asymptotics based on large deviations arguments [Section 9.6].

9.13 CONCLUSION

We have presented recent results on the large buffer asymptotics for an infinite capacity multiplexer with constant release rate and $M/G/\infty$ inputs. When the distribution of session durations has an exponential tail, these asymptotics are completely identified by means of large deviations arguments. The results are not as complete in the nonexponential case. Even in the more restricted setup where session durations have a subexponential distribution, only lower and upper bounds are available in general; they are derived through a variety of techniques, including the asymptotics of Pakes for the $GI/GI/1$ queue. These bounds have been shown to be tight for the generalized Pareto case. The issue is still open for other subexponential distributions (e.g., Weibull and lognormal), although the bounds are known to be tight in the lognormal case under some conditions on the release rate.

9.14 APPENDIX: THE BASIC UPPER BOUND

We begin the derivation of the companion upper bound (9.5) by establishing a basic asymptotic upper bound. For each $m = 1, 2, \ldots$ and $b > 0$, define the quantities

$$A(m, b) \equiv m \max_{n=1,\ldots,m} \mathbf{P}[S_n > b] \quad \text{and} \quad B(m, b) \equiv \mathbf{P}\left[\sup_{n=m+1,\ldots} S_n > b\right].$$

The representation (9.12) and a union bound argument lead to

$$\mathbf{P}[q_\infty > b] \leq A(m, b) + B(m, b) \leq 2\max(A(m, b), B(m, b)). \tag{9.127}$$

We now fix $y > 0$. Next, substitute $\lfloor b/y \rfloor$ for m in Eq. (9.127), take the logarithm and divide by $h(b)$. Letting b go to infinity in the resulting inequality, we obtain the basic asymptotic upper bound

$$\limsup_{b\to\infty} \frac{1}{h(b)} \ln \mathbf{P}[q_\infty > b] \leq \max(\mathcal{A}(y), \mathcal{B}(y)), \tag{9.128}$$

where we have used the notation

$$\mathcal{A}(y) \equiv \limsup_{b\to\infty} \frac{1}{h(b)} \ln A(\lfloor b/y \rfloor, b) \tag{9.129}$$

and

$$\mathcal{B}(y) \equiv \limsup_{b\to\infty} \frac{1}{h(b)} \ln B(\lfloor b/y \rfloor, b). \tag{9.130}$$

In the next two sections we show that $\mathcal{A}(y) \le -\alpha(y)$ and $\mathcal{B}(y) \le -\beta(y)$ with $\alpha(y)$ and $\beta(y)$ given by Eqs. (9.49) and (9.50), with the immediate consequence that Eq. (9.48) follows from Eq. (9.128).

9.15 APPENDIX: UPPERBOUNDING $\mathcal{A}(y)$ ($y > 0$)

Fix $y > 0$. It is plain that

$$\frac{1}{h(b)} \ln A(\lfloor b/y \rfloor, b) = \frac{1}{h(b)} \ln \lfloor b/y \rfloor + \frac{1}{h(b)} \max_{n=1,\ldots,\lfloor b/y \rfloor} \ln \mathbf{P}[S_n > b], \quad b > 0. \tag{9.131}$$

Under Condition (ii) and Eq. (9.36), it is a simple matter to check that

$$\limsup_{b\to\infty} \frac{1}{h(b)} \ln \lfloor b/y \rfloor = Kg(y). \tag{9.132}$$

For the second term of Eq. (9.131), we have

$$\begin{aligned}
\frac{1}{h(b)} \max_{n=1,\ldots,\lfloor b/y \rfloor} \ln \mathbf{P}[S_n > b] &\le \frac{1}{h(b)} \sup_{x>y} (\ln \mathbf{P}[S_{\lfloor b/x \rfloor} > b]) \\
&= \sup_{x>y} \frac{1}{h(b)} \ln \mathbf{P}\left[\frac{S_{\lfloor b/x \rfloor}}{\lfloor b/x \rfloor} > \frac{b}{\lfloor b/x \rfloor}\right] \\
&\le \sup_{x>y} \frac{1}{h(b)} \ln \mathbf{P}\left[\frac{S_{\lfloor b/x \rfloor}}{\lfloor b/x \rfloor} > x\right] \\
&= \sup_{x>y} \frac{v_{\lfloor b/x \rfloor}}{h(b)} \cdot \frac{1}{v_{\lfloor b/x \rfloor}} \ln \mathbf{P}\left[\frac{S_{\lfloor b/x \rfloor}}{\lfloor b/x \rfloor} > x\right], \quad b > 0,
\end{aligned}$$

and invoking Eq. (9.36) again we conclude

$$\limsup_{b\to\infty} \frac{1}{h(b)} \max_{n=1,\ldots,\lfloor b/y \rfloor} \ln \mathbf{P}[S_n > b] \le \limsup_{t\to\infty} \left(\sup_{x>y} \frac{v_t}{h(tx)} \cdot \frac{1}{v_t} \ln \mathbf{P}\left[\frac{S_t}{t} > x\right]\right). \tag{9.133}$$

At this stage, Duffield and O'Connell [10] invoke the assumed large deviations principle for the process $\{t^{-1}S_t,\ t = 1, 2, \ldots\}$, say, with good rate function $I\colon \mathbb{R} \to [0, \infty]$, to conclude the inequality

$$\limsup_{t\to\infty}\left(\sup_{x>y} \frac{v_t}{h(tx)} \cdot \frac{1}{v_t} \ln \mathbf{P}\left[\frac{S_t}{t} > x\right]\right) \le \limsup_{t\to\infty}\left(\sup_{x>y} \frac{v_t}{h(tx)} \cdot \left(\delta - \inf_{z>x} I(z)\right)\right) \tag{9.134}$$

for all $\delta > 0$. However, from Eq. (9.35) it is easy to conclude that for any $\delta > 0$, there exists t^* such that

$$\frac{1}{v_t} \ln \mathbf{P}\left[\frac{S_t}{t} > x\right] \le -\inf_{z>x} I(z) + \delta, \quad t > t^*. \tag{9.135}$$

Unfortunately, $t^* = t^*(x, \delta)$, and this dependence of t^* on x precludes taking the supremum $\sup_{x>y}$ on both sides of Eq. (9.135). Hence, Eq. (9.134) does not follow in a straightforward manner. To remedy this difficulty, we continue with the analysis by means of a slightly different approach; an alternative is offered by Duffield [11].

Lemma 9.15.1. *Under Condition (i), it holds that*

$$\limsup_{b\to\infty} \frac{1}{h(b)} \max_{t=1,\ldots,\lfloor b/y \rfloor} \ln \mathbf{P}[S_t > b]$$
$$\le -\sup_{\theta>0}\left(\liminf_{t\to\infty}\left(\inf_{x>y}\left(\frac{v_t}{h(tx)}(\theta x - \Lambda(\theta))\right)\right)\right), \quad y > 0. \tag{9.136}$$

Proof. Fix $y > 0$ and $x > 0$. For each $\theta > 0$, the usual Chernoff bound argument gives

$$\frac{1}{v_t} \ln \mathbf{P}\left[\frac{S_t}{t} > x\right] \le \frac{1}{v_t} \ln(\mathbf{E}[e^{\theta(v_t/t)S_t}]e^{-\theta x v_t})$$
$$= -\theta x + \Lambda_t(\theta), \quad t = 1, 2, \ldots. \tag{9.137}$$

Under Condition (i), if $\Lambda(\theta)$ is finite, then for each $\delta > 0$, there exists an integer $t^* = t^*(\theta, \delta)$ such that

$$\Lambda(\theta) - \delta \le \Lambda_t(\theta) \le \Lambda(\theta) + \delta, \quad t \ge t^*. \tag{9.138}$$

Reporting this fact into the Chernoff bound (9.137), we get

$$\frac{1}{v_t} \ln \mathbf{P}\left[\frac{S_t}{t} > x\right] \le -\theta x + \Lambda(\theta) + \delta, \quad t \ge t^*. \tag{9.139}$$

Consequently,

$$\begin{aligned}\sup_{x>y} \frac{1}{h(xt)} \ln \mathbf{P}\left[\frac{S_t}{t} > x\right] &\leq \sup_{x>y} \frac{v_t}{h(xt)} (\delta - \theta x + \Lambda(\theta)) \\ &\leq \sup_{x>y} \frac{v_t}{h(xt)} \delta - \inf_{x>y} \frac{v_t}{h(xt)} (\theta x - \Lambda(\theta)) \\ &\leq \frac{v_t}{h(yt)} \delta - \inf_{x>y} \frac{v_t}{h(xt)} (\theta x - \Lambda(\theta)), \quad t \geq t^*.\end{aligned}$$

It is now straightforward to conclude that

$$\limsup_{t\to\infty} \left(\sup_{x>y} \frac{1}{h(xt)} \ln \mathbf{P}\left[\frac{S_t}{t} > x\right]\right) \leq -\liminf_{t\to\infty} \left(\inf_{x>y} \left(\frac{v_t}{h(xt)} (\theta x - \Lambda(\theta))\right)\right) \quad (9.140)$$

because δ can be made arbitrarily small. The last inequality is automatically satisfied when $\Lambda(\theta) = \infty$ (in which case the right-hand side is ∞), and the least upper bound being the sharpest, we readily conclude Eq. (9.136) via Eq. (9.133). ■

Letting b go to infinity in Eq. (9.131) and making use of Eqs. (9.132) and (9.136) we indeed get $\mathcal{A}(y) \leq -\alpha(y)$.

9.16 APPENDIX: UPPERBOUNDING $\mathcal{B}(y)$ ($y > 0$)

Fix $y > 0$. If $g(y) = 0$, then $\beta(y) = 0$ and the bound $\mathcal{B}(y) \leq -\beta(y)$ immediately holds owing to the fact that $\mathcal{B}(y) \leq 0$. Hence, from now on we always assume that $g(y) > 0$.

Fix $b > 0$ and $\theta > 0$. By a simple Chernoff bound argument, we have

$$\mathbf{P}[S_t > b] \leq e^{-\theta(v_t/t)b} \mathbf{E}[e^{\theta(v_t/t)S_t}] \leq e^{v_t \Lambda_t(\theta)}, \quad t = 1, 2, \ldots. \quad (9.141)$$

Under Condition (i), if $\Lambda(\theta)$ is *finite*, then for each $\delta > 0$, there exists a finite integer $t^* = t^*(\theta, \delta)$ such that Eq. (9.138) holds. Hence, by a union bound argument, we get

$$\begin{aligned}\mathcal{B}(y) &= \limsup_{b\to\infty} \frac{1}{h(b)} \ln \left(\sum_{t=\lfloor b/y \rfloor+1}^{\infty} \mathbf{P}[S_t > b]\right) \\ &\leq \limsup_{b\to\infty} \frac{1}{h(b)} \ln \left(\sum_{t=\lfloor b/y \rfloor+1}^{\infty} e^{(\Lambda(\theta)+\delta)v_t}\right) \\ &= \limsup_{b\to\infty} \frac{v_{\lfloor b/y \rfloor}}{h(b)} \cdot \frac{1}{v_{\lfloor b/y \rfloor}} \ln \left(\sum_{t=\lfloor b/y \rfloor+1}^{\infty} e^{(\Lambda(\theta)+\delta)v_t}\right) \\ &= g(y) \limsup_{m\to\infty} \frac{1}{v_m} \ln \left(\sum_{t=m+1} e^{(\Lambda(\theta)+\delta)v_t}\right).\end{aligned} \quad (9.142)$$

Setting

$$L(\gamma) \equiv \limsup_{m\to\infty} \frac{1}{v_m} \ln\left(\sum_{t=m+1}^{\infty} e^{-\gamma v_t} \right), \quad \gamma \in \mathbb{R}, \tag{9.143}$$

we can rephrase Eq. (9.142) as

$$\mathcal{B}(y) \le g(y) L(-(\Lambda(\theta) + \delta)), \quad \delta > 0 \tag{9.144}$$

whenever $\Lambda(\theta)$ is finite for $\theta > 0$. The remainder of the discussion hinges on the following expression for Eq. (9.143), which shows that in Eq. (9.144) we need only be concerned with the situation $\Lambda(\theta) + \delta < -K$.

Lemma 9.16.1. *Under Condition (ii), we have*

$$L(\gamma) = \begin{cases} K - \gamma, & \text{if } \gamma > K, \\ \infty, & \text{if } \gamma < K. \end{cases} \tag{9.145}$$

When $K = 0$, Eq. (9.145) reads $L(\gamma) = -\gamma$ for all $\gamma > 0$. When $K > 0$, the boundary case $\gamma = K$ depends on the finer structure of the sequence $\{v_t / \ln t,\ n = 1, 2, \ldots\}$. The proof of Lemma 9.16.1 is elementary and is therefore omitted.

Lemma 9.16.2. *Under Conditions (i) and (ii), whenever $\Lambda^*(0) > K$, we have*

$$\mathcal{B}(y) \le g(y)(K - \Lambda^*(0)), \quad y > 0. \tag{9.146}$$

Proof. Fix $y > 0$ and $\delta > 0$. For each $\theta > 0$, Lemma 9.16.1 shows that

$$L(-(\Lambda(\theta) + \delta)) = \begin{cases} K + (\Lambda(\theta) + \delta), & \text{if } -(\Lambda(\theta) + \delta) > K, \\ \infty, & \text{if } -(\Lambda(\theta) + \delta) < K, \end{cases} \tag{9.147}$$

whenever $\Lambda(\theta)$ is finite. Hence, writing

$$\Theta(\delta) \equiv \{\theta > 0 : K + \Lambda(\theta) + \delta < 0\}, \tag{9.148}$$

we conclude from Eq. (9.144) that

$$\begin{aligned} \mathcal{B}(y) &\le g(y)\left(\inf_{\theta \in \Theta(\delta)} (K + \Lambda(\theta) + \delta)) \right) \\ &= g(y)\left(K + \delta + \inf_{\theta \in \Theta(\delta)} \Lambda(\theta) \right). \end{aligned} \tag{9.149}$$

We note the inclusions $\Theta(\delta) \subset \Theta(\delta')\ (0 < \delta' < \delta)$, together with the relations

$$-\Lambda^*(0) = \inf_{\theta \in \mathbb{R}} \Lambda(\theta) = \inf_{\theta > 0} \Lambda(\theta), \tag{9.150}$$

where the last equality made use of the fact $\mathbf{E}[\xi_1] < 0$. Consequently, under the condition $\Lambda^*(0) > K$, the set $\Theta(\delta)$ is not empty for δ in some nondegenerate interval $(0, \delta^*)$.

Therefore, letting δ go to zero in Eq. (9.149) we get

$$\mathcal{B}(y) \leq g(y)\left(K + \inf_{0<\delta<\delta^*}\left(\inf_{\theta \in \Theta(\delta)} \Lambda(\theta)\right)\right)$$

and the conclusion (9.146) readily follows from Eq. (9.150) by monotonicity. ∎

In short, when $K < \Lambda^*(0)$, Lemma 9.16.2 ensures $\mathcal{B}(y) \leq -\beta(y)$. On the other hand, if $\Lambda^*(0) \leq K$, then $\mathcal{B}(y) \leq 0 \leq -\beta(y)$ trivially. Note that in that case the sets $\Theta(\delta)$ are all empty.

ACKNOWLEDGMENTS

In the course of this work, we have benefited from interactions with several colleagues. Most notably we wish to thank: Marwan Krunz and Konstantin Tsoukatos for various discussions on the $M/G/\infty$ model, Nick Duffield for pointing out an error in Parulekar and Makowski [28], and Liu et al. [22] for making their paper available in preprint form. An early version of this manuscript was prepared while the first author was visiting INRIA—Sophia Antipolis; the hospitality of Projet MISTRAL is gratefully acknowledged. The work of the authors was supported partially through NSF Grant NSFD CDR-88-03012, NASA Grant NAGW77S, the Army Research Laboratory under Cooperative Agreement No. DAAL01-96-2-0002, and the Maryland Procurement Office under Grant No. MDA90497C3015.

REFERENCES

1. R. G. Addie, M. Zukerman, and T. Neame. Fractal traffic measurements, modeling and performance evaluation. In *Proc. IEEE INFOCOM'95*, pp. 985–992, Boston, April 1995.
2. F. Baccelli and P. Brémaud. *Elements of Queueing Theory: Palm–Martingale Calculus and Stochastic Recurrences*, Applications of Mathematics, Vol. 26. Springer-Verlag, Berlin, 1994.
3. J. Beran, R. Sherman, M. S. Taqqu, and W. Willinger. Long-range dependence in variable bit-rate video traffic. *IEEE Trans. Commun.*, **COM-43**:1566–1579, 1995.
4. V. P. Chistakov. A theorem on sums of independent positive random variables and its application to branching random processes. *Theory Probab. Appl.*, **9**:640–648, 1964.
5. G. L. Choudhury, D. M. Lucantoni, and W. Whitt. Squeezing the most out of ATM. *IEEE Trans. Commun.*, **COM-44**:203–217, 1996.

6. M. Crovella and A. Bestavros. Self-similarity in World Wide Web traffic: evidence and possible causes. *Perf. Eval. Rev.*, **24**:160–169, 1996; *Proc. ACM SIGMETRICS'96*, Philadelphia, May 1996.
7. D. R. Cox and V. Isham. *Point Processes*, Chapman and Hall, New York, 1980.
8. D. R. Cox. Long-range dependence: a review. In H. A. David and H. T. David, eds., *Statistics: An Appraisal*, pp. 55–74. The Iowa State University Press, Ames, 1984.
9. A. Dembo and O. Zeitouni. *Large Deviation Techniques and Applications*. Jones and Bartlett, Boston, 1993.
10. N. G. Duffield and N. O'Connell. Large deviations and overflow probabilities for the general single server queue, with applications. *Proc. Cambridge Philos. Soc.*, **118**:363–374, 1995.
11. N. G. Duffield. On the relevance of long-tailed durations for the statistical multiplexing of large aggregations. In *Proceedings of the 34th Annual Allerton Conference on Communications, Control and Computing*, pp. 741–750, Monticello (IL), October 1996,
12. P. Embrechts, C. Klüppelberg, and T. Mikosch. *Modeling Extremal Events*. Springer-Verlag, New York, 1997.
13. H. J. Fowler and W. E. Leland. Local area network traffic characteristics, with implications for broadband network congestion management. *IEEE J. Selected Areas Commun.*, **JSAC-9**:1139–1149, 1991.
14. V. S. Frost and B. Melamed. Traffic modeling for telecommunications networks. *IEEE Commun. Mag.*, **32**:70–81, 1994.
15. M. Garrett and W. Willinger. Analysis, modeling and generation of self-similar VBR video traffic. In *Proc. ACM SIGCOMM'94*, pp. 269–280, University College London, London, August 1994.
16. P. W. Glynn and W. Whitt. Logarithmic asymptotics for steady-state tail probabilities in a single-server queue. *J. Appl. Probab.*, **31**:131–159, 1994.
17. P. R. Jelenković and A. A. Lazar. Multiplexing on–off sources with subexponential on periods: Part II. In *Proceedings of the 15th International Teletraffic Congress*, pp. 965–974, Washington, DC, June 1997.
18. F. P. Kelly. *Reversibility and Stochastic Networks*. Wiley, New York, 1979.
19. G. Kesidis, J. Walrand, and C. S. Chang. Effective bandwidths for multiclass Markov fluids and other ATM sources. *IEEE/ACM Trans. Networking*, **1**:424–428, 1993.
20. W. Leland, M. Taqqu, W. Willinger, and D. Wilson. On the self-similar nature of Ethernet traffic (extended version). *IEEE/ACM Trans. Networking*, **TON-2**:1–15, 1994.
21. N. Likhanov, B. Tsybakov, and N. D. Georganas. Analysis of an ATM buffer with self-similar ("fractal") input traffic. In *Proc. IEEE INFOCOM'95*, pp. 985–992, Boston, April 1995.
22. Z. Liu, Ph. Nain, D. Towsley, and Z.-L. Zhang. Asymptotic behavior of a multiplexer fed by a long-range dependent process. *J. Appl. Probab.* **36**:105–118, 1999.
23. R. M. Loynes. The stability of a queue with non-independent inter-arrival and service times. *Proc. Cambridge Philos. Soc.*, **58**:497–520, 1962.
24. H. Michiel and K. Laevens. Teletraffic engineering in a broadband era. *Proc. IEEE*, **85**:2007–2033, 1997.
25. I. Norros. A storage model with self-similar input. *Queueing Syst. Theory Applic.*, **16**:387–396, 1994.
26. A. G. Pakes. On the tails distribution of waiting time distributions. *J. Appl. Probab.*, **12**:555–564, 1975.

27. M. Parulekar. *Buffer Engineering for $M/G/\infty$ Traffic Models*. Ph.D. Thesis, Electrical Engineering Department, University of Maryland, College Park, December 1999.

28. M. Parulekar and A. M. Makowski. Tail probabilities for a multiplexer with self-similar traffic. In *Proc. IEEE INFOCOM'96*, pp. 1452–1459, San Francisco, April 1996.

29. M. Parulekar and A. M. Makowski. $M/G/\infty$ input processes: a versatile class of models for traffic network. In *Proc. INFOCOM'97*, pp. 1452–1459, Kobe, Japan, April 1997.

30. M. Parulekar and A. M. Makowski. Tail probabilities for a multiplexer driven by $M/G/\infty$ input processes (I): preliminary asymptotics. *Queueing Syst. Theory Applic.*, **27**:271–296, 1997.

31. V. Paxson and S. Floyd. Wide area traffic: the failure of Poisson modeling. *IEEE/ACM Trans. Networking*, **TON-3**:226–244, 1993.

32. B. Tsybakov and N. D. Georganas. On self-similar traffic in ATM queues: definitions, overflow probability bound, and cell delay distribution. *IEEE/ACM Trans. Networking*, **TON-5**:397–409, 1997.

33. W. Willinger, M. S. Taqqu, W. E. Leland, and D. V. Wilson. Self-similarity in high-speed packet traffic: analysis and modeling of Ethernet traffic measurements. *Statist. Sci.*, **10**:67–85, 1995.

10

ASYMPTOTIC ANALYSIS OF QUEUES WITH SUBEXPONENTIAL ARRIVAL PROCESSES

P. R. JELENKOVIĆ

Department of Electrical Engineering, Columbia University, New York, NY 10027

10.1 INTRODUCTION

One of the major challenges in designing modern communication networks is providing quality of service to the individual users. An important part of this design process is understanding statistical characteristics of network traffic streams and their impact on network performance. Unlike the conventional voice traffic, modern data traffic exhibits an increased level of "burstiness" that spans over multiple time scales. It was observed that sample paths of these data sequences show evidence of self-similarity. Their autocorrelation structure is characterized by long-range dependency and the empirical distributions are easily matched with subexponential and long-tailed distributions. Early discovery of the self-similar nature of Ethernet traffic was reported in Leland et al. [42] (see also Leland et al. [43]). More recently, Crovella [22] attributed the long-range dependency of Ethernet traffic to the long-tailed file sizes that are transferred over the network. Long-range dependency of the variable bit rate video traffic was demonstrated by Beran et al. [9]. Long-tailed characteristics of the scene length distribution of MPEG video streams were explored in Heyman and Lakshman [30] and Jelenković et al. [37].

Practical importance, novelty, and the intriguing nature of these phenomena have attracted a great number of scientists to develop new traffic models and to understand the impact of these models on network performance. In this development there

Self-Similar Network Traffic and Performance Evaluation, Edited by Kihong Park and Walter Willinger
ISBN 0-471-31974-0

have been two basic approaches: self-similar (fractal) processes and fluid renewal models with long-tailed renewal distributions. In this presentation we focus on the latter. The investigation of queueing systems with self-similar arrival processes can be found in the literature [23, 24, 44, 47, 49, 51, 54, 55].

In this chapter some recent results are presented on the subexponential asymptotic behavior of queueing systems with subexponential arrival streams. The related references will be listed throughout the chapter. First, in Section 10.2 the classes of long-tailed and subexponential distributions are defined and some of their basic properties are presented. Section 10.3 begins with a presentation of a classical result on the subexponential asymptotics of a $GI/GI/1$ queue. That is followed by a brief discussion of various extensions of this result that can be found in the literature. The remainder of Section 10.3 contains two new results on this subject. In Section 10.3.1 a derivation is given for a straightforward asymptotic approximation for the loss rate in a finite buffer $GI/GI/1$ queue. It appears surprising that the derived asymptotic formula does not depend on the queue service process. However, a simple intuitive explanation of this insensitivity effect is provided. In Section 10.3.2 a $GI/GI/1$ queue with truncated heavy-tailed arrival sequences is analyzed. Explicit asymptotic characterization of a unique behavior of the queue length distribution is given. Informally, this distribution on the log scale resembles a *stair-wave* function that has steep drops at specific buffer sizes. This has important design implications, suggesting that negligible increases of the buffer size in certain buffer regions can decrease the overflow probabilities by orders of magnitude.

Section 10.4 describes a class of fluid queues and addresses the problem of multiplexing on/off sources with heavy-tailed on periods. A complete rigorous treatment of the subexponential asymptotic behavior of a fluid queue with a single on/off arrival process is presented in Section 10.4.1. Section 10.4.2 investigates multiplexing a heavy-tailed on/off process with a process that has a lighter (exponential) tail. It is shown that this queueing system is asymptotically equivalent to the queueing system in which the process with the lighter tail is replaced by its mean value. This has implications on multiplexing bursty data and video traffic with relatively smooth voice sources. Section 10.4.3 addresses the problem of multiplexing on/off sources with heavy-tailed on periods. Understanding of this problem is fundamental for achieving high network resource utilization and providing quality of service in the bursty traffic environment. Under a specific stability condition this problem admits an elegant asymptotic solution. A brief conclusion of the presentation is given in Section 10.5.

10.2 LONG-TAILED AND SUBEXPONENTIAL DISTRIBUTIONS

This section contains necessary definitions of long-tailed and subexponential distributions. An extensive treatment of subexponential distributions (and further references) can be found in Cline [17, 18] or in the recent survey by Goldie and Klüppelberg [27].

Definition 10.2.1. A distribution function F on $[0, \infty)$ is called *long-tailed* ($F \in \mathscr{L}$) if

$$\lim_{x \to \infty} \frac{1 - F(x - y)}{1 - F(x)} = 1, \quad y \in \mathbb{R}. \tag{10.1}$$

Definition 10.2.2. A distribution function F on $[0, \infty)$ is called *subexponential* ($F \in \mathscr{S}$) if

$$\lim_{x \to \infty} \frac{1 - F^{*2}(x)}{1 - F(x)} = 2, \tag{10.2}$$

where F^{*2} denotes the second convolution of F with itself, that is, $F^{*2}(x) = \int_{[0,\infty)} F(x - y)F(dy)$.

The class of subexponential distributions was first introduced by Chistakov [15]. The definition is motivated by the simplification of the asymptotic analysis of convolution tails. The best-known examples of distribution functions in $\mathscr{S}$ (and $\mathscr{L}$) are functions of regular variation $\mathscr{R}_{-\alpha}$ (in particular, Pareto family); $F \in \mathscr{R}_{-\alpha}$ if it is given by

$$F(x) = 1 - \frac{l(x)}{x^{\alpha}}, \quad \alpha \geq 0,$$

where $l(x)\colon \mathbb{R}_+ \to \mathbb{R}_+$ is a function of slow variation, that is, $\lim_{x \to \infty} l(\delta x)/l(x) = 1$, $\delta > 1$. These functions were invented by Karamata [38] (the main reference book is by Bingham et al. [10]). The other examples include lognormal and some Weibull distributions (see Jelenković and Lazar [36] and Klüppelberg [40]).

A few classical results from the literature on subexponential distributions follow. The general relation between $\mathscr{S}$ and $\mathscr{L}$ is presented in Lemma 10.2.3.

Lemma 10.2.3 [7]. $\mathscr{S} \subset \mathscr{L}$.

Lemma 10.2.4. *If $F \in \mathscr{L}$ then $(1 - F(x))e^{\alpha x} \to \infty$ as $x \to \infty$, for all $\alpha > 0$.*

NOTE 10.2.5. Lemma 10.2.4 clearly shows that for long-tailed distributions Cramér-type conditions are not satisfied.

One of the most basic properties of subexponential distributions is given in the following lemma. It roughly states that the sum of n i.i.d. random variables exceeds a large value x due to one of them exceeding x.

Lemma 10.2.6. *Let $\{X_n, n \geq 1\}$ be a sequence of i.i.d. random variables with a common distribution F and let $S_n = \sum_{i=1}^{n} X_i$. If $F \in \mathscr{S}$, then*

$$\mathbb{P}[S_n > x] \sim n\mathbb{P}[X_1 > x] \quad \text{as } x \to \infty. \tag{10.3}$$

Often in renewal theory it is of interest to investigate the *integrated tail* of a distribution function. To simplify the notation, for any distribution F we denote by $\bar{F}(x) = 1 - F(x)$, $\hat{F}(x) \stackrel{\text{def}}{=} \int_x^\infty \bar{F}(t)\,dt$, and $F_1(x) \stackrel{\text{def}}{=} m^{-1}(m - \hat{F}(x))$, where $m = \hat{F}(0)$. Throughout the text $F_1(x)$ will be referred as the integrated tail distribution of $F(x)$.

Definition 10.2.7. $F \in \mathscr{S}^*$ if

$$\int_0^x \frac{\bar{F}(x-y)}{\bar{F}(x)} \bar{F}(y)\,dy \to 2m_F < \infty, \qquad \text{as } x \to \infty,$$

where $m_F = \int_0^\infty yF(dy)$.

This class of distributions has the property that $F \in \mathscr{S}^* \Rightarrow F_1 \in \mathscr{S}$, and that $\mathscr{S}^* \subset \mathscr{S}$. Sufficient conditions for $F \in \mathscr{S}^*$ can be found in Klüppelberg [41], where it was explicitly shown that lognormal, Pareto, and certain Weibull distributions are in $\mathscr{S}^*$.

10.3 LINDLEY'S RECURSION AND *GI/GI/1* QUEUE

Let $\{A, A_n, n \in \mathbb{N}_0\}$ and $\{C, C_n, n \in \mathbb{N}_0\}$ be two independent sequences of i.i.d. random variables (on a probability space $(\Omega, \mathscr{F}, \mathbb{P})$). We term A_n and C_n as the arrival and service process, respectively. Then, for any initial random variable Q_0, the following Lindley's equation,

$$Q_{n+1} = (Q_n + A_{n+1} - C_{n+1})^+, \tag{10.4}$$

defines the *discrete-time queue length process* $\{Q_n, n \geq 0\}$. According to the classical result by Loynes [45] (see also Baccelli and Bremaud [8, Chap. 2]), there exists a unique stationary solution to recursion (10.4), and for all initial conditions the queue length process converges (in finite time) to this stationary process. In this chapter it is assumed that the queue is in its stationary regime, that is, $\{Q_n, n \geq 0\}$ is the stationary solution to recursion (10.4).

Recursion (10.4) also represents the waiting time process of the $GI/GI/1$ queue with C_n being interpreted as the interarrival time between the customer $n-1$ and n, A_n as the customer's n service requirement, and Q_n as the customer's n waiting time. For that reason the terms *waiting time distribution* for the $GI/GI/1$ queue and the *queue length distribution* for the discrete time queue will be used interchangeably.

Some of the first applications of long-tailed distributions in queueing theory were done by Cohen [20] and Borovkov [11] for the functions of regular variations. Cohen derived the asymptotic behavior of the waiting time distribution for the $M/GI/1$ queue. This result was extended by Pakes [48] to $GI/GI/1$ queue and the whole class of subexponential distributions. In Veraverbeke [56] the same result was rederived using a random walk technique. Let G and G_1 represent the distribution and its integrated tail distribution for A_n, respectively, $(G_1(x) = \int_0^x \mathbb{P}[A > u]\, du/\mathbb{E}A)$.

Theorem 10.3.1 (Pakes). *If $G_1 \in \mathcal{S}$ (or $G \in \mathcal{S}^*$), and $\mathbb{E}A_n < \mathbb{E}C_n$, then*

$$\mathbb{P}[Q_n > x] \sim \frac{1}{\mathbb{E}C_n - \mathbb{E}A_n} \int_x^\infty \mathbb{P}[A_n > u]\, du \text{ as } x \to \infty.$$

There are several natural avenues for extending this result. In Willekens and Teugels [58] and Abate et al. [1] asymptotic expansion refinements to Theorem 10.3.1 were investigated. For extensions of Theorem 10.3.1 to Markov-modulated $M/G/1$ queues see Asmussen et al. [4], and to Markov-modulated $G/G/1$ queues (equivalently random walks) see Jelenković and Lazar [36]. Further extension of these results to more general arrival processes was obtained in Asmussen et al. [6]. Recently, Asmussen et al. [5] established an asymptotic relationship between the number of customers in a $GI/GI/1$ queue and their waiting time distribution.

In the rest of this section recent results are presented on a $GI/GI/1$ queue with a finite buffer and truncated heavy-tailed arrival sequences.

10.3.1 Finite Buffer $GI/GI/1$ Queue

In engineering network switches it is very common to design them as loss systems. The main performance measures for these systems are loss probabilities and loss rates. Unfortunately, there are no asymptotic results in literature that address this problem under the assumption of long-tailed arrivals. Recently, I investigated this problem [31, 33].

Here, in Theorem 10.3.2, I present the main result from my earlier work [31]. The theorem gives an explicit asymptotic characterization of the loss rate in a finite buffer queue with long-tailed arrivals. This result, in combination with results from Jelenković and Lazar [35], yields a straightforward asymptotic formula for the loss rate in a fluid queue with long-tailed $M/G/\infty$ arrivals (for more details see Jelenković [31, 33]). In addition, I [31, 33] derived an explicit asymptotic approximation of buffer occupancy probabilities. This approximation is uniformly accurate for buffer sizes that are away from the buffer boundaries (zero and the maximum buffer size). Furthermore, as the maximum buffer size increases, the length of the buffer around the boundaries where the approximation does not apply stays constant. This precise knowledge of the buffer probabilities allows computation of various other functionals of the finite buffer queue.

The evolution of a finite buffer queue is defined with the following recursion:

$$Q_{n+1}^B = \min((Q_n^B + A_{n+1} - C_{n+1})^+, B), \quad n \ge 0,$$

where B is the buffer size. We assume that the queueing process is in its stationary regime. The loss rate is defined as

$$\lambda_{\text{loss}}^B \stackrel{\text{def}}{=} \mathbb{E}(Q_n^B + A_{n+1} - C_{n+1} - B)^+.$$

Theorem 10.3.2. *Let G_1 be the integrated tail distribution of A. If $G_1 \in \mathscr{S}$ and $\mathbb{E}A < \mathbb{E}C$, then*

$$\lambda_{\text{loss}}^B \stackrel{\text{def}}{=} \mathbb{E}(A - B)^+(1 + o(1)) \quad \text{as } B \to \infty.$$

HEURISTIC 10.3.4. Following the general heuristics for subexponential distributions the large buffer overflow is due to one (isolated) large arrival A_n. At the moment when this happens (say, time n) the queue length process is, because of the stability condition $\mathbb{E}A < \mathbb{E}C$, typically very small in comparison to B. Similarly, C_n is much smaller than B. Hence, the amount that is lost at the time of overflow is approximately $(Q_n^B + A_{n+1} - C_{n+1} - B)^+ \approx (A_{n+1} - B)^+$.

Accuracy of Theorem 10.3.2 was demonstrated [31, 33] with many numerical and simulation experiments. Here, an example is presented.

Example 10.3.5. Take $C_n \equiv 2$ and an arrival distribution $\mathbb{P}[A = 0] = \frac{1}{2}$, $\mathbb{P}[A = i] = 0.461969/i^4$, $i > 0$, $\mathbb{E}A = 0.5553$. Then, we numerically compute the loss rates λ_{loss}^B for the maximum buffer sizes $B = 100i$, $i = 1, \ldots, 7$. The results are presented with circles in Fig. 10.1. Note that for $B = 700$ we needed to solve a

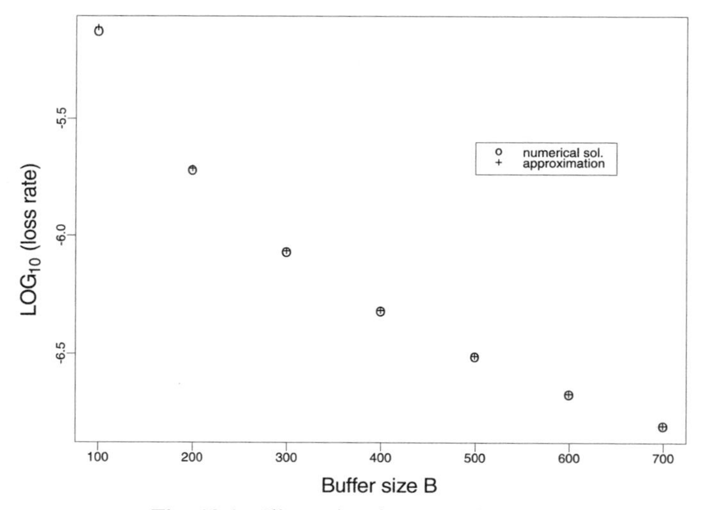

Fig. 10.1 Illustration for Example 10.3.5.

system of 700 linear equations! In contrast, Theorem 10.3.2 readily suggests an asymptotic approximation $\tilde{\lambda}_{\text{loss}}^{B} = 0.0767/B^2$. The approximation is presented on the same figure with "+" symbols. A precise match is apparent from the figure. In fact, relative error $|\tilde{\lambda}_{\text{loss}}^{B} - \lambda_{\text{loss}}^{B}|/\lambda_{\text{loss}}^{B}$ for all computed buffers was less than 4%.

10.3.2 Truncated Long-Tailed Arrival Distributions

In this section we investigate the queueing behavior when the distribution of the arrival sequence has a bounded (truncated) support [32, 34]. This arises quite frequently in practice when the arrival process distribution has a bounded support and inside that support is nicely matched with a heavy-tailed distribution (e.g., Pareto).

Our primary interest in this scenario is in its possible application to network control. More precisely, one can imagine network control procedure in which short network flows are separated from long ones. If the distribution of flows is long-tailed, this procedure will yield a truncated long-tailed distribution for the short network flows. Assume that long flows are transmitted separately using virtual circuits and short flows are multiplexed together. Intuitively, it can be expected that with short (truncated) flows one can obtain better multiplexing gains than with the original ones (before the separation). These gains are quantified in Theorem 10.3.6, which explicitly asymptotically characterizes a unique asymptotic behavior of the queue length distribution. Informally, this distribution on the log scale resembles a *stair-wave* function that has steep drops at specific buffer sizes (see Fig. 10.2). This has important design implications suggesting that negligible increases of the buffer

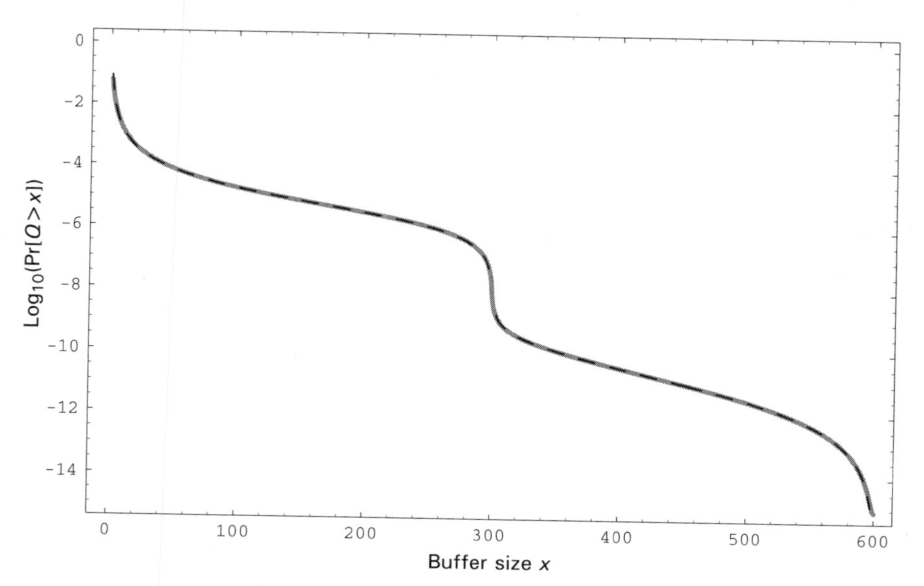

Fig. 10.2 Illustration for Example 10.3.9.

size in certain buffer regions can decrease the overflow probabilities by orders of magnitude.

Formally, for each $B > 0$ construct a sequence of truncated random variables

$$A_n^B = \min(A_{n_1} B).$$

Next, consider a single server queue with the arrival process $\{A^B, A_n^B, n \geq 0\}$, that is,

$$Q_{n+1}^B = (Q_n^B + A_{n+1}^B - C_{n+1})^+. \tag{10.5}$$

Assume that for all B, Q_n^B is in its stationary regime.

Theorem 10.3.6. *If $\mathbb{E}(A - C) < 0$, for all $n > 0$, $\mathbb{P}[C > x] \leq e^{-\eta x}$, $\eta > 0$, and A has a regularly varying distribution $\mathbb{P}[A > x] = l(x)/x^\alpha$, then*

$$\mathbb{P}[Q^B > (k + \delta)B] = \frac{h_k(\delta)}{(\mathbb{E}C - \mathbb{E}A)^{k+1}} \frac{l(B)^{k+1}}{B^{(k+1)(\alpha-1)}}(1 + o(1)) \quad \text{as } B \to \infty, \tag{10.6}$$

where $h_k(\delta)$, $0 < \delta < 1$, $k = 0, 1, 2, \ldots,$ are easily computable from

$$h_k(\delta) \stackrel{\text{def}}{=} \int\limits_{\substack{0 < x_i \leq 1, 1 \leq i \leq k+1 \\ x_1 + \cdots + x_{k+1} \geq \delta}} x_1^{-\alpha} \cdots x_{k+1}^{-\alpha} \, dx_1 \cdots dx_{k+1}. \tag{10.7}$$

HEURISTIC 10.3.7. In order that the queue exceeds a large buffer size $b = (k + \delta)B$ it is needed that exactly $k + 1$ large arrivals (of the order B) occur at approximately the same time. Since successive arrivals are independent this event is of the order $l(B)^{k+1}/B^{(k+1)(\alpha-1)}$. The detailed proof of this result can be found in Jelenković [32].

REMARK 10.3.8. (i) This result is related to Proposition 1 in Resnick and Samorodnitsky [50], where, under conditions similar to our theorem, a rough bound for the queue length increment during an activity period of the $M/GI/\infty$ arrival process was derived. (ii) Note that $h_0(\delta)$ is explicitly given by

$$h_0(\delta) = \frac{1}{(\alpha - 1)\delta^{\alpha-1}}(1 - \delta^{\alpha-1}). \tag{10.8}$$

Now, we illustrate Theorem 10.3.6 with the following example (for more examples see Jelenković [32, 34]).

Example 10.3.9. Parameterize the distribution of A_n^B as $a_0^B = 1 - p$, $a_i^B = pd/i^{\alpha+1}$, $1 \leq i \leq B - 1$, $a_B^B = 1 - \sum_{i=0}^{B-1} a_i$, where $d = 1/\zeta(\alpha + 1)$ and $\zeta(x)$ is a Zeta function. For the choice of arrival parameters $B = 300$, $\alpha = 2.8$, and $p = 0.3$ we compute $d = 1/\zeta(\alpha + 1) = 0.273345$, $a_0^B = 0.7$, $a_i^B = 0.0820/i^{\alpha+1}$, $1 \leq i \leq B - 1$, $\rho^B = 0.34086$. For these values we numerically invert the z-transform of the queue length distribution. These exact values of $\mathbb{P}[Q^B > x]$ are plotted with a gray line in Fig. 10.2. The values of approximation (10.6) are plotted on the same figure with dashed black lines. From the figure we can easily see that the approximation is almost identical to the exactly computed probabilities.

10.4 FLUID QUEUES AND MULTIPLEXING

Fluid queues with long-tailed characteristics have received significant attention in the recent queueing literature. The latest survey of the subject can be found in Boxma and Dumas [14]. In this section some results from Jelenković and Lazar [35] are presented.

The physical interpretation for a fluid queue is that, at any moment of time t, fluid is arriving to the system with rate a_t and is leaving the system with rate c_t. We term a_t and c_t to be the arrival and the service process, respectively. Then, the evolution of the amount of fluid Q_t (also called queue length) evolves according to

$$dQ_t = (a_t - c_t)\,dt \quad \text{if } Q_t > 0, \text{ or } a_t > c_t, \tag{10.9}$$

and $dQ_t = 0$, otherwise. It is not very difficult to see that, starting from $Q_0 = 0$, the solution Q_t, $t \geq 0$, to Eq. (10.9) is given by

$$Q_t = \sup_{0 \leq u \leq t} \int_u^t (a_u - c_u)\,du. \tag{10.10}$$

And, if a_t and c_t are stationary, Q_t is equal in distribution to

$$\mathbb{P}[Q_t \leq x] = \mathbb{P}\left[\sup_{0 \leq u \leq t} W_u \leq x\right],$$

where $W_t \stackrel{\text{def}}{=} \int_{-t}^{0} (a_u - c_u)\,du$, $t \geq 0$. Now, whenever the stability condition $\mathbb{E}a_t < \mathbb{E}c_t$ is satisfied (by Birkhoff's Strong Law of Large Numbers), $\mathbb{P}[Q_t \leq x]$ converges to a proper probability distribution; that is,

$$\mathbb{P}[Q \leq x] \stackrel{\text{def}}{=} \lim_{t \to \infty} \mathbb{P}[Q_t \leq x] = \mathbb{P}\left[\sup_{0 \leq u < \infty} W_u \leq x\right].$$

Furthermore, when the difference process $x_t \stackrel{\text{def}}{=} a_t - c_t$ is driven by a stationary and ergodic point process $\{T_n, -\infty < n < \infty\}$, that is,

$$x_t = x_{T_n}, \quad t \in [T_n, T_{n+1}),$$

then the fluid queue process evolves as

$$Q_t = (Q_{T_{n-}} + (t - T_n)x_{T_n})^+, \quad t \in [T_n, T_{n+1}), \tag{10.11}$$

where $q^+ = \max(q, 0)$. From the recursion above, it is clear that the process Q_t is essentially the same as the $G/G/1$ workload process. Hence, by the fundamental stability theorem of Loynes there exists a unique stationary solution to Eq. (10.11). We assume that $\{Q_t,\ -\infty < t < \infty\}$ is that stationary solution.

10.4.1 Fluid Queue with a Single On/Off Process

This section presents a complete asymptotic analysis of a fluid queue with a single subexponential on/off arrival process. A general storage model in a two-state random environment was investigated by Kella and Whitt [39].

More formally, consider two independent sequences of i.i.d. random variables $\{\tau_n^{\text{off}},\ n \geq 0\}$, $\{\tau_n^{\text{on}},\ n \geq 0\}$, $\tau_0^{\text{off}} = \tau_0^{\text{on}} = 0$. Define a point process $T_n^{\text{off}} \stackrel{\text{def}}{=} \sum_{i=0}^{n}(\tau_i^{\text{off}} + \tau_i^{\text{on}})$, $n \geq 0$; this process will be interpreted as representing the beginnings of off periods in an on/off process. Furthermore, define an on/off process a_t, with rate r, as

$$a_t = r \quad \text{if} \quad T_n^{\text{off}} - \tau_n^{\text{on}} \leq t < T_n^{\text{off}}, \qquad n \geq 1,$$

and $a_t = 0$ otherwise.

Then, if we observe the queue at the beginning of on periods, the queue length Q_n^{P} evolves as follows (P stands for Palm probability [8]).

$$Q_{n+1}^{\text{P}} = (Q_n^{\text{P}} + (r - c)\tau_n^{\text{on}} - c\tau_n^{\text{off}})^+, \qquad n \geq 0. \tag{10.12}$$

Let F and F_1 be the distribution and the integrated tail distribution, respectively, of τ^{on}.

Theorem 10.4.1. *If $r > c$, $(r - c)\mathbb{E}\tau_{\text{on}} < c\mathbb{E}\tau_{\text{off}}$, and $F_1 \in \mathcal{S}$ (or $F \in \mathcal{S}^*$), then*

$$\mathbb{P}[Q_n^{\text{P}} > x] \sim \frac{r - c}{c\mathbb{E}\tau_{\text{off}} - (r - c)\mathbb{E}\tau_{\text{on}}} \int_{x/(r-c)}^{\infty} \mathbb{P}[\tau^{\text{on}} > u]\, du \quad \text{as } x \to \infty. \tag{10.13}$$

Proof. Define $A_n = (r - c)\tau_n^{\text{on}}$ and $C_n = c\tau_n^{\text{off}}$ and apply Theorem 10.3.1. ■

10.4.1.1 Time Averages. At this point, we will compute queue time averages based on the queue Palm probabilities computed in Theorem 10.4.1. For this we need a stationary version a_t^s of the on/off arrival process a_t. Let T_n^{on}, $-\infty < n < \infty$, be a stationary point process that represents the beginnings of the on/off periods, with a convention that $T_0^{\text{on}} < 0 \leq T_1^{\text{on}}$. Then, according to Resnick and Samorodnitsky [51], the random variable T_0^{on} can be represented as $-T_0^{\text{on}} = B(\tau_{(0)}^{\text{off}} + \tau_0^{\text{on}}) + (1 - B)\tau_{(0)}^{\text{on}}$, where the random variables B, $\tau_{(0)}^{\text{on}}$, $\tau_{(0)}^{\text{off}}$ are independent of $\{\tau_n^{\text{on}}, \tau_n^{\text{off}}, n \leq -1\}$, τ_0^{off}, B is a Bernoulli random variable with $\mathbb{P}[B = 0] = 1 - \mathbb{P}[B = 1] = \mathbb{E}\tau^{\text{on}}/(\mathbb{E}\tau^{\text{on}} + \mathbb{E}\tau^{\text{off}})$, and $\tau_{(0)}^{\text{on}}$, $\tau_{(0)}^{\text{off}}$ are distributed as integrated tail distributions of τ^{on}, τ^{off}, respectively. Furthermore, the net increment of the load that comes to the queue in the interval $[T_0, 0]$ is given by the following equation:

$$\int_{T_0^{\mathrm{on}}}^{0} (a_t^s - c)\, dt = B[(r-c)\tau_0^{\mathrm{on}} - c\tau_{(0)}^{\mathrm{off}}] + (1-B)(r-c)\tau_{(0)}^{\mathrm{on}}. \tag{10.14}$$

Theorem 10.4.2. *If $r > c$, $(r-c)\mathbb{E}\tau_{\mathrm{on}} < c\mathbb{E}\tau_{\mathrm{off}}$, and $F_1 \in \mathcal{S}$ (or $F \in \mathcal{S}^*$), then*

$$\mathbb{P}[Q_t > x] \sim \mathbb{P}[Q_n^{\mathrm{P}} > x] + \frac{1}{\mathbb{E}\tau^{\mathrm{off}} + \mathbb{E}\tau^{\mathrm{on}}} \int_{x/(r-c)}^{\infty} \mathbb{P}[\tau^{\mathrm{on}} > u]\, du \tag{10.15}$$

$$\sim K \int_{x/(r-c)}^{\infty} \mathbb{P}[\tau^{\mathrm{on}} > u]\, du \quad \text{as } x \to \infty, \tag{10.16}$$

where

$$K = \frac{r-c}{c\mathbb{E}\tau_{\mathrm{off}} - (r-c)\mathbb{E}\tau_{\mathrm{on}}} + \frac{1}{\mathbb{E}\tau^{\mathrm{off}} + \mathbb{E}\tau^{\mathrm{on}}}. \tag{10.17}$$

REMARK 10.4.3. (i) This theorem improves on known results [16, 51] that were obtained under the assumption of τ^{on} being regularly varying. (ii) The following proof can be carried out to establish the relationship between the Palm and time averages in much more general settings like semi-Markov fluid queues.

Most of the results in this chapter can be found elsewhere and therefore these proofs are omitted. Here, as an illustration of a subexponential proving technique, the following proof of Theorem 10.4.2 is presented. This proof is taken from Jelenković and Lazar [35].

Proof. Let $\{Q_t, -\infty < t < \infty\}$ be a unique stationary solution to Eq. (10.12). Then, by using Eq. (10.14), and the independence of B of Q_{T_0}, $\tau_{(0)}^{\mathrm{off}}$, $\tau_{(0)}^{\mathrm{on}}$, τ_0^{off}, we obtain

$$\begin{aligned}
\mathbb{P}[Q_0 > x] &= \mathbb{P}[Q_0 > x, B = 1] + \mathbb{P}[Q_0 > x, B = 0] \\
&= \mathbb{P}[Q_{T_0} + \tau_0^{\mathrm{on}}(r-c) - c\tau_{(0)}^{\mathrm{off}} > x, B = 1] \\
&\quad + \mathbb{P}[Q_{T_0} + (r-c)\tau_{(0)}^{\mathrm{on}} > x, B = 0] \\
&= \frac{\mathbb{E}\tau^{\mathrm{off}}}{\mathbb{E}\tau^{\mathrm{on}} + \mathbb{E}\tau^{\mathrm{off}}} \mathbb{P}[Q_{T_0} + \tau_0^{\mathrm{on}}(r-c) - c\tau_{(0)}^{\mathrm{off}} > x] \\
&\quad + \frac{\mathbb{E}\tau^{\mathrm{on}}}{\mathbb{E}\tau^{\mathrm{on}} + \mathbb{E}\tau^{\mathrm{on}}} \mathbb{P}[Q_{T_0} + (r-c)\tau_{(0)}^{\mathrm{on}} > x].
\end{aligned} \tag{10.18}$$

(Note that $Q_0^{\mathrm{P}} = Q_{T_0}$). Since Q_{T_0} and $\tau_{(0)}^{\mathrm{on}}$ are independent and subexponential and have asymptotically proportional (equivalent) tails, by applying Lemma 5(ii)(A) of

Jelenković and Lazar [35], it follows that

$$\mathbb{P}[Q_{T_0} + (r-c)\tau_{(0)}^{\mathrm{on}} > x] \sim \mathbb{P}[Q_{T_0} > x] + \mathbb{P}[(r-c)\tau_{(0)}^{\mathrm{on}} > x] \quad \text{as } x \to \infty. \tag{10.19}$$

The independence of τ_0^{on} and $\tau_{(0)}^{\mathrm{off}}$, and $\tau_0^{\mathrm{on}} \in \mathscr{L}$, by the definition of long-tailed distributions it follows that $\mathbb{P}[\tau_0^{\mathrm{on}}(r-c) - c\tau_{(0)}^{\mathrm{off}} > x] \sim \mathbb{P}[\tau_0^{\mathrm{on}}(r-c) > x] = o(\mathbb{P}[Q_{T_0} > x])$ as $x \to \infty$. Subsequently, by applying Lemma 5(i)(A) of Jelenković and Lazar [35],

$$\mathbb{P}[Q_{T_0} + \tau_0^{\mathrm{on}}(r-c) - c\tau_{(0)}^{\mathrm{off}} > x] \sim \mathbb{P}[Q_{T_0} > x]) \quad \text{as } x \to \infty. \tag{10.20}$$

Finally, by replacing asymptotic relations (10.19) and (10.20) in Eq. (10.18), we obtain Eq. (10.15); combination of Eqs. (10.13) and (10.15) gives Eq. (10.16). This completes the proof. ■

10.4.2 Asymptotic Reduced Load Equivalence

In this section we consider multiplexing one long-tailed on/off process with exponential processes in a fluid queue. In Boxma [12, 13], a precise asymptotics of the embedded queue distribution was obtained for multiplexing on/off sources, one of which had regularly varying on periods, while the others had exponentially distributed on periods. A similar setting with intermediately varying on periods was investigated in Rolski et al. [52]. Jelenković and Lazar [35] observed that this queueing system is asymptotically interchangeable with a queueing system in which the on/off process is arriving alone and the exponential processes are replaced by their mean values. This result has been generalized in Agrawal et al. [2]. The title of this subsection is borrowed from the title of their paper.

In the remainder of this section, a result from Jelenković and Lazar [35] is presented. In order to state the result, the following definitions are introduced.

Definition 10.4.4. A distribution function F is *intermediate regular varying* $F \in \mathscr{IR}$ if

$$\lim_{\delta \downarrow 1} \liminf_{t\to\infty} \frac{\bar{F}(\delta t)}{\bar{F}(t)}.$$

REMARK 10.4.5. For recent results on distributions of intermediate regular variation we refer the reader to Cline [19]. Some basic properties of $\mathscr{IR}$ are: $\mathscr{IR} \subset \mathscr{S}$; $\mathscr{R} \subset \mathscr{IR}$. Also, it is not very difficult to see that $\mathscr{IR} \subset \mathscr{S}^*S$. Therefore, all of the results obtained in this chapter apply for $\mathscr{IR}$. In addition, directly from the definition it can be shown that $F \in \mathscr{IR}$ $\int_0^\infty \bar{F}(t)\,dt < \infty, \Rightarrow F_1 \in \mathscr{IR}$.

Under the general large deviation Gärtner–Ellis conditions (see Weiss and Shwartz [57]) on the arrival process, it can be proved that the queue length distribution is exponentially bounded. To avoid stating Gärtner–Ellis conditions, we will define an arrival process e_t to be *exponential*, if whenever this process is fed into a constant server fluid queue, the queue length distribution is exponentially bounded.

Definition 10.4.6. We say that a stationary and ergodic arrival process e_t is *exponential* if for any server capacity $c > \mathbb{E}e_t$ there exists $K \equiv K(c)$ and $\delta \equiv \delta(c) > 0$ such that

$$\mathbb{P}\left[\sup_{t\geq 0} \int_{-t}^{0} (e_u - c)\, du > x\right] \leq Ke^{-\delta x}.$$

REMARK 10.4.7. The main examples when the conditions of this definition are satisfied (i.e., Gärtner–Ellis conditions hold) are finite state space Markov chains or processes. Also, in terms of the on/off processes the conditions will hold whenever the distribution of on periods is exponentially bounded and off periods have a finite mean.

Recall that F and F_1 represent the distribution and the integrated tail distribution, respectively, of an on period. Then, we arrive at the following result (see Jelenković and Lazar [35]).

Theorem 10.4.8. *Consider a single server queue with a capacity c, and two independent arrival streams e_t and a_t. Assume that e_t is an exponential process and a_t is an on/off process with rate r, $F \in \mathcal{IR}$, and generally distributed off periods with a finite mean. If $\mathbb{E}(e_t + a_t) < c$, $r > c' \stackrel{\text{def}}{=} c - \mathbb{E}e_t$, then the queue asymptotics of this queueing system is equal to the queue asymptotics in which only the on/off process arrives and the server capacity is replaced by c', that is, it is given by Eq. (10.16) in which c is replaced by c'.*

REMARK 10.4.9. This result is true with exactly the same proof if the assumption of e_t being exponential is replaced with $\mathbb{P}[\sup_{t\geq 0} \int_{-t}^{0}(e_u - c)\, du > x] = o(F_1(x))$, for all $c > \mathbb{E}e_t$.

HEURISTIC 10.4.10. Large buildups in this fluid queue occur due to long and isolated on periods in a_t. During these long on periods the fluctuations in the exponential arrival stream e_t average out, and therefore its contribution to the asymptotic behavior is only through its mean value.

10.4.3 Multiplexing On/Off Sources

The problem of multiplexing on/off sources arises frequently as the basic model of contention in multimedia communication systems, as well as in some storage systems. The analysis of this problem dates back to Rubinovitch [53] and Cohen [21]. Cohen obtained a complete Laplace transform solution to this problem.

However, inverting the Laplace transform is usually a very tedious process. Hence, computationally tractable exact and approximate solution techniques are needed. For Markovian (fluid) on/off processes a thorough investigation of this problem was done in Anick et al. [3]. Many other results for multiplexing Markovian on/off processes followed. These led to the *equivalent bandwidth theory* for Markovian (or in general exponentially bounded) arrival processes; extensive references can be found in Duffield and O'Connell [24], Elwalid et al. [25], and Glynn and Whitt [26].

The analysis of a fluid queue in which more than one long-tailed process is multiplexed appears to be a very difficult problem. This is due to the fact that the renewal structure of an aggregate arrival process may be very complex, although the appearance of each individual process may be truly innocuous (like an on/off process). The complex autocorrelation structure of the aggregate process obtained by multiplexing long-tailed on/off processes has been examined in Heath et al. [28]. General bounds for multiplexing long-tailed fluid processes have been derived in Choudhury et al. [16]. In Poisson scaling the limiting case of an infinite number of on/off processes converges to the so-called $M/GI/\infty$ process. Asymptotic results for a fluid queue with a heavy-tailed $M/GI/\infty$ arrival process have been obtained in Boxma [12, 13] and Jelenković and Lazar [35]. Recently, new results on this model have been derived in Heath et al. [29] and Resnick and Samorodnitsky [50]. For various bounds in this context see Nain et al. [46].

10.4.3.1 Activity Period of an $M/GI/\infty$ Process. Let T_n, $n \geq 0$, $-\infty < n < \infty$, be a stationary Poisson process with rate Λ. Define $A_t^\infty = \sum_{n=-\infty}^{\infty} r1(T_n \leq t < T_n + r_n^{\text{on}})$, $r > 0$. Note that A_t^∞ represents the number of customers in an $M/GI/\infty$ queue; for that reason A_t^∞ is usually called an $M/GI/\infty$ process. An important observation is that this process represents a Poisson limit of a large number of on/off processes. Hence, it can be used as a good approximation of an aggregate process obtained by multiplexing a large (finite) number of on/off processes.

An important parameter that in many ways determines the fluid queue performance is the length of the arrival process activity period. Let $I^{\infty,\text{on}}$ be a generic activity period of an $M/GI/\infty$ process.

Theorem 10.4.11. *The asymptotics of the distribution of $I^{\infty,\text{on}}$ and its integrated tail are related as follows:*

(i) *If $F_1 \in \mathscr{S}$, then*

$$\int_t^\infty \mathbb{P}[I^{\infty,\text{on}} > u]\, du \sim e^{\Lambda \mathbb{E}\tau^{\text{on}}} \int_t^\infty \mathbb{P}[\tau^{\text{on}} > u]\, du \quad \text{as } t \to \infty.$$

(ii) *If in addition $F \in \mathscr{S}^*$, then*

$$\mathbb{P}[I^{\infty,\text{on}} > t] \sim e^{\Lambda \mathbb{E}\tau^{\text{on}}} \mathbb{P}[\tau^{\text{on}} > t] \quad \text{as } t \to \infty.$$

REMARK 10.4.12. For the case of τ^{on} being regularly varying $\mathbb{P}[\tau^{\text{on}} > t] = l(t)/t^{\alpha}$, $1 < \alpha < 2$, this result was obtained in Boxma [12] where Karamata's Tauberian/Abelian theorems were used to asymptotically relate $I^{\infty,\text{on}}$ and τ^{on}.

HEURISTIC 10.4.13. Cohen [21] shows that the expected number $\mathbb{E}N^{\infty}$ of on periods in one activity period is $e^{\Lambda\mathbb{E}\tau^{\text{on}}}$. Hence, by using the basic heuristics that a long period $I^{\infty,\text{on}}$ occurs due to exactly one long period we can jump into the conclusion

$$\mathbb{P}[I^{\infty,\text{on}} > t] \sim \mathbb{E}N^{\infty}\mathbb{P}[\tau^{\text{on}} > t] = e^{\Lambda\mathbb{E}\tau^{\text{on}}}\mathbb{P}[\tau^{\text{on}} > t].$$

However, it requires much more to rigorously prove this theorem (see Jelenković and Lazar [35]).

10.4.3.2 Queue Increment During an Activity Period. Let B_n, $n \geq 1$, be a sequence of random variables representing the total amount of fluid that is brought to the system during the nth activity period, that is, $B_n = \int_{t_n^b}^{t_n^e} A_t^{\infty}\, dt$, where t_n^b and t_n^e represent the beginning and end of the nth activity period, respectively. Furthermore, define $D_{c,n} \overset{\text{def}}{=} B_n - cl_n^{\text{on}}$, $0 < c \leq r$; note that $D_n \equiv D_{c,n}$ is a nonnegative random variable. If we imagine that A_t^{∞} represents the rate at which the fluid is arriving to a fluid queue, and that c is the constant rate at which the queue drains, then D_n represents the queue increment during the nth activity period. In order to derive the queueing asymptotics, we first have to understand the asymptotic behavior of D_n. A proof of the following result can be found in Jelenković and Lazar [35].

Theorem 10.4.14. *Consider an $M/GI/\infty$ arrival process with on periods being regularly varying $\mathbb{P}[\tau^{\text{on}} > x] = l(x)/x^{\alpha}$, $\alpha > 1$, where α is noninteger. If $0 < c \leq r$, then*

$$\mathbb{P}[D_n > x] \sim e^{\Lambda\mathbb{E}\tau^{\text{on}}}\mathbb{P}\left[\tau^{\text{on}} > \frac{x}{r + r\Lambda\mathbb{E}\tau^{\text{on}} - c}\right] \quad \text{as } x \to \infty. \tag{10.21}$$

REMARK 10.4.15. Recently in Resnick and Samorodnitsky [50] it was shown that this result holds under a more general condition of τ^{on} being intermediately regularly varying and $0 < c \leq r + r\Lambda\mathbb{E}\tau^{\text{on}}$.

10.4.3.3 Queueing Asymptotics. Let $Q_n^{\text{P},\infty}$ be the queue size observed at the beginning of the nth activity period of the $M/GI/\infty$ arrival process.

Theorem 10.4.16. *Let $\rho = \mathbb{E}A_t^{\infty} = \Lambda r\mathbb{E}\tau^{\text{on}} < c$. If $c \leq r$, and τ^{on} is regularly varying with noninteger exponent $\alpha > 1$, then*

$$\lim_{x\to\infty} \frac{\mathbb{P}[Q_t^{\text{P},\infty} > x]}{\int_{x/(\rho+r-c)}^{\infty} \mathbb{P}[\tau^{\text{on}} > u]\, du} = \Lambda\left(\frac{r}{c-\rho} - 1\right).$$

Proof. Denote with $A_n = D_{c,n}$, $C_n = cl_n^{\text{off}}$, use $\mathbb{E}(C_n - A_n) = e^{\Lambda \mathbb{E}\tau^{\text{on}}}(c - \rho)/\Lambda$ and apply Theorem 10.3.1. ■

In the next theorem, under more general assumptions, we obtain a tight lower bound for the fluid queue asymptotics with $M/GI/\infty$ arrivals (see Jelenković and Lazar [35]). For this fluid queue we denote its queue content process as Q_t^{∞}. It was conjectured [35] that the following bound represents actually the exact asymptotics.

Theorem 10.4.17. *Let* $\rho \overset{\text{def}}{=} \mathbb{E}A_t^{\infty,s} = \Lambda r \mathbb{E}\tau^{\text{on}} < c$. *If* $r + \rho > c$, *and* $\tau^{\text{on}} \in \mathcal{IR}$, *then*

$$\liminf_{x\to\infty} \frac{\mathbb{P}[Q_t^{\infty} > x]}{\int_{x/(r+\rho-c)}^{\infty} \mathbb{P}[\tau^{\text{on}} > u]\, du} \geq \frac{\Lambda r}{c - \rho}.$$

10.4.3.4 $M/G/\infty$ Approximation: Simulation Results. Based on Theorems 10.4.16 and 10.4.17, it is suggested that the queueing probabilities obtained by multiplexing N long-tailed on/off processes a_t^i, $1 \leq i \leq N$, are approximated as

$$\mathbb{P}[Q_t^N > x] \approx \frac{\Lambda_N r}{c_N} \int_{x/(r-c_N)}^{\infty} \mathbb{P}[\tau^{\text{on}} > u]\, du, \tag{10.22}$$

where $c_N \overset{\text{def}}{=} c - N\mathbb{E}a_t^i$, and $\Lambda_N \overset{\text{def}}{=} N\mathbb{E}a_t^i/(r\mathbb{E}\tau^{\text{on}})$. This approximation is termed an $M/G/\infty$ approximation. This approximation is to be used when the queue is stable and $r + (N-1)\mathbb{E}a_t^i > c$ *is satisfied.*

For simulation purposes we consider a discrete-time "fluid" queue. Correspondingly, we replace exponential off periods with geometrically distributed random variables $\mathbb{P}[\tau^{\text{off}} = t] = p(1-p)^{t-1}$, $t = 1, 2, 3, \ldots$. For on periods we consider the Pareto family $\mathbb{P}[\tau^{\text{on}} \geq t] = 1/t^{\alpha}$, $t = 1, 2, \ldots, \alpha > 0$. Here, for the discrete Pareto case we use

$$\mathbb{P}[Q_t^N = x] \approx \frac{\Lambda_N r}{c_N}(r - c_N)^{\alpha-1} x^{-\alpha}, \tag{10.23}$$

where c_N, and Λ_N are as defined earlier.

The efficacy of the approximation (10.23) is illustrated in the following simulation experiment (for additional experiments see Jelenković and Lazar [35]).

Example 10.4.18. Choose $p = 0.05$, $\alpha = 3$, $r = 2$, $c = 3$. This gives $\mathbb{E}\tau^{\text{on}} = 1.202$, and $\mathbb{E}a_t^i = 0.113$. Then, for $N = 20, 25$ processes, the approximations are given by β/x^3, $\beta = 4.14$, 48.04, respectively. The desirable closeness between the simulation results and the approximations is represented in Fig. 10.3. It is interesting to observe that in this case the peak rate of each individual process is smaller than the capacity of the server.

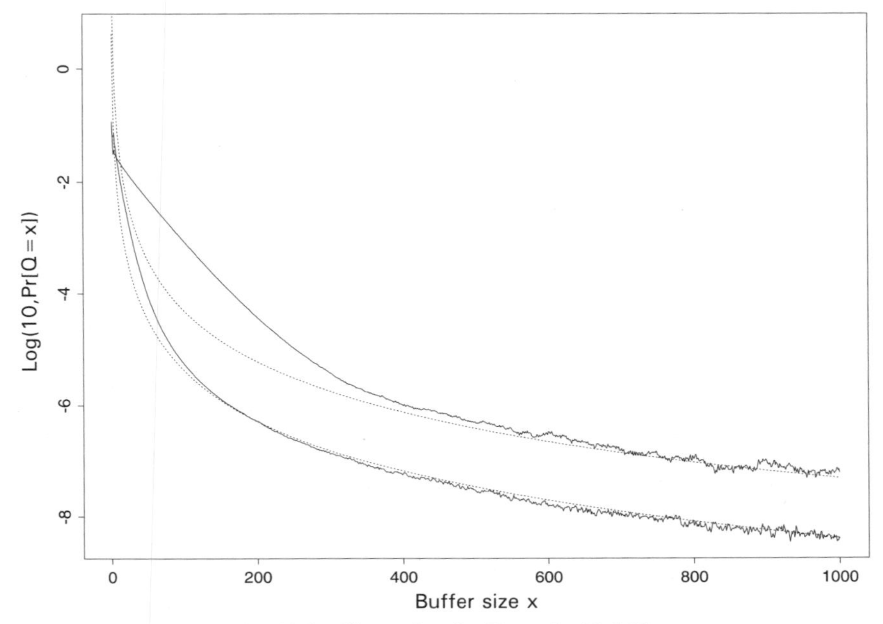

Fig. 10.3 Illustration for Example 10.4.18.

Note that in this experiment, for the case of $N = 20$ processes, the probabilities are very small ($\approx 10^{-8}$). Hence, in order to achieve reasonable simulation accuracy, we had to choose a very large number (10^9) of simulated on/off intervals. This means that the aggregate process was approximately 2×10^{10} samples long. The simulation of this case took *77 hours* on a modern (200 MIPS) IBM workstation. On the other hand, it is needless to say that the evaluation of Eq. (10.23), or Eq. (10.22), only takes a *negligible amount of time*!

10.5 CONCLUSION

In this chapter a variety of asymptotic results for queues with subexponential characteristics were presented. All of these results are *explicit*, *insightful*, and, as demonstrated with numerical examples, *accurate*. Due to these desirable characteristics, these results could be of practical use in designing future communication networks that will be able to carry efficiently and reliably bursty multimedia traffic.

ACKNOWLEDGMENTS

I am very grateful to all of my colleages who have sent me preprints of their papers.

REFERENCES

1. J. Abate, G. L. Choudhury, and W. Whitt. Waiting-time tail probabilities in queues with long-tail service-time distributions. *Queueing Syst.*, **16**(3/4):311–338, 1994.
2. R. Agrawal, A. M. Makowski, and P. Nain. On a reduced load equivalence for fluid queues under subexponentiality. *Queueing Syst.*, 1999, to appear.
3. D. Anick, D. Mitra, and M. M. Sondhi. Stochastic theory of a data handling system with multiple sources. *Bell Syst. Tech. J*, **61**:1871–1894, 1982.
4. S. Asmussen, L. F. Henriksen, and C. Klüppelberg. Large claims approximations for risk processes in a Markovian environment. *Stoch. Processes Applic.*, **54**:29–43, 1994.
5. S. Asmussen, C. Klüppelberg, and K. Sigman. Sampling at subexponential times, with queueing applications. Preprint, 1998.
6. S. Asmussen, H. Schmidli, and V. Schmidt. Tail probabilities for non-standard risk and queueing processes with subexponential jumps. *Adv. Appl. Probab.*, **32**(2), 1999.
7. K. B. Athreya and P. E. Ney, *Branching Processes*. Springer-Verlag, Berlin, 1972.
8. F. Baccelli and P. Brémaud. *Elements of Queueing Theory: Palm–Martingale Calculus and Stochastic Recurrence*. Springer-Verlag, Berlin, 1994.
9. J. Beran, R. Sherman, M. S. Taqqu, and W. Willinger. Long-range dependence in variable bit-rate video traffic. *IEEE Trans. Commun.*, **43**:1566–1579, 1995.
10. N. H. Bingham, C. M. Goldie, and J. L. Teugels. *Regular Variation*. Cambridge University Press, Cambridge, 1987.
11. A. A. Borovkov. *Stochastic Processes in Queueing Theory*. Springer-Verlag, Berlin, 1976.
12. O. J. Boxma. Fluid queues and regular variation. *Perf. Eval.*, **27,28**:699–712, 1996.
13. O. J. Boxma. Regular variation in a multi-source fluid queue. In *ITC 15*, pp. 391–402, Washington, DC, June 1997.
14. O. J. Boxma and V. Dumas. Fluid queues with long-tailed activity period distributions. Technical Report PNA-R9705, CWI, Amsterdam, April 1997.
15. V. P. Chistakov. A theorem on sums of independent positive random variables and its application to branching random processes. *Theor. Probab. Appl.*, **9**:640–648, 1964.
16. G. L. Choudhury and W. Whitt. Long-tail buffer-content distributions in broadband networks. *Perf. Eval.*, **30**:177–190, 1997.
17. D. B. H. Cline. Convolution tails, product tails and domains of attraction. *Probab. Theory Relat. Fields*, **72**(1):529–557, 1986.
18. D. B. H. Cline. Convolution of distributions with exponential and subexponential tails. *J. Austral. Math. Soc. Ser. A*, **43**:347–365, 1987.
19. D. B. H. Cline. Intermediate regular and π variation. *Proc. London Math. Soc.*, **68**(3):594–616, 1994.
20. J. W. Cohen. Some results on regular variation for distributions in queueing and fluctuation theory. *J. Appl. Probab.*, **10**:343–353, 1973.
21. J. W. Cohen. Superimposed renewal processes and storage with gradual input. *Stoch. Processes Applic.*, **2**:31–58, 1974.
22. M. Crovella. The relationship between heavy-tailed file sizes and self-similar network traffic. In *9th INFORMS Applied Probability Conferences*, Cambridge, MA, June 1997.
23. K. Debicki, Z. Michna, and T. Rolski. On the supremum for Gaussian processes over infinite horizon. Preprint, May 1997.

24. N. G. Duffield and N. O'Connell. Large deviations and overflow probabilities for the general single-server queue with applications. *Math. Proc. Cambridge Philos. Soc.*, **118**:363–374, 1995.

25. A. Elwalid, D. Heyman, T. V. Lakshman, D. Mitra, and A. Weiss. Fundamental bounds and approximations for ATM multiplexers with applications to video teleconferencing. *IEEE J. Select. Areas Commun.*, **13**(6):1004–1016, August 1995.

26. P. V. Glynn and W. Whitt. Logarithmic asymptotics for steady-state tail probabilities in a single-server queue. In J. Galambos and J. Gani, eds., *Studies in Applied Probability*, Vol. 31A (special issue of *J. Appl. Probab.*), pp. 131–156. Applied Probability Trust, Sheffield, England, 1994.

27. C. M. Goldie and C. Klüppelberg. Subexponential distributions. In M. S. Taqqu, R. Adler, and R. Feldman, eds., *A Practical Guide to Heavy Tails: Statistical Techniques for Analysing Heavy Tailed Distributions*. Birkhäuser, Basel, 1997.

28. D. Heath, S. Resnick, and G. Samorodnitsky. Heavy tails and long range dependence in on/off processes and associated fluid models. Preprint.

29. D. Heath, S. Resnick, and G. Samorodnitsky. How system performance is affected by the interplay of averages in a fluid queue with long range dependence induced by heavy tails. Preprint.

30. D. P. Heyman and T. V. Lakshman. Source models for VBR broadcast-video traffic. *IEEE J. Select. Areas Commun.*, **4**:40–48, February 1996.

31. P. R. Jelenković. Subexponential loss rates in a $GI/GI/1$ queue with applications. *Queueing Syst.*, 1999, to appear.

32. P. R. Jelenković. $GI/GI/1$ queue with truncated long-tailed service times. Submitted for publication, 1999.

33. P. R. Jelenković. Long-tailed loss rates in a single server queue. In *Proc. IEEE INFOCOM'98*, pp. 1462–1469, San Francisco, April 1998.

34. P. R. Jelenković. Network multiplexer with truncated long-tailed arrival streams. In *Proc. IEEE INFOCOM'99*, pp. 625–640, New York, NY, March 1999.

35. P. R. Jelenković and A. A. Lazar. Asymptotic results for multiplexing subexponential on–off processes. *Adv. Appl. Probab.*, **31**(2), 1999.

36. P. R. Jelenković and A. A. Lazar. Subexponential asymptotics of a Markov-modulated random walk with queueing applications. *J. Appl. Probab.*, **35**(2):325–347, June 1998.

37. P. R. Jelenković, A. A. Lazar, and N. Semret. The effect of multiple time scales and subexponentially of MPEG video streams on queueing behavior. *IEEE J. Select. Areas Commun.*, **15**(6):1052–1071, August 1997.

38. J. Karamata. Sur un mode de croissance réguilière des fonctions. *Mathematica (Cluj)*, **4**:38–53, 1930.

39. O. Kella and W. Whitt. A storage model with a two-state random environment. *Oper. Res.*, **40**:257–262, 1992.

40. C. Klüppelberg. Subexponential distributions and integrated tails. *J. Appl. Probab.*, **25**:132–141, 1988.

41. C. Klüppelberg. Subexponential distributions and characterizations of related classes. *Probab. Theory Relat. Fields*, **82**:259, 1989.

42. W. E. Leland, M. S. Taqqu, W. Willinger, and D. V. Wilson. On the self-similar nature of Ethernet traffic. In *Proc. ACM SIGCOMM'93*, pp. 183–193, 1993.

43. W. E. Leland, M. S. Taqqu, W. Willinger, and D. V. Wilson. On the self-similar nature of Ethernet traffic (extended version). *IEEE/ACM Trans. Networking*, **2**:1–15, 1994.

44. N. Likhanov, B. Tsybakov, and N. D. Georganas. Analysis of an ATM buffer with self-similar ("fractal") input traffic. In *Proc. IEEE INFOCOM'95*, pp. 985–991, Boston, April 1995.

45. R. M. Loynes. The stability of a queue with non-independent inter-arrival and service times. *Proc. Cambridge Philos. Soc.*, **58**:497–520, 1962.

46. P. Nain, Z. Liu, D. Towsley, and Z.-L. Zhang. Asymptotic behavior of a multiplexer fed by a long-range dependent process. *J. Appl. Probab.*, **36**(1).

47. I. Norros. A storage model with self-similar input. *Queueing Syst.*, **16**:387–396, 1994.

48. A. G. Pakes. On the tails of waiting-time distribution. *J. Appl. Probab.*, **12**:555–564, 1975.

49. M. Parulekar and A. M. Makowski. Tail probabilities for a multiplexer with self-similar traffic. In *INFOCOM'96*, pp. 1452–1459, San Francisco, March 1996.

50. S. Resnick and G. Samorodnitsky. Activity periods of an infinite server queue and performance of certain heavy tailed fluid queues. Preprint, 1997.

51. S. Resnick and G. Samorodnitsky. Performance decay in a single server queueing model with long range dependence. Preprint, 1996.

52. T. Rolski, S. Schlegel, and V. Schmidt. Asymptotics of Palm-stationary buffer content distribution in fluid flow queues. *Adv. Appl. Probab.*, **31**(1):235, 1999

53. M. Rubinovitch. The output of a buffered data communication system. *Stoch. Processes Applic.*, **1**:375–380, 1973.

54. B. K. Ryu and S. B. Lowen. Point process approaches to the modeling and analysis of self-similar traffic—part I: model construction. In *INFOCOM'96*, pp. 1468–1475, San Francisco, March 1996.

55. K. P. Tsoukatos and A. M. Makowski. Heavy traffic analysis for a multiplexer driven by $M/GI/\infty$ input processes. In *ITC 15*, pp. 497–506, Washington, DC, June 1997.

56. N. Veraverbeke. Asymptotic behavior of Wiener–Hopf factors of a random walk. *Stoch. Processes Applic.*, **5**:27–37, 1977.

57. A. Weiss and A. Shwartz. *Large Deviations for Performance Analysis: Queues, Communications, and Computing*. Chapman and Hall, New York, 1995.

58. E. Willekens and J. L. Teugels. Asymptotic expansion for waiting time probabilities in an $M/G/1$ queue with long-tailed service time. *Queueing Syst.*, **10**:295–312, 1992.

11

TRAFFIC AND QUEUEING FROM AN UNBOUNDED SET OF INDEPENDENT MEMORYLESS ON/OFF SOURCES

PHILIPPE JACQUET
INRIA, 78153 Le Chesnay Cédex, France

11.1 INTRODUCTION

11.1.1 Long-Term Dependence and Packet Loss in Telecommunication Traffic

In early literature about the performance of telecommunication systems, traffic was generally modeled as memoryless Poisson streams of packets. In these models the packet arrival processes show no time interdependence. Recent measurements on Web traffic show that this hypothesis is wrong and that Web traffic actually experiences what we now call *long-term* dependence.

Long-term dependence is interesting not only because it contradicts Poisson's law, but also because it significantly impacts the performance of networks. One effect is that it dramatically increases packet loss in data networks. For example, let us focus on an Internet router. In a simple model, the router can be seen as a buffer served by a single server. When the buffer overflows, some packets are lost. The lost packets must be re-sent following TCP/IP, thus adding extra delay and traffic.

If we simply model the router by a $M/M/1$ queue with an infinite buffer, input rate λ, and service rate 1, then the probability p_n that the queue length is greater than n is exactly λ^n. In a first-order approximation, quantity p_n can be identified with the packet loss rate in a buffer of size n. Therefore, to keep packet loss below some acceptable level ε, it suffices to make the buffer capacity greater than $(\log \varepsilon)/(\log \lambda)$,

Self-Similar Network Traffic and Performance Evaluation, Edited by Kihong Park and Walter Willinger
ISBN 0-471-31974-0

that is, a logarithmic function of $1/\varepsilon$. In general, telecommunication designers work on ε on the order of 10^{-6} and $\lambda < 0.8$.

Long-term dependent traffic can make loss and retry rates $p_n \asymp Bn^{-\beta}$ for some $\beta > 0$ [1]. In other words, the queue size distribution has a heavy tail. Under this condition it is clear that buffer capacity would need to be raised to $(\beta/\varepsilon)^{1/\beta}$ to keep packet loss rate under the acceptable level ε, which no longer leads to a logarithmic function of $1/\varepsilon$, but to a *polynomial* function of $1/\varepsilon$. Indeed, this minimal size would be several orders of magnitude higher than the capacity obtained with the Poisson model. In fact, actual router capacities are dangerously underestimated with regard to this new traffic condition.

11.1.2 Contribution of this Chapter

The Poisson law is the natural consequence of the law of large numbers, best describes the cumulated effect of several independent, identically distributed (i.i.d) sources in parallel. Assume, for example, N sources, each of them producing on average λ/N events per time unit according to a stationary random process. Then when N tends to infinity, the interevent times T tend to be i.i.d with a distribution function characterizing the Poisson law,

$$\Pr\{T > x\} = e^{-\lambda x}. \tag{11.1}$$

The convergence to a Poisson distribution still holds when the sources are not quite identical, as long as they have similar profiles. Feller [2] gave pretty general conditions for this convergence: basically, first moments of the source interevent generation time must be $O(1/N)$ and second moments must be $o(1/N)$.

In this chapter we are interested in the case where traffic is created from a large set of independent sources that *do not satisfy* Feller's conditions. In particular, we focus on certain sets of on/off sources that produce long-term dependence when their sizes tend to infinity. It will also be shown that queues submitted to such sets of sources will experience buffer occupation with a polynomially decaying tail distribution.

In the other chapters of this book it is assumed that some of the sources, taken individually, already produce long-term dependence. For example, some sources have heavy-tailed "on" periods. In this case the cumulated traffic shows long-term dependence and creates a heavy-tailed queue size distribution [3, 4]. The challenge in the present chapter is that none of the sources, taken separately, produces long-term dependence and a heavy-tailed queue size distribution, and that those phenomena eventually take place when the number of sources increases. To insist on this point, we will focus on individual on/off sources with memoryless profiles (exponentially distributed on periods and off periods). It has already been shown by Beran [5] and Jacquet [6] that such sources can create long-term dependence when their number increases. The contribution of the present chapter is to show that such sources can also create a polynomially decaying queue size distribution. We do not *claim* that pure memoryless on/off sources are necessarily realistic models for Web sources of traffic, but they have the advantage of being simple and analytically more tractable than more general models.

Our model stands halfway between the Boxma–Dumas models [7] with finite sets of on/off sources and the very exciting model of Tsybakov et al. [8], which assumes a *continuum* of on/off sources from which "on" periods are activated according to a Poisson process. A last technical point: we don't use fluid approximation in this chapter, although this feature has very interesting aspects.

11.2 TOOLS AND MODELS

11.2.1 The Mellin Transform Applied to Performance Analysis

The *technical* novelty of this chapter is in the extensive use of the Mellin transform. The Mellin transform is particularly well adapted to capturing the *polynomial* effects in function asymptotics with an unprecedented accuracy [9]. It has similar features to those of the Laplace transform, the latter being known for more than a century to be a good tool for capturing *exponential* effects in functions. The Mellin transform $f^*(s)$ of a function $f(x)$, defined for real $x > 0$, is

$$f^*(s) = \int_0^\infty f(x)x^{s-1}\, dx. \tag{11.2}$$

Note that the Mellin transform of function $f(x)$ is nothing more than the Laplace transform of $f(e^x)$. The Mellin transform is defined for s in the fundamental strip $\{s, \Re(s) \in]\alpha, \beta[\}$ of function $f(x)$. The constants α and β are the lower and upper bounds of the real numbers c such that $f(x) = o(x^{-c})$ for both $x \to 0$ and $x \to \infty$. The fundamental strip of function e^{-x} is $\{s, \Re(s) > 0\}$: in other words, $\alpha = 0$, $\beta = \infty$. In passing, the Mellin transform of e^{-x} is the celebrated Euler *Gamma* function, denoted $\Gamma(s)$. The fundamental strip of any polynomial function is the empty set: there is no Mellin transform of pure polynomial functions.

The inverse Mellin transform is

$$f(x) = \frac{1}{2i\pi}\int_{c-i\infty}^{c+i\infty} f^*(s)x^{-s}\, ds \tag{11.3}$$

valid for any c contained in the fundamental strip.

The Mellin transform has been introduced primarily to handle harmonic sums. A harmonic sum of function $f(x)$ is a series of the following kind:

$$\sum_{i\geq 0} a_i f(\omega_i x)$$

for some sequences a_i and ω_i, which make the sum properly convergent. The Mellin transform of the harmonic sum is

$$\left(\sum_{i \geq 0} a_i \omega_i^{-s}\right) f^*(s),$$

where $f^*(s)$ is the Mellin transform of function $f(x)$. This latter expression is sometimes much easier to handle than the harmonic sum itself.

The analysis of function $f(x)$ asymptotics, when x tends to both limit 0 and ∞, is equivalent to the singularity analysis of the Mellin transform $f^*(s)$ on the boundaries of its fundamental strip. The right boundary corresponds to the asymptotics when $x \to +\infty$; the left boundary corresponds to the asymptotics when $x \to 0$. For example, if $f^*(s)$ has one single pole on the right boundary at $s = \beta$, this pole has residue μ, and function $f^*(s)$ can be extended for $\text{Re}(s) \in [\beta, \beta + \varepsilon]$. Then in Eq. (11.3) the integration line can be moved on the right (see Fig. 11.1).

Applying the residue theorem, we find

$$f(x) = \mu x^{-\beta} + \frac{x^{-\beta-\varepsilon}}{2i\pi} \int_{-\infty}^{+\infty} x^{-it} f(\beta + \varepsilon + it)\, dt. \tag{11.4}$$

Using a majorization under the integral sign, we get a second-order estimate:

$$f(x) + \mu x^{-\beta} + O(x^{-\beta-\varepsilon})$$

when $x \to +\infty$.

The Mellin transform can provide a more accurate estimate and can even capture the case where the constant before the $x^{-\beta}$ estimate is a fluctuating function of x.

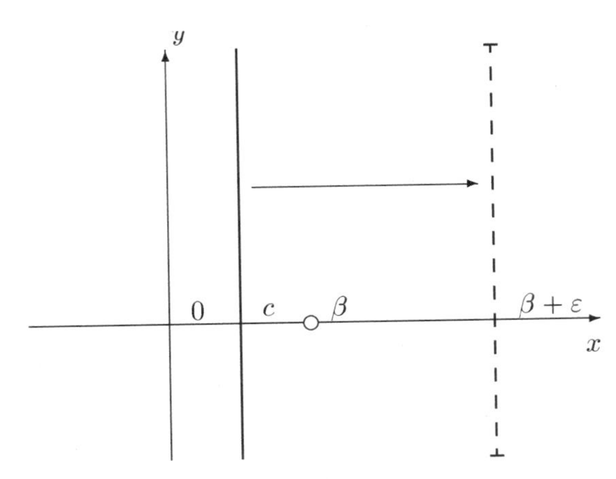

Fig. 11.1 Movement of the integration line in the reverse Mellin transform.

As an illustration, consider the following function $f(x)$ satisfying the functional equation:

$$f(x) = \tfrac{1}{2} f(x/2) + e^{-x}/2. \tag{11.5}$$

This equation arises in the analysis of *divide and conquer* algorithms [10]. The solution is obtained by iteration:

$$f(x) = \sum_{i \geq 0} 2^{-i} e^{x/2^i}. \tag{11.6}$$

By virtue of the harmonic sum expression, the Mellin transform $f^*(s)$ is immediately identified by $f^*(s) = 2^{-1+s} f^*(s) - \Gamma(s)$, or

$$f^*(s) = \frac{\Gamma(s)}{1 - 2^{-1+s}} \tag{11.7}$$

for s in the fundamental strip $\{s, \Re(s) \in]-0, 1[\}$. Note that there is a sequence of single poles $1 + 2ik\pi/(\log 2)$, for k integers, that are all located on the strip boundary $\{s, \Re(s) = 1\}$. Using the reverse Mellin transform and catching all the residues of these poles by moving the integration line on the right (see Fig. 11.2), we get

$$f(x) = P(\log x) x^{-1} + O(x^{-1-\varepsilon}) \tag{11.8}$$

with

$$P(y) = \sum_k \Gamma\left(1 + \frac{2ik\pi}{\log 2}\right) \exp\left(\frac{2ik\pi y}{\log 2}\right). \tag{11.9}$$

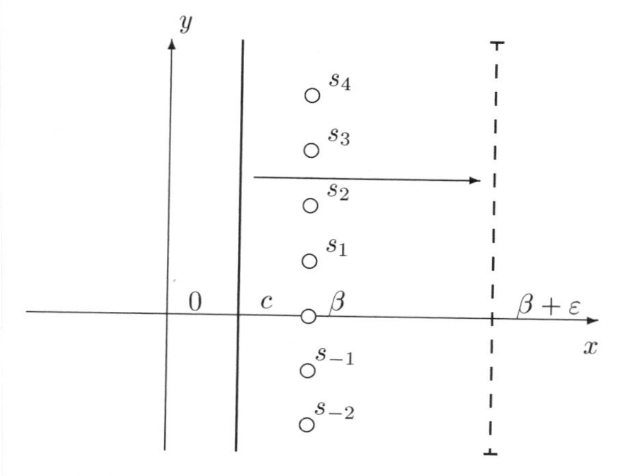

Fig. 11.2 Reverse Mellin transform with multiple poles.

It is clear that $P(x)$ is of period log 2 with Fourier coefficients identified in Eq. (11.9).

To make a more general statement, the Mellin transform is the ideal tool for capturing the slowly varying functions in front of polynomial factors. Here the term *analytical information theory* is used to describe problems of information theory that are solved by analytical methods borrowed from complex analysis [11, 12]. Refer to Szpankowski [12] for a survey and to Jacquet [6] for a detailed description of the tools and proofs used in the present chapter.

11.2.2 Memoryless On/Off Sources

As mentioned earlier we focus on the simplified case where each source has a memoryless profile. A memoryless on/off source is described by only two states—the "on" state and the "off" state—and has the following properties:

- In the "off" state, the source does not generate any packet.
- In the "on" state, the source generates packets as a Poisson stream with constant rate λ.
- The transition from the "on" state to the "off" state occurs with constant rate ν_0.
- The transition from the "off" state to the "on" state occurs with constant rate ν_1.

Note that the "on" periods and "off" periods are both exponentially distributed; that is, the state transition times follow a memoryless process.

We introduce the matrix $\mathbf{T}$, which we call the *transition* matrix:

$$\mathbf{T} = \begin{bmatrix} -\nu_1 & \nu_0 \\ \nu_1 & -\nu_0 \end{bmatrix} \tag{11.10}$$

The eigenvalues of $\mathbf{T}$ are 0 and $-(\nu_1 + \nu_0)$.

In most of this chapter we consider a general system where the arrival process comes from several independent memoryless on/off sources. We consider a denumerable set of on/off sources indexed from 1 to N, N tending to infinity in the analysis. See Fig. 11.3 for an illustration.

11.3 QUEUEING UNDER ON/OFF SOURCES

In the following, the network is modeled via several queues with single servers and an infinite buffer, which receive input from a set of on/off sources. The service time at each queue is exponential. Our aim is to find the steady-state distribution of the queue length and, in particular, the asymptotics of the probabilities p_n that the queue length exceeds n customers, when $n \to \infty$.

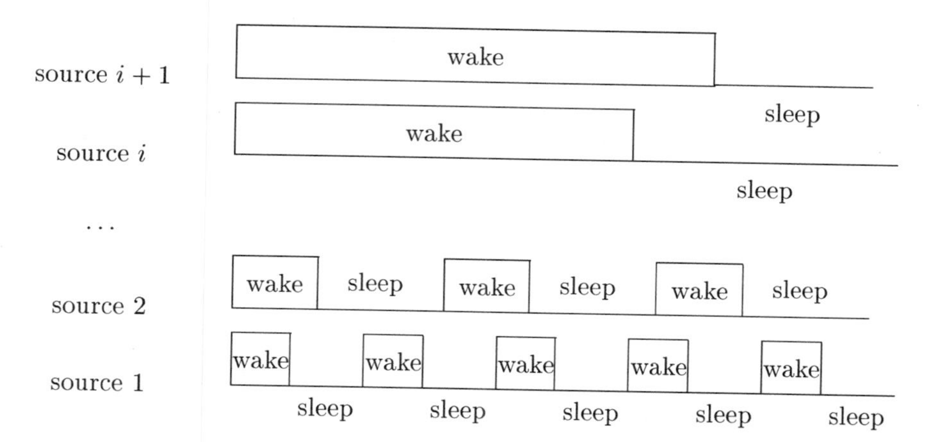

Fig. 11.3 Aggregation of a denumerable set of independent on/off sources.

11.3.1 Queueing with a Single On/Off Source

We consider a single queue with service mean equal to 1 time unit. We refer to Neuts [13] for the following theorem.

Theorem 11.3.1. *The queue length generating function with the exponential server and an on/off input source satisfies*

$$q(z) = 1 - \frac{(z-1)\lambda z_1 v_1}{(v_1 + v_0)(z - z_1)} \tag{11.11}$$

with

$$z_1 = \frac{1}{2}\left(\frac{1 + \lambda + v_1 + v_0 + \sqrt{(1-\lambda)^2 + 2(v_1 + v_0) + 2\lambda(v_1 - v_0) + (v_1 + v_0)^2}}{(1 + v_1)\lambda}\right). \tag{11.12}$$

Proof. This a straightforward adaptation of that in Neuts [13]. ■

Corollary 11.3.2. *When $n \to \infty$, quantities p_n exponentially decrease with*

$$p_n = \lambda \frac{v_1}{v_1 + v_o} z_1^{1-n}$$

11.3.2 Parallel Queues Under Self-Similar On/Off Sources

Here we consider several memoryless on/off sources, each of them served by a server. Each service time is exponential with a mean of 1 time unit, and the peak rate of each source is $\lambda > 1$. Therefore, the probability generating function $g(z)$ of the queue length distribution is simply

$$g(z) = \sum_{i=1}^{N} \frac{1}{N} q_i(z), \tag{11.13}$$

where $q_i(z)$ is the probability generating function of the queue assigned to source number i.

To simplify our analysis we look at a set of self-similar on/off sources. The vector (ν_1, ν_0) of each source is collinear to a fixed vector (v_1, v_0): for source i, $(\nu_1, \nu_0) = (\varepsilon_i v_1, \varepsilon_i v_0)$. Referring to Theorem 11.3.1, we see that the probability that queue i contains more than n packets is exactly

$$\frac{\lambda v_1}{v_1 + v_0} z_1^{1-n}(\varepsilon_i)$$

with $z_1(x)$ given by Eq. (11.12) with substitution of (ν_1, ν_0) by (xv_1, xv_0). If we denote by $r(\varepsilon)$ the asymptotic density of sources with vector (ν_1, ν_0) equal to $(\varepsilon v_1, \varepsilon v_0)$, then we have the following result [14].

Theorem 11.3.3. *Let* $0 < \beta < 1$. *If function* $r(x) = \mu x^{\beta-1} + O(x^{\beta-1+\varepsilon})$ *when* $x \to 0$, *then when* $N \to \infty$ *quantity* p_n *converges to the following expression:*

$$\frac{\lambda v_1}{v_1 + v_0}\left(\frac{v_1}{\lambda - 1} - v_0\right)^{-\beta} \mu n^{-\beta} + O(n^{-\beta-\varepsilon}). \tag{11.14}$$

Proof. Note that the conditions of the theorem are basically equivalent to the fact that the Mellin transform of function $r(x)$ has a single pole at $s = 1 - \beta$, with residue μ, and can be analytically continued on the strip $\{s, \Re(s) \in [1 - \beta - \varepsilon, 1 - \beta]\}$.

For any fixed N we have

$$p_n = \frac{1}{N} \sum_{i=1}^{N} \frac{\lambda v_1}{v_1 + v_0} z_1^{1-n}(\varepsilon_i). \tag{11.15}$$

At the limit we have

$$\lim_{M \to \infty} p_n = \frac{\lambda v_1}{v_1 + v_0} \int_0^{\infty} z_1(y)^{1-n} r(y)\, dy. \tag{11.16}$$

In the above expression, substituting a real number x for integer n, we obtain a function $p(x)$ whose Mellin transform $p^*(s)$ satisfies

$$p^*(s) = \frac{\lambda v_1}{v_1 + v_0} \int_0^\infty z_1(y) \log z_1(y))^{-s} r(y)\, dy. \tag{11.17}$$

Since

$$z_1(y) = 1 + \left(\frac{v_1}{\lambda - 1} - v_0\right) y + O(y^2),$$
$$(\log z_1(y))^{-s} = \left(\frac{v_1}{\lambda - 1} - v_0\right)^{-s} y^{-s}(1 + sO(y))$$

when $y \to 0$. Therefore, we have the estimate

$$\begin{aligned} p^*(s) &= \frac{\lambda v_1}{v_1 + v_0} \left(\frac{v_1}{\lambda - 1} - v_0\right)^{-s} \int_0^\infty r(y) y^{-s}(1 + sO(y))\, dy \\ &= \frac{\lambda v_1}{v_1 + v_0} \left(\frac{v_1}{\lambda - 1} - v_0\right)^{-s} [r^*(1 - s) + sO(r^*(2 - \Re(s)))], \end{aligned}$$

where function $r^*(s)$ is the Mellin transform of function $r(x)$. Therefore, the Mellin transform of $p(x)$ is related to the Mellin transform of $r(x)$. More precisely, if $r^*(s)$ has a single pole at $s = 1 - \beta$ with residue μ and can be extended further to the right, then the first encountered singularity of $p^*(s)$ is at $s = \beta$. Note that the singularity boundary of $r^*(2 - \Re(s))$ is shifted by 1 on the right from that of $r^*(1 - s)$,

$$p_n = \frac{\lambda v_1}{v_1 + v_0} \left(\frac{v_1}{\lambda - 1} - v_0\right)^{-\beta} \mu n^{-\beta} + O(n^{-\beta - \varepsilon}). \tag{11.18}$$

■

As an application, we choose $r(x) = \beta x^{\beta - 1}$ valid for $x \leq 1$ and $r(x) = 0$ for $x > 1$. In this case, $r^*(s) = \beta/(\beta - 1 + s)$ and we get

$$p_n = \frac{\lambda v_1}{v_1 + v_0} \left(\frac{v_1}{\lambda - 1} - v_0\right)^{-\beta} \beta n^{-\beta} + O(n^{-\beta - 1}).$$

Polynomial Tail with fluctuating Coefficients. By tuning system parameters, one can give rise to oscillating coefficients in the asymptotics of p_n when $n \to \infty$. Indeed, we can have $p_n \asymp P(n) n^{-\beta}$ with $P(x)$ oscillating between two values: lim inf $P(x) \neq$ lim sup $P(x)$.

For example, one can take $r(x) = \alpha x^{-2}[2 - \sum_{i \geq 9} 2^{-i/2} \exp(-2^i x^{2(1+\beta)})]$, the constant term α is here to make the density function $r(x)$ sum to 1 and is equal here to $-(2(1 + \beta)(1 - 2^{-\beta/2(1+\beta)})/\Gamma(-1/2(1 + \beta))$.

Therefore,

$$r^*(s) = -\frac{\alpha}{2(1+\beta)} \frac{\Gamma\left(\frac{s-2}{2(1+\beta)}\right)}{1 - 2^{[-\frac{1}{2}-(s-2)/2(1+\beta)]}}. \tag{11.19}$$

The Mellin transform $p^*(s)$ therefore has a singularity set made of simple poles $s_k = (\beta + 4i(1+\beta)k\pi)/2$, for k integer. As we saw in the tutorial section about the Mellin transform, this kind of set creates periodic fluctuating terms in front of the polynomial expansion. This periodic fluctuation is reflected in the asymptotics of p_n, namely:

$$\begin{aligned} p_n = {} & \frac{\lambda v_1}{v_1 + v_0}\left(\frac{v_1}{\lambda - 1} - v_0\right)^{-\beta} \frac{\alpha}{\log 2} \\ & \times \left\{\sum_k \Gamma(s_k) \exp\left[-4ik\pi\left(\frac{1+\beta}{\log 2}\right)\log n\right]\right\} n^{-\beta} + O(n^{-\beta-\varepsilon}). \end{aligned}$$

In other words, we have proved that $p_n \asymp P(\log n)n^{-\beta}$, where $P(\cdot)$ is a periodic function, of period $\log 2/(1-\beta)$, whose Fourier coefficients are proportional to $\Gamma(s_k)$. Since function $P(\cdot)$ is not constant (indeed Fourier coefficients are all nonzero), we don't have $\liminf P(x) = \limsup P(x)$.

Figure 11.4 displays function $r(x)$ computed for $\beta = 0.5$, which causes fluctuating polynomial coefficients.

11.3.3 A Multiplexer Queue Under an Infinite Number of On/Off Sources

In the previous section we considered on/off sources served by separate parallel queues. In this section we consider that all on/off sources forward their packet to a

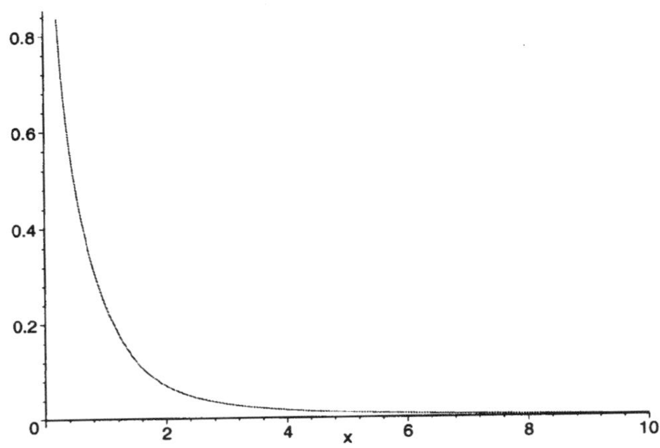

Fig. 11.4 Density function $r(x)$, which leads to fluctuating asymptotics with $\beta = 0.5$.

single server called the *mux* server. We assume that the mux server has an infinite buffer and performs exponential service times with mean 1. This will model a router, a switch, or a multiplexer device in the network.

We also assume that:

- for every i, the ith on/off source has peak rate λ_i;
- the λ_i are all identical and equal to a given $\lambda > 1$;
- for every on/off source we have $v_0 = \varepsilon_i^2$ and $v_1 = \varepsilon_i - \varepsilon_i^2$, for some sequence $\varepsilon_i > 0$.

To make the analysis relevant we need the system to be stable, which implies that the mean workload of the queue must be strictly smaller than 1:

$$\sum_{i=1}^{N} \varepsilon_i \lambda_i < 1. \tag{11.20}$$

Our target is still to give an asymptotic estimate of p_n, which is the probability that the mux queue length is greater than n.

We fix a parameter $\beta < 1$. In the sequel we suppose that the series $\sum \varepsilon_i$ is convergent. We call $\eta(s)$ the Dirichlet series $\sum_{j=1}^{\infty} \varepsilon_j^s$ and we assume that $\eta(s)$ is absolutely converent for all complex numbers s with real part strictly less than $1 - \beta$. For example, $\varepsilon_j = j^{1/(\beta-1)}$. Note that in this case the Dirichlet series $\eta(s) = \zeta(s/(1-\beta))$; that is, it can be identified with the Riemann *zeta* function. Figure 11.5 displays the v_1 (off/on rates) and v_0 (on/off rates) parameters of 60 such on/off sources with $\beta = 0.5$.

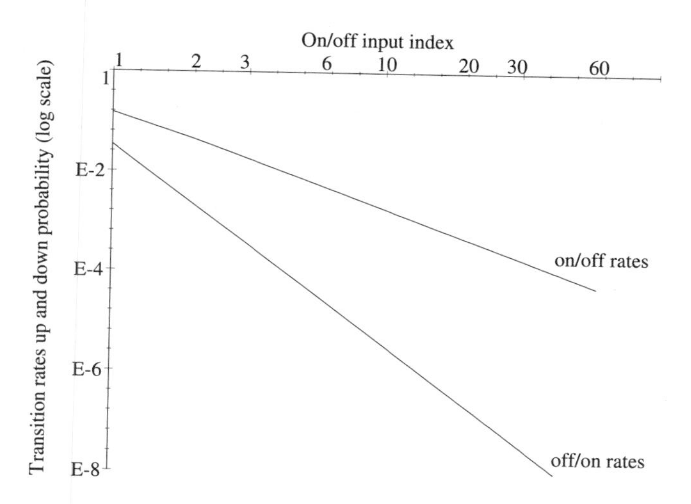

Fig. 11.5 Transition rates of 60 on/off sources with $\beta = 0.5$. Transition rates from on state to off state—v_i (on/off rates); transition rate from off state to on state—δ_i (off/on rates).

Theorem 11.3.4. *If the Dirichlet series $\sum_i \varepsilon_i^s$ has a simple pole on $s = 1 - \beta$ with residue μ, then the quantity p_n has a polynomial lower bound with exactly determined coefficients.*

Proof. To simplify, we assume that the Dirichlet series can be continued to a vertical strip $1 - \beta < \Re(s) < 1 - \beta + \varepsilon$.

Instead of looking directly at the queue length on the mux server mode, we consider again the parallel queues of the previous section, where each source has its own server. It is clear that the sum of the length of the parallel queues turns out to be always smaller than the length of the queue on the mux server.

Our aim is to show that the sum of the parallel queues has a polynomial tail of degree $-\beta$. Let $g(z)$ be the probability generating function of the sum of the queue lengths of the multiserver mode. We have the obvious identity $g(z) = \prod_{i=1}^{n} q_i(z)$, where $q_i(z)$ is the probability generating function of the queue size of the server attached to the on/off source number i.

Generating functions $q_i(z)$ are computed according to Theorem 11.3.1. Let z_i be the equivalent of z_1 in Theorem 11.3.1 for on/off source i. We have formally

$$g(z) = \prod_{i=1}^{\infty} \frac{(1 - \lambda\varepsilon_i z_i)z - (1 - \lambda\varepsilon)z_i}{z - z_i}. \tag{11.21}$$

Quantity $g(z)$ is defined as long as z does not belong to the set of singularities z_i. Under this restriction, the expression $g(z)$ converges because the series in $z_i - 1$ converges. Indeed, $z_n = 1 + \varepsilon_n/(\lambda - 1) + O(\varepsilon_1^2)$ when $n \to \infty$, and the series in ε_i converges (see Fig. 11.6). In passing, $g(z) = a + O(1/z)$ with $a = \prod_{i=1}^{\infty}(1 - \lambda\varepsilon_i z_i)$, when $|z| \to \infty$.

In the following we will establish that $g(1 - x) = 1 + \alpha x^{\beta} + O(x^{\beta+\varepsilon})$ when x converges to 0 with the condition that $|\arg(x)| < \pi - \theta$ for some α and $\theta < \pi/2$. We define $l(x) = \log(g(1 - x)) - \log(a)$. We have

$$l(x) = \sum_{i=1}^{\infty}\left[\log\left(1 + \frac{1 - \lambda\varepsilon_i z_i}{z_i - 1}x\right) - \log\left(1 + \frac{x}{z_i - 1}\right)\right] - \log(a). \tag{11.22}$$

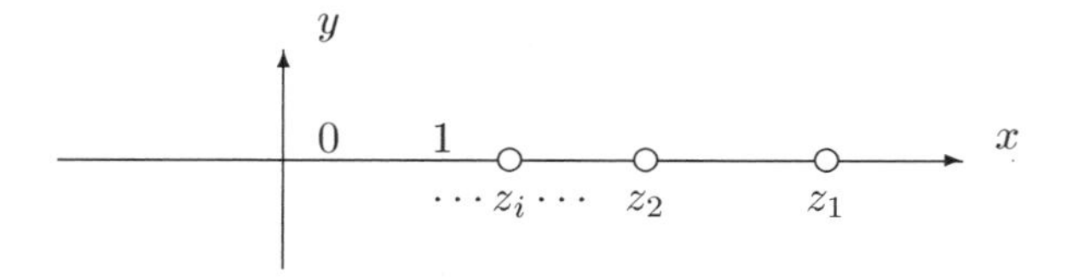

Fig. 11.6 Locations of the poles z_i in the complex plane.

Since $l(x) = -\log(a) + o(1)$ when $x \to 0$ and since $l(x) = O(1/x)$ when $x \to \infty$, the Mellin transform $l^*(s)$ of $l(x)$ is defined on the strip $0 < \Re(s) < 1$ and has expression (cf. [9])

$$\sum_{i=1}^{\infty}\left[\left(\frac{1-\lambda\varepsilon_i z_i}{z_i - 1}\right)^{-s} - \left(\frac{1}{z_i - 1}\right)^{-s}\right]\frac{-\pi}{s\sin\pi s}. \tag{11.23}$$

We identify in $(s\ \sin\pi s)^{-1}(-\pi)$ the Mellin transform of function $\log(1+x)$.

To get asymptotics of $l(x)$ when $x \to 0$, we identify the singularities of $l^*(s)$ located on the left of the definition domain, that is, for negative $\Re(s)$, in order to use the residue theorem in the reverse Mellin transform.

The first encountered pole is at $s = 0$ since $(s\sin\pi s)^{-1}\pi$ has a double root at $s = 0$. This pole is finally of degree 1 since the factor in front of it also has a root at $s = 0$; its residue is $\sum_{i=1}^{\infty}\log(1-\lambda\varepsilon_i z_i) = \log(a)$. There is also another pole at $s = -\beta$. Indeed,

$$\sum_{i=1}^{\infty}\left[\left(\frac{1-\lambda\varepsilon_i z_i}{z_i - 1}\right)^{-s} - \left(\frac{1}{z_i - 1}\right)^{-s}\right] = \sum_{i=1}^{\infty} s\varepsilon_i^{1+s}(1-\lambda)^{-s}(1+O(\varepsilon_i)) \tag{11.24}$$

since $z_1 = 1 + \varepsilon_i/(\lambda - 1) + O(\varepsilon_i^2)$. Therefore, the pole at $s = -\beta$ has residue $\mu\beta(1-\lambda)^{\beta}$.

Applying the residue theorem in the reverse Mellin transform by moving the line of integration from c to $-\beta - \varepsilon$, we find

$$\begin{aligned} l(x) &= \frac{1}{2i\pi}\int_{c-i\infty}^{c+i\infty} l^*(s)x^{-s}\,ds \\ &= -\log(a) - \mu\beta(1-\lambda)^{\beta}\pi(\beta\sin\pi\beta)^{-1}x^{\beta} \\ &\quad + \frac{1}{2i\pi}\int_{-\beta-\varepsilon-i\infty}^{-\beta-\varepsilon+i\infty} l^*(s)x^{-s}\,ds. \end{aligned}$$

A trivial majorization under the above integral translates immediately into the correct asymptotics of $g(z)$ around $z = 1$: $g(1-x) = 1 - \alpha x^{\beta} + O(x^{\beta+\varepsilon})$ with $\alpha = a\mu\pi\beta(1-\lambda)^{\beta}(\beta\sin\pi\beta)^{-1}$. Of course, this asymptotic is valid under the condition that x varies in an open set such that $|\arg(x)| < \pi - \theta$ for a θ arbitrarily chosen as $\theta < \pi/2$.

The latter conditions suffice to translate asymptotics of $g(z)$ into asymptotics over its coefficients. Let b_n be the coefficients of $g(z)$: $\sum_{n=0}^{\infty} b_n z^n = g(z)$. Since $g(1-x) = 1 - \alpha x^{\beta} + O(x^{\beta+\varepsilon})$, we immediately deduce from the theorems about singularity analysis of Flajolet–Odlyzko [15] that

$$b_n = \frac{\alpha}{\Gamma(-\beta)} n^{-1-\beta} + O(n^{-1-\beta-\varepsilon}). \tag{11.25}$$

The cumulative coefficients $p_n = b_n + b_{n+1} + \cdots$ observe similar asymptotics:

$$p_n = \frac{\alpha}{\Gamma(1-\beta)} n^{-\beta} + O(n^{-\beta-\varepsilon}). \tag{11.26}$$

The above expression proves the theorem. ■

Figure 11.7 displays the distribution of queue size. The results are obtained by sampling 20,000 windows with a time unit size of 10. If we select a logarithmic scale for the horizontal axis, the curve $\log(p_n)$ versus $\log n$ is asymptotically linear and illustrates of a polynomial tail distribution.

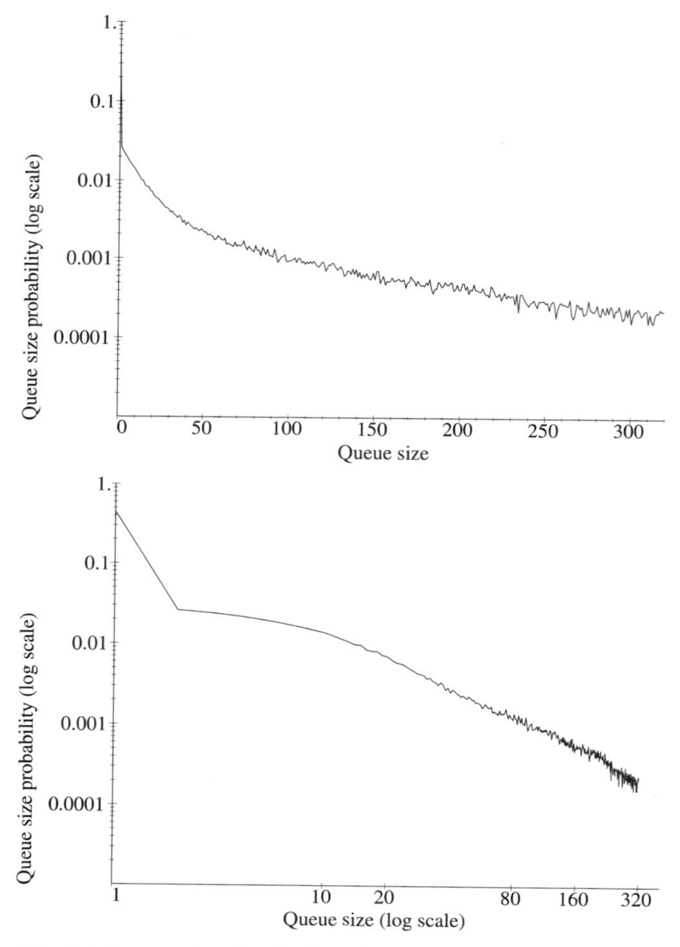

Fig. 11.7 Simulated queue size distribution (*bottom*, logarithmic scale), for on/off aggregation.

Remark. We can extend this result in a similar manner as for the parallel queue analysis, that is, making the singularity set a little more complicated. In principle, it is possible to obtain tail distributions with oscillating terms. We can also treat more general cases. The artifact condition $\lambda > 1$ can also be eliminated. But the most promising work item left in the present study is the estimation of a suitable *upper* bound for the queue length tail distribution, since the present work is limited to lower bounds.

11.4 CONCLUSION

A simple model is presented of arrivals based on the aggregation of independent memoryless on/off sources. It is shown that appropriate tuning of parameters produces long-term dependence and polynomial tail distributions of buffer occupancy. Such a model may give rise to a new source of interesting insight about long-term dependence in Web traffic. Future work can progress in two different directions. The first direction is to find less restrictive conditions on parameter tuning. The second direction is to find an exact estimate of the asymptotic tail distribution instead of a lower bound.

REFERENCES

1. W. Willinger, M. Taqqu, W. Leland, and D. Wilson. Self-similarity in high-speed packet traffic: analysis and modeling of Ethernet traffic measurement. *Statist. Sci.*, **10**(1):67–85, 1995.
2. W. Feller. *An Introduciton to Probability Theory and Its Applications*. Wiley, New York, 1971.
3. F. Baccelli. Effect of cross traffic on the performance of TCP. Workshop on the Modeling of TCP, École Normale Supérieure, 1998.
4. N. Duffield and N. O'Connell. Large deviations and overflow probabilities for the general single server queue, with applications. DIAS Technical Report No DIAS-STP-93-30, 1993.
5. J. Beran. *Statistics for Long Memory Processes*, Vol. 61 of Monographs on Statistics and Applied Probability. Chapman and Hall, New York, 1994.
6. P. Jacquet. Analytic information theory in service of queueing with aggregated exponential on/off arrivals. In *Proceedings of the 35th Annual Allerton Conference on Communication, Control and Computing*, pp. 242–251, 1997.
7. O. Boxma and V. Dumas. Fluid queues with long-tailed activity period distribution. In *Computer Communications*, invited survey in the special issue on stochastic analysis and optimization of communication systems, 1998.
8. B. Tsybakov and N. Georganaas. Overflow probability in ATM queue with self-similar input traffic. In *Proceedings IEEE International Communications Conference* (ICC '97), 1997.

9. P. Flajolet, X. Gourdon, and P. Dumas. Mellin transform and asymptotics: harmonic sums. *Theor. Comput. Sci.*, **144**(1–2):3–58, 1995.

10. P. Flajolet, M. Régnier, and D. Sotteau. Algebraic methods for trie statistics. *Ann. Discrete Math.*, **25**:145–188, 1985.

11. P. Jacquet and W. Szpankowski. Entropy computations via analytic depoisonization, *IEEE Trans. Inf. Theory*, 1072–1081, 1999.

12. W. Szpankowski. Techniques of the average case analysis of algorithms. In M. Atallah, ed., *Handbook on Algorithms and Theory of Computation*. CRC Press, Boca Raton, FL, 1997.

13. M. Neuts. *Matrix-Geometric Solutions in Stochastic Models: An Algorithmic Approach*. John Hopkins University Press, Baltimore, 1981.

14. P. Jacquet and N. Vvedenskaya. OnOff sources in an interconnection network: performance analysis when packets are routed to the shortest queue of two randomly selected nodes. INRIA Research Report RR-3570, 1998.

15. P. Flajolet and A. Odlyzko. Singularity analysis of generating functions. *SIAM J. Discuss. Math.*, **3**(2):216–240, 1990.

12

LONG-RANGE DEPENDENCE AND QUEUEING EFFECTS FOR VBR VIDEO

DANIEL P. HEYMAN
AT&T Labs, Middletown, NJ 07748
T. V. LAKSHMAN
Bell Laboratories, Holmdel, NJ 07733

12.1 INTRODUCTION

In this chapter we present some of our results concerning source models for H.261 coded VBR (variable bit rate) video. Video services have been forecasted to be a substantial portion of the traffic on emerging broadband digital networks. Statistical source models of video traffic are needed to design networks that delivery acceptable picture quality at minimum cost, and to control and shape the output rate of the coder. For example, one issue is deciding whether a new video connection can be admitted to a network, and the consequent determination of the bandwidth that must be allocated to the connection to ensure adequate quality of service. A model of the bandwidth that the connection will try to consume is required for this task. In addition to providing a good description of the bandwidth requirements, the source model should be usable in the connection-acceptance decision model.

Other chapters that contain related material are Chapters 9, 13, 16, and 17.

12.1.1 Special Properties of Video

There are some physical reasons why traces from video sources are special. Video is a succession of regularly spaced still pictures, called *frames*. Each still picture is represented in digital form by a coding algorithm, and then compressed to save

Self-Similar Network Traffic and Performance Evaluation, Edited by Kihong Park and Walter Willinger
ISBN 0-471-31974-0

bandwidth. See, for example, Netravili and Haskell [24] for full information about video coding. A common way to save bandwidth is to send a reference frame, and then send the differences of successive frames. This is called *interframe coding*. Since the adjacent pictures cannot be too different from each other (because most motion is continuous), this generates substantial autocorrelation in the sizes of frames that are near to each other. To protect against transmission errors, a full frame is sent periodically. Furthermore, when there is a scene change the frames no longer depend on the past frames, so functional correlation ends; this may also end the statistical correlation in the frame sizes. Scene changes require that a complete new picture be transmitted, so the scene lengths have an effect on the trace. For these and several other reasons that are too difficult to describe here, video traffic is different from broadband data traffic, and so the models and conclusions described in this chapter may not apply to other types of traffic.

Video quality degrades when information is lost during transmission or when the interarrival times of frames are either large or very variable. The latter is controlled by limiting buffer sizes; frames that arrive late might as well not arrive at all. Video engineers often describe the size of a buffer by the length of time it takes to empty it (which is the maximum delay a frame can incur). Current design objectives are for a maximum delay of between 100 and 200 ms. Since several buffers may be encountered from source to destination and there are other sources of delay (e.g., propagation time), some studies use 10 ms as the maximum buffer size.

Frames are transmitted in fixed size units that we call *cells*. The rate of information loss is the *cell-loss rate* or CLR. We are interested in situations where the cell losses occur because of buffer overflow. We consider models of a single station, where the buffer size and buffer drain rate are given. Under these conditions, the CLR is controlled by keeping the traffic intensity small enough to achieve a performance goal. A typical performance goal is to keep the CLR no larger than 10^{-k}, where k is usually between three and six.

These contraints on buffer size and CLR (and indirectly on traffic intensity) give rise to a practical region of operation where the constraints are satisfied. The notion of "high" traffic intensity is related to these constraints. When the design parameters (e.g., buffer size and processing speed) are specified, the traffic intensity is high when the constraints are just barely satisfied.

12.1.2 Source Modeling

The central problem of source modeling is to choose how to represent data traces by statistical models. A source model is sought for a purpose, which is usually as an input process to a performance model. We think that a source model is acceptable if it "adequately" describes the trace in the performance model at hand. By adequate we mean that when the source model is used in the performance model, the values of the operating characteristics of interest produced are "close enough" to the value produced by the trace. The definition of close enough may depend on the use to which the performance model will be put. For example, long-range network planning typically requires less accuracy for delay statistics and loss rates than equipment engineering does.

We don't regard good source models for a given trace to be unique. Different purposes may best be served by different models. For example, the DAR and GBAR models described in Sections 12.2.2 and 12.2.3 are designed for different purposes. A consequence of our emphasis on testing source models by how well they emulate the behavior of the trace they model in a performance model is that the confidence intervals we emphasize are on the operating characteristics of the performance models.

12.1.3 Outine

We divide VBR video into two classes, video conferences and entertainment video. Section 12.2 contains two models for video conferences, and Section 12.3 contains a model for entertainment video. The models are vetted by comparing the performance measures they induce in a simulation to the performance measures induced by data traces. All of these models are Markov chains, so they are short-range dependent (SRD). Hurst parameter estimates for the time series these models describe indicate the presence of long-range dependence. The reasons that short-range dependent models can provide good models for time series that exhibit long-range dependence are given in detail in Section 1.4. Our results are summarized in the last section.

12.2 VIDEO CONFERENCES

Video conferences show talking heads and may be the easiest type of video to model. The models developed for them will be expanded to describe entertainment video in Section 1.3.

12.2.1 Source Data

We have data from three different coders and four video teleconferences of about one-half hour in length. The data consists of the size of each still picture, that is, of each frame. All of the teleconferences show a head-and-shoulders scene with moderate motion and scene changes, and with little camera zoom or pan. All of the coders use a version of the H.261 video coding standard. The key differences in the sequences are that sequence A was recorded by a coder that uses neither discrete-cosine transform (DCT) nor motion compensation, sequence B was recorded by a coder that uses both DCT and motion compensation, and sequences C and D were recorded by a coder that used DCT but not motion compensation. The graphs in Figs. 12.1 and 12.2 show that the details (presence or absence of DCT or motion compensation) do not have a significant effect on the statistics of interest to us here. The summary statistics of these sequences are given in Table 12.1.

All of these sequences are adequately described by negative-binomial marginal distributions and geometric autocorrelation functions. Figure 12.1 shows Q-Q plots of the marginal distributions, which have been divided by their means; the fit is

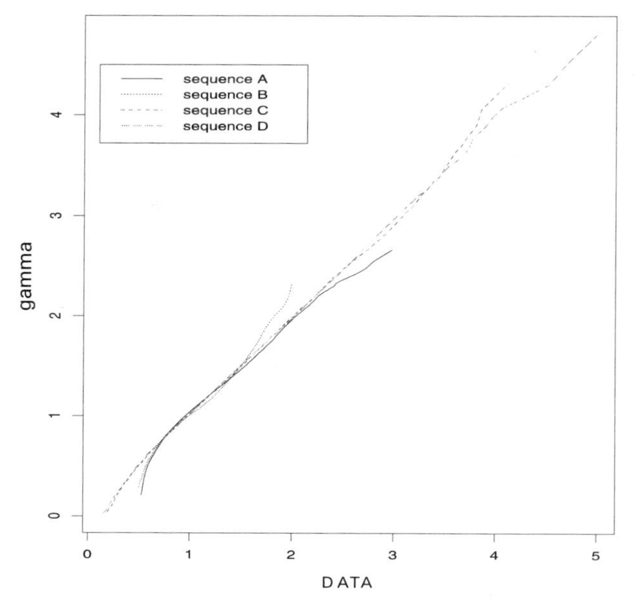

Fig. 12.1 Q-Q plots for four sequences.

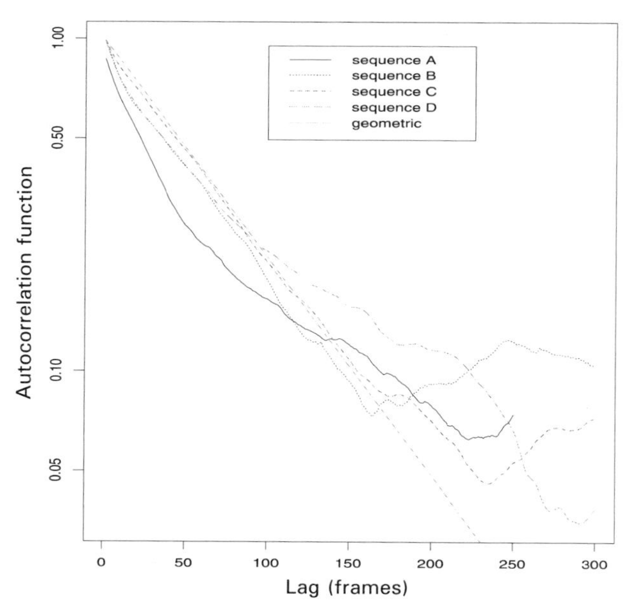

Fig. 12.2 Autocorrelation functions.

TABLE 12.1 Summary Statistics of Data Sequences

Sequence	Bytes per Cell	Mean (cells)	Standard Deviation (cells)	ρ
A	14	1506.4	512.7	0.981
B	48	104.9	29.7	0.984
C	64	130.3	74.4	0.985
D	64	170.6	107.6	0.970

excellent for sequences C and D, good for sequence A, and adequate for sequence B. The negative-binomial distribution is the discrete analog of the gamma distribution, and a discretized version of the latter can be used when it is more convenient to do so.

Figure 12.2 shows the autocorrelation functions. The ordinate has a log scale, so geometric functions will appear as straight lines. The geometric property holds for at least 100 lags (2.5 seconds) for sequences B, C, and D, and for 50 lags for sequence A. For lags larger than 250, the geometric function underestimates the autocorrelation function. We examined sequences A, B, and C and concluded they possess long-range dependence. Since the autocorrelation functions shown in Fig. 12.2 are so large for small lags, it seem intuitive (to us, at least) that the short-range correlations should be the important ones to capture in a source model. We propose using the geometric function ρ^k for the autocorrelation function.

Since the negative-binomial and gamma distributions are specified by two parameters, these parameters can easily be esimated from the mean and the variance of the number of cells per frame by the method of moments. Only those two moments and the correlation coefficient (ρ) are needed to specify the key properties of VBR teleconference traffic. The correlation coefficient can be estimated from the geometric portion of the autocorrelation function by taking logarithms and doing a linear regression.

12.2.2 The DAR Model

Our first investigations of these sequences with Tabatabai [16] and Heeke [18] focused on multiplexing issues. First, we established that the time series were stationary. This was done by examining plots of smoothed versions of the time series and boxplots of many partitions of the time series. Next, we showed a Markov chain provided a good description of the time series. This was done via simulations as described in Section 12.2.2.1. This means that the marginal distributions of the time series can be viewed as the steady-state distributions of the Markov chain. A Markov chain that has a geometric autocorrelation function and whose steady-state distribution can be specified is the DAR(1) process introduced by Jacobs and Lewis [20]. The only member of the DAR(k) family that is used here is the DAR(1), so the (1) will be deleted. The transition matrix is given by

$$P = \rho I + (1 - \rho)Q, \tag{12.1}$$

where ρ is the lag-one autocorrelation coefficient. I is the identity matrix, and each row of Q consists of the steady-state probabilities. In our case the steady-state probabilities are the negative-binomial probabilities described above, truncated at some convenient value at least as large as the peak rate (the missing probability is added to the last probability kept).

Equation (12.1) is convenient for analytical work, but it masks the simplicity of the DAR model. Let X_n be the size (in bits, bytes, or cells as appropriate) of the nth frame and ρ be as above; the DAR model is

$$X_n = \begin{cases} X_{n-1} & \text{with probability } \rho, \\ X' & \text{with probability } 1-\rho, \end{cases} \tag{12.2}$$

where X' is a sample from the marginal distribution (negative-binomial in our case). From Eq. (12.2) we see that the X-process maintains a constant value (cell rate) for a geometrically distributed number of steps (frames) with mean $1/(1-\rho)$, and then another value (possibly the same as the old value) is chosen. When ρ is close to one (it is about 0.98 in our examples, see Table 12.1), the mean time between cell rate changes is large (about 50 frames in our examples). This means that the sample paths are constant for long intervals. The data trace doesn't have this property, which is the reason the GBAR model described in Section 12.2.3 was introduced. This difference between the sample paths of the model and the data trace is mitigated when several sources are multplexed. The probability that $X_n' = X_{n-1}$ is small enough to be ignored in the following calculation. When k sources are multiplexed, $X_n = X_n - 1$ with probability ρ^k, so the mean time between potential cell rate changes with $\rho = 0.98$ and $k = 16$ is 3.6. Consequently, sample paths of the multiplexed cell streams from 16 sources are not constant for long intervals.

12.2.2.1 Validating the DAR Model We validate the DAR model by looking at performance models for multplexing gain and connection admission control. To estimate statistical multiplexing gain, we use cell-loss probabilities [16] from a simple model of a switch. The source model is a FIFO buffer that is drained at 45 Mb/s. The length of the buffer is expressed as the time to drain a full buffer; this is the maximum possible delay. The results of ten simulations of the DAR model for sequence C are given by 95% confidence intervals and are shown in Table 12.2. The

TABLE 12.2 Cell-Loss Rates for Trace and 95% Confidence Intervals for DAR Model of Sequence C

	Buffer Size (ms)				
Source	1	2	3	4	5
	Probability of Loss $\times 10^{-6}$ *for Various Buffer Sizes*				
Trace	2070.0	527.0	141.0	33.3	2.88
DAR model	(1738, 2762)	(433, 775)	(107.4, 212.6)	(15.1, 54.1)	(2.26, 9.34)

results of these simulations show that the DAR model does a good job of estimating the cell-loss rate when 16 sources are multiplexed. Similar results were obtained for the other sequences [18].

Now we consider connection admission control (CAC). Since the DAR model is a Markov chain model of the source, it conforms to one of the sets of conditions a source model must have for the effective bandwidth (EBW) theory of Elwalid and Mitra [8]. Moreover, the DAR model is a reversible Markov chain, and so it inspired a powerful extension of the EBW method, called the *Chernoff-dominated eigenvalue* (CCE) method [7]. Suppose we have a switch that can process at rate C (Mb/s) and has a buffer of size B (ms). We want to find the maximum number of statistically homogeneous sources that can be admitted while keeping the cell-loss rate no larger than 10^{-6}. The CDE method gives an approximate analytic solution with known error bounds; this solution is denoted by K_{CDE}. Another way to obtain the solution is to test candidate values by evaluating the cell-loss rate by simulation; we treat this as the exact solution and denote it by K_{sim}. Table 12.3 compares the results of the CDE method to the CAC found from simulations. The number admitted by the CDE method is a very close approximation to the "true" value obtained by simulation. This implies that the DAR model captures enough of the statistical properties of the trace to produce good admission decisions.

12.2.3 The GBAR Model

The DAR model may not be suitable for a single source (by a single source we mean a source that does not interact with other sources) as described above. Lucantoni et al. [22] give three areas where single source models are useful: studying what types of traffic descriptors make sense for parameter negotiation with the network at call setup, testing rate control algorithms, and predicting the quality of service degradation caused by congestion on an access link. For this reason, Heyman [11] proposed the GBAR model. Lakshman et al. [21] use the GBAR model to predict frame sizes in a rate control algorithm.

Lucantoni et al. [22] propose a Markov-renewal process model to describe a single source. This model has the advantage of being very general, and the disadvantage that it is not parameterized by some simple summary statistics of the data trace. The GBAR model exploits the properties enjoyed by teleconferencing traffic described in Section 12.2.2, the geometrically decaying autocorrelation

TABLE 12.3 CAC Performance for Video Conference A and Video Conference C

	Video Conference A										Video Conference B			
B	57	9	50	44	5	0.5	9	23	7	1	8.5	9.5	10	11
C	45	67	81	103	125	145	195	245	270	310	110	185	280	375
K_{sim}	20	30	40	50	60	70	98	128	139	156	16	30	49	66
K_{CDE}	16	25	33	44	53	63	90	120	130	150	15	30	50	70

function and the negative-binomial (or gamma) marginal distributions, to produce a simple model based on the three parameters that describe these features.

The GBAR(1) process was introduced by McKenzie [23], along with some other interesting autoregressive processes. (As with the DAR model we will drop the argument (1).) Two inherent features of this process are the marginal distribution is gamma and the autocorrelation function is geometric.

Toward defining the GBAR model, let $Ga(\beta, \lambda)$ denote a random variable with a gamma distribution with shape parameter β and scale parameter λ; that is, the density function is

$$f_G(t) = \frac{\lambda(\lambda t)^{\beta}}{\Gamma(\beta+1)} e^{-\lambda t}, \quad t > 0. \tag{12.3}$$

Similarly, let $Be(p, q)$ denote a random variable with a beta distribution with parameters p and q; that is, with density function

$$f_B(t) = \frac{\Gamma(p+q)}{\Gamma(p+1)\Gamma(q+1)} t^p (1-t)^q, \quad 0 < t < 1, \tag{12.4}$$

where p and q are both larger than -1. The GBAR model is based on two well-known results: the sum of independent $Ga(\alpha, \lambda)$ and $Ga(\beta, \lambda)$ random variables is a $Ga(\alpha + \beta, \lambda)$ random variable, and the product of independent $Be(\alpha, \beta - \alpha)$ and $Ga(\beta, \lambda)$ random variables is a $Ga(\alpha, \lambda)$ random variable. Thus, if X_{n-1} is $Ga(\beta, \lambda)$, A_n is $Be(\alpha, \beta - \alpha)$, and B_n is $Ga(\beta - \alpha, \lambda)$, and these three are mutually independent, then

$$X_n = A_n X_{n-1} + B_n \tag{12.5}$$

defines a stationary stochastic process X_n with a marginal $Ga(\beta, \lambda)$ distribution. Furthermore, the autocorrelation function of this process is given by

$$r(k) = \left(\frac{\alpha}{\beta}\right)^k, \quad k = 0, 1, 2, \ldots \tag{12.6}$$

The process defined by Eq. (12.5) is called the GBAR processes. The G and B denote gamma and beta, respectively, and the AR stands for autoregressive. Since the current value is determined by only one previous value, this is an autoregressive process of order one.

A possible physical interpretation of Eq. (12.5) is the following. Interpret A_n as the fraction of frame $n-1$ that is used in the predictor of frame n, so the first term on the left of Eq. (12.5) is the contribution of interframe prediction. In hybrid PCM/DPCM coding [24], for example, resistance to transmission error is accomplished by periodically setting some differential predictor coefficients to zero and sending a PCM value. We can think of B_n as the number of cells to do that. If the distributional and independence assumptions listed above Eq. (12.5) are valid, then the GBAR process will be formed.

Simulating the GBAR process only requires the ability to simulate independent and identically distributed gamma and beta random variables. This is easily done; for example, algorithms and Fortran programs are presented in Bratley et al. [3].

The GBAR process is used as a source model by generating noninteger values from Eq. (12.5) and then rounding to the nearest integer. It would be cleaner if a discrete process with negative-binomial marginals could be generated in the first place. McKenzie describes such a process (his Eq. (3.6)). Unfortunately, that process requires much more computation to simulate, and the extra effort does not appear to be worthwhile.

12.2.3.1 Validating the GBAR Model Ten sample paths of the GBAR process were generated and used as the arrival process (number of cells per frame with a fixed interframe time) in a simulation of a service system with a finite buffer and a constant-rate server. The cell-loss rates from these paths were averaged to obtain a point estimate for the GBAR model. The traffic intensity is varied by changing the service rate. The points produced by the simulations are denoted by an asterisk. In Fig. 12.3, we see that cell-loss rates computed from the GBAR model are close to the cell-loss rates computed from the data. Note that for each traffic intensity, the decrease in the cell-loss rate as the buffer size increases is very slight, for both the model and the data. This confirms the prediction of Hwang and Li [19] of buffer ineffectiveness. Since the GBAR model has only short-range dependence, this effect is not caused by long-range dependence here.

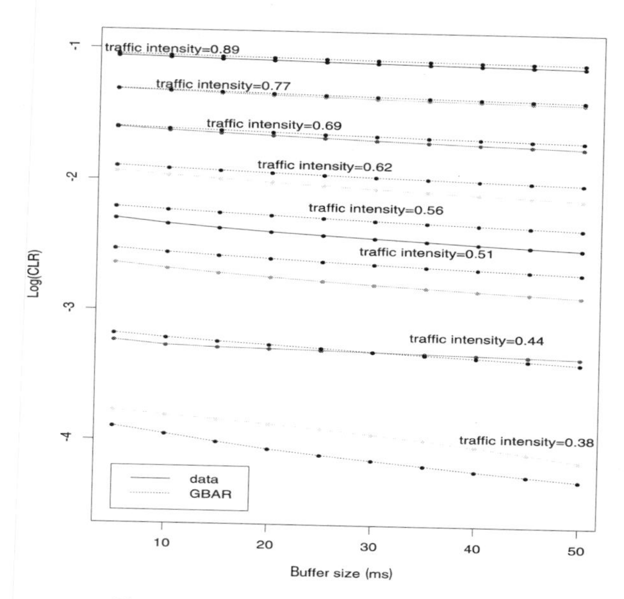

Fig. 12.3 Cell-loss rates for sequence A.

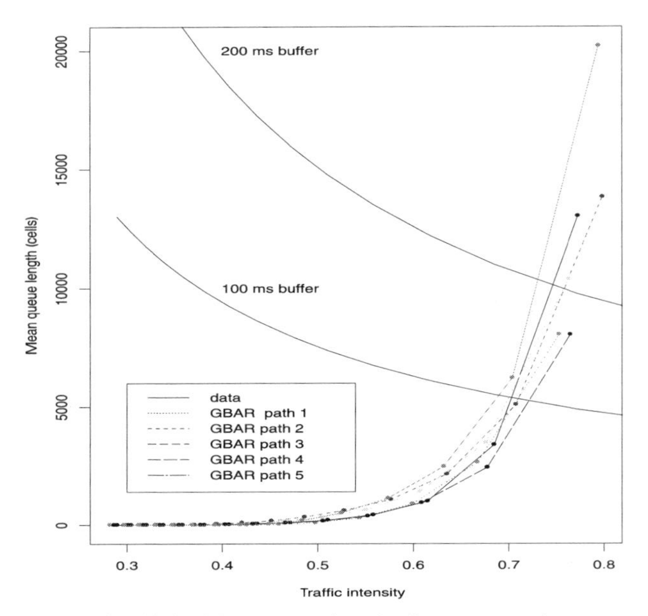

Fig. 12.4 Mean queue lengths for sequence A.

Figure 12.4 shows that the mean queue lengths in an infinite buffer computed from five GBAR paths are similar to the mean queue lengths computed using the data. In Fig. 12.4 the vertical axis on the left shows the mean queue length in cells. Video quality is poor when the cell delays are large; 100 ms is an upper bound on the acceptable delay at a node in a network that provides video services. Two buffer drain times are also shown; the practical region for the *maximum* is below and to the left of the 100 ms line. The range of the mean queue lengths shown exceeds the practical region for maximum queue length. In the practical region, the model and the trace give very similar mean delays. The differences between the mean queue length from the GBAR model and the mean queue lengths from the data would be even smaller if a finite buffer were imposed. (This is the truncating effects of finite buffers that is described in Section 12.4.2). The comparisons for sequences B, C, and D are qualitatively the same as for sequence A.

12.3 BROADCAST VIDEO

Now we turn to more dynamic sequences, such as films, news, sports, and entertainment television. Since the main purpose of the models is to aid in network performance evaluations, we are particularly interested in using the models to predict cell-loss rates.

Broadcast VBR-coded video has different bit-rate characteristics than VBR-coded video conferences. Video conference sequences consist of head-and-shoulders pictures with little or no panning, while broadcast video is characterized by a succession of scenes. With interframe coding, it is clear that scene changes require more bits than intrascene frames, so broadcast video will differ from video conferences in at least this respect; this was demonstrated by Yasuda et al. [32]. There are some other differences too, as demonstrated by Verbiest et al. [31] and further amplified by Verbiest and Pinnoo [30]. In these papers, a DPCM-based coding algorithm is used. In the latter paper, it is shown that the number of bits per frame has a different autocorrelation function for broadcast video than for video conferences or video telephony. The autocorrelation functions for the last two are similar to each other and decay geometrically to zero. For broadcast video, the autocorrelation function does not decay to zero. Moreover, the first frame after a scene change has significantly more bits than other frames in the scene. Ramamurthy and Sengupta [26] observe that the correlation function declines more rapidly at small lags than at large lags, and that the time series can be described by a semi-Markov process that has states identified by the bit rates for different types of scenes (and a state for scene changes). We build on this idea [12]; the simple DAR and GBAR models that described video conferences are not sufficient for broadcast video, although the DAR model is used as a building block in a more complex model.

12.3.1 Modeling Broadcast Video

We obtained several data sets giving the number of bits per frame for sequences encoded by an intrafield/interframe DPCM coding scheme without use of DCT or motion compensation. We did not have access to the actual video sequences. Hence, a visual identification of scene-change frames was not possible. Our modeling strategy was to first develop a way to identify scene changes, then construct models for the lengths of the scenes and the number of cells in a scene-change frame. Finally, models for the number of cells per frame for frames within scenes were developed.

12.3.1.1 Preliminary Data Analysis Before describing the statistical models, we report some elementary statistics about the sequences we examined. Figure 12.5 shows the peak and mean bit rates, and their ratios.

The peak-to-mean ratios vary from 1.3 to 2.4. By way of comparision, the peak-to-mean ratio for video-conference sequence with this codec (sequence A) is 3.2. Note that the larger peak-to-mean ratios are associated with the lower mean bit rates. The sequences *divers*, *film*, *Isuara 1*, and *Isaura 2*, which have a low mean rate and high peak-to-mean ratios, were different TV programs recorded from a Cable TV network (and designated as normal quality broadcast video). The sequences with low peak-to-mean ratios (such as *football*, *sport*, *news*, etc.) were taken directly from the TV studios (and designated high-quality broadcast video).

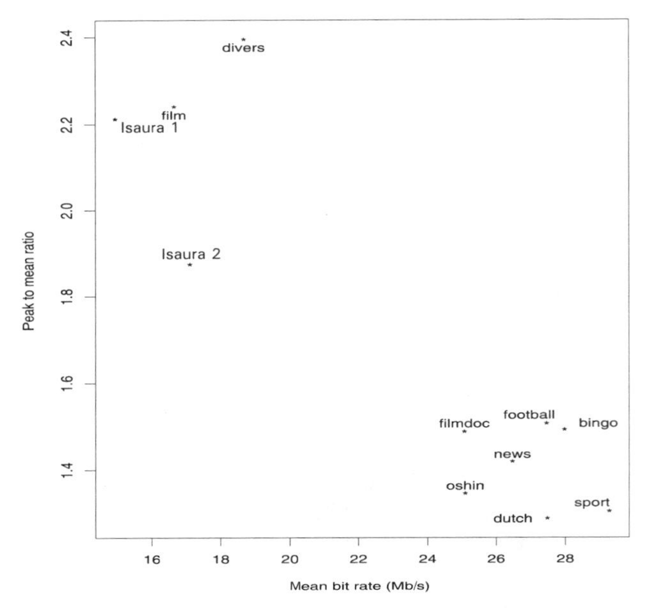

Fig. 12.5 Peak-to-mean ratios.

12.3.2 Identifying Scene Changes

Figure 12.6 shows two segments of the trace of *film*, and it can be seen that there are several spikes that are possibly due to scene changes. Since these spikes may be a dominant cause of cell losses, we need to model both their spacing and the magnitude. If we use merely fixed spacing at the correct rate but do not model the distribution of their spacing, multiplexed sources with nonidentical starting points will not have coincident spikes from time to time. This will underestimate cell losses.

Since we do not have a video record of the sequences, we will assume that a scene change occurs when a frame contains an abnormally large number of cells compared to its neighbors. We make this notion quantitative in the following way. Let X_i be the number of cells in frame i. At a scene change, the second difference $(X_{i+1} - X_i) - (X_i - X_{i-1})$ will be large in magnitude and negative in sign. To quantify what we mean by large, we divide the second difference by the average of the past few frames. We found that using 25 frames (1 second) in the average was about the same as using 6 frames (about $\frac{1}{4}$ second), and the latter was adopted. We chose -0.5 as the critical value; this choice is entirely subjective. It would be nice if the subjectivity could be replaced by an objective criterion. We examined the statistical theory of outlier identification for guidance and concluded that this is wishful thinking because objective tests need to have "outlier" specified externally.

To see if our criterion identified scene changes accurately, we looked at the time series X_i and the corresponding values of the scaled second differences. Some of the

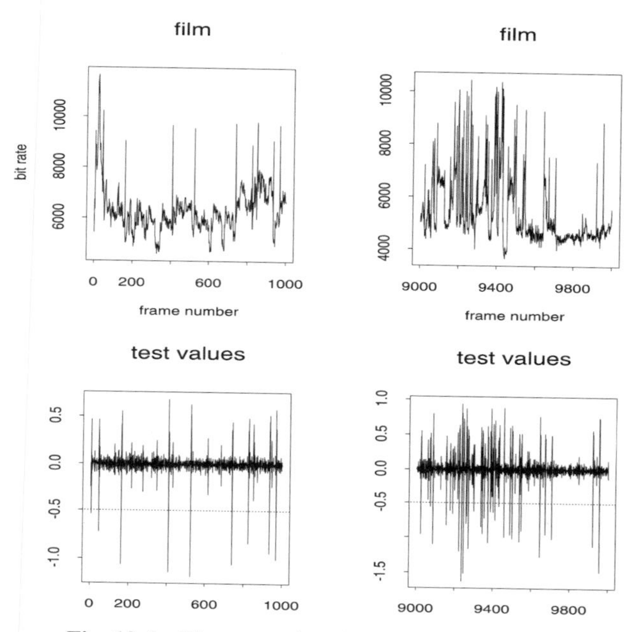

Fig. 12.6 Bit rates and scaled second differences.

data are shown in Fig. 12.6. The first 1000 frames are shown on the left. The test values identify the scene changes with no false positives. The choice of -0.5 as the critical value is not significant. Frames 9001 through 10,000 are shown on the right. This is a more active (in terms of bit-rate fluctuations) subsequence, and changing the critical value will affect the number of frames identified as scene changes. There are 317 scene changes when -0.5 is the critical value (the mean scene length is 6.5 seconds); there are 374 when -0.4 is used, and 283 when -0.6 is used. The density functions of the scene lengths produced by these critical values are shown in Fig. 12.7. We observe that the critical value does not have a large effect on the density function.

12.3.2.1 Scene Lengths Plots of the autocorrelation function showed that scene lengths are uncorrelated, so the main modeling issue is to characterize the distribution of the number of frames in a scene. The shape of the density in Fig. 12.7 was observed on all sequences except for *news*. (This is a news broadcast. There are 134 scene changes, and 75 of them are 3 frames long. Moreover, most of these occur consecutively. We do not have an explanation for this. The 3-frame scenes were deleted and the remainder are called *news.d.*)

This unimodal and long-tailed shape is characteristic of distributions used in reliability and insurance-loss models. We will use the following three distributions as candidates for describing scene lengths (and the number of cells per frame in the sequel).

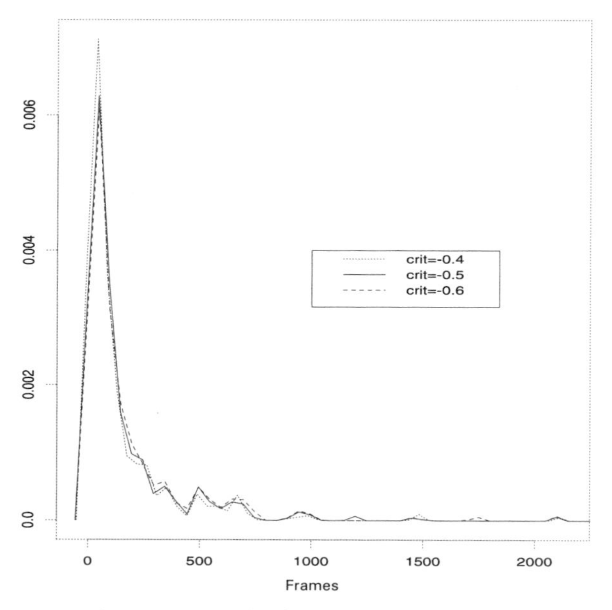

Fig. 12.7 Density functions of scene lengths.

The Gamma Distribution. The density function is given by Eq. (12.3).

The Weibull Distribution. It has the complementary distribution function (the probability that the random variable is larger than x)

$$S_{\mathrm{W}}(x) = 1 - e^{-\lambda x^{\beta}}, \quad x, \lambda, \beta > 0.$$

The Generalized Pareto Distribution. It has density function

$$f_{\mathrm{P}}(x) = \frac{\Gamma(\beta + k)\lambda^{\beta} x^{k-1}}{\Gamma(\beta)\Gamma(k)(\lambda + x)^{k+\beta}}, \quad x, \beta, \lambda, k > 0.$$

The Classical Pareto Distribution is the special case of $k = 1$. We will dispense with the adjective "general."

In all of these distributions, λ is a scale parameter and β is a shape parameter; the Pareto distribution has the second shape parameter k. The Pareto distribution contains the other two as special cases, but it is useful to maintain the distinction.

Figure 2.1 of Cox and Oakes [6] gives a nice way to compare these distributions. For a random variable X, let $\mu = E(X)$, $\sigma^2 = \mathrm{Var}(X)$, and $\mu_3 = E[(X - \mu)^3]$. Let γ_1 be the coefficient of variation; that is, $\gamma_1 = \sigma/\mu$. Let $\gamma_3 = \mu_3/\sigma^3$; it is the coefficient of skewness. For the distributions described above (and some others not mentioned here), γ_1 and γ_3 do not depend on the scale parameter λ. Plotting γ_3 versus γ_1 gives a

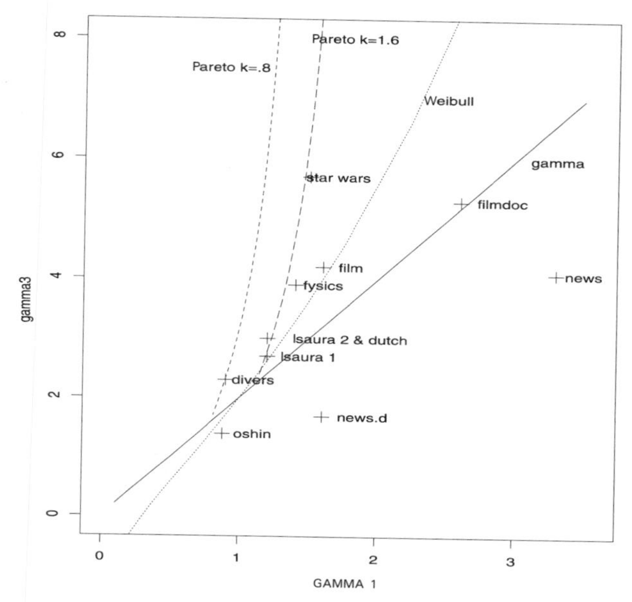

Fig. 12.8 Moments of scene lengths.

curve for the gamma and Weibull distributions, and a curve for each value of k for the Pareto distribution. Figure 12.8 shows these curves, and points for the sample values for the scene-length moments from our video sequences. (We excluded *bingo* from this and all subsequent figures because it contained only 12 scenes. We included data from a software intraframe coding [10] of the film "Star Wars.")

Except for sequences *news* and *news.d*, the moments of all the sequences plotted fall near the four curves traced out by the distributions; the distances from the curves are within the deviations we observed when samples from a true gamma distribution were compared with the gamma curves. We conclude that usually, but not always, scene lengths will follow a unimodal distribution that fits one of the common failure-time distributions.

12.3.3 Scene-Change Frames

Now we attempt to model the number of cells in the scene-change frames. It is clear that the frames that start a scene will have stochastically more cells than other frames because most of the picture has to be constructed afresh. Different models will be needed for the scene-change frames than for the intrascene frames. Plots of the autocorrelation function showed no correlation among frame sizes, so obtaining the distribution is the main modeling issue.

The distributions we use are on the interval $[0, \infty)$. The minimum number of cells in a scene-change frame is a few thousand, so we shifted the data by subtracting the

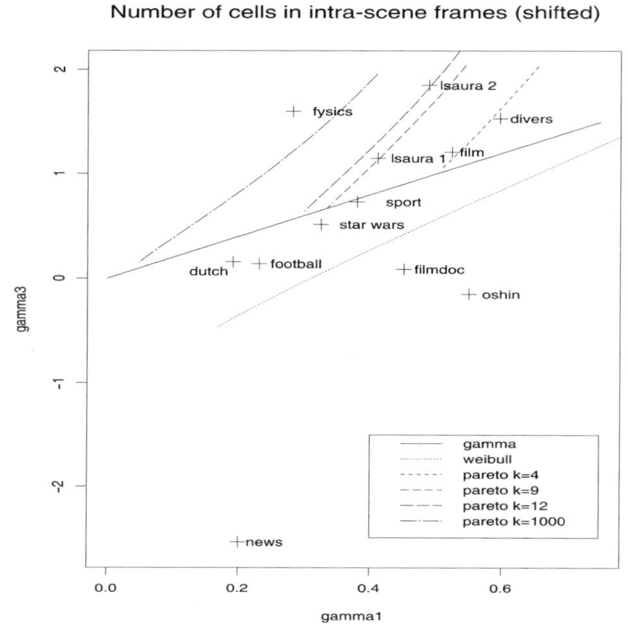

Fig. 12.9 Number of cells in scene-change frames.

smallest value. In other words, we model the amount larger than the minimum value in the data.

Figure 12.9 shows how the empirical values of γ_1 and γ_3 compare to the curves for the Gamma and Weibull distributions. Except for *fysics* and *football*, the data are far from the theoretical curves. Since the third moment has a large sampling error for long-tailed distributions [18, p. 26], the deviation from the theoretical curves may be due to sampling error. Based on Q-Q plots, two of the sequences (*Isaura 1* and *Isaura 2*) had good Weibull fits, and two (*dutch* and *star wars*) had good gamma fits. The lognormal distribution fits the *film* data well. We could not fit a model to the other sequences. We conclude that a known distribution may not always be a good fit to the number of cells in a scene-change frame, and that the same distribution is unlikely to be a good fit to all sequences.

12.3.4 Intrascene Frames

A close look at the bit-rate plot in Fig. 12.6 revealed that the effect of a scene change appears to last for two frames; the first frame after the scene change is also extra large. This was also observed by Ramamurthy and Sengupta [26] by looking at Fig. 12 in Verbiest and Pinnoo [30]. Since our data were produced by the same codec as the data in Verbiest and Pinnoo [30], this is not surprising. We examined *film, Isaura 1*, and *Isaura 2* and found that linear regression provided a good model for the number of cells in the first frame after a scene change. Letting Z_n be the number of cells in the nth scene-change frame, and Y_n be the number of cells in the succeeding frame, we have

$$Y_n = a + bZ_n + \epsilon_n, \tag{12.7}$$

where the ϵ_n are independent and identically distributed normal random variables with mean zero.

For the rest of the intrascene frames, the autocorrelation function and the distribution are important. Figure 13 in Verbiest and Pannoo [30] shows an autocorrelation function of a TV scene; it does not decline geometrically to zero as the video conference scenes do. We obtain similar functions when all of the intrascene frames (excluding the first two frames of each scene) are treated as a single time series. The left side of Fig. 12.10 shows this function for *flim* and *Isaura 1*. This method of estimating the autocorrelation function ignores scene boundaries. For example, the last frame of a scene and the first (counted) frame of the subsequent scene contribute to the lag one term in the autocorrelation function. For *film*, this causes the autocorrelation function to have periodic pulses, as shown in Fig. 12.10(c). These pulses are not present in the autocorrelation functions of the first six scenes and are just an artifact of the aggregation across scenes. A more refined estimate is obtained by dividing the data into scenes, calculating the empirical autocorrelation function for each scene, and then averaging over the scenes. To be

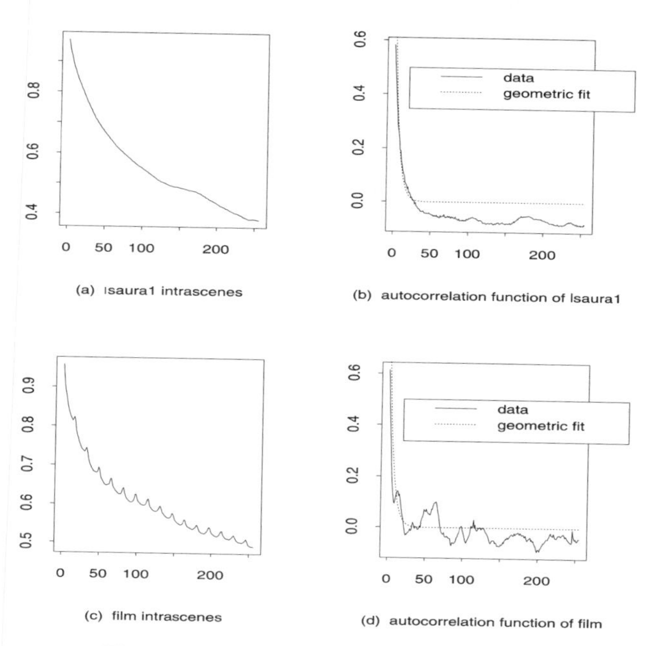

Fig. 12.10 Autocorrelation within scenes.

precise, first calculate $r_s(i)$, which is the lag i autocorrelation for scene s, for all i in scene s and all s. Then form the autocorrelation function $r(i)$ from

$$r(i) = \frac{1}{S}\sum_{s=1}^{S} R_s(i)\mathbf{1}_s(I), \quad i = 1, 2, \ldots,$$

where $\mathbf{1}_s(i)$ is one if scene s contains at least $i + 3$ frames (remember that the first two frames are treated separately and at least $i + 1$ observations are needed to estimate a lag i autocorrelation) and is zero otherwise. S is the number of scenes in the sequence. The autocorrelation functions obtained in this way are shown on the right side of Fig. 12.10. Now the autocorrelation functions decrease geometrically to zero; the dotted lines on the right-side of Fig. 12.10 show geometric functions fitted to the autocorrelation functions.

By using Q-Q plots we decided that the data indicate that *Isaura 1*, *Isaura 2*, *divers*, and *film* follow Pareto distributions, *dutch*, *sport*, and *star wars* follow gamma distributions, *football* and *filmdoc* can be modeled as either a gamma or a Weibull distribution, and *oshin*, *fysics*, and *news* do not follow any of the standard distributions. (*News* appears to be a mixture of five to seven of the standard distributions, but that is hard to fit.)

Since there is a wide variety of possible scenes, one might expect the distribution of the intrascene frames to be a mixture of unimodal distributions. It is perhaps surprising that except for *news*, we found that one distribution suffices. We were able to fit a distribution to the intrascene frames of some of the sequences, but not to all of them.

12.3.5 Validating the Models

We test our statistical models by using them as source models in a simulation of multiplexed video connections in an ATM network, and comparing the cell-loss probabilities from ten sample paths to the cell-loss probability obtained when the original sequences is used as the input stream. We arbitrarily selected *film* and *Isaura 1* as the sequences to use for the simulation tests.

12.3.5.1 Film From Fig. 12.8 we see that a Weibull distribution is a candidate for scene lengths of *film*. Frater et al. [9] use a Pareto distribution; this is also reasonable from Fig. 12.8. The parameters for this distribution are determined using the method of moments. We also see from Fig. 12.9 that the number of cells in scene-change frames follows a Weibull distribution. From Eq. (12.7), we know how to compute the number of cells in the frame succeeding a scene-change frame. The number of cells per frame in scene-change frames is uncorrelated between successive frames. So we need only construct a model for the number of cells per frame for frames within a scene. Using Fig. 12.11, we choose a Pareto distribution with $k = 4$ for the number of cells per frame in intrascene frames of *film*. We consider Markov chain and DAR models.

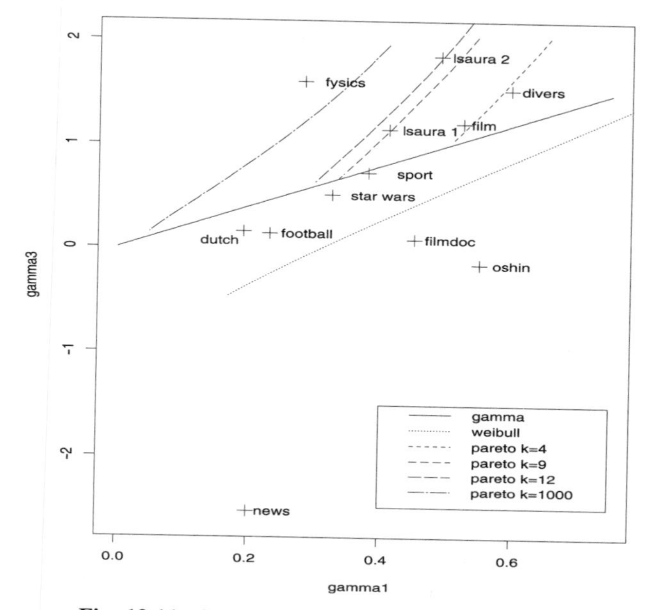

Fig. 12.11 Number of cells in intrascene frames.

The Markov chain model is created as follows. Let X_n be the number of cells in frame n, and $[X_n]$ be the integer part of $X_n/50$. (The choice of 50 is arbitrary; it makes the number of states about 100.) We propose to model $[X_n]$; $n = 1, 2, \ldots, N$ as a Markov chain with transition matrix $P = (p_{ij})$. We estimate (p_{ij}) in the usual way [2];

$$\hat{p}_{ij} = \frac{\text{number of transitions from } i \text{ to } j}{\text{number of transitions out of } i}$$

when the denominator is greater than zero.

The Markov chain model is used only for sequences where the DAR model is not sufficiently accurate. For *film*, the DAR model is a good model and the Markov chain model is not necessary for multiplexing studies; it is used in Section 12.4.3.1 as a model of a single source.

12.3.5.2 *Isaura 1* For *Isaura 1*, the models are generated in a similar manner. The distribution of scene lengths for *Isaura 1* is Weibull. Also, the distribution of the number of cells in scene-change frames is Weibull. The distribution of cells in frames within a scene is Pareto with $k = 9$. The Markov chain (here we modeled $[X_n] = X_n/100$ instead of $X_n/50$) and the DAR models are constructed in the same manner as for *film* except that Weibull distributions are used instead of Pareto.

12.3.6 Simulation Results

Figure 12.12 shows the simulation results evaluating the accuracy (as a predictor of cell losses) of the DAR model for the *film* sequence. This sequence has a mean rate of 16.7 Mb/s and a peak rate of 37.31 Mb/s. The multiplexer simulations were done with 8 sources being multiplexed. The output rate of the multiplexer was set to 155 Mb/s. The link utilization is 0.849. Simulations were done for maximum delays in the multiplexer equal to 0.5 ms, 1 ms, 2 ms, 3 ms, and 4 ms. The peak input rate from the source to the multiplexer was set to 43 Mb/s. This value only determines the cell transmission time to the multiplexer. The intercell spacing within a frame is equal and in an interframe interval the cell arrival rate is equal to the number of cells in the most recent frame divided by the interframe interval.

We generated 10 sample paths using the model and these are used in 10 runs of the simulation. In Fig. 12.12, the loss rates of the 10 sample paths for different buffer sizes are represented by their mean loss rate. The values corresponding to the mean plus one standard deviation and the mean minus one standard deviation are also plotted. For the actual traffic, the simulation was run not only using the trace of actual traffic but also with the rates in the actual traces scaled up and down by 1%. These are also shown in Fig. 12.12. The 1% scaling was used to provide a range of values against which the loss probabilities generated by the model may be compared. Since source rates may not be known very accurately, it is reasonable to compare the model's losses to actual losses with a 1% uncertainty in the source mean rate [9].

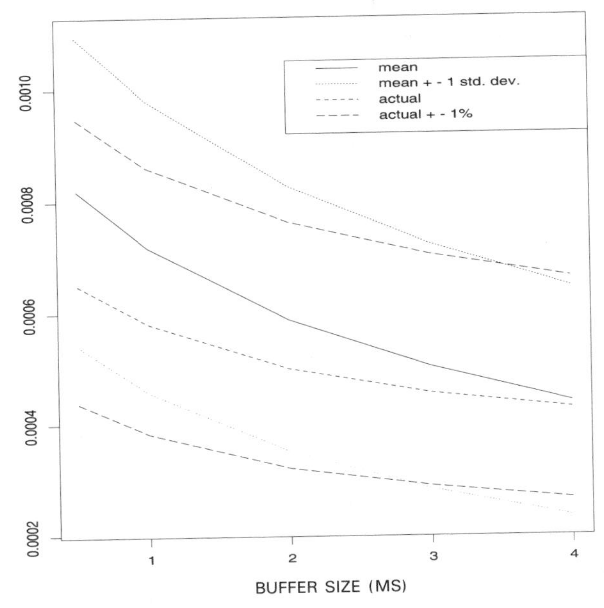

Fig. 12.12 Cell-loss probabilities for *film*.

As can be seen from Fig. 12.12, for all multiplexer buffer sizes studied, the mean loss rates of the 10 sample paths fall within the range of losses obtained for actual $\pm 1\%$. Also, the actual loss rate falls within the range of mean losses for the model ± 1 standard deviation. Hence, the DAR model is a sufficiently accurate model for the *film* sequence.

Frater et al. [9] successfully described this source with a simpler model. Their *scenic model* is a generalization of the basic DAR model of Eq. (12.1) that permits the times between changes of state to have any distribution on the positive integers.

Figure 12.13 shows the simulation results for the *Isaura 1* sequence. This sequence has a mean rate of 14.9 Mb/s and a peak rate of 33 Mb/s. The simulation was done with 8 sources being multiplexed. The buffer sizes and the multiplexer input and output rates are the same as those used for *film*. The output link utilization was 0.77. The losses resulting from use of the DAR and the Markov chain models are compared to the actual losses. As in Fig. 12.12, the actual losses and the losses resulting from a scaling of the mean rate of the actual sequence by $\pm 1\%$ is shown in the figure. The losses due to the models are represented using the mean losses from 10 sample paths and the mean ± 1 standard deviation values. From the figure, it can be seen that the DAR model overestimates cell losses and is not an acceptable model for this sequence unless conservative estimates are sufficient. However, as can be seen in Fig. 12.13, the Markov chain model is sufficiently accurate since the mean loss probabilities obtained using the Markov chain model are always lower than the losses obtained for actual rate plus 1% and always higher than the losses obtained for

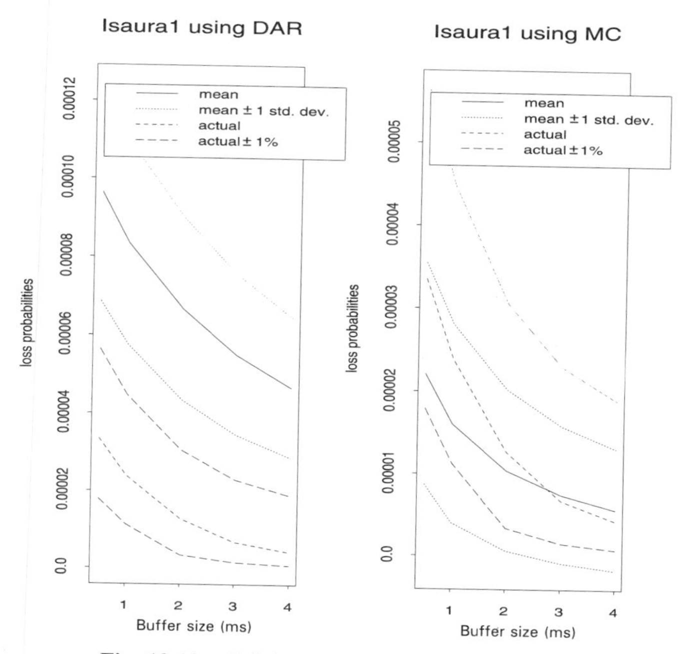

Fig. 12.13 Cell-loss probabilities for *Isaura 1*.

actual rate minus 1%. The drawback of the Markov chain model is that it has too many parameters.

12.3.7 Conclusions

The major conclusion that can be drawn from these experiments is that it is possible to construct a Markovian model for these video sequences, but a single model that can be used for all sequences does not seem to be possible. We use cell-loss rate as our performance measure here. Heyman and Lakshman [13] examined mean delay and also found that a Markov chain model could accurately represent these sequences.

12.4 WHY LONG-RANGE DEPENDENCE DOES NOT MATTER FOR VBR VIDEO

Beran et al. [1] observed long-range dependence in the video sequences we've just examined, and we found Markov chains (which have short-range dependence) provide good source models for several performance models. In this section we present two analytic explanations of this conundrum. The predictions of the analytic results are examined with numerical examples using the data traces described in the previous sections.

12.4.1 A Generic Buffer Model

We use a generic buffer model [13] to explore the effects of long-range dependence. The buffer has capacity c and receives inputs (e.g., frames consisting of cells) at equally spaced deterministic times $T_1, T_2, \ldots$. This assumption is appropriate for video sources because video frames are meant to be transmitted at equally spaced times. Let X_i be the number of arrivals (packets or cells) at time T_i, $i = 1, 2, \ldots$. Let d_i be the number of items that are processed during $[T_i, T_{i+1})$, which is called the ith interval. The length of an interval is the time to transmit one frame of video (40 ms for PAL video and 33.33 ms for NTSC video). This model applies to other types of traffic when packet or byte counts over a fixed time interval are used to describe the traffic. This is what Makowski and Parulekar do in Chapter 9 in this volume.

Let V_i be the buffer content at the end of the ith interval. We have the following law of motion for V_i, $i = 1, 2, \ldots$. V_0 is given, and

$$V_i = \min\{(V_{i-1} + X_i - d_i)^+, c\}, \quad i = 1, 2, \ldots, \tag{12.8}$$

where $x^+ = \max(x, 0)$. We will take $V_0 = 0$ to simplify the exposition. Thus, V_i is the smaller of the two numbers—the buffer capacity and the buffer content at the start of the ith interval, plus what arrived at the buffer, minus what went out.

We take d_i as a constant d, which is the number of packets or cells that can be processed in these intervals; for example, if 10,000 cells can be processed per second and the length of a frame interval is 40 ms, then $d = 10{,}000/25 = 400$. The arrivals are assumed to flow in such a way that the output is $\min(V_{i-1} + X_i, d)$. This will occur when the arrivals are spread uniformly over the frame interval (as suggested by Ott et al. [25]) and the cells are processed at a constant rate. When the X_i are stationary with common mean λ, say, the traffic intensity is λ/d.

12.4.1.1 Solution of the Model with an Infinite Buffer The model is easily solved when $c = \infty$, so we do that first. In this case, Eq. (12.8) becomes (recall that $d_i \equiv d$)

$$V_i = (V_{i-1} + X_i - d)^+, \quad i = 1, 2, \ldots, \tag{12.9}$$

which we recognize as Lindley's equation for the delay of the ith customer in a $GI/G/1$ queue, where the X_i's are interarrival times and the d's are service times. Define the partial sums

$$S_0 = 0 \quad \text{and} \quad S_m = \sum_{k=1}^{m} X_i, \quad m > 0.$$

Let $Y_i = X_i - d$; note that

$$\sum_{k=1}^{i} Y_k = S_i - di.$$

The solution of Eq. (12.9) is [15]

$$V_i = (S_i - di) - (S_{i_*} - di_*) = X_{i_*+1} + \cdots + X_i - (i - i_* - 1)d, \quad i = 1, 2, \ldots, \tag{12.10}$$

where

$$i_* = \arg\min_{0 \le j \le i} \{j : S_j - dj\}.$$

The arg min of a set is the argument that corresponds to the element that has the smallest value (if there is more than one smallest element, we will take the one with the largest argument). In the $GI/G/1$ queue, the index i_* corresponds to the last customer that left the server free, that is, the customer that starts the most recent busy period is i_{*+1}.

Equation (12.10) shows why the short-range correlations are important. Suppose $V_t = 0$ and $X_{t+1} > d$, so $V_{t+1} > 0$. As the lag one correlation increases, $P\{X_{t+2} > d\}$ tends to increase, and so Eq. (12.10) shows that V_{t+2} also increases stochastically. This effect gets repeated for $t + 2, t + 3, \ldots$ in turn, with the correlations at lags 2, 3, and so on coming into play. This causes buffer overflow in finite buffer models and large queue lengths in infinite buffer models.

For the DAR process in particular, X_{t+1} can be large by chance in Eq. (12.2). When ρ is near one, as it is for video conferences (see Table 12.1), X_{t+2}, $X_{t+3}, \ldots, X_{t+k}$ are likely to have the same large value for $k \gg 0$. This leads to large values of V_{t+2}, $V_{t+3}, \ldots, V_{t+k}$ and perhaps a few more values as the buffer empties. A similar argument applies to the GBAR model. This may explain why the DAR and GBAR models work so well for video conferences.

Equation (12.10) also shows that only those S_m that are formed by X's in the same busy period participate functionally in Eq. (12.10), although they are stochastically dependent on earlier X's and S_m's. (Since i_* is the start of a busy period, we know that the immediately previous X's are not too large. This means that they have some effect on the distribution of X_{i*} through the dependence structure. Since the marginal distribution of the X's has a long right tail in all data sets we've examined, this should not be a strong effect.) Since the X-process was taken to be stationary, the particular indices on the X's in Eq. (12.10) are (almost) irrelevant; it is the number of X's that are summed that is important. We call this the *resetting effect*. Thus, the effects of long-range dependence are significant only if long-range dependence causes the busy periods to be long enough for the long lags to come into play. Since VBR services carrying video traffic will be delay sensitive (to avoid jitter) and sensitive to cell losses (to avoid picture degradation), the traffic intensity for these services will not be large. This will make the busy periods short, so the resetting effect should be strong in practical operating regions. It might be weak when the traffic intensity is too large for practical operation.

12.4.2 Analytic Solution of the Model with a Finite Buffer

The solution of Eq. (12.8) with $c < \infty$ is complicated, and we look only at $i = 1$ and 2. We have

$$V_1 = \begin{cases} 0 & \text{if } Y_1 < 0, \\ Y_1 & \text{if } 0 \le Y_1 \le c, \\ c & \text{if } Y_1 > c. \end{cases}$$

This gives rise to three cases when $i = 2$.

Case 1: $Y_1 < 0$

$$V_2 = \begin{cases} 0 & \text{if } Y_2 < 0, \\ Y_2 & \text{if } 0 \le Y_2 \le c, \\ c & \text{if } Y_1 > c. \end{cases} \tag{12.11}$$

Case 2: $0 \leq Y_1 \leq c$

$$V_2 = \begin{cases} 0 & \text{if } Y_2 < -Y_1, \\ Y_1 + Y_2 & \text{if } -Y_1 \leq Y_2 \leq c - Y_1, \\ c & \text{if } Y_2 > c - Y_1. \end{cases} \tag{12.12}$$

Case 3: $c < Y_1$

$$V_2 = \begin{cases} 0 & \text{if } Y_2 < -c, \\ C + Y_2 & \text{if } -c \leq Y_2 \leq 0, \\ c & \text{if } Y_2 > 0. \end{cases} \tag{12.13}$$

In Case 1, the first busy period has length one, and we see the resetting effect as before. In Case 3, we see that when there is a buffer overflow at time 1, the value of Y_1 does not affect V_2. The effect of Y_1 on V_2 is in the fact that $Y_1 > c$; how much larger than c it is, is irrelevant. This is an enhancement to the resetting effect, which we call the *truncating effect of finite buffers.*

Suppose that a particular sequence $\{X_i\}$ and parameters c and $d = d_0$ produce the cell-loss rate ζ_0, say. It is intuitively obvious, and deducible from Eq. (12.8), that increasing d to $d_1 > d_0$ yields a cell-loss rate, say, ζ_1, that is smaller than ζ_0. A sample path argument will prove that the busy periods are stochastically shorter with $d = d_1$ than they are with $d = d_0$. This means that for a fixed buffer size, the truncating effect of finite buffers gets stronger as the cell-loss rate gets smaller. The simulation experiments we report have cell-loss rates larger than 10^{-6} because the data do not have enough cells to reliably estimate cell-loss rates any smaller. This result predicts that if we could do experiments with smaller cell-loss rates, the accuracy of Markov chain models would be better than the accuracy we are achieving now because the Markov chain models are at their best at small lags.

Consider two versions of the same buffer model, one with $c = \infty$ and the other with $c < \infty$. Another effect of the finite buffer is that *every busy period that contains an overflow is shorter than the corresponding busy period in an infinite buffer version*. By the corresponding busy period we mean the busy period that starts at the same time. There may be several busy periods in the finite buffer version before the busy period in the infinite buffer version ends. This result can formally be proved, but we give a heuristic "proof by picture." An intuitive reason for this result is that overflows reduce the number of cells that get into the buffer, and that should shorten the busy periods.

Figure 12.14 shows a portion of two buffer sample paths; one with an infinite buffer and the other with $c = 4$. The X's are the same for both paths, and $d = 1$. We need to emphasize the effect of the buffer size on V_i, so we will write V_i^c instead. The times when a stochastic process achieves a larger value than ever before are called [15] *ladder epochs*, and the record setting values are called *lader heights*. In Fig. 12.14 we see that overflows occur for V_i^4 at ladder heights of V_i^∞ that are larger than 4. The busy period of V_i^4 ends when V_i^∞ declines by 4 from the last ladder height in

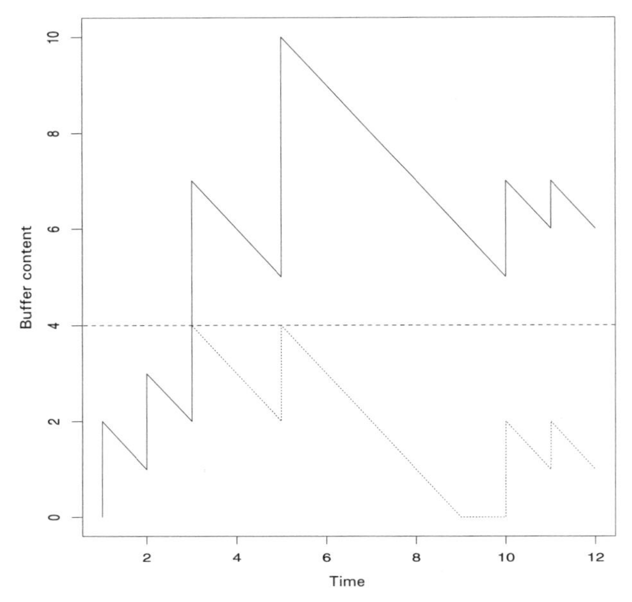

Fig. 12.14 Buffer sample paths for $c = \infty$ (solid line) and $c = 4$ (dotted line).

its busy period. This is necessarily before the end of the busy period for V_i^{∞}. Thus, the resetting effect is stronger for finite buffer models than it is for infinite buffer models.

The truncating effects given above are similar in spirit, but different in detail, to the *relevant time scale* notion introduced by Sriram and Whitt [29]. They make the observation that for large buffers, the queue lengths may be large and many arrivals interact in the queue. Then the effects of long-term positive covariances should cause more losses than models that do not have this property. However, when the buffer is fairly small, few arrivals interact in the queue and the effect of long-term positive covariances should be negligible.

12.4.3 Numerical Examples

The analytic results tell us that the resetting and truncation effects occur, but they don't tell us how strong they are. The reset effect should be weakest at the highest traffic intensity, since a slower drain rate should decrease the lengths of the busy periods. The truncating effect should get stronger as the buffer size decreases. Here we examine some numerical examples using the VBR video data to see if these effects have an impact on performance in realistic operating regions.

12.4.3.1 Video Conference D Data For video conference D, when the traffic intensity is 0.85, with an infinite buffer, 218 out of 14,5446 (1.50%) busy periods are

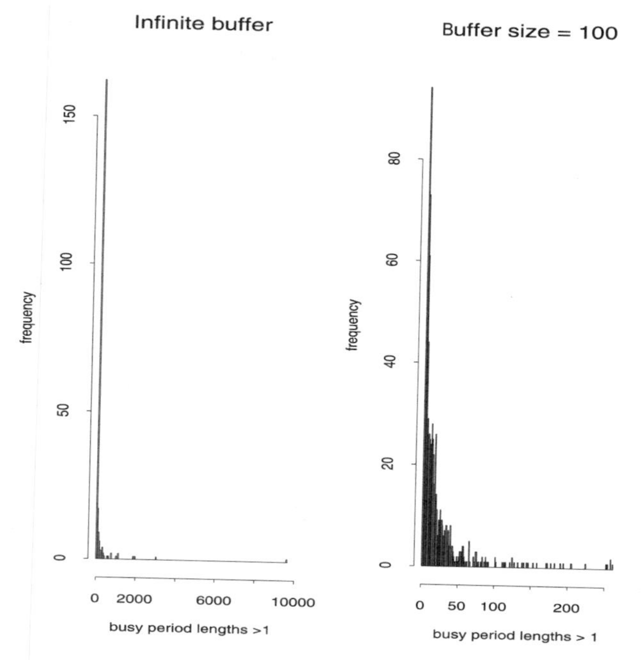

Fig. 12.15 Histograms for busy periods for video conference D.

longer than one. The median and upper-quartile point of these busy periods are 16 and 52; the maximum value is 9636. The histogram is displayed in Fig. 12.15.

There are six busy periods longer than 1000, and one of them has length 9636. These parameter conditions are well outside the operating range of practical interest; the cell-loss rate is 17% [13]. The mean buffer size is about 80,000 cells, or about 470 mean frames. There is close agreement between the mean buffer sizes from the trace and from the model at practical levels [13]. This can be accounted for by the resetting effect; when the buffer has capacity 100, the maximum busy period length is 262, which is in the region where the trace and the model yield similar S_m values [13, Fig. 7].

12.4.3.2 Film Data For this data set, when the traffic intensity is 0.66, 110 out of 48,104 (0.23%) of the busy periods have length greater than one. The median and upper-quartile points of these periods are 4 and 9, respectively. Only 4 are longer than 100 and the one that is longer than 200 has length 808. The cell-loss rate of the trace is 2.33×10^{-3}. The mean cell-loss rate from five model sample paths is 1.98×10^{-3}; the width of the range of the five cell-loss rates is 0.49×10^{-3}. Again, the data and the model agree well enough for engineering purposes. If the autocorrelations for long lags were important for determining cell-loss rates, agreement this close is unlikely. We think this example shows that the reset and truncating effects attenuate the effects of long-range dependence.

The truncating effect is particularly strong in this example. When the buffer is infinite, the expected queue lengths produced by the model underestimate the expected queue length produced by the data by a factor of about one-half for traffic intensities between 0.6 and 0.75. The same results occur when the buffer length is 12,000 cells. However, at DS 3 (45 Mb/s) rates, a 12,000 cell buffer would take about 30 ms to traverse. This is much larger than what we have been told VBR video buffers will be, because large buffers can cause substantial delay jitter. A 10 ms buffer is about as large as might be used; this corresponds to a 4000 cell buffer in this setting. With this size buffer, the model does a good job in estimating the mean queue length, as shown in Fig. 12.16. We attribute this improvement in estimating the mean queue length to the truncating effect of finite buffers.

12.4.3.3 *Isaura 1 Data* For this data set, when the buffer is 15,000 and $d = 7000$ (so the traffic intensity is 0.76), 372 out of 42,976 (0.87%) of the busy periods have length greater than one. The median and upper-quartile points of these periods are 5 and 8.5, respectively. Only 16 are longer than 100 and 2 are longer than 500. The longest is 664. The cell-loss rate of the data is 9.95×10^{-3}. The mean cell-loss rate from five model sample paths is 6.60×10^{-3}; the width of the range of the five cell-loss rates is 1.33×10^{-3}. Again, the data and the model agree well enough for engineering purposes. The mean queue lengths behave similarly to what occurs with *film*.

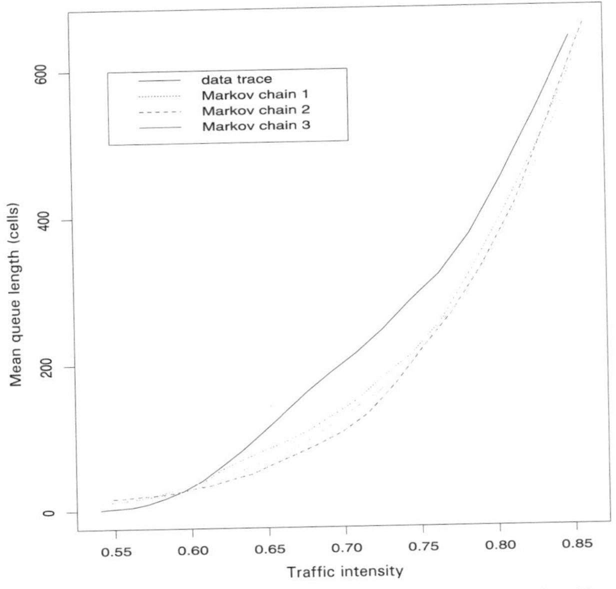

Fig. 12.16 Mean buffer size versus traffic intensity for *film*.

12.4.4 Large Deviations Estimate

In Eq. (12.8) assume that X_n is the superposition of N statistically identical and independent sources all of which have Gaussian $N(\lambda, \sigma^2)$ marginal distributions and autocorrelation function $r(k)$. The normal distribution is needed for the large deviations estimate of the steady-state solution of Eq. (12.8). We make it to gain insight into the effect of $r(k)$ on the solution and do not claim it is valid as a source model. We have been using d and c as the service rate and buffer capacity to which the source X_n is offered. In this section we emphasize that N sources are superposed. We take d and c as the service rate and buffer capacity *per source*, and we scale the service rate and the buffer size via

$$D = Nd \quad \text{and} \quad C = Nc,$$

which are the parameters in physical dimensions.

In the stationary regime, let V be the buffer content, and $\Phi(c) = \Phi(d, c, N) = P\{V = c\}$. The "large N asymptotic" is [5, 28]

$$\lim_{N\to\infty} = \frac{1}{N} \log \Phi(d, c, N) = -I(d, c), \tag{12.14}$$

where

$$I(d, c) = \inf_m \frac{(c + md - m\lambda)^2}{2V_m} \tag{12.15}$$

and

$$V_m = \text{Var}\left(\sum_{k=1}^{m} X_k\right) = \sigma^2\left(m + \sum_{k=1}^{m-1} (m-k) r(k)\right), \tag{12.16}$$

Let m_{c^*} be the infimizing m in Eq. (12.15); it is the value of m that maximizes the probability that the buffer overflows in m time periods, so Ryu and Elwalid call m_{c^*} the *the critical time scale.*

Ryu and Elwalid [28] make the following observations about m_{c^*}. Only those correlations smaller than the critical time scale enter Eq. (12.14) explicitly (through Eq. (12.16), so only the first m_{c^*} correlations are meaningful in evaluating $\Phi(d, c, N)$. Thus, if long-range dependence is to be an important effect, m_{c^*} must be large. They show that when $r(\cdot)$ is monotonically decreasing (which is a property observed in our video traces), then m_{c^*} is finite. It is easily established that $m_{0^*} = 1$, so a continuity argument suggests that m_{c^*} is small when c is small. When X_n is an *exact long-range dependent* (LRD) *process*, they prove that

$$\frac{m_{c^*}}{c} \approx \frac{H}{(1-H)(d-\lambda)}, \tag{12.17}$$

where H is the Hurst parameter. The traffic intensity is $a = \lambda/d$ and $\delta = c/d$ is the time to empty the buffer. Rewriting Eq. (12.17) as

$$m_{c^*} \approx \frac{H}{(1-H)} \frac{\delta}{(1-a)} \tag{12.18}$$

shows that long-range dependence can matter only when the buffer drain time is large and the traffic intensity is close to one. These conditions are not likely to occur in practice because δ is limited by maximum delay considerations and the traffic intensity is limited by packet-loss rate requirements.

12.4.5 Simulation Experiments

An empirical comparison of short- and long-range dependent processes is to use them to drive simulations of Eq. (12.8) and compare output statistics. Ryu and Elwalid [28] use the superposition of a fractal-binomial-noise-driven Poisson process and a DAR(1) process to emulate a process with large short-range correlations that decay geometrically and also possess long-range dependence. They find that the short-term correlations have the dominant impact on the value of the critical time scale and on the overflow probability.

Rao et al. [27] take a scaling of a fractional ARIMA process to be the "true" X-process and approximate it by autoregressive processes that match autocorrelation functions at certain points. They conclude that when these processes are such that the initial portions of their autocorrelation functions match, then the buffer overflow probabilities will be comparable.

A negative feature of these experiments is that the short- and long-range correlations cannot be varied independently, so the effect on the performance measures from changing one while holding the other fixed cannot be obtained. Heyman et al. [14] used the discrete-time $M/G/\infty$ queue to eliminate this feature. The arrival rate is λ and the mean service time is $1/\mu$. Makowski and Parulekar in Chapter 9 show that when $r(\cdot)$ is a given autocorrelation function that is monotonic, then the service time distribution

$$g_k = \frac{r(k-1) + r(k+1) - 2r(k)}{\mu}, \qquad k \geq 1, \tag{12.19}$$

will achieve $r(\cdot)$. In particular, we can choose $r(k)$ to be

$$r_k = \begin{cases} \alpha^k, & k \leq k_x, \\ \dfrac{\text{const}}{k^{2(1-H)}}, & k \geq k_x, \end{cases} \tag{12.20}$$

which has geometrically decaying short-term correlations and hyperbolically decaying long-term correlations. We can use Eq. (12.19) as the service time distribution in an $M/G/\infty$ queue and generate sample paths with autocorrelation function given by Eq. (12.20).

Suppose Eq. (12.20) is the autocorrelation function of the "true" $\{X_n\}$ in Eq. (12.18). The SRD version of the true process has a purely geometric autocorrelation function ($k_x = \infty$), and the LRD version has a purely hyperbolic autocorrelation function ($k_x = 1$). We chose parameters based on video conference C:

$$\lambda = 130, \quad \alpha = 0.985, \quad H = 0.7, \quad \text{and} \quad k_x = 250.$$

We take service rates that vary from 135 to 160 in steps of 5, so the traffic intensities range from 0.813 to 0.963 (cell losses are zero for smaller traffic intensities). Buffer sizes vary from 0 to 1 second. Each simulation was run for 500,000 iterations of Eq. (12.8). Figure 12.17 shows cell-loss rates averaged over 30 sample paths for each of the three input processes. The cell-loss rate increases with the traffic intensity, so the topmost LRD curve should be compared to the topmost source curve, and so on. It is clear that the SRD version is a very good representation of the true process for cell-loss calculations, and that the LRD version is a very poor representation. This could be predicted from the analytical results. The points to be learned from this example are that large correlations at short lags can be the dominant feature of the autocorrelation function, and that the presence of long-range dependence may not be relevant in a performance study. Note that an extreme case of the truncacting effect of finite buffers appears in Fig. 12.17; all three inputs yield the same loss rate when there is no buffer.

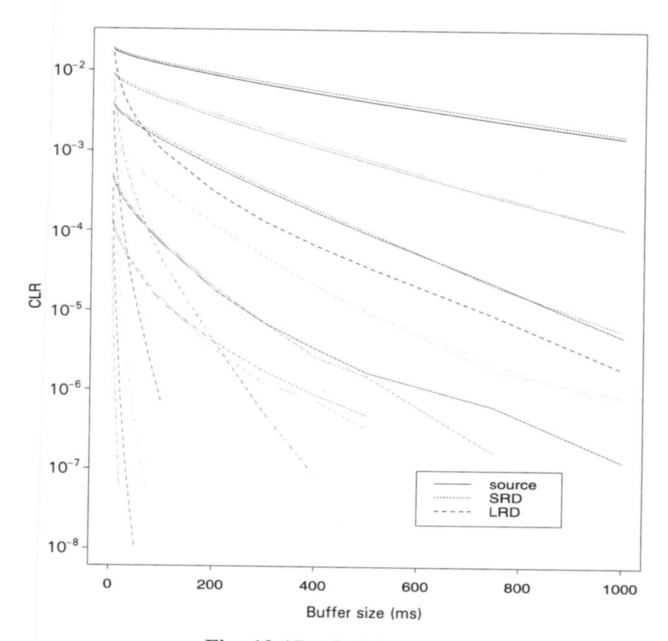

Fig. 12.17 Cell-loss rates.

Example 2 in Heyman et al. [14] shows that when $H = 0.8$, $\alpha = 0.8$, and $k_x = 10$, long-range dependence affects the cell-loss rates when the traffic intensity exceeds 0.90 and c/d exceeds 100, but not otherwise. These two examples verify the qualitative conclusions we drew from Eq. (12.18).

12.5 CONCLUSION

The VBR video sequences we have examined all have large autocorrelations at short lags as well as slowly decaying autocorrelations at long lags. For video conferences, the autocorrelation function is a geometric function for small lags, and source models based on this property when used in a performance model yield similar results to simulations using traces as the source. Entertainment video is more complex than video conferences because of the effects of scene changes. We find that the same class of models used for video conferences apply to scenes of entertainment video, and introducing the effects of scene changes and accounting for the random lengths of scenes we may obtain a usable source model. Such a model may not perform as well as an empirical Markov chain model.

We explain the success of these source models that allow only short-range dependence by deriving the *resetting effect* and the *trancating effect of finite buffers* from an analysis of a generic buffer model. These effects are shown to have a strong influence on the performance analysis of VBR video.

We have restricted our attention to VBR video coded according to the H.261 standard. The video community is currently exploring coding with MPEG standards. These codings generate a periodic structure to the frame sizes, so they are not directly amenable to modeling with stationary processes. The models we use here may provide building blocks for MPEG-encoded VBR video. Chandra and Reibman [4] and Heyman et al. [17] are first steps in this direction.

REFERENCES

1. J. Beran, R. Sherman, M. Taqqu, and W. Willinger. Long-range dependence in variable-bit-rate video traffic. *IEEE Trans. Commun.*, **43**:1566–1579, 1995.
2. P. Billingsley. *Statistical Inference for Markov Processes*. University of Chicago Press, Chicago, 1961.
3. P. Bratley, B. L. Fox, and L. E. Schrage. *A Guide to Simulation*. Springer-Verlag, Berlin, 1987.
4. K. Chandra and A. R. Reibman. Modeling two-layer MPEG-2 video traffic. In H. L. Bertoni, Y. Yang, S. P. Wang, and S. Panwar, eds., *Multimedia Communications and Video Coding*. Plenum Press, New York, 1996.
5. C. Courcoubetis and R. Weber. Buffer overflow asymptotics for a buffer handling many traffic sources. *J. Appl. Probab.*, **33**:886–903, 1996.

6. D. R. Cox and D. Oakes. *Analysis of Survival Data*. Chapman and Hall, New York, 1984.
7. A. Elwalid, D. Heyman, T. V. Lakshman, D. Mitra, and A. Weiss. Fundamental bounds and approximations for ATM multiplexing with applications to video teleconferencing. *IEEE J. Select. Areas Commun.*, **13**(6):1004–1016, 1995.
8. A. I. Elwalid and D. Mitra. Effective bandwidth of general Markovian traffic sources and admission control of high speed networks. *IEEE/ACM Trans. Networking*, **1**(3):329–343, 1993.
9. M. R. Frater, J. F. Arnold, and P. Tan. A new statistical model for traffic generated by VBR coders for television on the broadband ISDN. *IEEE Trans. Circuits Syst. Video Technol.*, **4**(6):521–526, December 1994.
10. M. W. Garrett and M. Vetterli. Congestion control strategies for packet video. In *Fourth International Workshop on Packet Video*, Kyoto, Japan, August 1991.
11. D. P. Heyman. The GBAR source model for VBR videoconferences. *IEEE/ACM Trans. Networking*, **5**(2):554–560, 1997.
12. D. P. Heyman and T. V. Lakshman. Source models for VBR broadcast-video traffic. *IEEE/ACM Trans. Networking*, **4**(1):40–48, February 1996.
13. D. P. Heyman and T. V. Lakshman. What are the implications of long-range dependence for VBR-video traffic engineering. *IEEE/ACM Trans. Networking*, **4**(3):301–317, 1996.
14. D. P. Heyman, T. V. Lakshman, and D. Liu. Assessing the effects of short-range and long-range dependence on overflow probabilities. In preparation, 1998.
15. D. P. Heyman and M. J. Sobel. *Stochastic Models in Operations Research, Vol. I.* McGraw-Hill, New York, 1982.
16. D. Heyman, A. Tabatabai, and T. V. Lakshman. Statistical analysis and simulation study of video teleconference traffic in ATM networks. *IEEE Trans. Circuits Syst. Video Technol.*, **2**(1):49–59, 1992.
17. D. P. Heyman, A. J. Tabatabai, and T. V. Lakshman. Statistical analysis of an MPEG2-coded VBR movie source. In H. L. Bertoni, Y. Yang, S. P. Wang, and S. Panwar, eds., *Multimedia Communications and Video Coding*, pp. 383–391. Plenum Press, New York, 1996.
18. D. Heyman, A. Tabatabai, T. V. Lakshman, and H. Heeke. Modeling teleconference traffic from VBR video coders. In *Proceedings of ICC '94*, pp. 1744–1748, 1994.
19. C. L. Hwang and S. Q. Li. On input state space reduction and buffer noneffective region. In *Proc. IEEE INFOCOM'94*, pp. 1018–1028, 1994.
20. P. A. Jacobs and P. A. W. Lewis. Time series generated by mixtures. *J. Time Series Anal.* **4**(1):19–36, 1983.
21. T. V. Lakshman, P. Mishra, and K. K. Ramakrishnan. Transporting compressed video over ATM networks with explicit rate feedback control. *IEEE/ACM Trans. Networking*, **7**: 1999.
22. D. M. Lucantoni, M. F. Neuts, and A. R. Reibman. Methods for performance evaluation of VBR video traffic models. *IEEE/ACM Trans. Networking*, **2**(2):176–180, 1994.
23. E. McKenzie. Autoregressive moving-average processes with negative-binomial and geometric marginal distributions. *Adv. Appl. Probab.* **18**: 679–705, 1986.
24. A. N. Netravili and B. G. Haskell. *Digital Pictures: Representation and Compression.* Plenum Press, New York, 1988.

25. T. J. Ott, T. V. Lakshman, and A. Tabatabai. A scheme for smoothing delay sensitive traffic offered to ATM networks. In *Proc. IEEE INFOCOM '92*, pp. 776–785, 1992.
26. G. Ramamurthy and B. Sengupta. Modeling and analysis of variable bit rate video multiplexer. In *7th Specialist Seminar, International Teletraffic Conference*, Morristown, NJ, 1990.
27. V. Venkateshwara Rao, K. R. Krishnan, and D. P. Heyman. Performance of finite-buffer queues under traffic with long-range dependence. In *Proc. IEEE Globecom '96*, pp. 3–14, London, 1996.
28. B. K. Ryu and A. Elwalid. The importance of long-range dependence of VBR video traffic in ATM traffic engineering: myths and realities. *Proc. ACM SIGCOMM'96*, Stanford, CA, 1996.
29. K. Sriram and W. Whitt. Characterizing superposition arrival processes in packet multiplexers for voice and data. *IEEE J. Select. Areas Commun.*, **SAC-4**(6):833–846, 1986.
30. W. Verbiest and L. Pinnoo. A variable bit rate video codec for asynchronous transfer mode networks. *IEEE J. Select. Areas Commun.*, **7**:761–770, 1989.
31. W. Verbiest, L. Pinnoo, and B. Vosten. The impact of the ATM concept on video coding. *IEEE J. Select. Areas Commun.*, **6**:1623–1632, 1988.
32. Y. Yasuda, H. Yasuda, H. Ohta, and F. Kishino. Packet video transmission through ATM networks. In *Proc. IEEE Globecom*, pp. 25.1.1–25.1.5, 1989.

13

ANALYSIS OF TRANSIENT LOSS PERFORMANCE IMPACT OF LONG-RANGE DEPENDENCE IN NETWORK TRAFFIC

GUANG-LIANG LI AND VICTOR O. K. LI

Department of Electrical & Electronic Engineering, The University of Hong Kong, Pokfulam, Hong Kong, China

13.1 INTRODUCTION

To support multimedia applications, high-speed networks must be able to provide quality-of-service (QoS) guarantees for connections with drastically different traffic characteristics. Some of the characteristics fall beyond the conventional framework of Markov traffic modeling. For instance, recent studies have demonstrated convincingly that there exists long-range dependence or self-similarity in packet video, which is an important traffic component in high-speed networks. Essentially, long-range dependence cannot be captured by Markov traffic models. Although long-range dependence in network traffic has been widely recognized [1, 2, 4, 8, 11, 15, 17, 18], QoS impact of long-range dependence is still an open issue. For example, there are different opinions regarding whether Markov traffic models can still be used to predict loss performance in the presence of long-range dependence. This and other related issues are also discussed in Chapter 12 of this book. QoS guarantee for long-range dependent (LRD) traffic is the topic of Chapters 16 and 19 as well. The issue of congestion control for self-similar traffic is addressed in Chapter 18.

Self-Similar Network Traffic and Performance Evaluation, Edited by Kihong Park and Walter Willinger
ISBN 0-471-31974-0

In this chapter, we present an analysis of transient loss performance impact of long-range dependence in network traffic. This work is only the first step of our exploration. But we hope that it will still be helpful for understanding loss performance impact of long-range dependence in the transient state, although much further work needs to be done in the future. A different transient analysis in the context of capacity planning and recovery time is given in Chapter 17.

In general, the transient analysis of queueing models is very challenging. A transient solution to a queueing model is a function of a time index, either continuous or discrete, defined over an infinite range. It is very difficult to find an explicit, closed-form solution if the system is not Markovian. For a Markov model, the probabilistic evolution of the system state is governed by the Chapman–Kolmogorov equation. In the presence of long-range dependence, the model is essentially non-Markovian. So the Chapman–Kolmogorov equation does not hold anymore. Due to this and other difficulties involved in transient analysis, most of the existing work on QoS impact of long-range dependence is limited to asymptotic analysis in the steady state [4, 6, 9, 10, 15–17] (see also Chapters 4–10 in this volume). Although certain insights have been gained through investigation carried out in steady-state, we feel that it is still necessary to extend the investigation beyond the region of steady-state.

First, due to the high variability caused by long-range dependence, the convergence of an LRD traffic process toward steady-state can be very slow. In contrast, Markov traffic processes converge to steady-state exponentially fast. Consequently, for a link carrying LRD traffic, the discrepancy between steady-state performance and transient performance can be more significant, compared with the situations in which traffic can be modeled by Markov processes. Second, to guarantee QoS for traffic with the high variability caused by long-range dependence, dynamic and adaptive resource allocation may be necessary to account for the effect of the current system state. Performance analysis based on the steady state may not be appropriate for this purpose, since in steady-state, any initial effect will disappear eventually. Because steady-state performance may differ from transient performance significantly for LRD traffic, the image regarding loss behavior of LRD traffic in the transient state still largely remains vague.

In this chapter, an approach different from conventional transient analysis is used, which allows us to investigate the transient performance impact of long-range dependence in traffic without first seeking a closed-form transient solution. That is, we limit our analysis to some short period of time, and even to a single state of a traffic process. The reasons for us to adopt this approach are as follows. First, it is relatively easy, and may also be sufficient, to consider transient solutions defined only for a relatively short time period, since a short time period may actually cover the time span in which we are interested for transient performance analysis. A large time span may be less interesting from a point of view of transient analysis, since the difference between the steady state and the transient state may diminish significantly after a long time has elapsed. Second, if the arrival process is renewal type, then the behavior of the system is probabilistically periodic. So it may be sufficient to focus only on a "typical" probabilistic period to study the transient performance of the system. Finally, for a multiple-state arrival process, we may further limit the analysis to an arbitrary single state of the arrival process and compare transient performance measures computed under different modeling assumptions. With this approach, we have gained some insights into QoS impact of long-range dependence in network traffic.

The rest of the chapter is organized as follows. In Section 13.2, we first introduce a framework for traffic modeling that captures the essential property of long-range dependence. Within this framework, traffic is modeled by multistate, fluid-type stochastic processes. When such a process is in a given state, the underlying traffic source generates traffic at a constant rate. The time spent by the process in a state is a random variable. For the purpose of this chapter, we let the distribution of the random variable be arbitrary. As a result, we can construct Markov and LRD traffic models as we wish. Then we define loss performance measures in the transient state. In Section 13.3, we compare transient loss performance between the traditional Markov models and the LRD models. To keep the comparison reasonable, for the Markov and LRD models, except for the distributions of the times spent by the traffic processes in their respective states, we let all other traffic parameters be the same. By doing so, the difference in loss behavior between Markov and LRD traffic is only due to the modeling assumption on the underlying traffic process. We then compare transient loss of Markov and LRD traffic for two cases. In the first case, we assume that both traffic processes are in the same state with the same initial condition characterized by the amount of traffic left in the system when the processes enter the state. In the second case, we consider two-state Markov and LRD fluids. To examine whether it is appropriate to predict loss performance computed according to Markov models in steady state for LRD traffic, in Section 13.4, we show how to compute steady-state limits of transient loss measures for general two-state fluids, and compare transient loss against loss in steady state. In Section 13.5, we discuss the impact of long-range dependence in network traffic, based on the analytical and numerical results obtained. We conclude this chapter in Section 13.6, with a summary of the findings of our study, and a brief discussion on the challenge posed by transient performance guarantee in the presence of long-range dependence and some extension of this work. Section 13.7 contains two appendixes.

13.2 TRAFFIC MODELS AND TRANSIENT LOSS MEASURES

We adopt general fluid-type stochastic processes with multiple states as a framework for traffic modeling. The state of such a process is associated with the bit rate of the underlying traffic source. When the process is in a given state, the source generates traffic at a constant bit rate. The bit rates are different for different states. Such fluid-type traffic models have been used in many previous studies for traffic engineering. A well-known example is the Markov-modulated fluid model [5]. However, the traffic model in our study is essentially different from traditional fluid traffic models: our traffic model is not necessarily Markovian, which allows us to capture the property of long-range dependence in traffic. A special case of our traffic model is the general two-state fluid, which can capture the most important traffic properties such as long-range dependence and burstiness. An important part of this work is based on the two-state fluid model. Various on/off fluid models are special cases of the general two-state fluid and have widely been used for traffic modeling. For example, on/off sources with heavy-tailed on/off periods are proposed to explain long-range dependence or self-similarity in traffic [18]. For an on/off fluid, no traffic is generated in the off state. In this book, on/off traffic models are also considered in Chapters 5, 7, 11, and 17.

Let us denote a fluid-type traffic process by $R(t)$. The physical meaning of $R(t)$ is the time-dependent bit rate of the underlying traffic source. Denote $R(t)$ by $R(n)$ for $t \in [t_n, t_{n+1})$, where t_n is the instant at which the nth transition of the state of $R(t)$ occurs. Accordingly, $[t_n, t_{n+1})$ is an interval during which $R(t)$ remains unchanged. Suppose that the bit rate of the traffic source is r during the interval, that is, $R(n) = r$. Denote the length of the interval $[t_n, t_{n+1})$ by Δt_n. Clearly Δt_n is a random variable, representing the time spent by $R(t)$ in the state in which the bit rate of the traffic source is r. Suppose that Δt_n obeys a distribution $F_{\Delta t_n}(s) = P\{\Delta t_n \leq s\}$. We assume that the distributions of Δt_n are the same when the traffic process is in the same state but may differ for different states. For a two-state process, we use on and off to refer to the states. When the state is on, the bit rate is denoted by r_1, and the bit rate corresponding to the off state is r_0, where $r_1 > r_0 \geq 0$. Denote the lengths of the nth on and off intervals, respectively, by S_n and T_n. We assume that for $n \geq 1$, S_n are independent and identically distributed (i.i.d.) random variables as are T_n. Since both S_n and T_n are i.i.d., we can drop the subscript n in S_n and T_n. The general two-state fluid model is appealing from an analysis point of view, since it can capture the essential property of long-range dependence in network traffic while still permitting an exact analysis without approximation.

For a Markov fluid, Δt_n is of course exponentially distributed. To capture the property of long-range dependence in traffic, we can assume that Δt_n obeys some heavy-tailed distribution. Readers can find a simple formal proof in Grossglauser and Bolot [9] for a special case of the general fluid traffic model, which shows that if for all $n \geq 1$, Δt_n are i.i.d. with respect to both n and $R(n)$, and are drawn from a common heavy-tailed distribution, then the corresponding traffic process $R(t)$ is an LRD process or, more exactly, an asymptotically second-order self-similar process, with autocorrelation function $C(t) \sim t^{-\alpha+1}$ as $t \to \infty$, where the symbol $\sim$ represents an asymptotic relation.

To define transient loss measures, we assume that traffic loss is caused only by buffer overflow. For a multistate fluid, the loss measures are the expected traffic loss ratio and the probability that loss of traffic occurs in interval $[t_n, t_{n+1})$, conditioned on w, the amount of traffic left in the system at t_n, where $n \geq 1$. The quantity w is a random variable. In general, for a multistate fluid, it is difficult to obtain the distribution of w. Therefore, we have to treat w as a given condition. However, in the special case of a two-state fluid, we only need to treat w_1 as a given condition, where w_1 is the initial amount of traffic in the system when $n = 1$. For any $n > 1$, we can compute the distribution of w by recurrence. So for the special case of a two-state fluid, it is not necessary to treat w for $n > 1$ as a given condition in the transient loss measures. Instead, we can account for the impact of w by its distribution.

In the next section, we discuss how to compute the above transient loss measures, and compare the transient loss behavior of fluid traffic based on the loss measures under different modeling assumptions on the traffic process, which will provide useful insights into loss performance impact of long-range dependence in the transient state.

13.3 TRANSIENT LOSS OF MARKOV AND LRD TRAFFIC

Now let us consider a link with finite buffer B and bandwidth C. Suppose that the link carries a fluid traffic process $R(t)$. Recall that $[t_n, t_{n+1})$ is the nth interval between transitions of the state of $R(t)$. The modeling assumption on $R(t)$ is determined by the distribution of Δt_n, the length of the interval $[t_n, t_{n+1})$. We are curious about the transient loss behavior of $R(t)$ at the link under two conflicting assumptions in traffic modeling:

- $R(t)$ is a Markov process. Consequently, Δt_n is exponentially distributed.
- $R(t)$ is an LRD process, which implies that Δt_n obeys some (asymptotically) heavy-tailed distribution.

To compare the transient loss behavior of Markov and LRD traffic, we consider the following two cases.

13.3.1 Loss Behavior in Single States

To compare the loss behavior of multistate Markov and LRD fluids in single states, we use the traffic loss probability and the expected traffic loss ratio defined in the interval $[t_n, t_{n+1})$ where $n \geq 1$, as the transient loss measures. Both the above loss measures are conditioned on the amount of traffic left in the system at t_n.

Suppose $R(n) = r$. for convenience of exposition, we simply let $t_n = 0$ and $t_{n+1} = S$, where S is a random variable with distribution $F_S(s) = P\{S \leq s\}$ that may depend on r. The amount of traffic in the system at time t is represented by $w(t)$, where $t \in [0, S)$. Denote by $w(0)$ the initial amount of traffic left in the system at the beginning of interval $[0, S)$. Accordingly, the loss probability and the expected loss ratio are denoted, respectively, by $P\{\text{loss}|R(n) = r, w(0) = w\}$ and $E[l|R(n) = r, w(0) = w]$, where l is the fraction of traffic lost in $[0, S)$. The following two lemmas show how to compute the loss measures.

Lemma 13.3.1

$$P\{\text{loss}|R(n) = r, w(0) = w\} = \begin{cases} 0, & r \leq C, \\ 1, & r > C, w = B, \\ P\{S > \tau(w)\}, & r > C, w < B, \end{cases} \tag{13.1}$$

where

$$\tau(w) = \frac{B - w}{r - C}. \tag{13.2}$$

Proof. Clearly, if $r \leq C$, then $P\{\text{loss}|R(n) = r,\ w(0) = w\} = 0$, and $r > C$ together with $w = B$ implies that $P\{\text{loss}|R(n) = r,\ w(0) = w\} = 1$. On the other hand, if $r > C$ and $w < B$, then the random event that traffic loss due to buffer overflow occurs in the interval is equivalent to existing $\tau \in (0, S)$ such that $w(t) = B$ for $t \in [\tau, S)$ and hence $P\{\text{loss}|R(n) = r, w(0) = w\} = P\{S > \tau\}$. It is easy to see that

$$\tau = \frac{B - w}{r - C}.$$

Since τ depends on w, we denote τ by $\tau(w)$. ■

COMMENT 13.3.2. The physical meaning of $\tau(w)$ is the instant in $[0, S)$ after which traffic loss due to buffer overflow begins immediately in the interval. Since $\tau(w)$ depends on bandwidth and buffer allocated to the underlying traffic, it can be viewed as a control parameter.

Lemma 13.3.3

$$E[l|R(n) = r, w(0) = w] = \begin{cases} 0, & r \leq C, \\ \displaystyle\int_0^{u_0} P\left\{S > \frac{u_0 \tau(w)}{u_0 - u}\right\} du, & r > C, \end{cases} \tag{13.3}$$

where $\tau(w)$ is given by Eq. (13.2) and

$$u_0 = 1 - \frac{C}{r}. \tag{13.4}$$

Proof. From Lemma 13.3.1 we know that traffic loss due to buffer overflow will not occur during $[0, S)$ if $r \leq C$, so $l = 0$ when $r \leq C$, and as a result,

$$E[l|R(n) = r, w(0) = w] = 0, \quad r \leq C.$$

In the following, we consider $r > C$. If $S \leq \tau(w)$, then $l = 0$, since in $[0, S)$, traffic loss due to buffer overflow begins only after $t \in [0, S)$ reaches $\tau(w)$. If $S > \tau(w)$, then

$$\begin{aligned} &\text{the amount of traffic lost in } [0, S) \\ &\quad = (\text{the amount of traffic arrived in } [0, S)) \\ &\qquad - (\text{the amount of traffic accepted in } [0, S)) \\ &\quad = rS - (CS + B - w) = (r - C)[S - \tau(w)]. \end{aligned}$$

For $S > \tau(w)$, we have

$$l = \frac{(r - C)[S - \tau(w)]}{rS} = \left(1 - \frac{C}{r}\right)\left[1 - \frac{\tau(w)}{S}\right].$$

Therefore,

$$l = \begin{cases} 0, & S \leq \tau(w), \\ u_0\left[1 - \frac{\tau(w)}{S}\right], & S > \tau(w), \end{cases} \tag{13.5}$$

where u_0 is given by Eq. (13.4). Since l depends on S, we can express l by $l(S)$. Recall that $F_S(s)$ is the distribution of S. We have

$$\begin{aligned} E[l(S)|R(n) = r, w(0) = w] &= \int_{\tau(\omega)}^{\infty} l(s)\, dF_S(s) \\ &= [l(s)F_S(s)]_{s=\tau(w)}^{\infty} - \int_{\tau(w)}^{\infty} F_S(s)\, dl(s) \\ &= u_0 - \int_{\tau(w)}^{\infty} [1 - P\{S > s\}]\, dl(s) \\ &= u_0 - \int_{\tau(w)}^{\infty} dl(s) + \int_{\tau(w)}^{\infty} P\{S > s\}\, dl(s) \\ &= \int_{\tau(w)}^{\infty} P\{S > s\}\, dl(s). \end{aligned}$$

In the second line of the above equations, since $F_S(\infty) = P\{S \leq \infty\} = 1$ by definition, $\lim_{s\to\infty} l(s) = u_0$, and $\lim_{s\to\tau(w)} l(s) = 0$ from Eq. (13.5), we see that $[l(s)F_S(s)]_{\tau(w)}^{\infty} = u_0$. The proof is completed by changing the integral variable as follows. Denote $l(s)$ by u; we have

$$u = u_0\left[1 - \frac{\tau(w)}{s}\right]$$

and

$$s = \frac{u_0\tau(w)}{u_0 - u}.$$

When $s = \tau(w)$, we have $u = 0$, and $s = \infty$ is equivalent to $u = u_0$. ■

COMMENT 13.3.4. If $P\{\text{loss}|R(n) = r,\ w(0) = w\} = 1$, then it only means that traffic loss due to buffer overflow will occur for certain in $[0, S)$, while not necessarily implying that all traffic arrived in $[0, S)$ is lost. As we can see from Lemmas 13.3.1 and 13.3.3, when $P\{\text{loss}|R(n) = r,\ w(0) = w\} = 1$, we still have $E[l|R(n) = r,\ w(0) = w] = 1 - C/r < 1$, given $r > C$.

COMMENT 13.3.5. For loss behavior of fluid traffic in single states, the only nontrivial case is $r > C$ and $w < B$. Otherwise, the loss probability equals either 0 or 1, and the expected loss ratio is either 0 or a constant equal to u_0. So it is sufficient to consider only $r > C$ and $w < B$ for the purpose of this study.

As shown above, both the conditional loss probability and the conditional expected loss ratio depend explicitly on the distribution of S, which is in turn determined by the assumption on the underlying traffic process $R(t)$. For example, if we assume that $R(t)$ is a Markov process, then S is exponentially distributed. On the other hand, if we assume that $R(t)$ is an LRD process, then the distribution of S is heavy-tailed. To compare the two conflicting modeling assumptions, we consider the following scenario.

Suppose that a Markov fluid model, denoted by $M(t)$, is used for modeling a fluid-type traffic process $R(t)$. But, in fact, the underlying traffic process $R(t)$ is an LRD process, denoted by $L(t)$, which has the same state space as that of the Markov process $M(t)$. The essential difference between $M(t)$ and $L(t)$ lies in the way to characterize Δt_n, the length of the time interval $[t_n, t_{n+1})$ for arbitrary $n \geq 1$. We still use $[0, S)$ to represent $[t_n, t_{n+1})$, so the interval length Δt_n can be denoted simply by S. As we have already mentioned, for Markov model $M(t)$, S is exponentially distributed, but for LRD model $L(t)$, the distribution of S is heavy tailed or asymptotically heavy tailed; that is, the functional form of the distribution possesses the property of heavy tail if the value of S is sufficiently large. For an asymptotically heavy-tailed distribution, it is only necessary for us to consider the case that the value of S is large enough to be in the heavy tail, since only the heavy-tailed effect appears essentially different from Markov traffic modeling and hence is of great interest for the purpose of this study. To be specific, we consider the following heavy-tailed distribution:

$$P\{S \leq s\} = 1 - (\gamma s + 1)^{-\alpha}, \quad 0 \leq s < \infty, \gamma > 0, \quad 1 < \alpha < 2, \tag{13.6}$$

which is a variant of the conventional Pareto distribution. The reason for us to consider this variant is that the range of the random variable of interest in our study is $[0, \infty)$ while for the conventional Pareto distribution, the range of the random variable is $[\omega, \infty)$, where $\omega > 0$. As we can see, the tail of the distribution becomes heavier and heavier as α decreases toward 1. In fact, a smaller α corresponds to a stronger LRD effect [18].

We are concerned with transient loss performance of the underlying traffic process $R(t)$ predicted by $M(t)$. In other words, we want to know the impact of long-range dependence on the transient loss performance predicted by the Markov model. For convenience of exposition, when necessary, M and L, representing respectively the Markov and LRD traffic models, will substitute for R in the notation

$R(n)$ for distinction of the use of the notation. For example, $L(n)$ represents the bit rate of $L(t)$ in the nth interval between transitions of the state of $L(t)$ and $P\{\text{loss}|L(n)=r,\ w(0)=w\}$ is the conditional loss probability of $L(t)$.

Our approach is to compare $P\{\text{loss}|M(n)=r,\ w(0)=w\}$ and $E[l|M(n)=r,\ w(0)=w]$ with $P\{\text{loss}|L(n)=r,\ w(0)=w\}$ and $E[l|L(n)=r,\ w(0)=w]$, respectively, for given $B<\infty$, $C<\infty$, $r>C$, and $w<B$, under the assumption that $E[S]$ is the same for Markov model $M(t)$ and LRD model $L(t)$. That is, the comparison is made such that $M(t)$ and $L(t)$ are in the same state with the same initial amount of traffic left in the buffer. With such a comparison, we believe that the difference in transient loss performance predicted by the Markov model and the LRD model is only due to the different modeling assumptions on the underlying traffic process $R(t)$. According to Lemmas 13.3.1 and 13.3.3, the comparison is straightforward. To exclude the trivial cases, we consider only $r>C$ and $w<B$.

Theorem 13.3.6. *Supose that S obeys the Pareto distribution (13.6) for $L(t)$, and $E[S]$ is the same for $L(t)$ and $M(t)$. For given $B<\infty$, $C<\infty$, $r>C$, and $w<B$, we have $P\{\text{loss}|M(n)=r,\ w(0)=w\}\geq P\{\text{loss}|L(n)=r,\ w(0)=w\}$ if $\tau(w)\leq\gamma^{-1}z$ and $P\{\text{loss}|M(n)=r,\ w(0)=w\}<P\{\text{loss}|L(n)=r,\ w(0)=w\}$ otherwise, where $\tau(w)$ is given by Eq. (13.2) and $z>0$ is the solution of $e^{\beta x}=x+1$ for $x\in(0,\infty)$ and $\beta=1-\alpha^{-1}$.*

Proof. Under the assumptions that S obeys the Pareto distribution (13.6) for $L(t)$, and $E[S]$ is the same for $M(t)$ and $L(t)$, we have $E[S]=(\alpha-1)^{-1}\gamma^{-1}$ for both $L(t)$ and $M(t)$, and according to Eq. (13.1),

$$\frac{P\{\text{loss}|L(n)=r,\ w(0)=w\}}{P\{\text{loss}|M(n)=r,\ w(0)=w\}}=\frac{e^{(\alpha-1)\gamma\tau(w)}}{[\gamma\tau(w)+1]^{\alpha}}=\frac{[e^{(\alpha-1)\gamma\tau(w)/\alpha}]^{\alpha}}{[\gamma\tau(w)+1]^{\alpha}}=\left[\frac{e^{(1-\alpha^{-1})\gamma\tau(w)}}{\gamma\tau(w)+1}\right]^{\alpha}.$$

Let $\beta=1-\alpha^{-1}$ and denote $\gamma\tau(w)$ by x; then $P\{\text{loss}|L(n)=r,\ w(0)=w\}\leq P\{\text{loss}|M(n)=r,\ w(0)=w\}$ is equivalent to $e^{\beta x}\leq x+1$. Define $y(x)\stackrel{\text{def}}{=}e^{\beta x}-x-1$ for $x\in[0,\infty)$. We see that $y(x_0)$ is the only extreme of $y(x)$, where $x_0=\beta^{-1}\ln\beta^{-1}>0$ satisfying $dy/dx=\beta e^{\beta x}-1=0$. In fact, $y(x_0)$ is a minimum of $y(x)$ since $d^2y/dx^2(x_0)=\beta>0$, and $y(x_0)$ cannot be nonnegative since $y(0)=0$ and $y(x_0)<y(0)$. So we have $y(x_0)<0$. On the other hand, it is evident that $y(x)$ will become and remain positive after x reaches a sufficiently large value. Thus, there must exist one and only one zero $z>x_0$ of $y(x)$ for $x\in(0,\infty)$ such that $y(x)\leq 0$ for $x\leq z$ and $y(x)>0$ otherwise. The result to be proved then follows. ∎

Theorem 13.3.6 shows that if an LRD fluid is modeled by a Markov process, then the Markov model may indeed underestimate the loss probability of the underlying LRD traffic. A similar result holds for the conditional expected loss ratio.

Theorem 13.3.7. *Suppose that S obeys the Pareto distribution (13.6) for $L(t)$, and $E[S]$ is the same for $L(t)$ and $M(t)$. For given $B<\infty$, $C<\infty$, $r>C$, and $w<B$,*

we have $E[l|M(n) = r,\ w(0) = w] \geq E[l|L(n) = r,\ w(0) = w]$ *if* $\theta \leq \gamma^{-1}z$ *and* $E[l|M(n) = r, w(0) = w] < E[l|L(n) = r, w(0) = w]$ *otherwise, where*

$$\theta = \frac{u_0 \tau(w)}{u_0 - \xi},$$

$\xi \in (0, u_0)$, $\tau(w)$ *is given by Eq. (13.2),* u_0 *is given by Eq. (13.4), and* $z > 0$ *is the solution of* $e^{\beta x} = x + 1$ *for* $x \in (0, \infty)$ *and* $\beta = 1 - \alpha^{-1}$.

Proof. We first recall a well-known result (the generalized mean value theorem) in elementary analysis. Suppose that $F(u)$ and $G(u)$ are continuous on $[a, b]$ and differentiable on (a, b), and $G'(u) \neq 0$ for $a < u < b$, where $'$ indicates derivation with respect to u. Then there exists at least one $\xi \in (a, b)$ such that

$$\frac{F(b) - F(a)}{G(b) - G(a)} = \frac{F'(\xi)}{G'(\xi)}.$$

Now let $a = 0, b = u_0$ and denote

$$\int P\left\{S > \frac{u_0 \tau(w)}{u_0 - u}\right\} du$$

by $F(u)$ if S obeys the Pareto distribution (13.6) and by $G(u)$ if S is exponentially distributed. Both $F(u)$ and $G(u)$ are continuous on $[0, u_0]$ and differentiable on $(0, u_0)$. According to the above result and Eq. (13.3),

$$\frac{E[l|L(n) = r, w(0) = w]}{E[l|M(n) = r, w(0) = w]} = \frac{F(u_0) - F(0)}{G(u_0) - G(0)} = \frac{F'(\xi)}{G'(\xi)},$$

where $\xi \in (0, u_0)$,

$$F'(\xi) = \left[\gamma \frac{u_0 \tau(w)}{u_0 - \xi} + 1\right]^{-\alpha},$$

and

$$G'(\xi) = \exp\left[-(\alpha - 1)\gamma \frac{u_0 \tau(w)}{u_0 - \xi}\right] > 0$$

for $0 < \xi < u_0$. Letting

$$\theta = \frac{u_0 \tau(w)}{u_0 - \xi},$$

we have $F'(\xi) = (\gamma\theta + 1)^{-\alpha}$ and $G'(\xi) = e^{-(\alpha-1)\gamma\theta}$. Replacing $\tau(w)$ in Theorem 13.3.6 by θ, then using the same arguments as that used in the proof of Theorem 13.3.6, we see that the result to be proved follows directly. ■

COMMENT 13.3.8. The assumption of Pareto distribution is not restrictive. Similar results hold for any other heavy-tailed distributions. One numerical example is given in Section 13.3.2.

COMMENT 13.3.9. The above theorems imply that if an LRD fluid is modeled by a Markov fluid, then in all nontrivial cases, that is, for each state with a duration interval $[t_n, t_{n+1})$ such that $W_n < B$ and $R(n) > C$, there exists some critical value of $\tau(w)$ and θ, beyond which the Markov model underestimates traffic loss of LRD fluid characterized by both the loss probability and the expected loss ratio.

COMMENT 13.3.10. The critical value of both $\tau(w)$ and θ is $\gamma^{-1}z$, where γ is one of the two parameters of the Pareto distribution (13.6), and z is the solution of equation $e^{\beta x} = x + 1$. Since $\beta = 1 - \alpha^{-1}$, we see that z is determined by α, while α is another parameter of the Perato distribution. Therefore, the critical value of $\tau(w)$ and θ is independent of buffer and bandwidth allocated to the traffic source and determined only by the property of LRD traffic. Applications with stringent loss constraints require more bandwidth and buffer, which can cause $\tau(w)$ and θ to increase beyond the critical value, and therefore result in underestimation of loss performance degradation in the transient state for LRD traffic modeled by Markov processes.

13.3.2 Loss Behavior of General Two-State Fluids

We have just shown that long-range dependence can affect transient loss performance of fluid traffic in single states. To extend the analysis beyond single states, we examine further transient loss behavior of general two-state fluids [12]. We still denote the two-state traffic process by $R(t)$ and refer to the states as on and off. The lengths of both on and off intervals are i.i.d. random variables denoted, respectively, by S and T. Such traffic models are appealing from a mathematical point of view, since they can capture the essential property of long-range dependence while still keeping the analysis tractable. In fact, as we can see later, general two-state traffic models permit a complete analysis by which we can obtain exact results without approximation.

Let us consider a link carrying two-state fluid traffic with bit rates $r_1 > r_0$. We can model the link by a finite buffer queueing system with buffer B and service rate C corresponding to the link bandwidth. The link transmits traffic in the queue unless the queue is empty. We are concerned with transient loss performance defined by the traffic loss probability and the expected traffic loss ratio in the nth on interval for $n \geq 1$.

If C is greater than r_0 but less than r_1, then loss of traffic due to buffer overflow can only occur when the traffic process is in the on state, that is, $R(t) = r_1$. In fact, this is the only nontrivial case that should be considered. If $C \leq r_0$, then loss performance will be out of control in the transient state. On the other hand, if $C \geq r_1$, then loss will never occur for certain. Therefore, in the following, we assume $r_0 < C < r_1$ and consider only loss measures defined for on intervals. Since transient performance of the queueing system depends on temporal behavior of the system state, before we derive transient loss measures, we shall first investigate the stochastic evolution of system state variables.

13.3.2.1 Stochastic Dynamics of the Queueing Model

The state variables of interest regarding the queueing model are W_n and Q_n, which represent the amounts of traffic in the system at the beginning and the end of the nth on interval of the traffic process, respectively. Suppose that initially the traffic process is on with

$$W_1 = w_1,$$

where w_1 is a given constant between 0 and B. According to flow balance, the evolution of W_n and Q_n is as follows:

$$W_n = \begin{cases} -(C - r_0)T + Q_{n-1}, & T < Q_{n-1}/(C - r_0), \\ 0, & T \geq Q_{n-1}/(C - r_0), \end{cases}$$

where $n \geq 2$ and

$$Q_n = \begin{cases} (r_1 - C)S + W_n, & S < (B - W_n)/(r_1 - C), \\ B, & S \geq (B - W_n)/(r_1 - C), \end{cases}$$

where $n \geq 1$. Since S and T are random variables, W_n and Q_n are also random variables. Denote by $\varphi_n(w)$ and $\psi_n(q)$ the probability density functions of W_n and Q_n, respectively. Evidently, $\varphi_n(w)$ and $\psi_n(q)$ can be viewed as a solution of the above evolution equations in the probabilistic sense, which can readily be obtained by recurrence as shown below.

$$\psi_1(q) = \begin{cases} \dfrac{1}{r_1 - C} f_S\left(\dfrac{q - w_1}{r_1 - C}\right) + \delta(q - B)P\left\{S \geq \dfrac{q - w_1}{r_1 - C}\right\}, & 0 < q \leq B, \\ 0, & \text{otherwise}, \end{cases}$$

$$\psi_n(q) = \begin{cases} \displaystyle\int_0^B \psi_n(q|w)\varphi_n(w)\, dw, & 0 < q \leq B, \\ 0, & \text{otherwise}, \end{cases} \qquad n \geq 2,$$

and

$$\varphi_n(w) = \begin{cases} \displaystyle\int_0^B \varphi_n(w|q)\psi_{n-1}(q)\, dq, & 0 \leq w < B, \\ 0, & \text{otherwise}, \end{cases} \qquad n \geq 2,$$

where for $n \geq 2$

$$\psi_n(q|w) = \begin{cases} \dfrac{1}{r_1 - C} f_S\left(\dfrac{q - w}{r_1 - C}\right) + \delta(q - B)P\left\{S \geq \dfrac{q - w}{r_1 - C}\right\}, & 0 < q \leq B, \\ 0, & \text{otherwise}, \end{cases}$$

is the probability density function of Q_n conditioned on $W_n = w$, $f_S(s)$ is the probability density function of S,

$$\varphi_n(w|q) = \begin{cases} \dfrac{1}{c - r_0} f_T\left(\dfrac{q-w}{C-r_0}\right) + \delta(w)P\left\{T \geq \dfrac{q-w}{C-r_0}\right\}, & 0 \leq w < B, \\ 0, & \text{otherwise}, \end{cases}$$

is the probability density function of W_n conditioned on $Q_{n-1} = q$, $f_T(t)$ is the probability density function of T, and $\delta(\cdot)$ is the Dirac delta function. The Dirac delta function is defined by

$$\delta(x - x_0) = \begin{cases} 0, & x \neq x_0, \\ \infty, & x = x_0. \end{cases}$$

An important property of the Dirac delta function is

$$\int_{-\infty}^{\infty} f(x)\delta(x - x_0)\, dx = f(x_0),$$

where $f(x)$ is a function defined for $x \in (-\infty, \infty)$. It is easy to verify that the above probability density functions are nonnegative and normalizable. The process W_n has significant impact on transient loss performance of the queueing system. To reach the steady state for realizing steady-state performance, the system needs time to forget its history characerized by the initial condition $W_1 = w_1$ and the distributions of W_n for $1 < n < \infty$.

*13.3.2.2 **Transient Loss Measures*** To compare the loss behavior of two-state Markov and LRD traffic in the transient state, we consider two transient loss measures defined for on intervals. Recall that $R(n)$ denotes $R(t)$ for $t \in [t_n, t_{n+1})$. As we have pointed out already, for a two-state process, it is only necessary to consider on intervals, that is, $[t_n, t_{n+1})$ during which $R(n) = r_1$.

Our first transient loss measure is the probability that loss of traffic occurs during the nth on interval due to buffer overflow, denoted by $P\{\text{loss}|R(n) = r_1\}$. The second transient loss measure is the expected traffic loss ratio of the nth interval, denoted by $E[l|R(n) = r_1]$. Since $W_1 = w_1$ is a constant, we can use $P\{\text{loss}|R(1) = r_1, W_1 = w_1\}$ and $E[l|R(1) = r_1, W_1 = w_1]$ directly as transient loss measures for $n = 1$. Using Lemmas 13.3.1 and 13.3.3, we can easily prove the following theorem, which shows how to compute the loss measures.

Theorem 13.3.11

$$P\{\text{loss}|R(n) = r_1\} = \begin{cases} P\{S > \tau(w_1)\}, & n = 1, \\ \displaystyle\int_0^B P\{S > \tau(w)\}\varphi_n(w)\, dw, & n > 1, \end{cases}$$

and

$$E[l|R(n) = r_1] = \begin{cases} \displaystyle\int_0^{u_0} P\left\{S > \frac{u_0 \tau(w_1)}{u_0 - u}\right\} du, & n = 1, \\ \displaystyle\int_0^B \int_0^{u_0} P\left\{S > \frac{u_0 \tau(w)}{u_0 - u}\right\} \varphi_n(w)\, du\, dw, & n > 1. \end{cases}$$

COMMENT 13.3.12. For $n \geq 2$, the loss probability and the expected loss ratio depend on the distributions of both S and T, since $\varphi_n(w)$ depends on the distributions of not only S but also T when $n \geq 2$.

13.3.2.3 Investigating Transient Loss Behavior We can investigate the transient loss behavior of a general two-state fluid-type traffic process, at least in principle, by computing the loss measures for any given $n \geq 1$. By doing so, in fact, we have already broken, although in a "brute-force" way, the barrier of the Markov assumption in transient performance analysis of queueing systems. Recall that the basis of transient performance analysis for a Markov model is the Chapman–Kolmogorov equation governing the probabilistic evolution of the system state. If the model is not Markovian, then the Chapman–Kolmogorov equation does not necessarily hold. In this case, some auxiliary Markov model may be constructed such that the original system can be analyzed approximately. In contrast to this conventional approach, our method does not rely on the Markov assumption, so it can analyze non-Markov systems without approximation. In addition, it can also analyze Markov systems without using the Chapman–Kolmogorov equation.

However, if we compute the transient performance measures for each $n \geq 1$ directly, then the computational complexity will gradually become prohibitive as n increases. But in the context of transient analysis, performance measures corresponding to a large n may be less interesting, since when n is large, the difference between transient and steady states may be less significant. Consequently, for transient performance analysis, it may be sufficient to focus only on a relatively short time periods spanned by a number of on and off intervals. By doing so, the computation involved will be feasible.

In fact, there is a simple way to explore the transient loss performance impact of long-range dependence in two-state fluid traffic. Recall that a two-state fluid process can be viewed as an alternating renewal process. By taking advantage of the *renewal* nature of two-state fluids, we can simply focus only on a series of a small number of on and off intervals. Since both on and off intervals have impact on transient loss performance, such a series must contain at least two on intervals and one off interval between them, as shown in Fig. 13.1.

Now let us consider such an on–off–on series. Without loss of generality, the two on intervals in the series are indexed by $n = 1$ and 2. Since loss behavior of the queueing system during the first on interval in the series cannot capture the impact of the off interval, we examine loss behavior of the system in the second on interval.

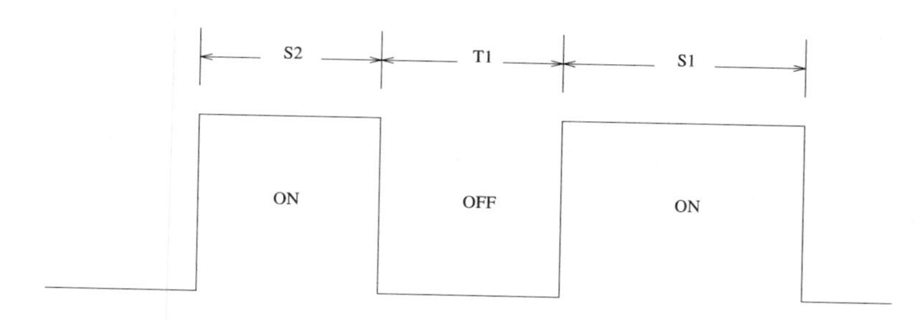

Fig. 13.1 An on–off–on series of a two-state traffic process.

Before the on–off–on series begins, the history of the system is summarized by $W_1 = w_1$. Due to the renewal nature of two-state fluids, as time evolves, the on–off–on series will probabilistically duplicate itself; that is, this traffic pattern will appear repeatedly with lengths of on and off intervals drawn from the same distributions. At the beginning of each such duplication, there is an amount of traffic w_1 left in the buffer, reflecting the impact of the history of the system.

Therefore, in order to investigate transient loss performance of the queueing system, it is sufficient to consider $P\{\text{loss}|R(n) = r_1\}$ or $E[l|R(n) = r_1]$ only for $n = 2$ and examine the behavior of the loss measures as w_1 varies between 0 and B. In fact, by focusing only on this simple traffic pattern, we can observe any possible transient loss behavior of the queueing system. Therefore, we use the transient loss measures for $n = 2$, that is, $P\{\text{loss}|R(2) = r_1\}$ and $E[l|R(2) = r_1]$, to define two performance functions, denoted respectively by P and E. The two functions depend on w_1, so we express P and E as functions of w_1. For convenience of exposition, w_1 is simply denoted by w in P and E. That is,

$$P(w) \stackrel{\text{def}}{=} P\{\text{loss}|R(2) = r_1\}$$

and

$$E(w) \stackrel{\text{def}}{=} E[l|R(2) = r_1].$$

The above arguments are based on the fact that a two-state stochastic fluid can be modeled by an alternating renewal process. Essentially, the behavior of a renewal process is probabilistically periodic in the sense that interarrival times between renewals are independent and identically distributed. Each time a renewal occurs, the process restarts itself probabilistically. This property allows us to focus on some basic traffic patterns. The on–off–on series is the minimum of such traffic patterns that can take the impact of both on and off intervals into account. The renewal nature makes the investigation of transient loss behavior for two-state fluid traffic simple. Based on the results presented previously, the derivation of the performance functions is straightforward. The final expressions of the performance functions are given in Section 13.7. Although at first sight the expressions look tedious, the computation of $P(w)$ and $E(w)$ consists of only straightforward evaluation of probability distribution and density functions, as well as multiple integrals, and therefore is tractable. By means of numerical analysis, we can obtain accurate results without approximation. Accurate results are desirable for investigating the transient loss performance impact of long-range dependence in traffic, which can help us see what happens exactly in the transient state if LRD traffic is modeled by Markov models.

13.3.2.4 Numerical Examples With the performance functions defined above, we can now investigate numerically the transient loss behavior for two-state Markov and LRD fluids. For simplicity, we let S and T be identically distributed and all traffic processes considered have the same $E[S]$ and $E[T]$. We consider exponential distribution as well as two heavy-tailed distributions for S and T. The first heavy-tailed distribution is given by Eq. (13.6). Another heavy-tailed distribution is defined by the following probability density function:

$$f(t) = \begin{cases} \alpha A^{-1} e^{-\alpha t/A}, & t \leq A, \\ \alpha e^{-\alpha} A^{\alpha} t^{-(\alpha+1)}, & t > A, \end{cases} \quad A > 0,\ 1 < \alpha < 2. \tag{13.7}$$

The traffic processes determined by Eqs. (13.6) and (13.7) are LRD processes [1, 6, 11], henceforth referred to as LRD process I and LRD process II, respectively. We use Eq. (13.7) as an example to show that the assumption of the Pareto distribution in Theorems 13.3.6 and 13.3.7 is not restrictive as far as the impact of long-range dependence is concerned. What is essential here is that the distribution is heavy tailed, regardless of the specific form of the distribution. Clearly, the process determined by exponentially distributed S and T is a Markov process. Therefore, the queueing system with such a Markov arrival process can also be modeled by a Markov process. However, the queueing system with an LRD arrival process is not Markovian anymore.

To be specific, we let $\alpha = 1.6$, $\gamma = 0.5$, $r_1 = 1$, and $r_0 = 0$. Since all traffic processes considered have the same $E[S]$ and $E[T]$, and S and T are identically distributed, the expectations determined by Eqs. (13.6) and (13.7) are the same. So the parameter A in Eq. (13.7) can be determined by the above parameters.

As shown in Section 13.3.2.3, by exploiting the renewal property of two-state fluids, we can investigate the transient loss behavior of the queueing system by examining the values of performance functions $P(w)$ and $E(w)$ as w varies in $[0, B)$. We consider three traffic processes determined by the distributions of S and T as described above. The assumption that all traffic processes considered have the same $E[S]$ and $E[T]$ implies that the marginal state distributions of the traffic processes are the same. We first let buffer size $B = 1$ and service rate $C = 0.8$ for all the traffic processes, and compare transient loss behaviors of the traffic processes based on the functions $P(w)$ and $E(w)$ computed, respectively, according to Eqs. (13.8) and (13.9) in Section 13.7.1.

We observe that even though the marginal state distributions of the traffic processes are the same, the transient loss measures of the traffic processes can be

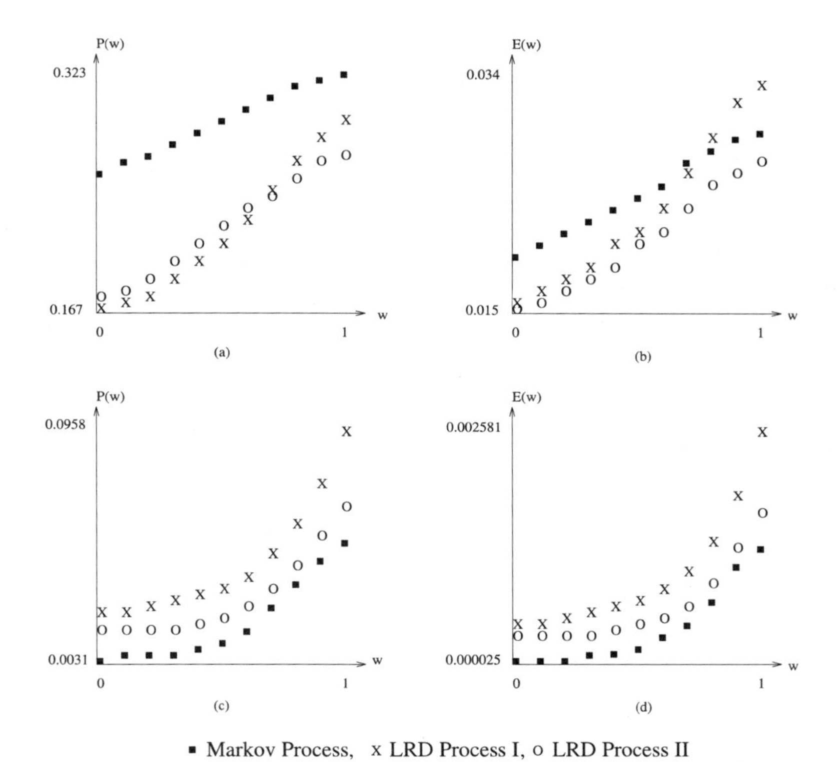

Fig. 13.2 Transient loss behaviors of different traffic processes. (a) The loss probability with $B = 1$ and $C = 0.8$. (b) The expected loss ratio with $B = 1$ and $C = 0.8$. (c) The loss probability with $B = 1$ and $C = 0.95$. (d) The expected loss ratio with $B = 1$ and $C = 0.95$.

significantly different (Fig. 13.2). The difference in transient loss behavior is only due to the distributions of S and T, which characterize the stochastic properties of the basic time scales in the underlying traffic. We note in Fig. 13.2(b) that if w exceeds some critical value, then the expected loss ratio of the Markov process is less than that of LRD process I. This is due to the effect of the heavy-tailed distributions of S and T. Such an effect becomes more significant if C (or B) is large. For example, suppose that we keep $B = 1$ but let $C = 0.95$; see Fig. 13.2(c) and (d). Note that a large C implies a stringent loss performance requirement. In this case, we observe that for any $w \in [0, B)$, both the loss measures of the Markov process are less than those of LRD processes I and II. This is because for the Markov process, the transient loss measures decay exponentially fast as C or B increases, but for the LRD processes, the loss measures decay only hyperbolically, due to the LRD effect caused by the heavy-tailed distributions of S and T. The numerical results are consistent with the analytical results regarding transient loss of LRD traffic in single states. As C or B increases, $\tau(\omega)$ and θ also increase. According to Theorems 13.3.6

and 13.3.7, beyond some critical value of $\tau(w)$ and θ, Markov models can underestimate the loss performance degradation of LRD traffic in on intervals.

13.4 STEADY-STATE LIMITS OF TRANSIENT LOSS MEASURES

According to some previous studies carried out in steady state, Markov models may still be used for traffic engineering in the presence of long-range dependence. Even though the above conclusion holds in steady state, we still need to examine whether it is valid to extend the conclusion beyond the region of the steady state. To this end, we must know whether loss behavior in the steady state is significantly different from that in the transient state.

To make an exact comparison between transient loss and loss in the steady state, we consider two-state fluid traffic [12]. We have already shown how to compute transient loss measures for two-state fluid traffic. Now we show how to compute the steady-state limits of the loss measures, which are defined by

$$P\{\text{loss}\} \stackrel{\text{def}}{=} \lim_{n\to\infty} P\{\text{loss}|R(n) = r_1\}$$

and

$$E[l] \stackrel{\text{def}}{=} \lim_{n\to\infty} E[l|R(n) = r_1].$$

13.4.1 Analysis

For a given stationary two-state traffic process, we have the following limiting conditional probability density functions,

$$\psi(q|w) \stackrel{\text{def}}{=} \lim_{n\to\infty} \psi_n(q|w), \quad \varphi(w|q) \stackrel{\text{def}}{=} \lim_{n\to\infty} \varphi_n(w|q),$$

and limiting probability density functions,

$$\psi(q) \stackrel{\text{def}}{=} \lim_{n\to\infty} \psi_n(q), \quad \varphi(w) \stackrel{\text{def}}{=} \lim_{n\to\infty} \varphi_n(w),$$

regarding the corresponding state variables. The following theorem and its corollary give a procedure for computing the steady-state loss measures.

Theorem 13.4.1. *For a two-state fluid process, we have*

$$\varphi(w) = \varphi_0(w) + v(w)\delta(w),$$

where $\delta(w)$ is the Dirac delta function,

$$\varphi_0(w) = v(0)\bar{\varphi}_0(w), \quad v(0) = \frac{1}{1 + \int_0^B \bar{\varphi}_0(w)\, dw},$$

and $\bar{\varphi}_0(w)$ *is the solution of the following integral equation:*

$$\bar{\varphi}_0(w) = K_0(w, 0) + \int_0^B K_0(w, x)\bar{\varphi}_0(x)\, dx$$

with kernel

$$K_0(w, x) = \frac{1}{(C - r_0)(r_1 - C)} \int_{\max\{w,x\}}^B f_T\left(\frac{y - w}{C - r_0}\right) f_S\left(\frac{y - x}{r_1 - C}\right) dy$$
$$+ \frac{1}{C - r_0} f_T\left(\frac{B - w}{C - r_0}\right) P\left\{S \geq \frac{B - x}{r_1 - C}\right\}.$$

The proof of the theorem is given in Section 13.7.2. Due to the property of the Dirac function $\delta(w)$, only $v(0)$ rather than the whole $v(x)$ will appear in the steady-state loss measures. So in Theorem 13.4.1, we shown how to compute $v(0)$ instead of the whole $v(w)$. From Theorem 13.3.11, the steady-state limits of the transient loss measures can be expressed as follows:

$$P\{\text{loss}\} = \int_0^B P\{S > \tau(w)\}\varphi(w)\, dw$$

and

$$E[l] = \int_0^B \int_0^{u_0} P\left\{S > \frac{u_0\tau(w)}{u_0 - u}\right\}\varphi(w)\, du\, dw.$$

According to Theorem 13.4.1, the steady-state loss measures now can be computed with $\bar{\varphi}_0(w)$, which is obtained by solving the integral equation.

Corollary 13.4.2

$$P\{\text{loss}\} = v(0)\left[P\{S > \tau(0)\} + \int_0^B P\{S > \tau(w)\}\bar{\varphi}_0(w)\, dw\right]$$

and

$$E[l] = v(0)\left[\int_0^{u_0} P\left\{S > \frac{u_0\tau(0)}{u_0 - u}\right\} du + \int_0^B \int_0^{u_0} P\left\{S > \frac{u_0\tau(w)}{u_0 - u}\right\}\bar{\varphi}_0(w)\, du\, dw\right].$$

COMMENT 13.4.3. The integral equation in Theorem 13.4.1 is the Fredholm integral equation of the second kind. We have proved that similar equations hold for general queueing models with a single server, including the queue $G/G/1$ and $G/D/1$. Therefore, we have actually developed a novel solution technique for the steady-state analysis of general single server queueing sytems [14].

13.4.2 Numerical Examples

Based on the analysis results, we can now use numerical examples to compare transient and steady-state loss measures for different traffic processes with different buffer size B and bandwidth C. We still adopt the traffic parameters given in Section 13.3.2.4. The transient loss performance is characterized by the performance functions defined in Section 13.3.2.3. We first fix B and let C take on several different values (Fig. 13.3). Then we fix C and let B take on different values (Fig. 13.4).

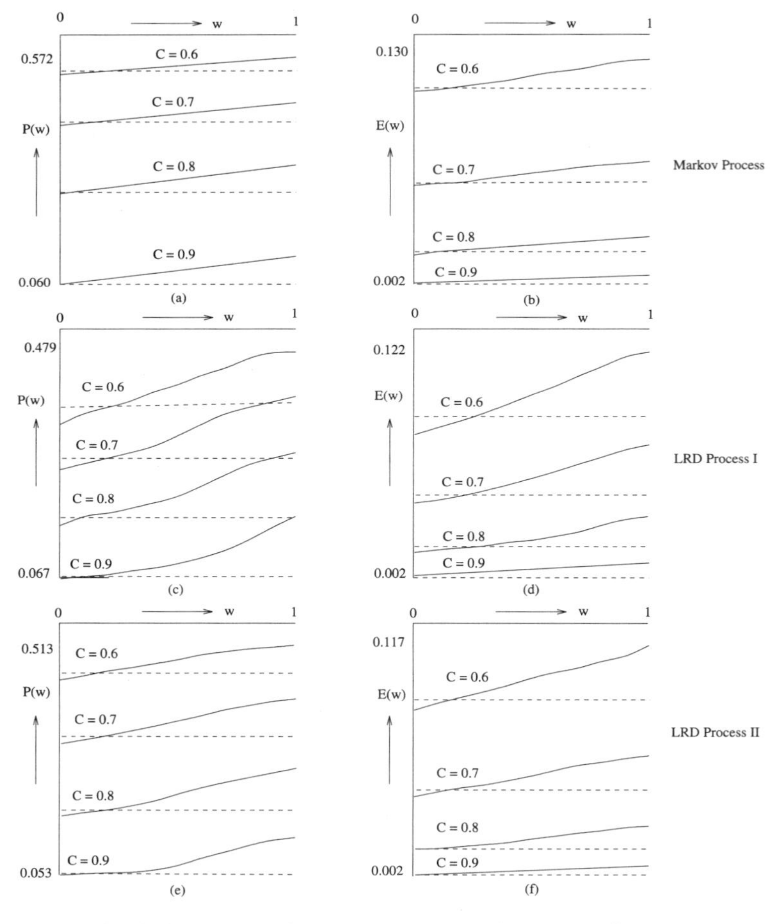

Fig. 13.3 Transient loss versus steady-state loss for $B = 1$ and $C = 0.6, 0.7, 0.8, 0.9$. (a) The loss probability of the Markov process. (b) The expected loss ratio of the Markov process. (c) The loss probability of LRD process I. (d) The expected loss ratio of LRD process I. (e) The loss probability of LRD process II. (f) The expected loss ratio of LRD process II.

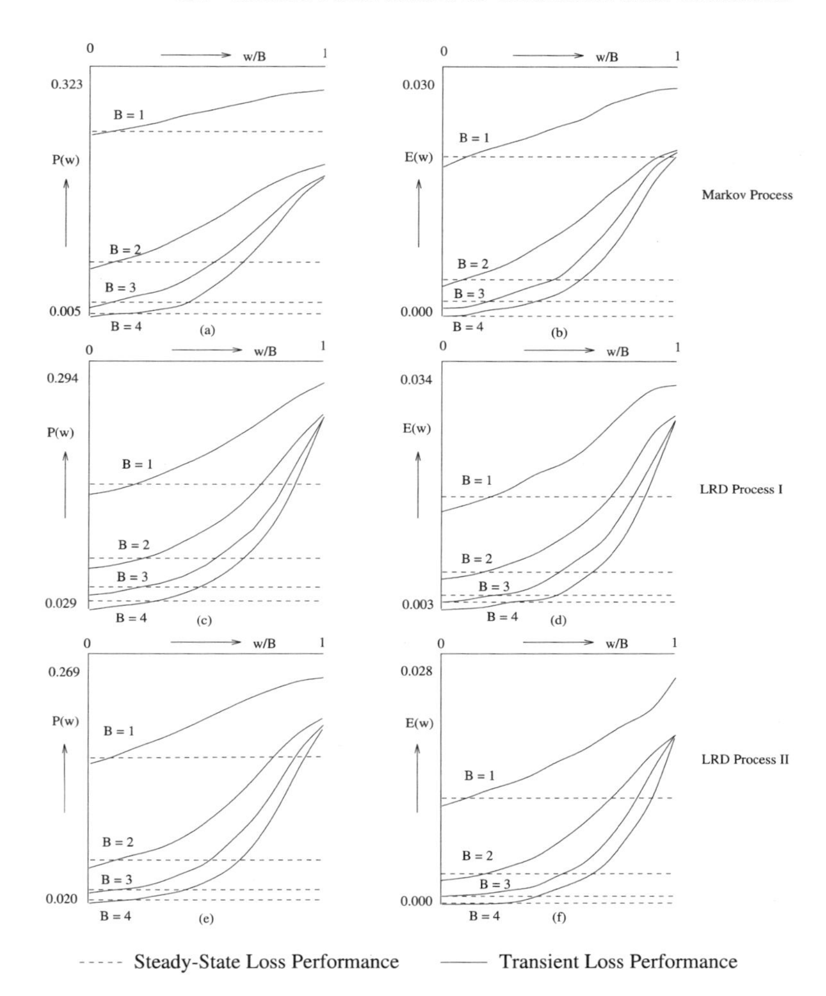

Fig. 13.4 Transient loss versus steady-state loss for $C = 0.8$ and $B = 1, 2, 3, 4$. The horizontal axis represents w scaled by B. (a) The loss probability of the Markov process. (b) The expected loss ratio of the Markov process. (c) The loss probability of LRD process I. (d) The expected loss ratio of LRD process I. (e) The loss probability of LRD process II. (f) The expected loss ratio of LRD process II.

From the numerical results, we see that, in the steady state, the loss behavior of each of the above two-state traffic processes is indeed significantly different from its loss behavior in the transient state. Since steady-state loss measures are weighted by the marginal distributions of the underlying traffic process, the random variance in transient loss measures is completely averaged out in the steady state. As a result, the steady-state loss measures are simply some constants.

Unlike the loss measures in the steady state, the transient loss measures are functions of the initial value w that summarizes the history of the system. For all the traffic processes considered, we observe that as the initial value varies, the transient loss measures will become greater than the corresponding steady-state loss measures. That is, steady-state loss measures can underestimate actual loss in the transient state. As the buffer size B or the service rate C increases, we can still observe such underestimation. Although increasing C can decrease the probability for w to take larger values, due to the randomness of on and off periods, it is still possible for w to take values greater than some critical value beyond which the steady-state loss measures will underestimate transient loss. We even observe that if C (or B) is large enough, then for all $w \in [0, B)$, the steady-state loss measures are strictly less than the corresponding transient loss measures; for example, see the case of $C = 0.9$ in Fig. 13.3(d).

13.5 THE IMPACT OF LONG-RANGE DEPENDENCE IN NETWORK TRAFFIC

If bandwidth allocated to a traffic source equals the peak bit rate of the source, then whether there exists long-range dependence in the underlying traffic has no impact at all on traffic modeling and engineering. So in the following discussions, we assume that the bandwidth allocated to an LRD traffic source is less than its peak rate.

First of all, perhaps the most important issue is whether long-range dependence is relevant to traffic engineering [6, 9, 10, 16, 17]. It has been shown that in the steady state, when the buffer size is small, long-range dependence may not have significant impact [10, 17]. However, the above conclusion is drawn in the steady state, so it may not be necessarily valid beyond the region of steady state. Our analysis and numerical results have shown that, in the transient state, there exists significant difference in traffic loss behavior between traditional Markov models and LRD models. Analysis results presented in Section 13.3 describe a scenario regarding transient loss performance impact of long-range dependence in multistate fluid traffic.

In nontrivial cases, after the fluid enters a state, it will pour more traffic into the buffer than what the bandwidth can transmit. However, according to Lemma 13.3.1, traffic loss will not take place until the time spent by the fluid in the state exceeds the value of $\tau(w)$, which is the instant after which loss of traffic begins. From Theorems 13.3.6 and 13.3.7, we see that there exists some critical value of $\tau(w)$ and θ determined only by the property of the underlying LRD traffic, beyond which Markov models can underestimate loss performance degradation for LRD traffic. Since large values of $\tau(w)$ and θ imply stringent loss performance requirements, the above results in fact say that, in the transient state, long-range dependence is more likely to have significant impact on applications with stringent loss constraints.

Although a small buffer may alleviate the impact of long-range dependence in the steady state [9, 10], it does not help in the transient state if the loss constraint is stringent. Let us take the conditional loss probability as an example. Since $\tau(w) = (B - w)/(r - C)$ depending on buffer and bandwidth allocated to the fluid, it can be viewed as a control variable. From Lemma 13.3.1, in order to meet a stringent loss performance requirement, the value of $\tau(w)$ should be sufficiently large. On the other hand, however, as shown by Theorem 13.3.6, if Markov models are used for traffic engineering in the presence of long-range dependence, then increasing the value of $\tau(w)$ can result in underestimation of loss performance degradation for LRD traffic in the transient state. Clearly, a large buffer can of course increase the value of $\tau(w)$, and hence lead to underestimation of loss performance degradation of LRD traffic, if loss performance is predicted based on Markov models. However, the buffer is not the only factor causing such underestimation. If the buffer size is small, then the bandwidth must be large enough to keep the value of $\tau(w)$ sufficiently large, which can also result in underestimation of traffic loss in the presence of long-range dependence. Similar results hold also for the conditional expected loss ratio. In this case, we have

$$\lim_{C\to r} \theta = \lim_{C\to r} \frac{u_0 \tau(w)}{u_0 - \xi} = \frac{\lim_{C\to r} \tau(w)}{1 - \lim_{C\to r} \xi/u_0} = \infty,$$

since from Eq. (13.2), $\tau(w) \to \infty$ as C increases toward r, and $0 \le \lim_{C\to r} \xi/u_0 \le 1$ since $\xi \in (0, u_0)$. As we know, the critical value of $\tau(w)$ and θ is independent of C. Consequently, increasing C will eventually lead to sufficiently large $\tau(w)$ and θ exceeding the critical value, and therefore result in underestimation of loss performance degradation for LRD traffic. In fact. Such underestimation is intrinsic if loss performance requirements are stringent. Therefore, buffer size is not an essential issue regarding the impact of long-range dependence on traffic loss in the transient state. According to our analysis, the difference in transient loss behavior of different traffic processes is determined by the distributions of times spent by the traffic processes in their states. Concerning the impact long-range dependence in traffic, what is essential then is the heavy-tailed distribution of the random variables that represent the time scales in the underlying LRD traffic, which is the most important property of LRD traffic.

In addition, our analysis and numerical results based on the fluid traffic model with two states have demonstrated that the loss probability as well as the expected loss ratio in the steady state can be less than those in the transient state. This is because steady-state analysis cannot capture actual loss behavior of traffic processes in the transient state. In an idealized steady state, the system has forgotten its history. As a result, the steady-state loss measures are merely some constants. In contrast, transient loss measures depend on the history of the system and therefore are variables. Within the framework of steady-state performance analysis, loss performance is weighted by the marginal state distribution of the traffic process. However, the marginal distribution is not important to transient loss performance. Our numerical results have shown that different two-state fluid-type traffic processes with the same marginal state distribution but different distributions of S and T can experience significantly different transient loss. What is important here again is the distributions of the random variables regarding the time scales in the underlying traffic. Since steady-state loss performance can be much better than that in the transient state even for Markov traffic, and transient loss performance of Markov traffic can also be much better than that of LRD traffic, prediction of loss performance based on Markov models in the steady state can then lead to even worse underestimation of loss performance degradation for LRD traffic in the transient state.

13.6 CONCLUSION

In this chapter, we have presented an analysis of the transient loss performance impact of long-range dependence in network traffic. We have examined transient loss behavior for multistate fluid traffic in single states and for two-state fluid traffic, under two fundamentally different assumptions, that is, Markov versus LRD, on the underlying traffic process. Numerical results are given to illustrate the analysis. To sum up, the main findings of this study are as follows.

- In the transient state, different traffic processes exhibit different loss behaviors. The essential cause of the difference is the distributions of the times spent by the traffic processes in their states. Such distributions characterize the stochastic properties of the basic time scales in the traffic processes. Heavy-tailed distributions appear to be a core factor of LRD traffic.
- For LRD traffic modeled by Markov processes, if loss constraints are stringent, then Markov models are more likely to underestimate loss performance degradation of the underlying LRD traffic. If loss performance is predicted based on Markov models in the steady state, then such underestimation can be even worse.

Based on the above findings, we believe that in the transient state, long-range dependence can indeed have significant impact on traffic modeling and engineering.

Since traffic loss is sensitive to long-range dependence in the transient state, how to guarantee transient loss performance in the presence of long-range dependence then poses a significant challenge to traffic engineering. The challenge lies mainly in two aspects. First, beyond the region of steady state, marginal state distributions of traffic sources have only limited use. On the other hand, it is not feasible, if not impossible, to use transient state distributions for traffic engineering. Second, in a network, traffic processes will lose any renewal properties that they might possess at the edge of the network, making transient loss performance analysis and guarantee even more difficult. Transient loss performance guarantee in the presence of long-range dependence may be an important issue in future research.

Another related issue is transient call blocking analysis. Due to the same reason that motivated the work presented in this chapter, call blocking performance should also be analyzed in the transient state. Recent studies show that it may be more appropriate to model call duration times in wide-area networks by heavy-tailed distributions [7]. If this is indeed the case, then due to the high variability caused by heavy-tailed distributions, it will take much longer for the system to approach the steady state, and hence call blocking probabilities in steady and transient states can be significantly different. We have developed a technique to compute transient call blocking probabilities explicitly and exactly for arbitrarily distributed call interarrival and holding times and analyzed the impact of the distribution of the call holding times [13]. Our analysis shows that in the transient state, the distribution of the call holding times indeed has a significant impact on call blocking performance, and this impact cannot be captured by steady-state call blocking analysis.

13.7 APPENDIXES

13.7.1 Transient Performance Functions of Two-State Fluid

By recurrence, we can compute $\varphi_n(w)$ for any $n > 1$, according to the formulas given in Section 13.3.2.1. Then with straightforward computation, for $n = 2$, the final expressions of the transient performance functions for general two-state fluids defined in Section 13.3.2.3 can be obtained as follows:

$$
\begin{aligned}
P(w) \stackrel{\text{def}}{=} P\{\text{loss}|R(2) = r_1\} &= \frac{1}{(C - r_0)(r_1 - C)} \\
&\times \int_0^B P\{S > \tau(x)\} \int_{\max\{x,w\}}^B f_T\left(\frac{q-x}{C-r_0}\right) f_S\left(\frac{q-w}{r_1-C}\right) dq\, dx \\
&+ \frac{P\{S \geq \tau(w)\}}{C - r_0} \int_0^B P\{S > \tau(x)\} f_T\left(\frac{B-x}{C-r_0}\right) dx \\
&+ \frac{P\{S > \tau(0)\}}{r_1 - C} \int_w^B f_S\left(\frac{q-w}{r_1-C}\right) P\left\{T \geq \frac{q}{C-r_0}\right\} dq \\
&+ P\{S \geq \tau(w)\} P\left\{T \geq \frac{B}{C-r_0}\right\} P\{S > \tau(0)\}
\end{aligned}
\tag{13.8}
$$

and

$$
\begin{aligned}
E(w) \stackrel{\text{def}}{=} E[l|R(2) = r_1] &= \frac{1}{(C - r_0)(r_1 - C)} \\
&\times \int_0^B \int_0^{u_0} P\left\{S > \frac{u_0 \tau(x)}{u_0 - u}\right\} \int_{\max\{x,w\}}^B f_T\left(\frac{q-x}{C-r_0}\right) f_S\left(\frac{q-w}{r_1-C}\right) dq\, du\, dx \\
&+ \frac{P\{S \geq \tau(w)\}}{C - r_0} \int_0^B \int_0^{u_0} P\left\{S > \frac{u_0 \tau(x)}{u_0 - u}\right\} f_T\left(\frac{B-x}{C-r_0}\right) du\, dx \\
&+ \frac{1}{r_1 - C} \int_0^{u_0} P\left\{S > \frac{u_0 \tau(0)}{u_0 - u}\right\} du \int_w^B f_S\left(\frac{q-w}{r_1-C}\right) P\left\{T \geq \frac{q}{C-r_0}\right\} dq \\
&+ P\{S \geq \tau(w)\} P\left\{T \geq \frac{B}{C-r_0}\right\} \int_0^{u_0} P\left\{S > \frac{u_0 \tau(0)}{u_0 - u}\right\} du.
\end{aligned}
\tag{13.9}
$$

13.7.2 Proof of Theorem 13.4.1

Since in the steady state we have by definition

$$\varphi(w) = \int_0^B \varphi(w|y)\psi(y)\, dy$$

and

$$\psi(y) = \int_0^B \psi(y|x)\varphi(x)\, dx,$$

we see that $\varphi(w)$ satisfies a homogeneous Fredholm integral equation of the second kind,

$$\varphi(w) = \int_0^B K(w, x)\varphi(x)\, dx \tag{13.10}$$

with kernal $K(w, x) = \int_0^B \varphi(w|y)\psi(y|x)\, dy$. We can see readily that

$$K(w, x) = K_0(w, x) + H(w, x)\delta(w), \tag{13.11}$$

where

$$K_0(w, x) = \frac{1}{(C - r_0)(r_1 - C)} \int_{\max\{w,x\}}^B f_T\left(\frac{y - w}{C - r_0}\right) f_S\left(\frac{y - x}{r_1 - C}\right) dy + \frac{1}{C - r_0} f_T\left(\frac{B - w}{C - \tau_0}\right) P\left\{S \geq \frac{B - x}{r_1 - C}\right\}$$

and

$$H(w, x) = \frac{1}{r_1 - C} \int_x^B f_S\left(\frac{y - x}{r_1 - C}\right) P\left\{T \geq \frac{\max\{0, y - w\}}{C - r_0}\right\} dy + P\left\{T \geq \frac{B - w}{C - r_0}\right\} P\left\{S \geq \frac{B - x}{r_1 - C}\right\}.$$

In general, the Fredholm integral equation can be solved numerically. However, due to the singularity of $K(w, x)$ caused by the Dirac delta function $\delta(w)$ at $w = 0$, we cannot solve Eq. (13.10) directly. The singularity caused by the Dirac delta function $\delta(w)$ will also appear in $\varphi(w)$. Since this singularity in $\varphi(w)$ is intrinsic and cannot be removed from $\varphi(w)$, the functional form of $\varphi(w)$ will be

$$\varphi(w) = \varphi_0(w) + v(w)\delta(w), \tag{13.12}$$

where $\varphi_0(w)$ and $v(w)$ are unknown functions. Substituting Eqs. (13.12) and (13.11) into (13.10), we have

$$\varphi_0(w) + v(w)\delta(w) = v(0)K_0(w, 0) + \int_0^B K_0(w, x)\varphi_0(x)\, dx + \left[v(0)H(w, 0) + \int_0^B H(w, x)\varphi_0(x)\, dx\right]\delta(w).$$

Comparing both sides of the above equation, we see that

$$\varphi_0(w) = v(0)K_0(w, 0) + \int_0^B K_0(w, x)\varphi_0(x)\, dx \tag{13.13}$$

and

$$v(w) = v(0)H(w, 0) + \int_0^B H(w, x)\varphi_0(x)\, dx. \tag{13.14}$$

Note that

$$v(0) = \lim_{\eta\to 0}\int_0^\eta \varphi(w)\, dw = \lim_{n\to\infty} P\{W_n = 0\},$$

where η is a small positive variable. To avoid the trivial case of $v(0) = 0$, we assume $v(0) > 0$. Since $v(0)$ is still unknown, we let

$$\bar{\varphi}_0(w) \stackrel{\text{def}}{=} \frac{\varphi_0(w)}{v(0)} \tag{13.15}$$

and divide both sides of Eq. (13.13) by $v(0)$, so

$$\bar{\varphi}_0(w) = K_0(w, 0) + \int_0^B K_0(w, x)\bar{\varphi}_0(x)\, dx. \tag{13.16}$$

The above equation is a nonhomogeneous Fredholm integral equation of the second kind in $\bar{\varphi}_0(w)$ with kernel $K_0(w, x)$ and can be solved numerically with the standard method [3].

With $\bar{\varphi}_0(w)$ obtained by solving Eq. (13.16), we can determine $v(0)$ as follows. Since by definition

$$\bar{\varphi}_0(w) = \frac{\varphi_0(w)}{v(0)} = \frac{\varphi(w) - v(w)\delta(w)}{v(0)},$$

integrating both sides of the above equation from 0 to B, we have

$$\int_0^B \bar{\varphi}_0(w)\, dw = \frac{\int_0^B \varphi(w)\, dw - \int_0^B v(w)\delta(w)\, dw}{v(0)} = \frac{1 - v(0)}{v(0)}.$$

Therefore,

$$v(0) = \frac{1}{1 + \int_0^B \bar{\varphi}_0(w)\, dw}. \tag{13.17}$$

Substituting Eq. (13.17) into Eqs. (13.15) and (13.14), we obtain finally

$$\varphi_0(w) = \frac{\bar{\varphi}_0(w)}{1 + \int_0^B \bar{\varphi}_0(x)\, dx}$$

and

$$v(w) = \frac{H(w, 0) + \int_0^B H(w, x)\bar{\varphi}_0(x)\, dx}{1 + \int_0^B \bar{\varphi}_0(x)\, dx}$$

Comparing $v(w)$ with $v(0)$, we should have

$$H(0, 0) + \int_0^B H(0, x)\bar{\varphi}_0(x)\, dx = 1. \tag{13.18}$$

The above equation indeed holds. We can see this by letting $w = 0$ in Eq. (13.14) and dividing both sides of Eq. (13.14) by $v(0)$. Note also that Eq. (13.18) ensures $\int_0^B \varphi(w)\, dw = 1$.

ACKNOWLEDGMENTS

The authors would like to thank Ms. Jun-Hong Cui for performing the numerical calculations. This work was supported in part by the National Natural Science Foundation of China.

REFERENCES

1. J. Beran, R. Sherman, and W. Willinger. Long range dependence in variable bit rate video traffic. *IEEE Trans. Commun.*, **43**(3):1566, 1995.
2. M. E. Crovella and A. Bestavros. Self-similarity in World Wide Web traffic: evidence and possible causes. In *Proc. ACM SIGMETRICS '96*, p. 160, 1996.
3. L. M. Delves and J. L. Mohamed. *Computational Methods for Integral Equations*. Cambridge University Press, Cambridge, England, 1988.
4. N. G. Duffield, J. T. Lewis, N. O'Connell, R. Russell, and F. Toomey. Predicting quality of service for traffic with long-range fluctuations. In *Proc. IEEE ICC*, Seattle, WA, 1995.
5. A. I. Elwalid and D. Mitra. Effective bandwidth of general Markovian sources and admission control of high speed networks. *IEEE/ACM Trans. Networking*, **1**:329, 1993.

6. A. Erramilli, O. Narayan, and W. Willinger. Experimental queueing analysis with long-range dependent packet traffic. *IEEE/ACM Trans. Networking*, **4**(2):209, 1996.
7. A. Feldmann, A. C. Gilbert, and W. Willinger. Data networks as cascades: investigating the multifractal nature of internet WAN traffic. In *ACM SIGCOMM '98*, p. 42, 1998.
8. M. W. Garret and W. Willinger. Analysis, modeling and generation of self-similar VBR video traffic. In *Proc. ACM SIGCOMM '94*, p. 269, 1994.
9. M. Grossglauser and J.-C. Bolot. On the relevance of long-range dependence in network traffic. In *Proc. ACM SIGCOMM '96*, p. 15, 1996.
10. D. P. Heyman and T. V. Lakshman. What are the implications of long-range dependence for VBR-video traffic engineering? *IEEE/ACM Trans. Networking*, **4**(3):301, 1996.
11. W. E. Leland, M. S. Taqqu, W. Willinger, and D. V. Wilson. On the self-similar nature of Ethernet traffic (extended version). *IEEE/ACM Trans. Networking*, **2**((1):1, 1994.
12. G.-L. Li, J.-H. Cui, F.-M. Li, and B. Li. Transient loss performance of a class of finite buffer queueing systems. In *ACM/IFIP SIGMETRICS/PERFORMANCE '98*, p. 111, 1998.
13. G.-L. Li and V. O. K. Li. Transient call blocking probabilities: computation and applications. Submitted for publication.
14. G.-L. Li and V. O. K. Li. Explicit steady-state solutions of general queueing models with a single server. Technique Report, Department of Electrical and Electronic Engineering, The University of Hong Kong, 1999.
15. N. Likhanov, B. Tsybakov, and N. D. Georganas. Analysis of an ATM buffer with self-similar ("fractal") input traffic. In *Proc. IEEE INFOCOM*, p. 985, 1995.
16. I. Norros. On the use of fractional Brownian motion in the theory of connectionless networks. *IEEE J. Select. Areas Commun.*, **13**(6):953, 1995.
17. B. K. Ryu and A. Elwalid. The importance of long-range dependence of VBR video traffic in ATM traffic engineering: myths and realities. In *Proc. ACM SIGCOMM '96*, p. 3, 1996.
18. W. Willinger, M. S. Taqqu, R. Sherman, and D. V. Wilson. Self-similarity through high-variability: statistical analysis of Ethernet LAN traffic at the source level. In *Proc. ACM SIGCOMM '95*, p. 100, 1995.

14

THE PROTOCOL STACK AND ITS MODULATING EFFECT ON SELF-SIMILAR TRAFFIC

KIHONG PARK
Network Systems Lab, Department of Computer Sciences, Purdue University, West Lafayette, IN 47907

GITAE KIM AND MARK E. CROVELLA
Department of Computer Science, Boston University, Boston, MA 02215

14.1 INTRODUCTION

Recent measurements of local-area and wide-area traffic [14, 22, 28] have shown that network traffic exhibits variability at a wide range of scales. Such scale-invariant variability is in strong contrast to traditional models of network traffic, which show variability at short scales but are essentially smooth at large time scales; that is, they lack long-range dependence. Since self-similarity is believed to have a significant impact on network performance [2, 15, 16], understanding the causes and effects of traffic self-similarity is an important problem.

In this chapter, we study a mechanism that induces self-similarity in network traffic. We show that self-similar traffic can arise from a simple, high-level property of the overall system: the heavy-tailed distribution of file sizes being transferred over the network. We show that if the distribution of file sizes is heavy tailed—meaning that the distribution behaves like a power law thus generating very large file transfers with nonnegligible probability—then the superposition of many file transfers in a client/server network environment induces self-similar traffic, and this causal

Self-Similar Network Traffic and Performance Evaluation, Edited by Kihong Park and Walter Willinger
ISBN 0-471-31974-0

mechanism is robust with respect to changes in network resources (bottleneck bandwidth and buffer capacity), topology, interference from cross-traffic with dissimilar traffic characteristics, and changes in the distribution of file request interarrival times. Properties of the transport/network layer in the protocol stack are shown to play an important role in mediating this causal relationship.

The mechanism we propose is motivated by the *on/off model* [28]. The on/off model shows that self-similarity can arise in an idealized context—that is, one with independent traffic sources and unbounded resources—as a result of aggregating a large number of 0/1 renewal processes whose on or off periods are heavy tailed. The success of this simple, elegant model in capturing the characteristics of measured traffic traces is surprising given that it ignores nonlinearities arising from the interaction of traffic sources contending for network resources, which in real networks can be as complicated as the feedback congestion control algorithm of TCP. To apply the framework of the on/off model to real networks, it is necessary to understand whether the model's limitations affect its usefulness and how these limitations manifest themselves in practice.

In this chapter, we show that in a "realistic" client/server network environment—that is, one with bounded resources leading to the coupling of multiple traffic sources contending for shared resources—the degree to which file sizes are heavy tailed directly determines the degree of traffic self-similarity. Specifically, measuring self-similarity via the Hurst parameter H and file size distribution by its power-law exponent α, we show that there is a linear relationship between H and α over a wide range of network conditions and when subject to the influence of the protocol stack. The mechanism gives a particularly simple structural explanation of why self-similar network traffic may be observed in many diverse networking contexts.

We discuss a traffic-shaping effect of TCP that helps explain the modulating influence of the protocol stack. We find that the presence of self-similarity at the link and network layer depends on whether reliable and flow-controlled communication is employed at the transport layer. In the absence of reliability and flow control mechanisms—such as when a UDP-based transport protocol is used—much of the self-similar burstiness of the downstream traffic is destroyed when compared to the upstream traffic. The resulting traffic, while still bursty at short ranges, shows significantly less long-range correlation structure. In contrast, when TCP (Reno, Tahoe, or Vegas) is employed, the long-range dependence structure induced by heavy-tailed file size distributions is preserved and transferred to the link layer, manifesting itself as scale-invariant burstiness.

We conclude with a discussion of the effect of self-similarity on network performance. We find that in UDP-based non-flow-controlled environment, as self-similarity is increased, performance declines drastically as measured by packet loss rate and mean queue length. If reliable communication via TCP is used, however, packet loss, retransmission rate, and file transmission time decline gracefully (roughly linearly) as a function of H. The exception is mean queue length, which shows the same superlinear increase as in the unreliable non-flow-controlled case. This graceful decline in TCP's performance under self-similar loads comes at a cost: a disproportionate increase in the consumption of buffer space. The sensitive dependence of mean queue length on self-similarity is consistent with previous works [2, 15, 16] showing that queue length distribution decays more slowly for long-range dependent (LRD) sources than for short-range dependent (SRD) sources. The aforementioned traffic-shaping effect of flow-controlled, reliable transport transforming a large file transfer into an on-average "thin" packet train (stretching-in-time effect) suggests, in part, why the on/off model has been so successful despite its limitations—a principal effect of interaction among traffic sources in an internetworked environment lies in the generation of long packet trains wherein the correlation structure inherent in heavy-tailed file size distributions is sufficiently preserved.

The rest of the chapter is organized as follows. In the next two sections, we discuss related work, the network model, and the simulation setup. This is followed by the main section, which explores the effect of file size distribution on traffic self-similarity, including the role of the protocol stack, heavy-tailed versus non-heavy-tailed interarrival time distribution, resource variations, and traffic mixing. We conclude with a discussion of the effect of traffic self-similarity from a performance evaluation perspective, showing its quantitative and qualitative effects with respect to performance measures when both the degree of self-similarity and network resources are varied.

14.2 RELATED WORK

Since the seminal study of Leland et al. [14], which set the groundwork for considering self-similar network traffic as an important modeling and performance evaluation problem, a string of work has appeared dealing with various aspects of traffic self-similarity [1, 2, 7, 11, 12, 15, 16, 22, 28].

In measurement based work [7, 11, 12, 14, 22, 28], traffic traces from physical network measurements are employed to identify the presence of scale-invariant burstiness, and models are constructed capable of generating synthetic traffic with matching characteristics. These works show that long-range dependence is an ubiquitous phenomenon encompassing both local-area and wide-area network traffic.

In the performance evaluation category are works that have evaluated the effect of self-similar traffic on idealized or simpified networks [1, 2, 15, 16]. They show that long-range dependent traffic is likely to degrade performance, and a principal result is the observation that queue length distribution under self-similar traffic decays much more slowly than with short-range-dependent sources (e.g., Poisson). We refer the reader to Chapter 1 for a comprehensive survey of related works.

Our work is an extension of the line of research in the first category, where we investigate causal mechanisms that may be at play in real networks responsible for generating the self-similarity phenomena observed in diverse networking contexts.

[1] H-estimates and performance results when an open-loop flow control is active can be found in Park et al. [17].

The relationship between file sizes and self-similar traffic was explored in Park et al. [18], and is also indicated by the work described in Crovella and Bestavros [7], which showed that self-similarity in World Wide Web traffic might arise due to the heavy-tailed distribution of file sizes present on the Web.

An important question is whether file size distributions in practice are in fact typically heavy-tailed, and whether file size access patterns can be modeled as randomly sampling from such distributions. Previous measurement-based studies of file systems have recognized that file size distributions possess long tails, but they have not explicitly examined the tails for power-law behavior [4, 17, 23–25]. Crovella and Bestavros [7] showed that the size distribution of files found in the World Wide Web appears to be heavy-tailed with α approximately equal to 1, which stands in general agreement with measurements reported by Arlitt and Williamson [3]. Bodnarchuk and Bunt [6] show that the sizes of reads and writes to an NFS server appear to show power-law behavior. Paxson and Floyd [22] found that the upper tail of the distribution of data bytes in FTP bursts was well fit to a Pareto distribution with $0.9 \leq \alpha \leq 1.1$. A general study of UNIX file systems has found distributions that appear to be approximately power law [13].

14.3 NETWORK MODEL AND SIMULATION SETUP

14.3.1 Network Model

The network is given by a directed graph consisting of n nodes and m links. Each output link has a buffer, link bandwidth, and latency associated with it. A node v_i $(i = 1, 2, \ldots, n)$ is a *server node* if it has a probability density function $p_i(X)$, where $X \geq 0$ is a random variable denoting file size. We will call $p_i(X)$ the *file size distribution* of server v_i. v_i is a *client node* (it may, at the same time, also be a server) if it has two probability density functions $h_i(X)$, $d_i(Y)$, $X \in \{1, \ldots, n\}$, $Y \in \mathbf{R}_+$, where h_i is used to select a server, and d_i is the *interarrival time* (or *idle time distribution*), which is used in determining the time of next request. In the context of reliable communication, if T_k is the time at which the kth request by client v_i was

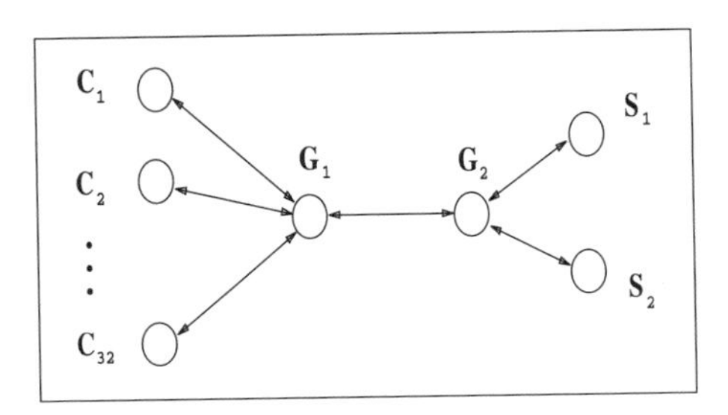

Fig. 14.1 Network configuration.

reliably serviced, the next request made by client v_i is sheduled at time $T_k + Y$, where Y has distribution d_i. Requests from individual clients are directed to servers randomly (independently and uniformly) over the set of servers. In unreliable communication, this causal requirement is waived. A 2-server, 32-client network configuration with a bottleneck link between gateways G_1 and G_2 is shown in Fig. 14.1. This network configuration is used for most of the experiments reported below. We will refer to the total traffic arriving at G_2 from servers as *upstream* traffic and the traffic from G_2 to G_1 as *downstream* traffic.

A file is completely determined by its size X and is split into $\lceil X/M \rceil$ packets, where M is the maximum segment size (1 kB for the results shown in this chapter). The segments are routed through a packet-switched internetwork with packets being dropped at bottleneck nodes in case of buffer overflow. The dynamical model is given by all clients independently placing file transfer requests to servers, where each request is completely detemined by the file size.

14.3.2 Simulation Setup

We have used the LBNL Network Simulator (`ns`) as our simulation environment [8]. `Ns` is an event-driven simulator derived from S. Keshav's REAL network simulator supporting several flavors of TCP (in particular, the TCP Reno's congestion control features—Slow Start, Congestion Avoidance, Fast Retransmit/Recovery) and router scheduling algorithms.

We have modified the distributed version of `ns` to model our interactive client/server environment. This entailed, among other things, implementing our client/server nodes as separate application layer agents. A UDP-based unreliable transport protocol was added to the existing protocol suite, and an aggressive opportunistic UDP agent was built to service file requests when using unreliable communication. We also added a TCP Vegas module to complement the existing TCP Reno and Tahoe modules.

Our simulation results were obtained from several hundred runs of `ns`. Each run executed for 10,000 simulated seconds, logging traffic at 10 millisecond granularity. The result in each case is a time series of one million data points; using such extremely long series increases the reliability of statistical measurements of self-similarity. Although most of the runs reported here were done with a 2-server/32-client bottleneck configuration (Fig. 14.1), other configurations were tested including performance runs with the number of clients varying from 1 to 132. The bottleneck link was varied from 1.5 Mb/s up to OC-3 levels, and buffer sizes were varied in the range of 1–128 kB. Non-bottleneck links were set at 10 Mb/s and the latency of each link was set to 15 ms. The maximum segment size was fixed at 1 kB for the runs reported here. For any reasonable assignment to bandwidth, buffer size, mean file request size, and other system parameters, it was found that by adjusting either the number of clients or the mean of the idle time distribution d_i appropriately, any intended level of network contention could be achieved.

14.4 FILE SIZE DISTRIBUTION AND TRAFFIC SELF-SIMILARITY

14.4.1 Heavy-Tailed Distributions

An important characteristic of our proposed mechanism for traffic self-similarity is that the sizes of files being transferred are drawn from a heavy-tailed distribution. A distribution is *heavy tailed* if

$$P[X > x] \sim x^{-\alpha} \quad \text{as } x \to \infty,$$

where $0 < \alpha < 2$. That is, the asymptotic shape of the distribution follows a power law. One of the simplest heavy-tailed distributions is the *Pareto* distribution. The Pareto distribution is power law over its entire range; its probability density function is given by

$$p(x) = \alpha k^{\alpha} x^{-\alpha-1},$$

where $\alpha, k > 0$, and $x \geq k$. Its distribution function has the form

$$F(x) = P[X \leq x] = 1 - (k/x)^{\alpha}.$$

The parameter k represents the smallest possible value of the random variable.

Heavy-tailed distributions have a number of properties that are qualitatively different from distributions more commonly encountered such as the exponential or normal distribution. If $\alpha \leq 2$, the distribution has infinite variance; if $\alpha \leq 1$ then the distribution has also infinite mean. Thus, as α decreases, a large portion of the probability mass is present in the tail of the distribution. In practical terms, a random variable that follows a heavy-tailed distribution can give rise to extremely large file size requests with nonnegligible probability.

14.4.2 Effect of File Size Distribution

First, we demonstrate our central point: the interactive transfer of files whose size distribution is heavy-tailed generates self-similar traffic even when realistic network dynamics, including network resource limitations and the interaction of traffic streams, are taken into account.

Figure 14.2 shows graphically that our setup is able to induce self-similar link traffic, the degree of scale-invariant burstiness being determined by the α parameter of the Pareto distribution. The plots show the time series of network traffic measured at the output port of the bottleneck link from the gateway G_2 to G_1 in Fig. 14.1. This *downstream traffic* is measured in bytes per unit time, where the aggregation level or time unit varies over five orders of magnitude from 10 ms, 100 ms, 1 s, 10 s, to 100 s. Only the top three aggregation levels are shown in Fig. 14.2; at the lower aggregation levels traffic patterns for differing α values appear similar to each other. For α close to 2, we observe a smoothing effect as the aggregation level is increased, indicating a

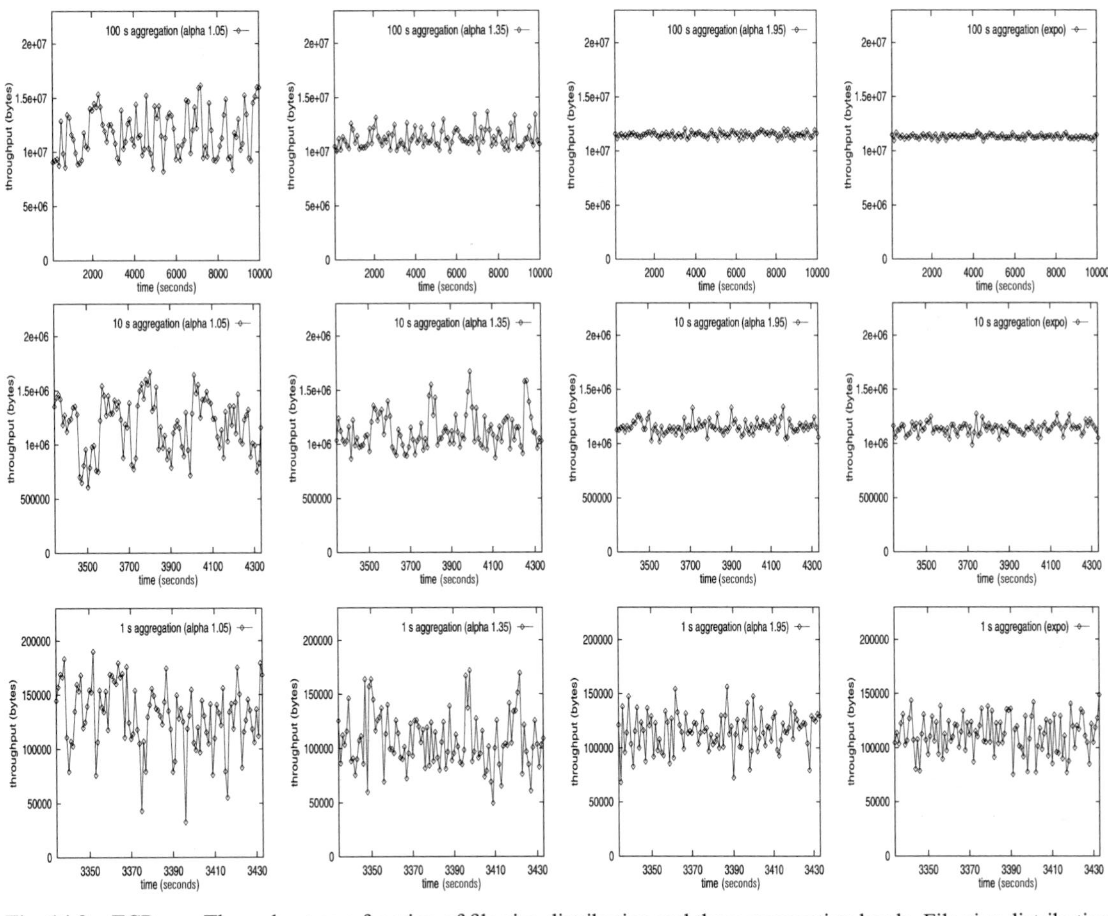

Fig. 14.2 TCP run. Throughput as a function of file size distribution and three aggregation levels. File size distributions constitute Pareto with $\alpha = 1.05$, 1.35, 1.95, and exponential.

weak dependency structure in the underlying time series. As α approaches 1, however, burstiness is preserved even at large time scales, indicating that the 10 ms time series possesses long-range dependence. The last column depicts time series obtained by employing an exponential file size distribution at the application layer with the mean normalized so as to equal that of the Pareto distributions. We observe that the aggregated time series between exponential and Pareto with $\alpha = 1.95$ are qualitatively indistinguishable.

A quantitative measure of self-similarity is obtained by using the *Hurst parameter* H, which expresses the speed of decay of a time series' autocorrelation function. A time series with long-range dependence has an autocorrelation function of the form

$$r(k) \sim k^{-\beta} \quad \text{as } k \to \infty,$$

where $0 < \beta < 1$. The Hurst parameter is related to β via

$$H = 1 - \frac{\beta}{2}.$$

Hence, for long-range dependent time series, $\frac{1}{2} < H < 1$. As $H \to 1$, the degree of long-range dependence increases. A test for long-range dependence in a time series can be reduced to the question of determining whether H is signficantly different from $\frac{1}{2}$.

In this chapter, we use two methods for testing self-similarity.[2] These methods are described more fully in Beran [5] and Taqqu et al. [23], and are the same methods used in Leland et al. [12]. The first method, the *variance–time plot*, is based on the slowly decaying variance of a self-similar time series. The second method, the R/S plot, uses the fact that for a self-similar data set, the *rescaled range* or R/S statistic grows according to a power law with exponent H as a function of the number of points included. Thus the plot of R/S against this number on a log–log scale has a slope that is an estimate of H. Figure 14.3 shows H-estimates based on variance–time and R/S methods for three different network configurations. Each plot shows H as a function of the Pareto distribution parameter for $\alpha = 1.05$, 1.15, 1.25, 1.35, 1.65, and 1.95.

Figure 14.3(a) shows the results for the baseline TCP Reno case in which network bandwidth and buffer capacity are both limited (1.5 Mb/s and 6 kB), resulting in an 4% packet drop rate for the most bursty case ($\alpha = 1.05$). The plot shows that the Hurst parameter estimates vary with file size distribution in a roughly linear manner. The $H = (3 - \alpha)/2$ line shows the values of H that would be predicted by the on/off model in an idealized case corresponding to a fractional Gaussian noise process. Although their overall trends are similar (nearly coinciding at $\alpha = 1.65$), the slope of the simulated system with resource limitations and reliable transport layer running TCP Reno's congestion control is consistently less than -1, with an offset below the

[2] A third method based on the periodgram was also used. However, this method is believed to be sensitive to low-frequency components in the series, which led in our case to a wide spread in the estimates; it is omittted here.

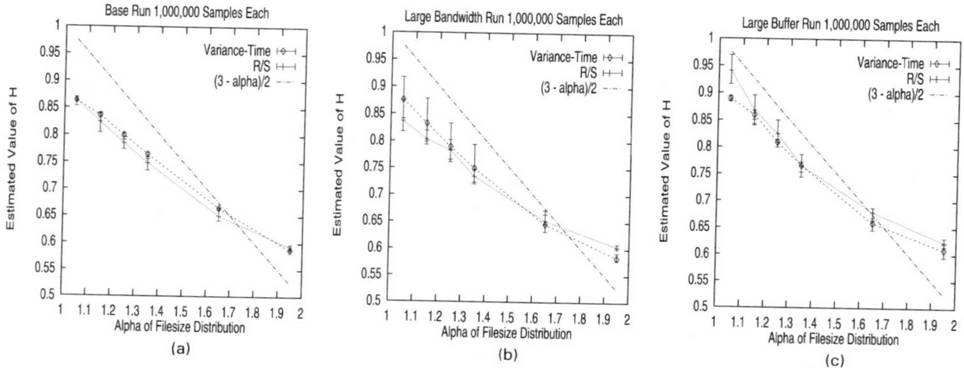

Fig. 14.3 Hurst parameter estimates (TCP run): R/S and variance–time for $\alpha = 1.05, 1.35, 1.65$ and 1.95. (a) Base run, (b) large bandwidth/large buffer, and (c) large buffer.

idealized line for α close to 1, and above the line for α close to 2. Figure 14.3(b) shows similar results for the case in which there is no significant limitation in bandwidth (155 Mb/s) leading to zero packet loss. There is noticeably more spread among the estimates, which we believe to be the result of more variability in the traffic patterns since traffic is less constrained by bandwidth limitations. Figure 14.3(c) shows the results when bandwidth is limited, as in the baseline case, but buffer sizes at the switch are increased (64 kB). Again, a roughly linear relationship between the heavy-tailedness of file size distribution (α) and self-similarity of link traffic (H) is observed.

To verify that this relationship is not due to specific characteristics of the TCP Reno protocol, we repeated our baseline simulations using TCP Tahoe and TCP Vegas. The results, shown in Figure 14.4, were essentially the same as in the TCP Reno baseline case, which indicates that specific differences in implementation of

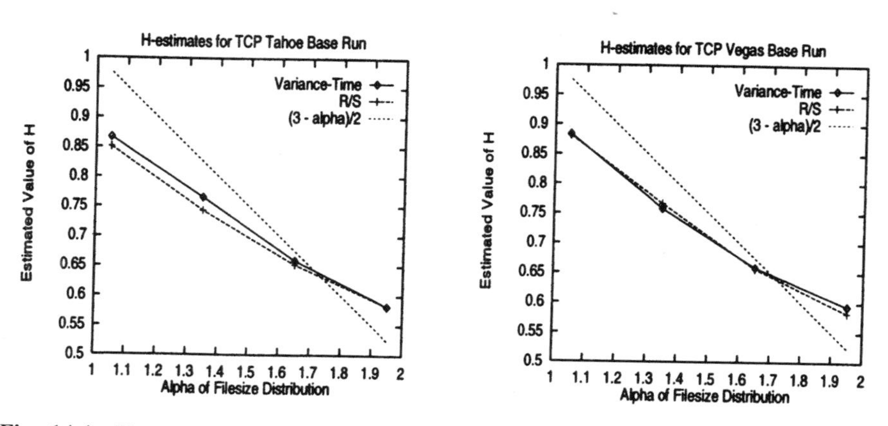

Fig. 14.4 Hurst parameter estimates for (a) TCP Tahoe and (b) TCP Vegas runs with $\alpha = 1.05, 1.35, 1.65, 1.95$.

TCP's flow control between Reno, Tahoe, and Vegas do not significantly affect the resulting traffic self-similarity.

Figure 14.5 shows the relative file size distribution of client/server interactions over the 10,000 second simulation time interval, organized into file size buckets (or bins). Each file transfer request is *weighted* by its size in bytes before normalizing to yield the relative frequency. Figure 14.5(a) shows that the Pareto distribution with $\alpha = 1.05$ generates file size requests that are dominated by file sizes above 64 kB. On the other hand, the file sizes for Pareto with $\alpha = 1.95$ (Fig. 14.5(b)) and the exponential distribution (Fig. 14.5(c)) are concentrated on file sizes below 64 kB, and in spite of fine differences, their aggregated behavior (cf. Figure 14.2) is similar with respect to self-similarity.

We note that for the exponential distribution and the Pareto distribution with $\alpha = 1.95$, the shape of the relative frequency graph for the weighted case is analogous to the *nonweighted* (i.e., one that purely reflects the frequency of file size requests) case. However, in the case of Pareto with $\alpha = 1.05$, the shapes are "reversed" in the sense that the total number of requests are concentrated on small file sizes even though the few large file transfers end up dominating the 10,000 second simulation run. This is shown in Figure 14.6.

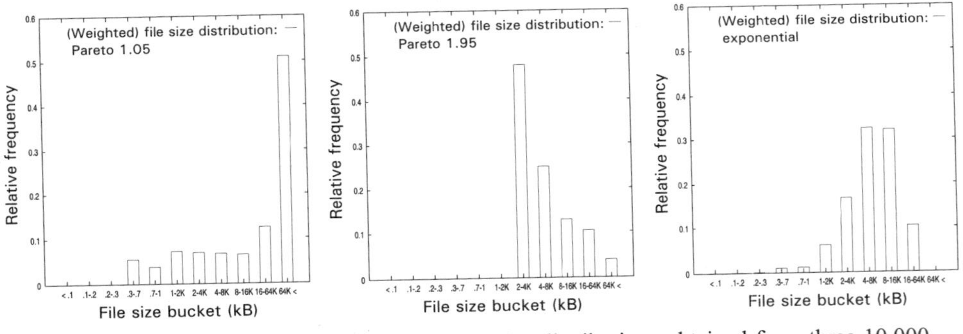

Fig. 14.5 Relative frequency of *weighted* file size distributions obtained from three 10,000 second TCP runs—Pareto (a) with $\alpha = 1.05$ and (b) with $\alpha = 1.95$; (c) exponential distribution.

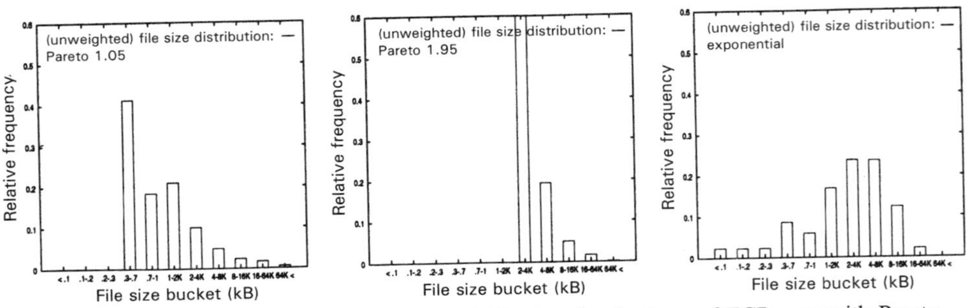

Fig. 14.6 Relative frequency of unweighted file size distributions of TCP runs with Pareto (a) with $\alpha = 1.05$ and (b) with $\alpha = 1.95$; (c) exponential distribution.

14.4.3 Effect of Idle Time Distribution

All the runs thus far were obtained with an exponential idle time distribution with mean 600 ms. Figure 14.7(a) and (b) show the H-estimates of the baseline configuration when the idle time distribution is exponential with mean 0.6 s and Pareto with $\alpha = 1.05$ and mean 1.197 s. The file size distribution remained Pareto. As the H-estimates show, the effect of a Pareto-modeled heavy-tailed idle time distribution is to boost long-range dependence when α is close to 2, decreasing in effect as α approaches 1.

This phenomenon may be explained as follows. For file size distributions with α close to 2, the correlation structure introduced by heavy-tailed idle time is significant relative to the contribution of the file size distribution, thus increasing the degree of self-similarity as reflected by H. As α approaches 1, however, the tail mass of the file size distribution becomes the dominating term, and the contribution of idle time with respect to increasing dependency becomes insignificant in comparison.

Figure 14.7(c) shows the Hurst parameter estimates when the file size distribution was exponential with mean 4.1 kB, but the idle time distribution was Pareto with α ranging between 1.05 and 1.95 and mean 1.197 s at $\alpha = 1.05$. As the idle time distribution is made more heavy-tailed ($\alpha \to 1$), a positive trend in the H-estimates is discernible. However, the overall level of H-values is significantly reduced from the case when the file size distribution was Pareto, indicating that the file size distribution is the dominating factor in determining the self-similar characteristics of network traffic.

14.4.4 Effect of Traffic Mixing

Figure 14.8 shows the effect of making one of the file size distributions heavy-tailed ($\alpha = 1.05$) and the other one exponential in the 2-server system. Downstream throughput is plotted against time where the aggregation level is 100 seconds. Figure 14.8(a) shows the case when both servers are Pareto with $\alpha = 1.05$. Figure 14.8(c)

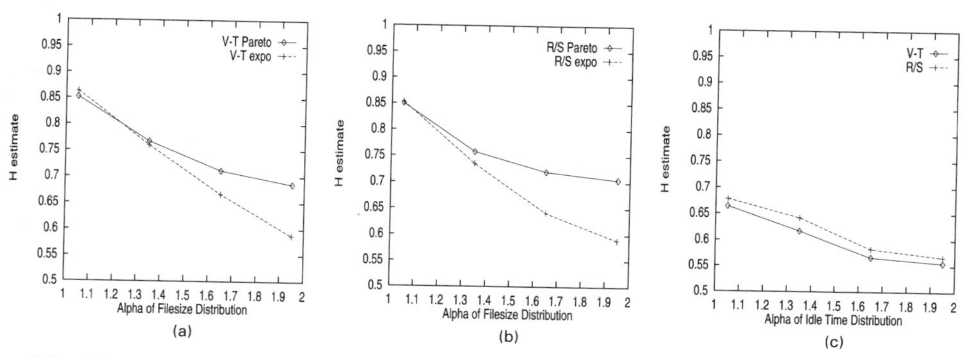

Fig. 14.7 TCP run: exponential idle time versus Pareto idle time with Pareto file size distributions—(a) variance–time, (b) R/S, (c) Pareto idle times with exponential file size distribution (right).

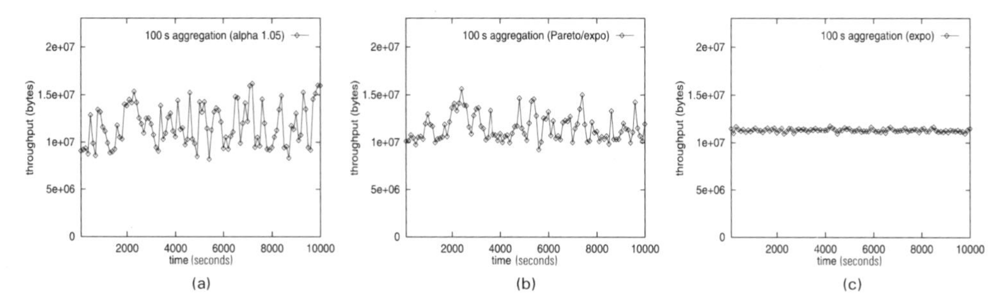

Fig. 14.8 Traffic mixing effect for two file size distributions: Pareto $\alpha = 1.05$ and exponential at 100 second aggregation level. (a) Both servers are Pareto, (b) one server is Pareto, the other one is exponential; and (c) both servers are exponential.

shows the case when both servers have exponential file size distributions. Figure 14.8(b) is the combined case, where one server has a Pareto distribution with $\alpha = 1.05$ and the other server has an exponential distribution. Figure 14.8 shows that the mixed case is less "bursty" than the pure Pareto case but more bursty than the pure exponential case. Performance indicators such as packet drop rate and retransmission rate (not shown here) exhibit a smooth linear degradation when transiting from one extreme to the other. That is, the presence of less bursty cross-traffic does not drastically smooth out the more bursty one, nor does the latter swallow up the smooth traffic entirely. Traffic mixing was applied to all combination pairs for $\alpha = 1.05$, 1.35, 1.65, and 1.95, keeping one server fixed at $\alpha = 1.05$. The H-values for the three cases shown are 0.86, 0.81, and 0.54, respectively.

14.4.5 Effect of Network Topology

Figure 14.9 shows a variation in network topology from the base configuration (Fig. 14.1) in which the 32 clients are organized in a caterpillar graph with 4 articulation points (gateways G_3, G_4, G_5, G_6), each containing 8 clients. The traffic volume intensifies as we progress from gateway G_6 to G_2 due to the increased multiplexing effect. Link traffic was measured at the bottleneck link between G_3 and G_2, which was set at 1.544 Mb/s. All other links were set at 10 Mb/s. The Hurst parameter estimates for various values of α (not shown here) indicate that for both V–T and

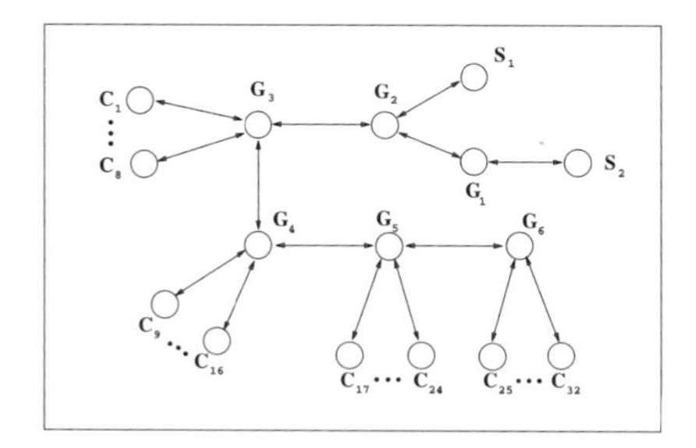

Fig. 14.9 Variation in network topology.

R/S, the degree of self-similarity measured across both topologies is almost the same.

14.4.6 Effect of the Protocol Stack

In this section, we explore the role of the protocol stack with respect to its effect on traffic self-similarity. We concentrate on the functionality of the transport layer and its modulating influence on the characteristics of downstream traffic via its two end-to-end mechanism: reliable transport and congestion control.

14.4.6.1 Unreliable Communication and Erosion of Long-Range Dependence Figure 14.10 shows the Hurst parameter estimates for a 32-client/2-server system with exponential idle time distribution and Pareto file size distributions for $\alpha = 1.05$, 1.35, 1.65, and 1.95. In these simulations, communication is unreliable; they use a UDP-based transport protocol, which is driven by a greedy application whose output rate, upon receiving a client request, was essentially only bounded by the local physical link bandwidth. (The flow-controlled case is described in Park et al. [20].) The H-estimates show that as source burstiness is increased, the estimated Hurst parameter of the downstream traffic decreases relative to its value in the upstream traffic.

Another interesting point is the already low Hurst estimate of the upstream traffic for Pareto $\alpha = 1.05$. We believe this is due to a *stretching-in-space* effect: given an exponential idle time distribution, the extremely greedy nature of the UDP-based application encourages traffic to be maximally stretched out in space, and stretching-in-time is achieved only for very large file size requests. The concentration of its mass on a shorter time interval decreases the dependency structure at larger time scales, making the traffic less self-similar.

14.4.6.2 Stretching-in-Time In contrast to the unreliable non-flow-controlled case, reliable communication and flow control, together, act to preserve the long-range dependence of heavy-tailed file size distributions, facilitating its transfer and

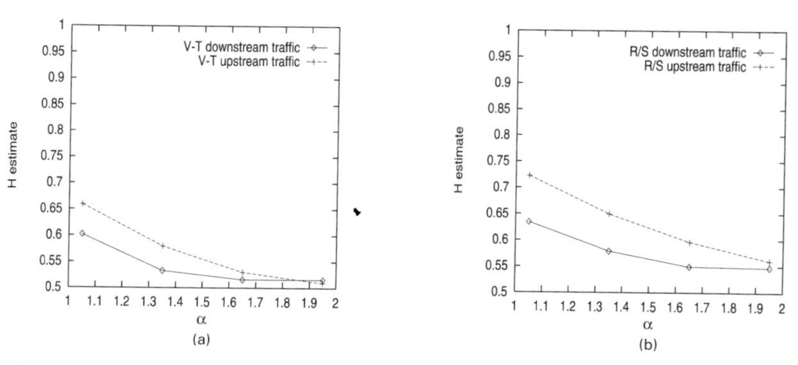

Fig. 14.10 UDP run: erosion of long-range dependence through excessive buffer overflow. (a) Variance–time and (b) R/S.

ultimate realization as self-similar link traffic. Efficiency dictates that file transmissions, including retransmission of lost packets, complete in a short amount of time. Subject to the limitations of congestion control in achieving optimally efficient transfer [21], this has the effect of stretching out a large file or message transfer in time into an on-average, "thin" packet train. This also suggests why the linear on/off model may have been successful in modeling the output characteristics of a complicated nonlinear system, which real networks undoubtedly are. In some sense, the effect of the unaccounted-for nonlinearity is reflected back as a stretching-in-time effect, thus conforming to the model's original suppositions.

14.4.7 Network Performance

In this section, we present a summary of performance results evaluating the effects of self-similarity. A comprehensive study of the performance implications of self-similarity including quality-of-service (QoS) issues, resource trade-offs, and performance comparisons between TCP Reno, Tahoe, and Vegas can be found in Park et al. [20].

14.4.7.1 Performance Evaluation Under Reliable Communication We evaluated network performance when both traffic self-similarity (α of Pareto file size distribution) and network resources (bottleneck bandwidth and buffer capacity) were varied. Figure 14.11(a) shows packet loss rate as a function of α for buffer sizes in the range 2–128 kB. We observe a gradual increase in the packet loss rate as α approaches 1, the flatness of the curve increasing as buffer capacity is decreased. The latter is due to an overextension of buffer capacity whereby the burstiness associated with $\alpha = 1.95$ traffic is already high enough to cause significant packet drops. The added burstiness associated with highly self-similar traffic ($\alpha = 1.05$) bears little effect. The same gradual behavior is also observed for packet retransmission and throughput.

Figure 14.11(b) shows mean queue length as a function of α for the same buffer range. In contrast to packet loss rate, queueing delay exhibits a superlinear dependence on self-similarity when buffer capacity is large. This is consistent

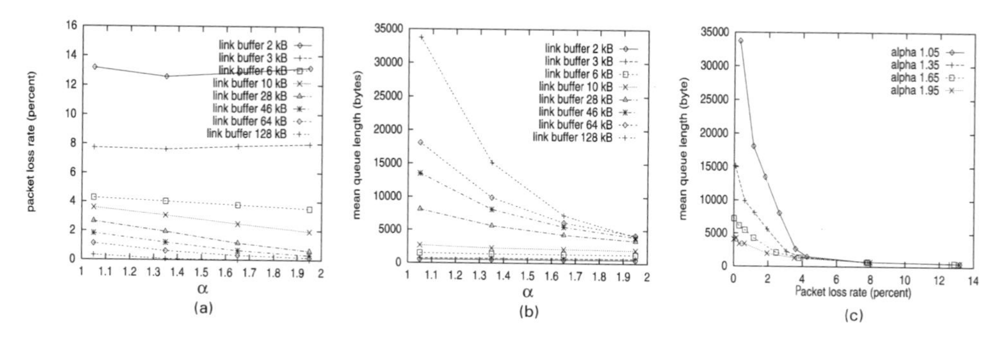

Fig. 14.11 TCP run. (a) Packet loss rate and (b) mean queue length as a function of α; (c) queueing delay-packet loss trade-off curve.

with performance evaluation works [8, 15, 16] which show that queue length distribution decays more slowly for long-range dependent sources than for short-range dependent sources.

Figure 14.11(c) shows the queueing delay-packet loss trade-off curve for four levels of α. The individual performance points were obtained by varying buffer size while keeping bandwidth fixed at the baseline value. The performance curves show that under highly self-similar traffic conditions, the negative effects of self-similarity are significantly amplified in the packet loss rate regime below 4%. A similar trade-off relation exists for queueing delay and throughput. The effect of varying bandwidth to obtain the trade-off graphs and evaluation of the marginal benefit of network resources is shown in Figure 14.12. We observe that bandwidth affects a smooth, well-behaved performance curve with respect to self-similarity. It is, in part, due to this behavior that a "small buffer capacity/large bandwidth" resource provisioning policy is advocated.

14.4.7.2 Performance Evaluation Under Unreliable Communication Performance evaluations under unreliable, non-flow-controlled transport yield performance results that are significantly worse than their reliable, flow-controlled counterparts. In particular, the dependence of throughput-related measures such as effective throughput, packet loss, and packet retransmission is no longer gradual—their shapes exhibit a superlinear dependence similar to the mean queue length relation in the reliable communication case. The superlinear dependence of queueing delay on the degree of self-similarity is further amplified, and so are trade-off relations between queueing delay and throughput. This is shown in Figure 14.13.

14.5 CONCLUSION

In this chapter, we have shown that self-similarity in network traffic can arise due to a particularly simple cause: the reliable transfer of files drawn from heavy-tailed

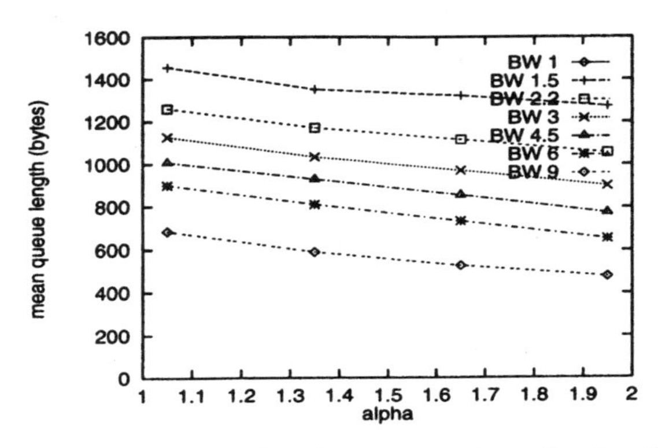

Fig. 14.12 TCP run. Mean queue length as a function of α for different bottleneck bandwidths.

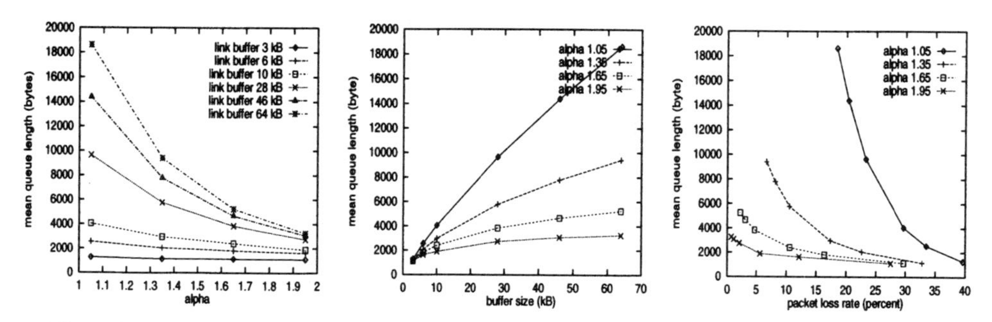

Fig. 14.13 UDP run. Mean queue length (left) as a function of α, mean queue length (middle) as a function of buffer size, and mean queue length vs. packet loss rate trade-off (right).

distributions. Such a high-level explanation of the self-similarity phenomenon in network traffic is appealing because there is evidence that file systems indeed possess heavy-tailed file size distributions [3, 7, 13, 22]. It also relates a networking problem—traffic characterization and performance—to a system-wide cause that has traditionally been considered outside the networking domain. The growth and prevalence of multimedia traffic only aggravates the situation by facilitating the structural conditions for inducing self-similar network traffic via long transfers, and our work supports recent efforts directed at managing network resources in a more integrated way ("middleware" research) in which issues such as caching and server selection may turn out to be relevant in formulating effective solutions for congestion control. We refer the reader to Chapter 18 for a feedback traffic control approach to managing self-similarity.

We have shown that the relationship between file size distribution and traffic self-similarity is not significantly affected by changes in network resources, topology, traffic mixing, or the distribution of interarrival times. We have also shown that reliability and flow control mechanisms in the transport layer of the protocol stack give rise to a traffic-shaping effect that preserves self-similarity in network traffic. This helps explain why the on/off model [28], in spite of ignoring traffic interactions through resource limitations and feedback control, has been successful in modeling observed traffic characteristics. The coupling between traffic sources sharing and contending for common network resources leads to a stretching-in-time effect, which reflects back to the on/off model by conforming, at a qualitative level, to its simplifying suppositions.

Finally, we have shown that network performance, as measured by packet loss and retransmission rate, declines smoothly as self-similarity is increased under reliable, flow-controlled packet transport. The only performance indicator exhibiting a more sensitive dependence on self-similarity was mean queue length, and this concurs with the observation that queue length distribution under self-similar traffic decays more slowly than with Poisson sources. In contrast, we showed that performance declines drastically with increasing self-similarity when a UDP-based unreliable transport mechanism was employed. This gives a sense of the moderating effect of TCP on network performance in the presence of highly bursty traffic. Lastly, this chapter has focused on the large time scale or long-range structure of nerwork traffic and its performance effects. In recent work [9], multiplicative scaling has been discovered with respect to short time scale structure, which is conjectured to stem from TCP's feedback congestion control.

REFERENCES

1. A. Adas and A. Mukherjee. On resource management and QoS guarantees for long range dependent traffic. In *Proc. IEEE INFOCOM '95*, pp. 779–787, 1995.
2. R. Addie, M. Zukerman, and T. Neame. Fractal traffic: measurements, modelling and performance evaluation. In *Proc. IEEE INFOCOM '95*, pp. 977–984, 1995.
3. M. F. Arlitt and C. L. Williamson. Web server workload characterization: the search for invariants. In *Proc. ACM SIGMETRICS '96*, pp. 126–137, May 1996.
4. M. G. Baker, J. H. Hartman, M. D. Kupfer, K. W. Shirriff, and J. K. Ousterhout. Measurements of a distributed file system. In *Proceedings of the Thirteenth ACM Symposium on Operating System Principles*, Pacific Grove, CA, October 1991, pp. 198–212.
5. Jan Beran. *Statistics for Long-Memory Processes*. Monographs on Statistics and Applied Probability. Chapman and Hall, New York, 1994.
6. R. R. Bodnarchuk and R. B. Bunt. A synthetic workload model for a distributed system file server. In *Proceedings of the 1991 SIGMETRICS Conference on Measurement and Modeling of Computer Systems*, pp. 50–59, 1991.
7. M. Crovella and A. Bestavros. Self-similarity in World Wide Web traffic: evidence and possible causes. In *Proc. ACM SIGMETRICS '96*, pp. 151–160, 1996.
8. N. G. Duffield and N. O'Connell. Large deviations and overflow probabilities for the general single server queue, with applications. *Mathematical Proc. of the Cambridge Phil. Soc.* **118**, pp. 363–374, 1995.
9. A. Feldman, A. C. Gilbert, P. Huang, and W. Wilinger. Dynamics of IP traffic: A study of the role of variability and the impact of control. In *Proc. ACM SIGCOMM '99*, pp. 301–313, 1999.
10. S. Floyd. Simulator tests. Available in `ftp://ftp.ee.lbl.gov/papers/simtests.ps.Z.` `ns` is available at `http://www-nrg.ee.lbl.gov/nrg.`, July 1995.
11. M. Garreet and W. Willinger. Analysis, modeling and generation of self-similar VBR video traffic. In *Proc. ACM SIGCOMM '94*, pp. 269–280, 1994.
12. C. Huang, M. Devetsikiotis, I. Lambadaris, and A. Kaye. Modeling and simulation of self-similar variable bit rate compressed video: a unified approach. In *Proc. ACM SIGCOMM '95*, pp. 114–125, 1995.
13. G. Irlam. Unix file size survey—1993. Available at `http://www.base.com/gordoni/ufs93.html`, September 1994.
14. W. E. Leland, M. S. Taqqu, W. Willinger, and D. V. Wilson. On the self-similar nature of Ethernet traffic (extended version). *IEEE/ACM Trans. Networking*, **2**:1–15, 1994.

15. N. Likhanov and B. Tsybakov. Analysis of an ATM buffer with self-similar ("fractal") input traffic. In *Proc. IEEE INFOCOM '95*, pp. 985–992, 1995.
16. I. Norros. A storage model with self-similar input. *Queueing Syst.*, **16**:387–396, 1994.
17. J. K. Ousterhout, H. Da Costa, D. Harrison, J. A. Kunze, M. Kupfer, and J. G. Thompson. A trace-driven analysis of the UNIX 4.2 BSD file system. In *Proceedings of the Tenth ACM Symposium on Operating System Principles*, Orcas Island, WA, pp. 15–24, December 1985.
18. K. Park, G. Kim, and M. Crovella. On the relationship between file sizes, transport protocols, and self-similar network traffic. In *Proc. IEEE International Conference on Network Protocols*, pp. 171–180, 1996.
19. K. Park, G. Kim, and M. Crovella. On the relationship between file sizes, transport protocols, and self-similar network traffic. Technical Report 96-016, Boston University, Computer Science Department, 1996.
20. K. Park, G. Kim, and M. Crovella. On the effect of traffic self-similarity on network performance. Technical Report CSD-TR 97-024, Purdue University, Department of Computer Sciences, 1997.
21. K. Park. Warp control: a dynamically stable congestion protocol and its analysis. In *Proc. ACM SIGCOMM '93*, pp. 137–147, 1993.
22. V. Paxson and S. Floyd. Wide-area traffic: the failure of Poisson modeling. In *Proc. ACM SIGCOMM '94*, pp. 257–268, 1994.
23. K. K. Ramakrishnan, P. Biswas,and R. Karedla. Analysis of file I/O traces in commercial computing environments. In *Proceedings of the 1992 SIGMETRICS Conference on Measurements and Modeling of Computer Systems*, pp. 78–90, June 1992.
24. M. Satyanarayanan. A study of file sizes and functional lifetimes. In *Proceedings of the Eighth ACM Symposium on Operating System Principles*, pp. x–x, December 1981.
25. A. J. Smith. Analysis of long term file reference patterns for applications to file migration algorithms. *IEEE Trans. Software Eng.*, **7**(4):403–410, July 1981.
26. M. S. Taqqu, V. Teverovsky, and W. Willinger. Estimators for long-range dependence: an empirical study. Preprint, 1995.
27. B. Tsybakov and N. D. Georganas. Self-similar traffic and upper bounds to buffer overflow in an ATM queue. *Performance Evaluation*, **36**(1):57–80, 1998.
28. W. Willinger, M. Taqqu, R. Sherman, and D. Wilson. Self-similarity through high-variability: statistical analysis of Ethernet LAN traffic at the source level. In *Proc. ACM SIGCOMM '95*, pp. 100–113, 1995.

15

CHARACTERISTICS OF TCP CONNECTION ARRIVALS

ANJA FELDMANN
AT&T Labs–Research, Shannon Laboratories, Florham Park, NJ 07932

15.1 INTRODUCTION

Packets are the basic unit of the Internet. Yet, most user operations involve more than one packet and user experience depends on the performance of the network on a set of packets. Thus it is not surprising that in today's Internet protocol (IP) networks sets of packets are starting to be used as the basis for network operations. The goals of the network operators include improving traffic engineering by better load balancing and moving beyond best effort services. Operating on sets of packets is more efficient for network elements. In fact, per packet decisions may lead to undesired or even unstable solutions. At the edge of the Internet the transmission control protocol (TCP) offers the abstraction of a TCP connection, each consisting of a set of packets. Within the center of the network, or for non-TCP traffic, a related connection abstraction is provided by an IP flow, a group of related IP packets that are close in time.

Examples of network resource allocation problems that arise on a per connection level include signaling to reserve buffers and bandwidth. Signaling could be initiated by the end system to achieve explicit quality of service (QoS), using protocols such as RSVP. Alternatively, signaling could be initiated by a network element to achieve better load balancing or to provide QoS for certain types of traffic. In this approach, edge routers detect flows of related packets and implicitly establish dedicated connections through the network to carry this traffic. Finally, individual network

Self-Similar Network Traffic and Performance Evaluation, Edited by Kihong Park and Walter Willinger
ISBN 0-471-31974-0

routers could allocate resources for a flow to improve routing and forwarding performance. For example, a router could cache next-hop routes for common destination addresses to reduce the number of routing computations and to forward related packets along a single route. The router could even establish dedicated connections through the switching fabric for long-lived flows to avoid software processing of subsequent packets.

Each of these network mechanisms performs operations and allocates resources on the time scale of connection arrivals. Therefore, the burstiness of the connection arrival process affects two separate provisioning tasks: the central processing unit (CPU) resources necessary to perform the algorithm and the network resources required to acheive a desired level of blocking. The burstier the arrival process, the more CPU resources are necessary to execute the algorithm and the more network resources are needed to maintain a given level of blocking.

In this chapter we demonstrate that the TCP connection arrival process is bursty. We show that the arrival process is asymptotic self-similar. Self-similarity of the TCP connection arrival process implies that the use of standard models in evaluating the performance of resource allocation methods can yield misleading results. Therefore, we characterize TCP connection interarrival times[1] using heavy-tailed distributions. We present statistical evidence that such distributions, especially the Weibull distributions, yield a better model for the interarrival times of TCP connections than exponential models. Intuitively, a heavy-tailed interarrival time means that if no connection arrived for some time it becomes more and more unlikely that one will arrive soon. This holds even if the underlying arrival process is nonstationary. Finally, based on a simple resource allocation problem, we show that there are advantages to using Weibull distributions to model TCP connection interarrival times over a nonstationary Poisson process.

Our results are based on extensive analyses of multiple traces collected at Carnegie Mellon University in 1995, at AT&T Bell Laboratories in 1995 and 1996, and at AT&T Labs–Research in 1996. In addition we augment our results with the analysis of traces collected at Lawrence Berkeley Laboratories (LBL) in 1993 and 1995, at Digital Equipment Corporation (DEC) in 1995, and within AT&T WorldNet in 1997 and 1998.

The rest of the chapter is organized as follows. In Section 15.2 we present, on an intuitive level, why TCP connection arrivals are bursty and what impact this may have on resource allocation problems. A more detailed description of the TCP/IP traces on which we base our traffic characterizations is given in Section 15.3. The results in Section 15.4 indicate the self-similar nature of the arrival process. Section 15.5 outlines the methods used to analyze the interarrival time distribution and then shows that the Weibull distribution yields a good fit for connection interarrival times. Section 15.6 contrasts the use of the Weibull model to that of a nonstationary Poisson process using an example application. Finally we conclude with a brief summary.

[1] Given two consecutive timestamps of observed TCP connections from a TCP connection arrival process, the interarrival time is the time difference between the two timestamps.

15.2 BURSTINESS OF TCP CONNECTION ARRIVAL PROCESSES AND ITS IMPLICATIONS

In this section we discuss, on an intuitive level, the self-similarity of arrival processes and the reason why TCP connection arrival processes are self-similar. The fact that TCP connection arrivals are bursty leads to the question of how to characterize the distribution of time between arrivals of TCP connections, or the TCP connection interarrival time distribution. Finally, we discuss the implications of the burstiness on resource allocation problems.

15.2.1 Self-Similarity of Packet Arrival Processes

In the context of the Internet, the most studied arrival process is the packet arrival process. Prior to the work by Leland et al. [27], the Poisson model was the most commonly used model for network traffic. Leland and co-workers point out that on a local-area network (LAN) the packet arrival process shows self-similar behavior. An obvious pitfall of the Poisson model, compared to a self-similar model, is that its aggregation behavior differs substantially from the self-similar model; aggregated self-similar traffic stays burstier than aggregated Poisson traffic. If one assumes a Poisson arrival process of packets, the amount of buffering needed in an ATM switch should be fairly small. Yet, performance of ATM switches, in terms of loss rate improved significantly when adding larger buffers [19, 38]. Willinger et al. [42] explain the self-similar nature of Ethernet traffic observed on the packet level as the result of a superposition of many on/off sources (also referred to as packet trains [25]), where the lengths of the on and off periods are drawn from heavy-tailed distributions.

Paxson and Floyd [35] show that wide-area network (WAN) traffic at the packet level is of asymptotically self-similar nature, that is, self-similar behavior over large time scales. They propose a structural model to explain the asymptotic self-similarity in terms of the characteristics of the main applications. Their model is based on an $M/G/\infty$ model originally due to Cox [9]: session arrivals are assumed to be Poisson or, more generally, of renewal-type; session duration (in seconds) or session size (in bytes) is required to be heavy tailed. Accordingly, traffic can be seen as the superposition of user activities that arrive according to a Poisson process but carry an amount of work that is drawn from a heavy-tailed distribution. The authors find that for wide-area traffic the arrival process of sessions is consistent with Poisson processes and that their durations are heavy tailed. Park et al. [36] and Crovella and Bestavros [8] showed a possible causal relationship between heavy-tailed transfer durations and file sizes on UNIX file systems.

15.2.2 Self-Similarity of Connection Arrival Processes

To be able to study the arrival process of sessions, Paxson and Floyd equate sessions with TCP connections of applications such as TELNET and FTP connections, which, at the time of the study (1995), were among the most popular applications on the

Internet. They point out that other arrival processes including FTP DATA connection arrivals and SMTP connection arrivals are not consistent with Poisson processes.

A Poisson process arises when users initiate actions more or less independently. While this may have been the case for FTP and TELNET sessions in 1995, the Internet has changed since then. It has grown beyond expectation so that an enormous amount of information is available, and its dominant application is the World Wide Web (or the Web). The Web contributes more than half of the packets and more than two-thirds of all TCP connections to the traffic of commercial ISPs such as AT&T's WorldNet. During a Web user session, a user is likely to download not just one single Web object, since most Web pages consist of 3 to 4 embedded images [4, 18], but also different Web pages. Indeed, he/she is very likely to download a set of Web pages from different Web sites or to initiate multiple FTP downloads. Therefore, equating a user session that includes Web browsing with a TCP connection is questionable. Indeed, with the growth of the Internet, a person using the Internet is more likely to initiate more than one operation during a session (independent of the kind of application FTP, TELNET, or Web). This breaks the Poisson paradigm and suggests a new look at the burstiness of the TCP connection arrival process. In fact, we show that WEB connection arrivals show self-similar behavior. This change is an illustration of how the characteristics of the Internet can change over time.

While it is still true that users arrive in a more or less memoryless fashion, this does not translate into TCP connections arriving in a memoryless way. Feldmann et al. [16] show that Cox's model is still applicable but should be applied to a higher level abstraction of a user session. Feldmann et al. [16] explain that the connection arrival process is self-similar using Cox's model by showing that the number of TCP connections initiated by a user session (in this case a modem call) is heavy-tailed.

15.2.3 Distribution of TCP Connection Interarrival Times

TCP connection interarrival times are derived from the TCP arrival process by considering the distribution of the times between consecutive arrivals of TCP connections. The fact that the TCP connection arrival process is self-similar indicates that distributions with heavy tails such as Weibull, Pareto, and lognormal distributions may yield better models for TCP connection interarrival times than exponential models. We augment previous results with statistical evidence that this is indeed the case. We show that the Weibull distribution yields a good fit for the connection interarrival times of Web connections over time periods of an hour or two. Deng [11], independent of our results, also found that the Weibull distribution gives a good fit.

Indeed, the Weibull distribution gives a good fit not just for Web connections but for all connections of a specific type (such as FTP and TELNET) that arrived within a time period of an hour, two hours, or even days. The same holds true even if we include connections of all applications, when calculating the interarrival times over a specific time period, or if we consider all connections in a given trace. While the Weibull distribution gives a good fit when the set of connections is enlarged to include more connections, it also gives a good fit if the set of connections is reduced.

Indeed, if we isolate traffic sources (e.g., machines that serve a number of users, such as computer servers) or consider only connections from a subset of the sources, we observe that the Weibull distribution still results in better fits than the other models.

A possible explanation for this somewhat surprisingly good fit is that the Weibull distribution is a rather flexible distribution that seems to capture small to large interarrival times. Applications such as the Web create interarrival times from very short to very long. Very short interarrival times are created when a Web page has multiple embedded images. Currently, each of these images will be downloaded using a separate TCP connection and typical browsers open up to four parallel TCP connections. Persistent connections [37] may reduce the number of TCP connections from four to two and may transfer multiple Web objects over the same TCP connections. Still the Web protocol is responsible for creating some very short interarrival times. Once a user starts browsing the Web, he/she usually visits a set of different Web pages. This browsing and reading of Web pages generates interarrival times that are of intermediate size. Once the browsing session is done a user is likely to take a long break, lets say for lunch, dinner, or even a vacation, thus creating long interarrival times. As such, it is no surprise that Web connection interarrival times are heavy tailed.

We are by no means claiming that the arrival process is a stationary process that can be completely described by an independent identical distribution (i.i.d.) arrival process. The traces show a substantial dependence on the time of day. Yet, depending on the application, it might be more important to understand the small time scale behavior than matching a longer time scale behavior. To this extent we present evidence that using an i.i.d. Weibull model can yield more accurate performance prediction than a nonstationary Poisson process that matches the number of arrivals over time periods as small as minutes. Our example application is signaling on a single edge network.

15.2.4 Implications of the Burstiness of TCP Connections

The burstiness of connection interarrival times has implications on the design and evaluation of algorithms such as signaling and dynamic routing as well as resource allocation for Web servers.

Signaling Signaling consists of two subproblems: connection-admission control and routing. Connection-admission control decides which connections to accept or reject while routing chooses along which (multihop) path through the network to reserve resources. The connection arrival process affects not only the CPU resources necessary to perform the algorithms, but also the network resources required to achieve a desired level of blocking.

Depending on a router's CPU resources, it may not be feasible to do an implicit setup for every connection. In this case one has to find the most beneficial subset and, in addition, a different way of grouping the packets [18]. While different groupings may reduce the burstiness, they do not eliminate it.

Connection-level simulations [14] show that an increase in burstiness of the connection arrival stream increases the level of blocking in connection-admission control. This can be even more severe for multihop paths if several links on the path are affected.

Dynamic Routing Currently, routing in the Internet is completely static. Dynamic routing (e.g., on a per connection level) is a candidate for taking advantage of shifts in the traffic matrix to improve traffic engineering. Yet, if arrivals are too bursty, this may not lead to much improvement [40].

Benchmarking of Web Servers Web servers often allocate resources, such as processes, on a per TCP connection basis. SURGE [3] is a realistic Web workload generation tool that mimics a set of real users accessing a Web server. It creates a bursty connection arrival process, which exercises the resource allocation policies of Web servers. In this way it is able to detect significant performance problems that would have gone unnoticed using other benchmarks that do not generate a bursty arrival process.

15.3 OVERVIEW OF NETWORK TRAFFIC TRACES

Our analysis of TCP connections is based on trace analysis of transmission control protocol/Internet protocol (TCP/IP) internetwork traffic. The traces were collected on three different Ethernet segments at Carnegie Mellon University (CMU), AT&T Bell Laboratories, and AT&T Labs–Research using the *tcpdump* packet capture tool [24] running the Berkeley Packet Filter [28]. The number of packets that tcpdump reported dropped by the kernel was negligible.

The first set of traffic data was collected on an Ethernet segment of the School of Computer Science at CMU that is one of a total of 18 Ethernet segments that are bridged through a backbone with an aggregate 0.5 Gbit/s throughput. The workstation we used for the trace collection (DEC Alpha 400/300 with 64 Mbytes RAM) is connected to a segment that connects approximately 120 systems, including UNIX workstations, MacIntoshs, and PCs. The second traffic data set was collected on an Ethernet segment at AT&T Bell Laboratories. The third traffic data set was collected on the same Ethernet segment shortly after the split of AT&T Bell Laboratories into AT&T Labs–Research and Lucent Bell Laboratories. The workstation we used for the second set of traces (SGI with one 134 MHz MIPS R4600 processor and 64 Mbytes RAM) was connected to AT&T Bell Laboratories' internal network while the workstation for the third set of traces (SGI with one 100 MHz MIPS R4000 processor and 64 Mbytes RAM) was connected to an Ethernet segment outside AT&T's firewall. During some trace collection periods all TCP connections (including all World Wide Web access) to the external Internet passed this Ethernet segment. We refer to the second set of traces as the internal AT&T traces and to the third set as the external AT&T traces.

Traces were restricted to TCP traffic only; all user datagram protocol (UDP) traffic was discarded. Thus roughly one-third of all traffic at CMU and on the internal AT&T network was discarded but almost none on the external AT&T network. Note that neither of these networks carried any substantial amount of MBONE traffic.

To be able to collect data over a reasonable time period only those TCP packets that are involved in the TCP connection establishment handshakes between the source and destination pair were collected, that is, those packets that either have their synchronize sequence numbers flag (SYN) or their finish flag (FIN) set in the TCP header (ignoring RST packets).

From the traces one can derive the arrival time of TCP connections, their source and destination, their application, their durations, and the number of bytes transferred (excluding TCP/IP overhead) using the *tcp-conn* tool [22]. The CMU datasets cover nearly 161 hours over 8 different days and the AT&T datasets more than 1290 hours over 52 days.

The packets with SYN, FIN flags are classified according to the application that generated them. We distinguished the traffic classes shown in Table 15.1. This classification to applications is based on the port numbers of the packets. All packets with port numbers that fit none of these applications are collected in a separate class.

Table 15.2 presents a breakdown of the observed TCP connection arrivals according to application classes. First we note that the traffic volume and traffic mix are highly dependent on the kind of network we were monitoring and when we were monitoring them. The observed change in the CMU traffic mix between December 1994 and June 1995 reflects the increasing popularity of the World Wide

TABLE 15.1 Traffic Classes

HTTP:	Packets generated by World Wide Web applications such as Netscape and Mosaic.
X:	Packets generated by the X11 window system, for example, by xterms.
RFS:	CMU uses three remote file systems: the Brunhoff remote file system, the Andrew file system (AFS), and the network file system (NSF). Of these three the Brunhoff remote file system is the only one that uses TCP connections.
SMTP:	Packets generated by the simple mail transfer protocol (smtp).
FTP:	Packets generated by the file transfer protocol (ftp). This includes ftp-control connections as well as ftp- data connections. Sometimes it is desirable to distinguish between these two types of connections. In this case we refer to them as FTP.CONTROL and FTP.DATA.
POP:	Packets generated by the post office protocol.
TELNET:	Packets generated by the remote terminal protocol.
NNTP:	Packets generated by the network news transfer protocol.
FINGER:	Packets associated with the user information query application.
CUSTOM:	Packets generated by custom, CMU, or AT&T-specific applications.
PRINTER:	Packets generated by spooling files to networked printers.
PROTO:	Packets generated by a collection of protocols, including domain name service protocol and echo protocol.

Web and the fact that CMU is phasing out RFS. The trace collected on the internal network at AT&T Bell Laboratories shows a very low overall volume but contains a very high percentage of X connections. Given that most machines connected to this Ethernet are X-terminals and that most people work on the machine to which their X-terminal is connected, this is not surprising. Comparing the three external AT&T traces one can see the effects of certain network changes. Most traffic observed in the period between 18 November and 8 December are accesses to AT&T Bell Laboratories' World Wide Web server that is connected to this Ethernet segment. While the January dataset still contains such HTTP accesses it is dominated by HTTP traffic that originates at AT&T Labs–Research but accesses other machines in the Internet. The final dataset also contains all traffic between Lucent Bell Laboratories and AT&T Labs–Research's Web server and between AT&T and Lucent Web server (using a different port number). After we accounted for this, the percentage of HTTP traffic is 65.83% and the percentage of SMTP traffic is 28.55%. This explains the huge difference in the number of collected packets.

TABLE 15.2 Breakdown of TCP Connection Packets According to Application Type for a Subset of the Traces

	CMU	CMU	Internal AT&T	External AT&T	External AT&T	External AT&T
Start	9 Dec. 94 12 a.m.	29 June 95 12 a.m.	8 Dec. 95 12 a.m.	18 Nov. 95 12 a.m.	15 Jan. 96 9 p.m.	15 Mar. 96 1:30 p.m.
End	10 Dec. 94 3 p.m.	31 June 95 12 a.m.	23 Dec. 95 12 a.m.	8 Dec. 95 12 a.m.	19 Jan. 96 10 a.m.	31 Mar. 96 2:30 p.m.
Type	in %	in %	in %	in %	in %	in %
http	10.26	25.14	2.86	93.94	84.48	7.26
X	3.44	6.95	42.05	0.02	0.08	0.00
rfs	31.17	14.86	0.00	0.00	0.00	0.00
smtp	13.92	11.14	11.78	0.12	3.25	3.15
ftp	3.71	3.73	0.28	2.62	1.75	0.09
pop	6.60	9.26	0.01	0.06	0.37	0.00
telnet	1.24	2.57	1.45	0.46	0.66	0.03
nntp	5.77	4.10	0.01	0.03	0.02	0.00
finger	3.79	2.00	0.07	0.09	0.01	0.00
custom	2.16	3.65	0.00	0.69	4.35	0.00
printer	9.62	1.45	2.23	0.00	0.00	0.00
proto	2.65	8.68	11.69	0.05	2.80	89.05
other	5.63	6.38	27.58	1.92	2.23	0.42
Number of packets with SYN/FIN flags	98,852	135,119	49,643	1,327,278	2,856,079	3,993,416

All in all the traffic mixes observed on three Ethernets (two carrying mostly local-area traffic, and one, at AT&T, carrying mostly wide-area traffic) reflect different, but not uncommon, TCP connection usage patterns.

We have applied the same analysis techniques to some of the LBL and DEC traces collected prior to 1996 that were used for the study by Paxson and Floyd [35] (available through the Internet Traffic Archive [21]). Most of the results are similar to the results we get for the CMU and AT&T datasets. In the cases where the results differ we will briefly comment on how and why they differ. The key to understanding the difference is that the use of the network has changed with more and more people using the Web and better and better wide-area networking connectivity. The main difference between the traces is that the amount of HTTP traffic in some of the traces considered here is substantially larger than in the LBL and DEC traces we considered.

In addition, we selectively analyze more recent traces collected from a commercial ISP, AT&T's WorldNet. For these traces the Web is the dominant application. HTTP transfers are responsible for more than 85% of all of the TCP connections and more than 50% of all the packets. The difference between the percentages is because Web downloads on average involve a smaller number of packets than some of the other applications.

15.4 SELF-SIMILARITY OF CONNECTION ARRIVAL PROCESS

Intuitively, an arrival process is consistent with a Poisson process if it is a superposition of many independent sources whose activity is more or less memoryless. Before the Web became the dominant application of the Internet, this was mostly true for the then dominant applications: TELNET and FTP CONTROL. Just because a user had initiated one telnet session did not necessarily imply that he/she was going to be more likely to initiate another telnet session. The Web changed this principle. Once a user starts browsing he/she is much more likely to download another set of Web pages than to stop after just one Web page. After all, finding the "right" information on the Web is not always easy. This implies that the arrival process may not be consistent with a Poisson process anymore. Indeed, the arrival process of TCP connections shows self-similar behavior.

To support this claim Fig. 15.1 shows a series of plots of the arrival process (i.e., number of connection arrivals per time unit) for HTTP connection arrivals for three different choices of time units for the external AT&T trace from 18 November to 8 December 1995. Starting with a time unit of 312.5 minutes (Fig. 15.1(a)), the subsequent plot is obtained from the previous one by decreasing the time unit by a factor of 25, increasing the time resolution by the same factor, and concentrating on a random subinterval of the previous plot. All plots look very "similar" to each other regardless of the chosen time scale. The burstiness of the connection arrival does not seem to decrease as the time resolution is decreased. Rather we see the same degree of variability for all time resolution. Similar plots can be obtained for other datasets as well as other applications or shorter time intervals.

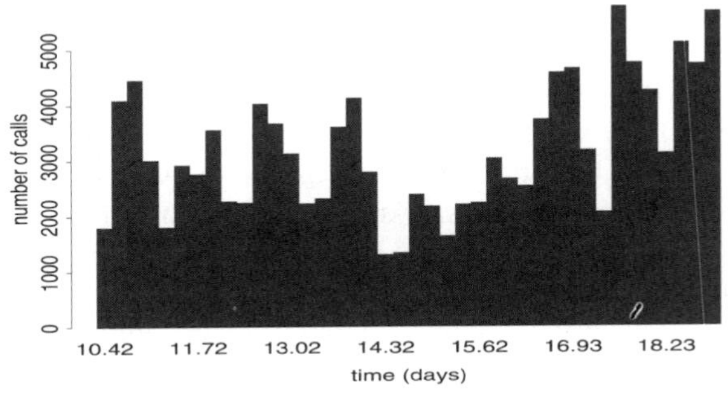

(a) bin size: 312.5 minutes

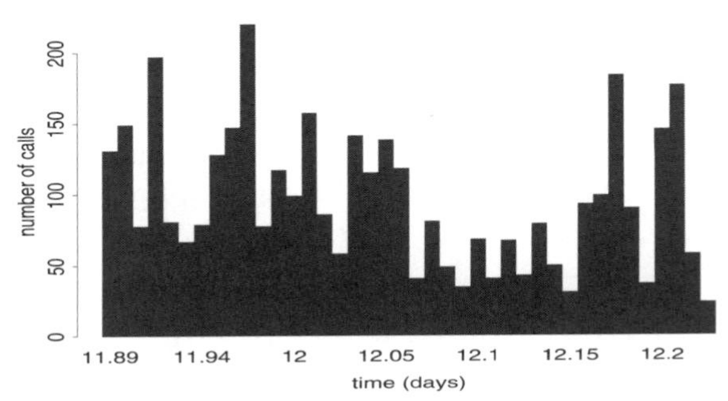

(b) bin size: 12.5 minutes

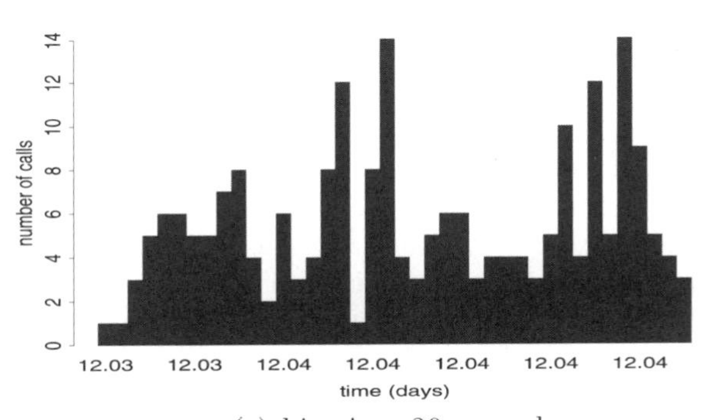

(c) bin size: 30 seconds

Fig. 15.1 Pictorial indication of self-similarity: number of connection arrivals over time for traffic class HTTP on three different time scales.

The degree of burstiness over different time scales or the extent of self-similarity can be expressed with just one single parameter, the Hurst parameter [5]. For self-similar processes its value is between 0.5 and 1 and the degree of self-similarity increases as the Hurst parameter approaches 1. More formally, a covariance-stationary process $X = (X_k : k \geq 1)$ is called *asymptotically self-similar* (with self-similarity parameter H, $0 < H < 1$), if for all large enough m,

$$X \approx m^{1-H} X^{(m)},$$

where $X^{(m)} = (X^{(m)}(k) : k \geq 1)$ is the *aggregated* process of order m, given by

$$X^{(m)}(k) = \frac{1}{m}(X_{(k-1)m+1} + \cdots + X_{km}), \quad k \geq 1.$$

The process under consideration is the number of connection arrivals per time unit.

In the past [5, 7, 20, 27, 35, 41] various graphical tools, such as "variance–time plots," "pox plots of R/S," and "periodogram plots," and statistical tools, such as "periodogram-based MLE estimate," have been used to estimate the Hurst parameter. Using these tools on the HTTP connection arrival from the busy hour of the external AT&T dataset from March, the estimates are $\hat{H} = 0.749$ for the variance–time plot; $\hat{H} = 0.764$ for the pox plot of R/S; $\hat{H} = 0.796$ for the periodogram plot (not restricted to the lower frequencies); $\hat{H} = 0.737$ with a 95% confidence interval from 0.715 to 0.758 for the MLE Whittle estimate based on the periodogram. Similar estimates have been derived for other subsets of the traces.

More recently, Abry and Veitch [1, 2] proposed a wavelet-based technique for analyzing long-range dependent data and for estimating the associated Hurst parameter. Yet, more important, their method allows the identification of scaling regions, breakpoints, and nonscaling behavior and yields an unbiased estimator for the Hurst parameter (see Feldmann et al. [16] for examples). Abry and Veitch's method utilizes the ability of wavelets to "localize" a signal in both time and scale (see Kaiser [26] for an introduction to wavelets, and Daubechies [10] for a more mathematical treatment of the subject).

Given a process X, the discrete wavelet transform of the process will result in a set of wavelet coefficients $d_{j,k}$ that capture the contribution of the process at scale j and time $2^j k$. If X is a self-similar process with Hurst parameter $H \in (\frac{1}{2}, 1)$, then Abry and Veitch [1] have shown that the expectation of the energy E_j that lies within a given bandwidth 2^{-j} around frequency $2^{-j}\lambda_0$ is given by

$$\mathbf{E}[E_j] = \mathbf{E}\left[\frac{1}{N_j}\sum_k |d_{j,k}|^2\right] = c|2^{-j}\lambda_0|^{1-2H},$$

where c is a prefactor that does not depend on j, and where N_j denotes the number of wavelet coefficients at scale j. One can study the scaling of a process by plotting $\log_2 E_j$ against scale j (where $j = 1$ is the finest scale and $j = N > 1$ is the coarsest). The scaling analysis of a signal that is exactly self-similar will yield a linear plot for all scales; for a Poisson trace (i.e., $H = 0.5$) the plot will be a horizontal line. On the other hand, for an asymptotically self-similar signal a linear relationship between $\log_2 E_j$ and scale j will be apparent only for the larger times or scales.

If we apply the wavelet technique to our datasets, we confirm our suspicion that the HTTP connection arrival process is asymptotically self-similar. Figure 15.2 shows the scaling plots for some of the busier two hour periods of the datasets. All plots show a clear nonhorizontal scaling region from about scale 4–6 to scale 12, verifying the assumption of self-similar nature of the arrival process. The corresponding Hurst parameter estimation results, based on the larger scales, are all around $\hat{H} \approx 0.7$. To estimate simple trends (e.g., linear and x^2 trends) we used Daubechies wavelets [10] with three vanishing moments.

To contrast these results with the analysis of Paxson and Floyd [35], Fig. 15.3 shows the results of the scaling analysis for two subsets of the DEC-PKT- 4 dataset. The first dataset contains all FTP CONTROL connection arrivals while the second one contains all HTTP connection arrivals. In line with the analysis results reported by Paxson and Floyd [35], the arrival process of FTP CONTROL connections gives a Hurst

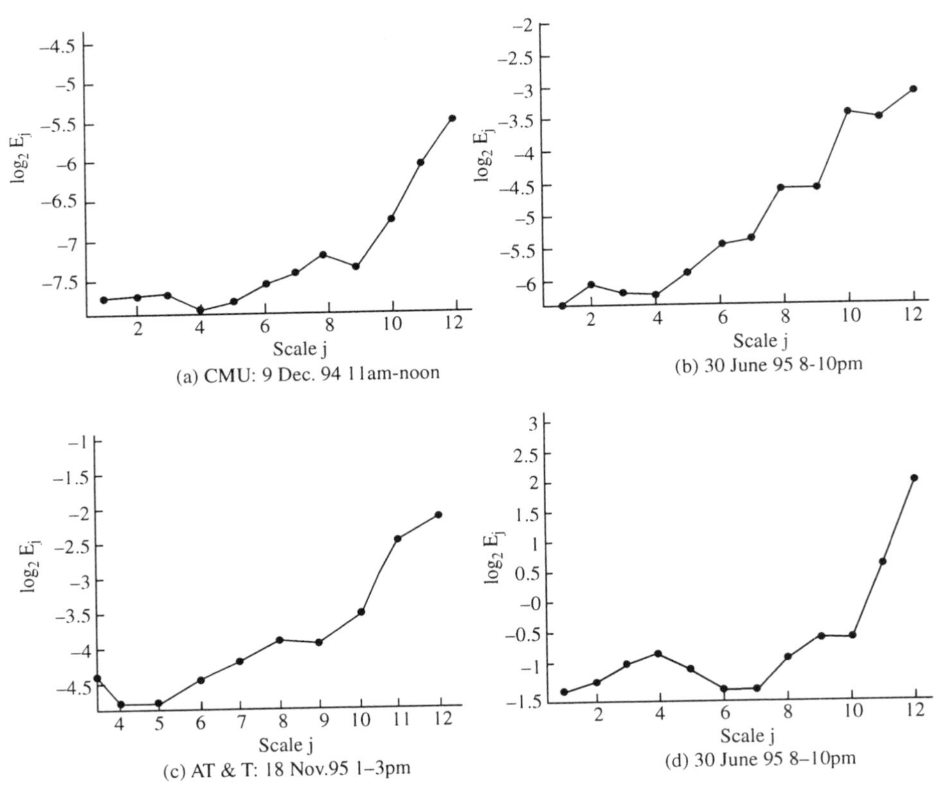

Fig. 15.2 Scaling analysis of arrival process of HTTP connection arrival for subsets of our datasets. Note the different labeling at the bottom and the top of the plots: scale j (bottom, $j = 1$ is finest scale); actual time (top, in seconds). The vertical bars at each scale represent 95% confidence intervals for $\log_2(E_j)$.

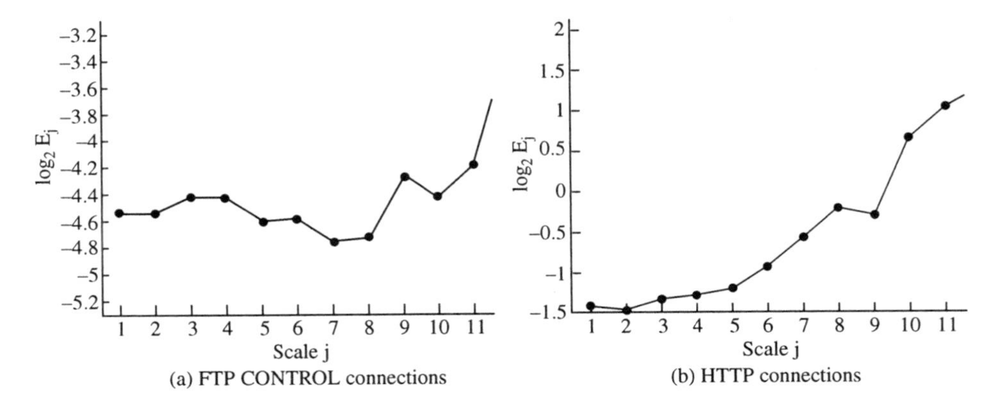

Fig. 15.3 Scaling analysis of arrival process of FTP CONTROL and HTTP connection arrival for dataset DEC-PKT-4.

parameter estimation of 0.5, which is consistent with a Poisson process. Yet, even in this dataset, the process that counts the number of HTTP connections is consistent with a self-similar process with Hurst parameter of $\hat{H} \approx 0.65$.

To underline the observations that the number of Web connection arrivals is consistent with asymptotic self-similar behavior, Fig. 15.4 shows the results of the scaling analysis for data collected on AT&T WorldNet's network. The first dataset was collected on 14 August 1998 at 10:30 p.m. on a T3 backbone link while the second dataset was collected on 23 July 1997 at 7 p.m. on a FDDI ring carrying traffic from roughly 420 modems used by WorldNet dialup customers (for more

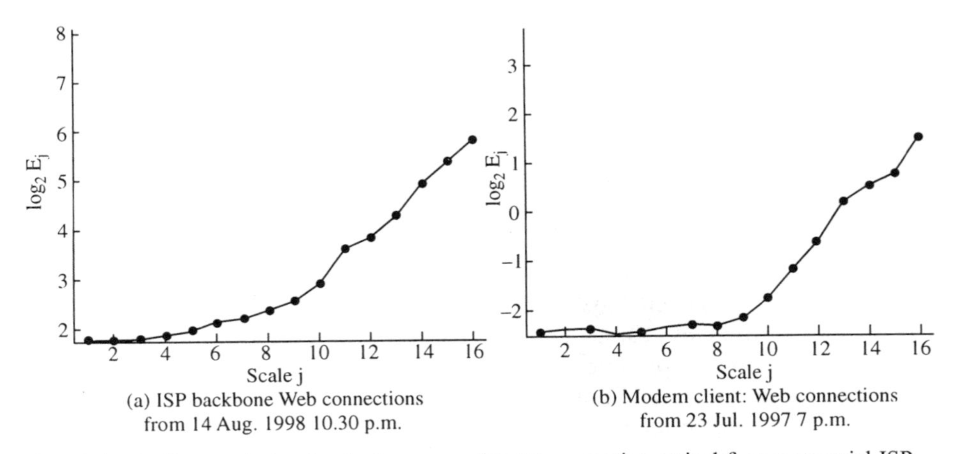

Fig. 15.4 Scaling analysis of arrival process of HTTP connection arrival for commercial ISP (AT&T WorldNet).

details see Feldmann et al. [13, 15]). In the first dataset 87.6% of the TCP connections are Web connections, of which there are a total of 1,187,866. In the dial dataset at least 85.8% of all TCP connections are Web connections, of which there are a total of 51,027 Web connections. Again both dataset as well as many others show that the arrival process is consistent with self-similarity with a Hurst parameter of $\hat{H} \approx 0.7$.

15.5 CHARACTERIZATION OF CONNECTION INTERARRIVAL TIMES

In this section we fit distributions with heavy tails to the empirical distribution of TCP connection interarrival times and explore their goodness of fit. To this end we first outline the methods used to analyze interarrival time distribution. Next, we show that the Weibull distribution in particular gives a good fit for TCP connection interarrival times and finish by discussing recent results of using heavy-tailed distributions in traffic modeling.

15.5.1 Modeling of Empirical Distribution

Until recently [33–35, 42] traffic has almost exclusively been modeled by exponential distributions. In the last few years it has been shown that this is insufficient and that heavy-tailed distributions are more appropriate and can be used to explain self-similar behavior or long-range dependent behavior. Given a random variable X with distribution function $F(x) = P(X \leq x)$, its distribution function $F(x)$ is called heavy tailed if its tail $1 - F(x) = P(X > x)$ decreases subexponentially for large values of x. This means that if a random variable represents waiting time, then for a random variable with a heavy-tailed distribution the longer the already accumulated waiting time is the lower in the likelihood of the next arrival within the following time interval.

The models we consider are based on the exponential, the Weibull, the Pareto, or the lognormal probability distributions. Past work has concentrated on the exponential, the lognormal, and the Pareto distribution, but not the Weibull distribution. Often, if the conditions of "strict randomness" of the exponential distribution are not satisfied, the Weibull distribution is a suitable alternative [23]. The Weibull distribution is a generalization if the exponential distribution in the sense that a variable x has a Weibull distribution of $y = (x/a)^c$ has an exponential distribution with probability density function $p(y) = e^{-y}$. As the value of c decreases the probability of longer as well as shorter values increases, and the burstiness of the traffic increases. The definitions and the maximum likelihood estimators for all the probability distributions can be found in Tables 15.3 and 15.4. For more details see Johnson and Kotz [23]. Note that all but the exponential distributions are two-parameter distributions and can have heavy tails.

For the analysis of the traffic traces described above we follow the approach of using goodness-of-fit measures suggested by Paxson [34, 35] for the analysis of wide-area TCP connections. To judge if the fit of one model is better than the other

TABLE 15.3 Definition of Several Probability Distributions

Distribution	Probability Density $p(x)$	Cumulative Probability $F(x)$	Mean $E(x)$
Exponential	$\frac{1}{\rho}e^{-x/\rho}$	$1-e^{-x/\rho}$	ρ
Weibull	$\frac{1}{a}\left(-\frac{x}{a}\right)^{c-1}e^{-(x/a)^c}$	$1-e^{-(x/a)^c}$	$a\Gamma\left(\frac{1}{c}+1\right)$
Pareto $(k>0, a>0; x\geq k)$	$\frac{ak^a}{x^{a+1}}$	$1-\left(\frac{k}{x}\right)^a$	$\frac{ak}{a-1}$ if $a>1$ ∞ if $a\leq 1$
Lognormal with trans $z=(\log x-\zeta)/\sigma$	$\frac{1}{x\sqrt{2\pi}\sigma}e^{-[\log(x)-\zeta]^2/2\sigma^2}$	No closed form	$e^{\zeta+\sigma^2/2}$

TABLE 15.4 Maximum Likelihood Estimators for Several Probability Distributions

Probability Distribution	Maximum Likelihood Estimator
Exponential	$\hat{\rho}=\frac{1}{n}\sum_{i=1}^{n}x_i=\bar{x}$
Weibull	$\hat{a}=\left[\frac{1}{n}\sum_{i=1}^{n}x_i^{\hat{c}}\right]^{1/\hat{c}}$ $\hat{c}=\left[\left(\sum_{i=1}^{n}x_i^{\hat{c}}\log x_i\right)\left(\sum_{i=1}^{n}x_i^{\hat{c}}\right)^{-1}-\frac{1}{n}\sum_{i=1}^{n}\log x_i\right]^{-1}$ Solve for $\hat{a}$ and $\hat{c}$
Pareto	$\hat{a}$, $\hat{k}$ estimated via least squares fit to $\log(1-F(x))=a\log k-a\log x$
Lognormal	$\hat{\zeta}=\overline{\log(x)}$ $\hat{\rho}=\sqrt{\text{variance }(\log(x))}$

we use a discrepancy estimate $\hat{\lambda}^2$, which has an estimated variance of $\hat{v}(\hat{\lambda}^2)$ [31, 34] and say that one model is better than another if the value of its estimated discrepancy minus its variance is larger than the other model's estimated discrepancy plus its variance.

The discrepancy $\hat{\lambda}^2$ is computed by estimating the values of the experimental cumulative probability for a set of bins and comparing it to the values of the actual cumulative probability function. Therefore, the computation depends on the choice of a number of bins and the spacing of the bins. Following suggestions by Scott [39] and Mann and Wald [32], and Paxson [34], we space the bins logarithmically where the number of bins is $w=3.49\hat{\sigma}n^{-4/9}$ if n is the number of observations. Adjacent bins are combined if the number of observations in the bins is less than five.

15.5.2 Modeling Interarrival Times of TCP Connections

The first indication that the Weibull distribution might be a good model is obtained by plotting the standardized skewness versus the coefficient of variation for both models and datasets and observing that the points of the datasets are clustered around the Weibull distribution. Figure 15.5(a) demonstrates this for eight complete datasets, Fig. 15.5(b) for eight hour periods of the external AT&T dataset from 18 November to 8 December 1995, and Fig. 15.5(c) for one hour periods of all Web requests of the external AT&T dataset from 18 November to 8 December 1995.

The arrival processes are by no means stationary. Indeed, Fig. 15.6(a) shows how the number of Web connections changes over time for the CMU trace from 29 June. Each point represents the number of Web connections that arrived within a 15 minute interval. For the longer external AT&T trace from 18 November, Fig. 15.6(b) shows how many Web connections arrived within each four hour time period. These graphs show clear time of day and day of week dependencies.

Nevertheless, we will show in this section that the Weibull distribution provides a good fit for the connection interarrival times of TCP connections over all different measurement periods. We start by first examining the distribution of all connection

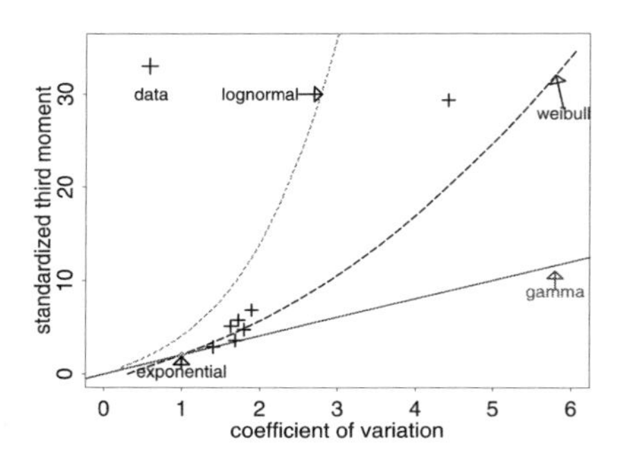

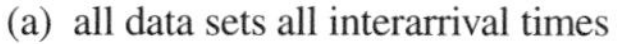

(a) all data sets all interarrival times

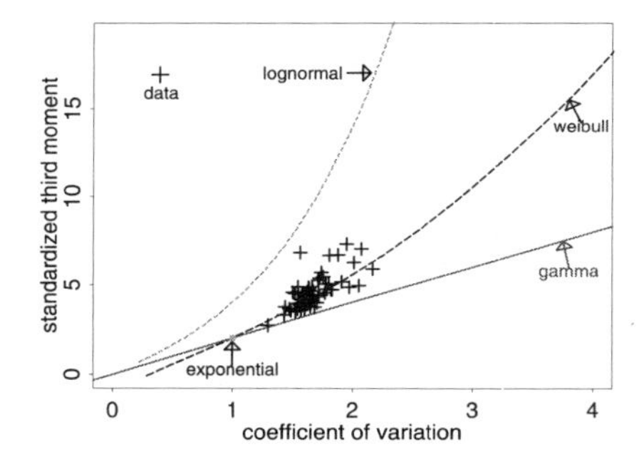

(b) 18 Nov. AT&T 8 hours of data

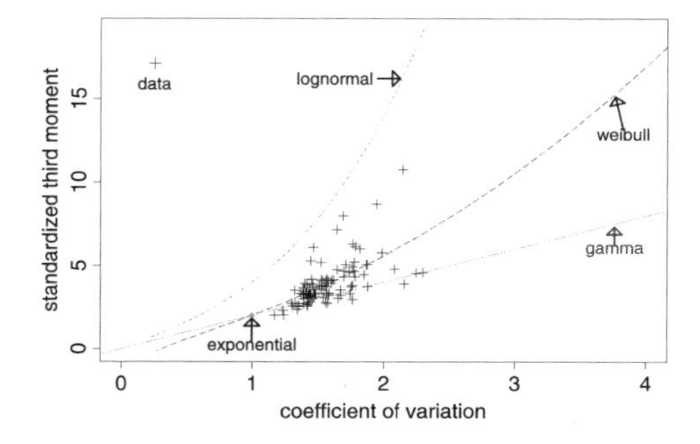

(c) 18 Nov. AT&T one hour of Web data

Fig. 15.5 Skewness.

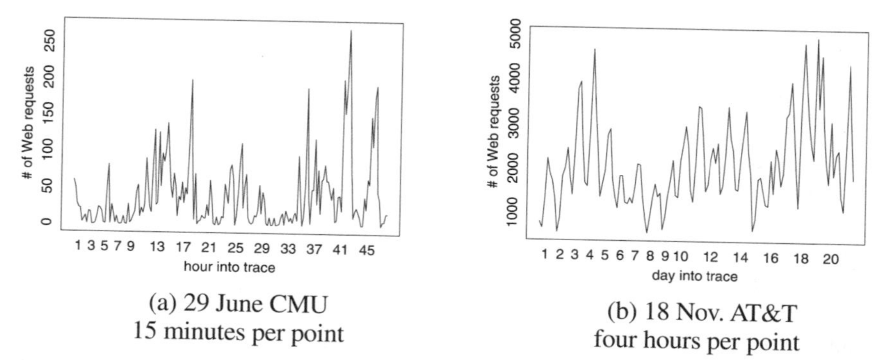

(a) 29 June CMU
15 minutes per point

(b) 18 Nov. AT&T
four hours per point

Fig. 15.6 (a) Time of day effects for Web traffic in June 1995 CMU dataset. (b) Time of day and day of week effects for Web traffic in external AT&T dataset.

requests, then distinguishing between the application types, next considering application-specific arrivals over time periods of various length, and finally we present time-dependent models for only a subset of the sources. While one might argue that we are planning to present the most useful models last (after all, arrivals are dependent on the time of day), we choose this order to familiarize the reader with the data and some graphical tools.

15.5.3 All Interarrival Times

The first quantity of interest is the shape and the characteristics of the empirical cumulative distribution of all interarrival times. Figure 15.7 shows the empirical density function for the CMU trace from 29 to 30 June 1995 and for the external AT&T trace from 18 November to 8 December 1995. Although the empirical density is fairly smooth, there are quite a few spikes, especially for the small interarrival times. The spikes in the small interarrival times are due to timer resolution problems and simultaneous arrival of connections.

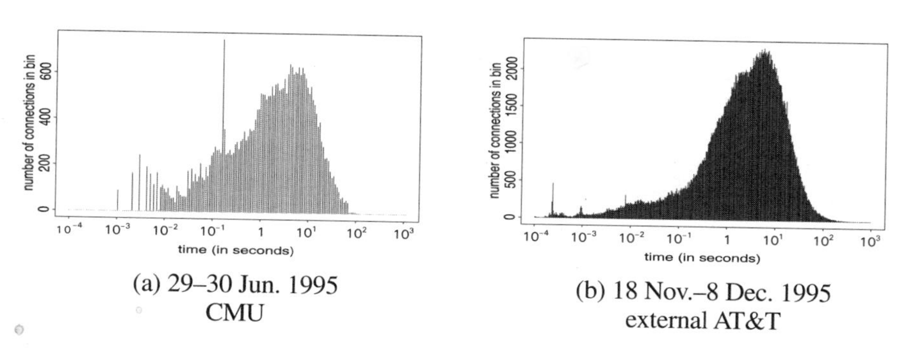

(a) 29–30 Jun. 1995
CMU

(b) 18 Nov.–8 Dec. 1995
external AT&T

Fig. 15.7 Histogram (empirical density function) for interarrival time of one CMU and one external AT&T dataset.

The obvious spikes around 0.146 s for the CMU dataset and around 0.000234 s for the external AT&T dataset are due to protocol dependencies. The spike in the CMU dataset is caused almost entirely by the KSHELL protocol. Each KSHELL invocation establishes two TCP connections within a time period of roughly 0.146 s. Since, on this Ethernet segment, it is quite likely that no other TCP connection request is initiated within such a small time window, this creates a spike in the empirical cumulative distribution. The spike in the external AT&T dataset is caused almost entirely by World Wide Web accesses that correspond to an HTTP client establishing parallel TCP connections to load inlined HTTP objects such as gif files. Since this external AT&T dataset is dominated by HTTP traffic, this spike is much more prevalent here than in the CMU dataset. For the purpose of the analysis discussed in this chapter we did not exclude any of these observations.

Next, consider the shape and characteristics of the empirical cumulative distribution of all interarrival times. Figure 15.8(a) shows the empirical cumulative distribution resulting from the empirical density function of Fig. 15.7(a) for a CMU trace together with the fitted cumulative distributions of the Weibull, Pareto, lognormal (lnorm), and exponential distributions. Figure 15.8(b) shows the cumulative distributions for the external AT&T trace. It is immediately obvious that the exponential, Pareto, and lognormal distributions result in very poor fits, while the agreement between the fitted Weibull distribution and the measured interarrival times is quite good. It is interesting that the exponential model underestimates short as well as long interarrival times, while the lognormal model overestimates both. The model based on the Pareto distribution overestimates the medium length interarrival times and yields a very poor fit for the shorter interarrival times.

The observation that the fitted Weibull distribution yields a much better fit than any of the other ones can be confirmed using the discrepancy measure λ^2. Table 15.5 summarizes $\hat{\lambda}^2 - \sqrt{\hat{v}(\hat{\lambda}^2)}$ and $\hat{\lambda}^2 + \sqrt{\hat{v}(\hat{\lambda}^2)}$ for all fits and three data sets. This corresponds to a 68.26% confidence interval. Figure 15.9 visualizes the data from Table 15.5 for all datasets in a bar plot where the height of the bars corresponds to the value of $\hat{\lambda}^2$ and the error bars to the standard deviation of $\hat{\lambda}^2$ or the 68.26% confidence intervals. To visualize the discrepancy better, all values above a threshold of five are cut off.

The Weibull distribution yields statistically significant better models for the interarrival times of all ten datasets. For all datasets except the CMU dataset from November 1994 the lognormal model is an order of magnitude better than both the exponential model and the Pareto model, while the three cannot be distinguished for the CMU November dataset. This difference between the datasets is most likely due to the fact that they span significantly different time periods and capture different traffic mixes.

15.5.4 Application-Specific Interarrival Times

While the overall times are of interest, one expects that different protocols have different characteristics. We compute the discrepancies on a per application basis. Figure 15.10 visualizes their values in a bar plot. Again the height of the bars

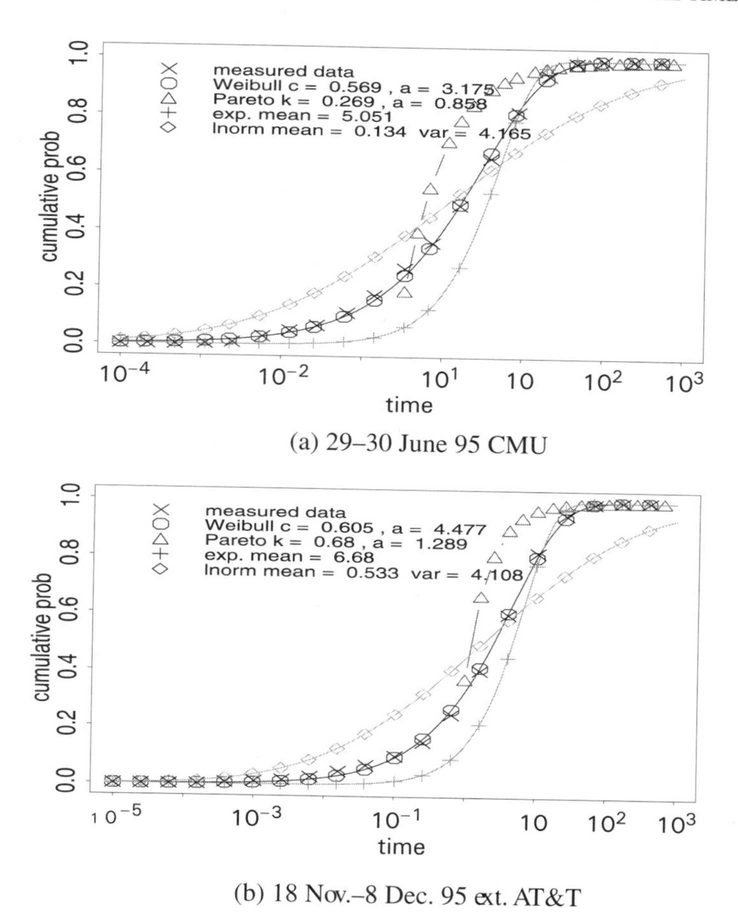

(a) 29–30 June 95 CMU

(b) 18 Nov.–8 Dec. 95 ext. AT&T

Fig. 15.8 Empirical and fitted cumulative interarrival time distributions of one CMU and one external AT&T dataset.

TABLE 15.5 The 68.26% Confidence Intervals for $\hat{\lambda}^2$ for the Fitted Models of Some Datasets and the Parameters of the Fitted Weibull model

	Dataset		
Model	CMU 29–30 June	External AT&T 18 Nov.–8 Dec.	Internal AT&T 8–23 Dec.
Weibull	0.06–0.07	0.08–0.09	0.18–0.20
Lognormal	0.79–0.82	0.93–0.95	1.48–1.58
Exponential	2.08–2.17	1.12–1.14	2.34–2.53
Pareto	2.25–2.42	3.25–3.41	13.6–14.9
Weibull c	0.569	0.600	0.419
Weibull a	3.175	3.894	32.582

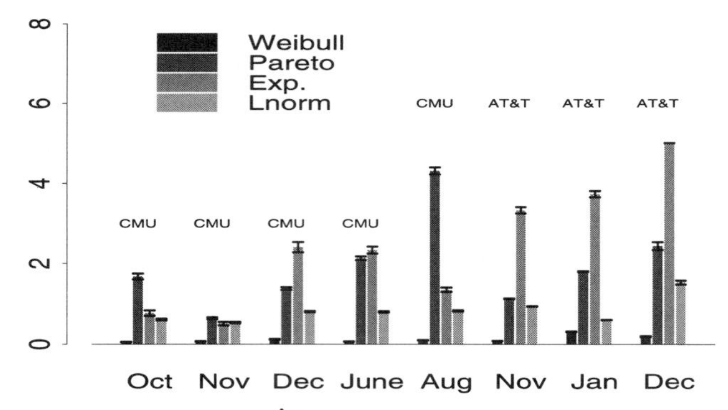

Fig. 15.9 Confidence intervals for $\hat{\lambda}^2$ for the fitted Weibull, lognormal, exponential, and Pareto models of all interarrival times for the eight datasets.

corresponds to the value of $\hat{\lambda}^2$ and the error bars to the standard deviation of $\hat{\lambda}^2$ or the 68.26% confidence intervals. We see that for our datasets the Weibull model does substantially better than the others (except for FTP.DATA, consistent with Paxson and Floyd [35]) although the value of the discrepancy is quite substantial for some of the application types and some datasets. In fact, for all CMU datasets the Weibull model is statistically better than the other models in 203 of 240 cases, it is indistinguishable in 25 cases, and is beaten by one of the other models in only 12 instances. Actually the Pareto model is better in eight cases, while the exponential model is better in three cases, and the lognormal model in one case. For the AT&T datasets the Weibull model is statistically better than the other models in 90 of 95 instances, it is indistinguishable from the Pareto or the exponential model in three instances, and it is beaten by the Pareto model in only two instances.

For the LBL and DEC datasets we observe that the Weibull distribution is capable of modeling some but not all applications. For example, the Weibull distribution performs very well for SMTP, NNTP, and FTP.CONN connections but poorly for HTTP and TELNET connections for the 1993 LBL-7 dataset. An explanation for HTTP is that very few people were using it at the time of the collection of the trace. Therefore the sample may not be appropriate. For TELNET it seems that a nonstationary approach is

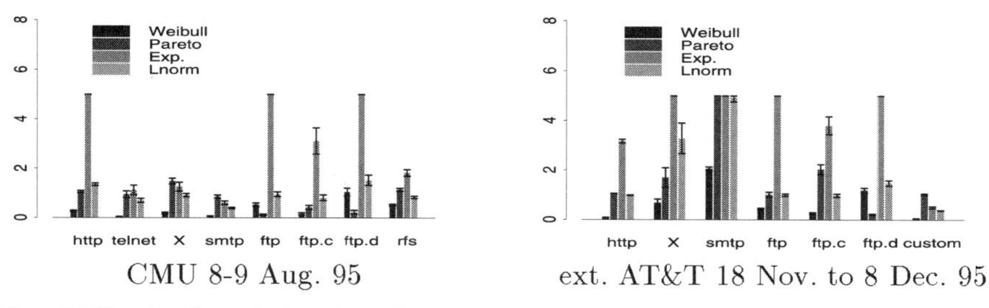

Fig. 15.10 Quality of fit of application-specific connection interarrival times for two datasets.

necessary to model this data. For the one hour long 1995 LBL-5 dataset, the distribution of HTTP interarrival times is modeled best using a lognormal distribution while the total dataset is explained best via an exponential or Weibull model. For the one hour long 1995 DEC-4 dataset the distribution of HTTP interarrival times is modeled best using the Weibull distribution. The overall distribution is also explained best by a Weibull distribution. Yet, with a shape parameter of 0.83, the distribution is relatively close to an exponential distribution, which is not too surprising given that most connections are either TELNET or FTP connections. Indeed, because more than one-fifth of all TCP connections are HTTP connections, a distribution with a heavier tail than the exponential distribution appears appropriate.

While most of the application-specific Weibull models have small discrepancies, we observe that the discrepancies are quite large for some applications such as NNTP, POP, or PROTO. We identified two major causes for such large discrepancies: bimodality and timer-driven applications. Since neither of our models is able to fit such distorted distributions, the resulting large discrepancy values are no surprise.

15.5.5 Application-Specific Interarrival Times over Shorter Time Periods

Although the traces where collected over time periods of varying length (from less than one day to three weeks), the Weibull distribution produces good models for the interarrival times of the various applications. This is a first indication that the appropriateness of the fitted Weibull distribution is not adversely affected by the considered time interval. Indeed, the observations regarding the goodness of the fitted distribution hold true over almost all time periods that are long enough to allow sufficient observations. It even holds if we select just one compute server over a one hour time period or aggregate over a number of single user machines within one hour time periods.

We start by considering the HTTP connection interarrival times of the external AT&T dataset from 18 November. Figure 15.11 shows the discrepancy measures λ^2 for time periods of different length. For periods of one day (Fig. 15.11(a)), one hour (Fig. 15.11(b)) and even 20 minutes (Fig. 15.11(c)), the Weibull distribution yields the best fit.

If we isolate specific traffic sources within our traces (e.g., a compute server at AT&T), the Weibull distribution still gives the statistically best model. Figure 15.12(a) depicts the arrival time series of the individual sources as well as the total arrival time series for all HTTP connections of the external AT&T dataset from 16 to 31 March over some arbitrary one hour time period. Figure 15.12(b) and (c) show the empirical and fitted cumulative interarrival time distribution of the interarrival time distribution of two specific sources from the same data used for Fig. 15.12(a). Again the fit of the Weibull distribution is far superior to all the others—an observation that is underlined by the discrepancy measure for other sources and time periods.

The time-dependent discrepancies for the interarrival times of another application, TELNET, are shown in Fig. 15.13 for the 29–30 June 1995 dataset. Here the first

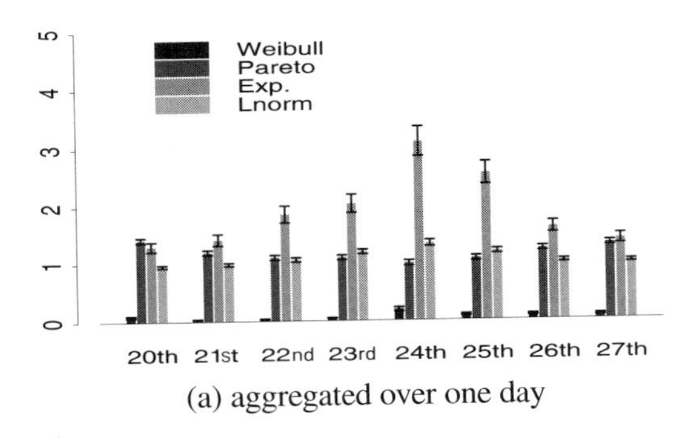

(a) aggregated over one day

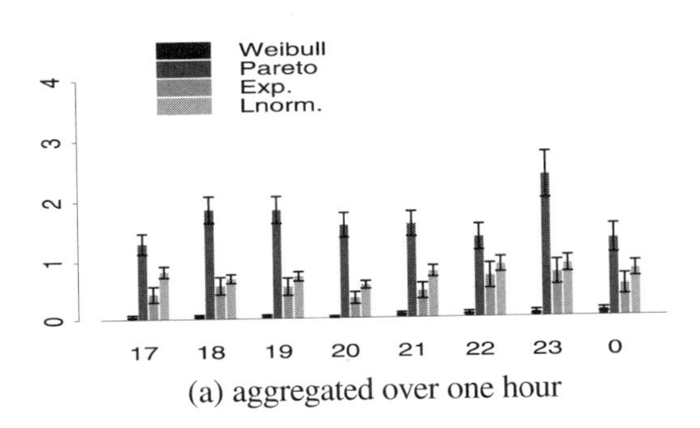

(a) aggregated over one hour

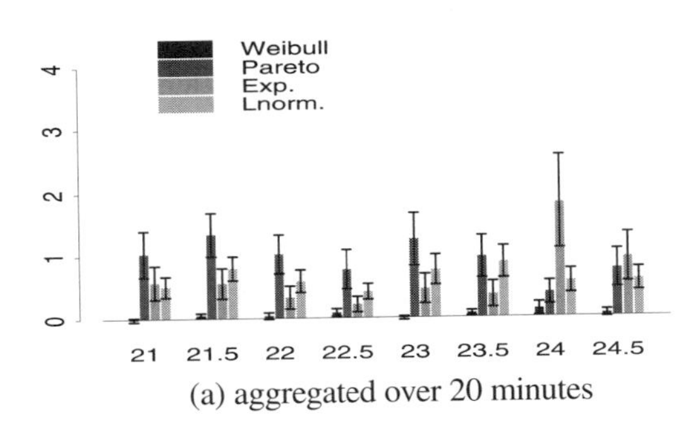

(a) aggregated over 20 minutes

Fig. 15.11 Quality of fit of application-specific (HTTP) and time-dependent connection interarrival times for the external AT&T 18 November to 8 December 1995 dataset.

four columns correspond to models fitted over 12 hour time periods while the fifth column corresponds to models fitted over the whole dataset. It is rather interesting to note the huge variations in the value of the discrepancy for all models except the Weibull one. This is another indication that only the Weibull and none of the other models yields a satisfying fit even over the different time periods.

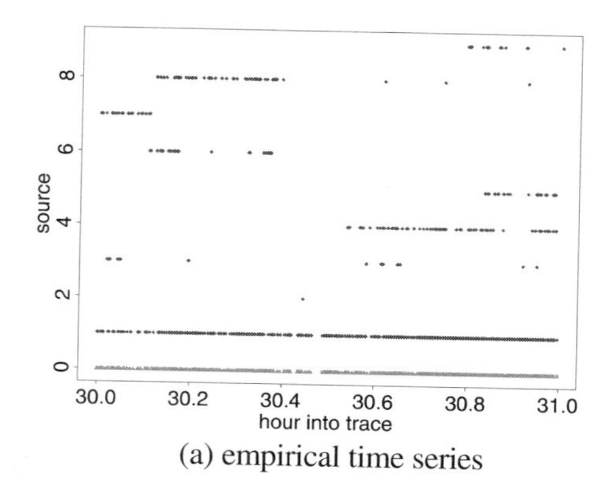

(a) empirical time series

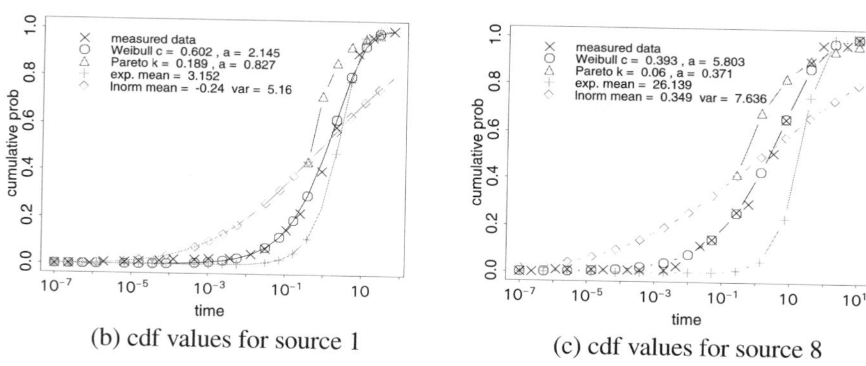

(b) cdf values for source 1

(c) cdf values for source 8

Fig. 15.12 Fit of source- and application-specific (HTTP) interarrival times over a one hour period of the external AT&T dataset from 16 to 31 March 1996. (a) The arrivals versus the source. (b) The empirical and the fitted cumulative distributions of sources 1 and 8.

Within the models for the 12 hour time period a clear periodicity is recognizable, a periodicity that has also been observed within other datasets. During the first half of a day the Pareto and lognormal models give a better fit than the exponential one, while the situation is reversed for the second half of the day. An explanation for this is that during night hours many fewer actions are initiated and, as such, the number of long interarrival times is larger, thus explaining why the heavy-tailed Pareto distribution yields a better fit than the exponential distribution. The Weibull distribution yields an even better fit than either the Pareto or the exponential one.

We also consider the interarrival time distribution of all HTTP connections over all 1204 possible one hour time periods that contain sufficient observations. The Weibull models are statistically better than the Pareto models in 92.9% of the instances, better than the exponential models in 92.3% of the instances, and better than the lognormal in 94.5% of the instances. In only 0.7% of the instances is the Pareto model better than the Weibull one and in only 0.2% is the exponential model better than the Weibull model. (Even though the Weibull model contains the

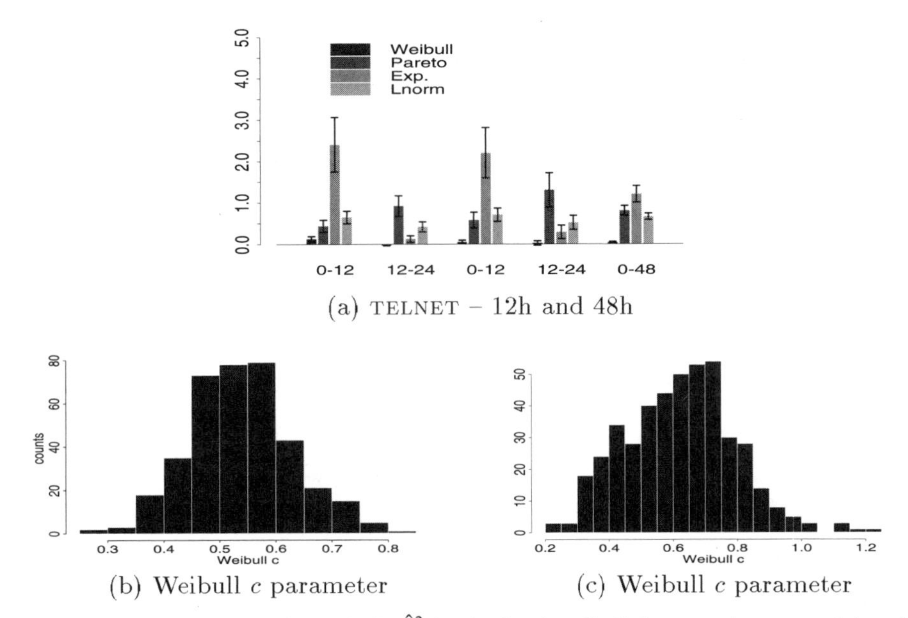

Fig. 15.13 (a) Confidence intervals for $\hat{\lambda}^2$ for the fitted Weibull, lognormal, exponential, and Pareto model of TELNET interarrival times within each 12 hour time frame for the 29–30 June 1995 dataset. (b) Histogram of Weibull shape parameter c of the fitted Weibull distributions for the HTTP application for one hour time periods for the external AT&T dataset from 16 to 30 March. (c) Histogram of Weibull shape parameter c of the fitted Weibull distributions for the applications TELNET, X, FTP, HTTP, RFS, SMTP, and FINGER and two hour time periods for all CMU datasets.

exponential model, the exponential model can sometimes result in a better fit than the Weibull model. This is possible since the parameter estimation for the Weibull model is more complex and can be less precise than for the exponential model.) Similar observations apply to the CMU datasets and other applications.

Given that the Weibull model is superior for many applications and that its shape parameter c may vary quite substantially from one time period to the next, it is of interest to look at the distribution of the shape parameter c of the fitted Weibull distributions. Figure 15.13(b) gives the histogram of Weibull shape parameter c for the fitted Weibull distribution of the external AT&T dataset from 16 to 31 March 1996 and Fig. 15.13(c) does the same for the fitted Weibull distributions for the applications TELNET, X, FTP, HTTP, RFS, SMTP, and FINGER over two hour time periods for all CMU datasets. That the number of Weibull models that have a shape parameter around one is small is another indication that the exponential model is inappropriate to model these interarrival times. That more than 50% of the shape parameters are less than 0.65 indicates the inherent burstiness of the request sequences.

To underscore the observation that the Weibull distribution gives good fits especially for HTTP TCP connection interarrival times, we again consider traces from AT&T's WorldNet. The first trace was collected on the 23 July 1997 starting at 7 p.m. for a duration of seven hours on an FDDI ring carrying traffic from roughly 420 modems used by WorldNet dialup customers (for more details see Feldmann et al. [13, 15]). For this dataset we consider the interarrival time distribution of HTTP connections over all 36 possible 12 minute time periods. In all instances the Weibull models are statistically better than the Pareto and exponential models, and better than the lognormal in 91.6% of the case. Never is any other model statistically better than the Weibull one. Similar observations apply to other subsets. One of the problems in this dataset is that it includes 371,972 HTTP connections. With this many connections arriving in a bursty fashion the timer resolution can become a potential problem. This can disturb small interarrival times and therefore undersample them. Figure 15.14(a) shows the histogram of the interarrival times for a two hour sample of the data. The spikes in the range of 0.0001–0.001 second have two origins: timer inaccuracy and browsers almost simultaneously opening multiple TCP connections. Nevertheless, the data, except for the spikes, is still well modeled by a Weibull distribution. Figure 15.14(b) shows the histogram of Weibull shape parameter c for the fitted Weibull distributions. The backbone AT&T WorldNet dataset contains so many HTTP requests (1,187,865 over one hour) that the timer resolution on the monitoring machine is not sufficient to resolve the small time scale effects. Therefore we do not attempt to model the interarrival times for this dataset.

15.5.6 Heavy-Tailed Distributions in Traffic Modeling

In the following we summarize related work in the area of using heavy-tailed distributions in traffic modeling.

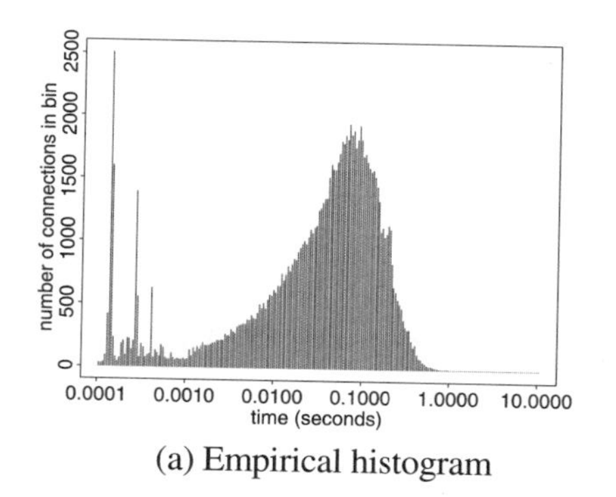

(a) Empirical histogram

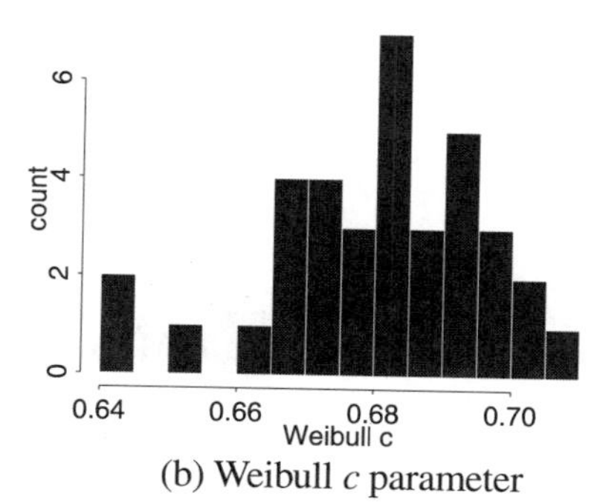

(b) Weibull c parameter

Fig. 15.14 (a) Histogram (empirical density function) for interarrival times during two hours of the AT&T WorldNet modem user trace. (b) Histogram of Weibull shape parameter c for the AT&T WorldNet modem user trace over 12 minute time periods.

Bolotin [6] studied a year's worth of complete call records for phone calls and found that their call duration is better described by a mixture of lognormal distributions than exponential distributions. Duffy et al. [12] suggest that heavy-tailed distributions with possibly infinite variance/mean might be best suited for describing the durations of phone calls.

Marshall and Morgan [29] note that the empirical statistical distributions of local-area network traffic have longer tails than an exponential distribution. They observe that some of the interarrival time histograms look as if they could be well fitted by a lognormal distribution or a mixture of two lognormal ones.

Fowler and Leland [17] showed that local-area network traffic has a high level of variability on every time scale that they measured. Such observations are inconsistent with the assumption of exponential distributions.

As mentioned earlier, Willinger et al. [42] explain the self-similar nature of Ethernet traffic as the result of superposition of many on/off sources where the length of the on and off periods are heavy tailed. Fractal-like behavior has also been observed in VBR video traffic [20].

Paxson and Floyd [35] find that for wide-area traffic Poisson processes are only valid for modeling the arrival of user sessions (TELNET and FTP connections) and that they fail to model other arrival processes. Traffic processes that are better modeled using self-similar processes include FTP data connection arrivals, FTP.DATA bytes, and TELNET packet interarrival times.

Paxson [33, 34] examines the behavior of TELNET, NNTP, SMTP, and FTP wide-area TCP connections. His results show that lognormal, Pareto, and extreme distributions yield statistically better models for the tails of a variety of parameters. His study presents models for the number of bytes sent by the originator and responder of TELNET connections and their durations, the number of originator bytes in NNTP and SMTP connections, and the number of bytes and the burst bytes in FTP connections.

Mogul [30] considered the arrival of World Wide Web requests at a busy World Wide Web server. His conclusions include that HTTP request arrivals approximately follow Poisson curves for interarrival times below the mean but have much larger tails.

Crovella and Bestavros [7, 8] and Park et al. [36] examine World Wide Web requests using a different data collection scheme. They point out that durations of HTTP connections follow heavy-tailed distributions. They explain this observation by pointing out that file size distributions on Web servers are also heavy-tailed.

15.6 INTERARRIVAL TIMES—NONSTATIONARY OR WEIBULL

To address the question of modeling interarrival times of TCP connections by either a stationary renewal process with interarrival times drawn from an i.i.d. Weibull distribution (Weibull process) or a nonstationary Poisson process, we present in this section simulation results for one application—connection admission for a single unit capacity edge—that indicate that a Weibull process might capture certain aspects of the arrival process more appropriately than a nonstationary Poisson process.

A connection, characterized by arrival time, duration, and bandwidth requirement, is accepted if the edge can support the connection; otherwise it is rejected. The measure of interest is the blocking probability that depends on the network load. Given that we want to compare different arrival sequences, we use simple models for the durations and the bandwidth requirements of the connections; namely, bandwidth requirements are chosen uniformly from $[0.05, 0.45]$ or $[0.005, 0.195]$ and durations have an exponential distribution. The load of the network is varied by adjusting the mean of the duration.

Figure 15.15 shows the results of the simulations for various interarrival time models. The curve labeled "orig" shows the blocking probability of the arrival sequence derived from the trace. The label "weib" refers to i.i.d. Weibull interarrival times. Here, the Weibull parameters are determined by fitting a single Weibull model to the interarrival times of the trace. The label "poisson (DUR)" means that arrivals follow a nonstationary Poisson process. First, the number of arrivals within each segment of length DUR is counted. Next, these counts are used to generate a corresponding number of arrivals within each interval of length DUR by choosing the interarrival times according to an exponential distribution. The label "weib (DUR)" refers to a similar construction except that the interarrival times are drawn from a Weibull distribution. In this sense the label "weib (DUR)" denotes a nonstationary Weibull process with segment length DUR.

A standard assumption is that a process should be fairly stationary over a time period of less than one hour. Therefore our first set of simulations involves a set of 5214 arrivals from a one hour period of the 16–31 March external AT&T dataset (see Fig. 15.15(a)). The simulation results show that the Weibull arrival process yields a much better prediction of the actual blocking probability than any of the various nonstationary Poisson models. This is the case even though the segment length of the nonstationary model covers values as small as 0.5 minute. Note that the accuracy of the nonstationary Poisson models improves only slightly as the segment length of the models is reduced from 60 minutes to 0.5 minute. This indicates that the Weibull models capture some aspect of the arrival sequence—that is, its burstiness—that cannot be captured by the nonstationary Poisson models.

Next, Fig. 15.15(b) and (c) show results that are based on the complete CMU trace for 29–30 August. In Fig. 15.15(b) the average bandwidth requirement is one-quarter of the edge bandwidth. In this case the stationary Weibull model predicts the blocking probability more accurarately than any of the nonstationary Poisson models. Once the average bandwidth requirements of the connection are reduced to one-tenth (Fig. 15.15(c)), the performance of the pure Weibull model is not that impressive. Still, it does as well as the nonstationary process with a time segment of one hour. This, in itself, is an accomplishment considering that we compare a model that is based on one set of two parameters over a 48 hour time period to a model that is based on 48 parameters. Even more impressive is that a nonstationary Weibull model with segments of one hour clearly outperforms the nonstationary Poisson model with segments as small as 6 minutes.

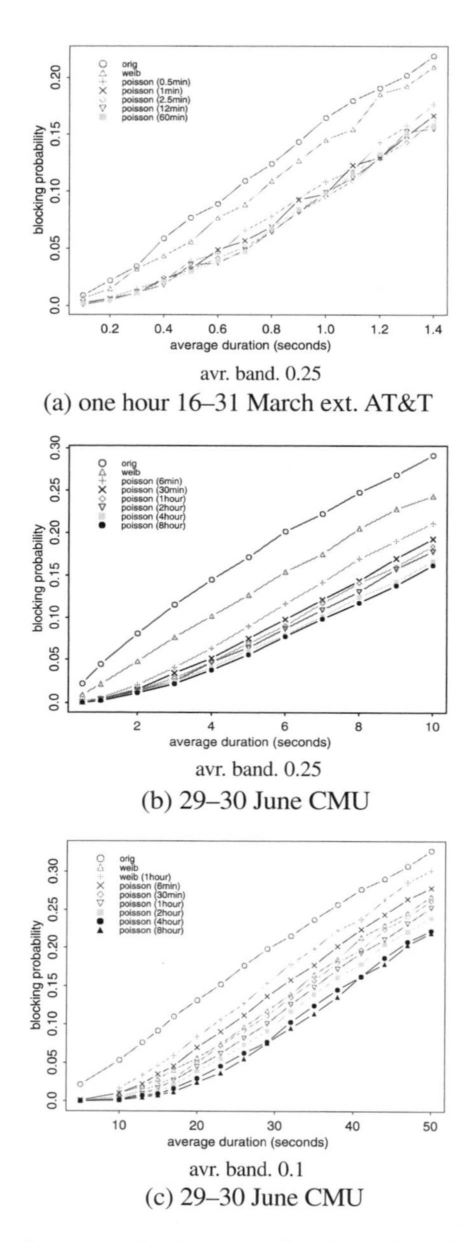

Fig. 15.15 Results of a connection-by-connection simulation of a connection-admission strategy for a single unit capacity edge contrasted for different interarrival time models. (a) 5214 arrivals taken from a one hour time period from an AT&T dataset. (b) 34,152 arrivals from a CMU dataset.

This means that the Weibull distribution is modeling long breaks in the arrival process as least as well as a nonstationary Poisson process taken over a reasonable period. In addition, the Weibull process is capable of modeling the short interarrival times much more accurately. Figure 15.16 reveals some interesting comparisons of a nonstationary Poisson process to the stationary Weibull process. Figure 15.16(a) shows the actual arrival times of the connections for the arrivals derived from the June dataset. Figure 15.16(b) shows the empirical cumulative distributions of the original interarrival time distribution, the fitted Weibull distribution, the fitted exponential distribution, and the empirical distribution of the nonstationary Poisson process with period 6 minutes for the June dataset. While Fig. 15.16 shows the result for a specific nonstationary Poisson process, the results are almost identical for other nonstationary Poisson processes.

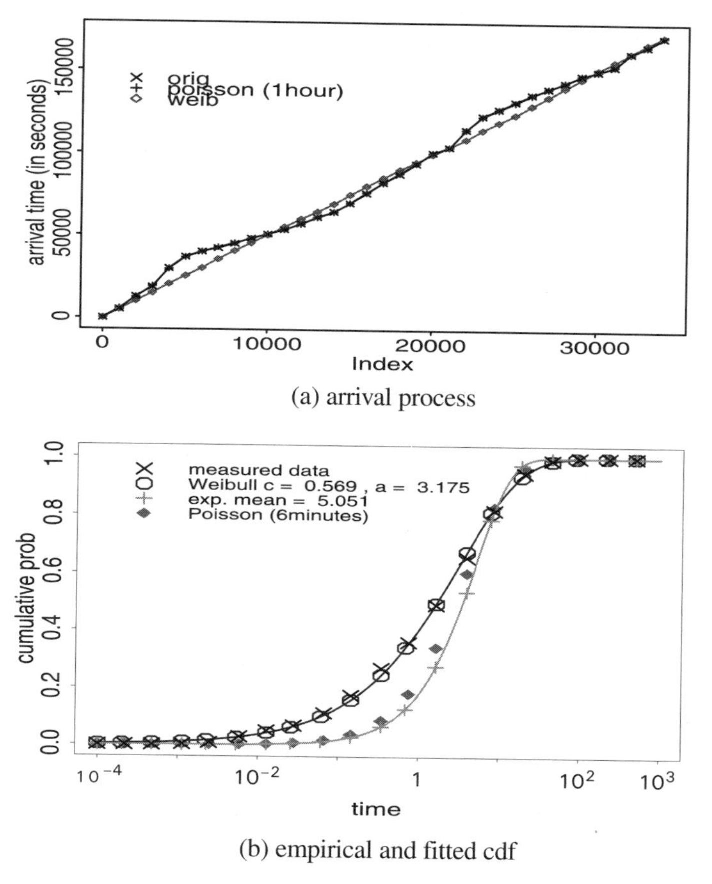

Fig. 15.16 (a) Arrival sequence of 34,152 arrivals from a CMU dataset. (b) Empirical and fitted cumulative distributions of the interarrival times of the simulations.

By design the nonstationary Poisson process follows the original arrival sequence very accurately; this is not the case for the Weibull time series. This would indicate that the stationary Weibull model is not appropriate. Yet, given that the simulation results[2] indicate otherwise, consider the implications of Fig. 15.16(b). The plot indicates that the cumulative distribution of the interarrival times of the total nonstationary Poisson process is much closer to the fitted exponential distribution than to the one of the original data (that it is trying to model). This implies that the nonstationary Poisson process is not modeling the short interarrivals very accurately. Therefore, since exactly these interarrival times are crucial for the burstiness of the arrival process, the stationary Weibull model can outperform the nonstationary Poisson process for the connection-admission problem.

In addition, it explains why the stationary Weibull model loses prediction accuracy as the average bandwidth requirements of the connections are decreased. With decreased bandwidth requirements large time scale load variations can dominate load variations on smaller time scales and the nonstationary Poisson process may match the relevant time scale of interest. Therefore, a decrease of the segment sizes leads to only a small improvement in predicting the blocking probability.

In summary, the Weibull distribution matches the interarrival time distribution over more time scales than the nonstationary Poisson model. If short time scale interactions within the arrival process are relevant to the network management problem, then a Weibull arrival process may well be an appropriate model to capture the burstiness of the arrival process.

15.7 CONCLUSION

Traditional models such as exponential models are insufficient to explain the observed distributions of interarrival times of TCP connections. Instead, we find that most parameters are statistically better fitted using distributions with heavy tails, such as Weibull, Pareto, and lognormal distributions. The Weibull distributions, especially, yield superior fits for all interarrival times, all application-specific interarrival times that are not timer driven, and time-dependent interarrival times, implying that the number of long and short interarrival times and therefore the number and intensity of bursts are higher than traditional traffic models suggest. Indeed, we observed that the increased burstiness of the arrival process for some applications is consistent with a self-similar process. This means it is scale-invariant behavior for application types such as Web and FTP. Simulation results for a connection-admission problem let us suggest the Weibull model as an alternative model for analysis and simulations of this and other network resource allocation problems.

[2] While the plots show only one simulation result, we evaluated several simulations, each using a different seed for the random number generator. Each simulation run resulted in quantitatively similar results.

ACKNOWLEDGMENTS

The author would like to thank Anna Gilbert, Jennifer Rexford, and Walter Willinger for many helpful discussions.

Many of the traces considered in this chapter were collected using the *tcp-dump* packet capture tool developed by V. Jacobson, C. Leres, and S. McCanne and are available via anonymous ftp to `ftp.ee.lbl.gov`. To extract TCP connection information from the traces, we relied on V. Paxson's *tcp-conn* tool, which is available from `http://ita.ee.lbl.gov/index.html`.

The DEC traces were collected by J. Mogul of Compaq's Western Research Lab (WRL), the LBL traces were gathered by V. Paxson and are available from `http://www.acm.org/sigcomm/ITA`, and we thank S. Alexander and S. Gao from AT&T Labs–Research for making the MH and FP traffic collection possible. We also acknowledge the help of many of our colleagues at AT&T Labs, especially of J. Friedmann and A. Greenberg, with the data collection effort within WorldNet.

Finally, we would like to thank P. Abry and D. Veitch for making their programs to perform the wavelet-based scaling analysis available to us.

REFERENCES

1. P. Abry and D. Veitch. Wavelet analysis of long-range dependent traffic. *IEEE Trans. Inf. Theory*, **44**:2–15, 1998.
2. P. Abry and D. Veitch. Wavelet for the analysis, estimation and synthesis of scaling data. In K. Park and W. Willinger, eds., *Self-Similar Network Traffic and Performance Evaluation*. Wiley, New York, 2000.
3. P. Barford and M. E. Crovella. Generating representative web workloads for network and server performance evaluation. In *Proceedings of Performance'98/ACM SIGMETRICS '98*, pp. 151–160, 1998.
4. P. Barford and M. E. Crovella. A performance evaluation of hyper text transfer protocols. In *Proceedings of Performance'99/ACM SIGMETRICS'99*, 1999.
5. J. Beran. *Statistics for Long-Memory Processes*. Chapman and Hall, New York, 1994.
6. V. A. Bolotin. Modeling call holding time distributions for CCS network design and performance analysis. *IEEE J. Select. Areas Commun.*, **12**(3):433–438, 1994.
7. M. E. Crovella and A. Bestavros. Explaining World Wide Web traffic self-similarity. Technical Report BU-CS-95-015, Boston University, Boston, 1995.
8. M. E. Crovella and A. Bestavros. Self-similarity in World Wide Web traffic—evidence and possible causes. *IEEE/ACM Trans. Networking*, **5**(6):835–846, 1997.
9. D. R. Cox. Long-range dependence: a review. In H. A. David and H. T. David, eds., *Statistics: An Appraisal*, pp. 55–74. Iowa State University Press, Ames, 1984.
10. I. Daubechies. *Ten Lectures on Wavelets*. SIAM, Philadelphia, 1992.
11. S. Deng. Empirical model of WWW document arrivals at access link. In *Proceedings of ICC/SUPERCOMM'96*, pp. 1197–1802, 1996.
12. D. E. Duffy, A. A. McIntosh, M. Rosenstein, and W. Willinger. Statistical analysis of

CCSN/SS7 traffic data from working CCS subnetworks. *IEEE J. Select. Areas Commun.*, **12**(3):544–551, 1994.

13. A. Feldmann, R. Cáceres, F. Douglis, and M. Rabinovich. Performance of web proxy caching in heterogeneous bandwidth environments. In *Proceedings of the Conference on Computer Communications (IEEE INFOCOM)*, pp. 107–116 April 1999.
14. A. Feldmann. Impact of non-Poisson arrival sequences for call admission algorithms with and without delay. In *GLOBECOM'96. IEEE Global Telecommunications Conference*, London, UK, 1996.
15. A. Feldmann, A. Gilbert, and W. Willinger. Data networks as cascades: explaining the multifractal nature of internet WAN traffic. In *Proc. ACM SIGCOMM*, pp. 42–55, September 1998.
16. A. Feldmann, A. Gilbert, W. Willinger, and T. G. Kurtz. The changing nature of network traffic: scaling phenomena. *Comput. Commun. Rev.*, **28**(2):5–29, 1998.
17. H. J. Fowler and W. E. Leland. Local area network traffic characteristics, with implications for broadband network congestion management. *IEEE J. Select. Areas Commun.*, **9**(7):1139–1149, 1991.
18. A. Feldmann, J. Rexford, and R. Cáceres. Reducing overhead in flow-switched networks: an empirical study of web traffic. *IEEE/ACM Trans. Networking*, **6**:673–685, December 1998.
19. M. Grossglauser and J.-C. Bolot. On the relevance of long-range dependence in network traffic. In *SIGCOMM Symposium on Communications Architectures and Protocols*, pp. 15–24, 1996.
20. M. W. Garrett and W. Willinger. Analysis, modeling, and generation of self-similar VBR video traffic. In *SIGCOMM Symposium on Communications Architectures and Protocols*, pp. 269–280, 1994.
21. Internet traffic archive. `http://www.acm.org/sigcomm/ITA/index.html`.
22. V. Jacobson. tcp-conn, 1996. Available through Internet Traffic Archive.
23. N. L. Johnson and S. Kotz. *Distributions in Statistics: Continuous Univariate Distributions—1*. Wiley-Interscience, New York, 1975.
24. V. Jacobson, C. Leres, and S. McCanne. tcpdump, June 1989. Available via anonymous ftp to `ftp.ee.lbl.gov`.
25. R. Jain and S. A. Routhier. Packet trains: measurements and a new model for computer network traffic. *IEEE J. Select. Areas Commun.*, **4**:986–995, 1986.
26. G. Kaiser. *A Friendly Guide to Wavelets*. Birkhauser, Boston, 1994.
27. W. E. Leland, M. S. Taqqu, W. Willinger, and D. V. Wilson. On the self-similar nature of Ethernet traffic (extended version). *IEEE/ACM Trans. Networking*, **2**(1):1–15, 1994.
28. S. McCanne and V. Jacobson. The BSD packet filter: a new architecture for user-level packet capture. In *Proceedings of the 1993 Winter USENIX Conference*, pp. 259–269, San Diego, CA, 1993.
29. W. T. Marshall and S. P. Morgan. Statistics of mixed data traffic on a local area network. *Comput. Networks ISDN Syst.*, **10**:185–195, 1985.
30. J. Mogul. Network behavior of a busy web server and its clients. Technical Report 95/5, Digital Equipment Corp. Western Research Laboratory, 1995.
31. D. S. Moore. Measures of lack of fit from tests of chi-squared type. *J. Statist. Planning Inference*, **10**:151–166, 1984.

32. H. Mann and A. Wald. On the choice of the number of class intervals in the application of the chi-square test. *Ann. Math. Statist.*, **13**:306–317, 1942.
33. V. Paxson. Empirically derived analytic models of wide-area TCP connections: extended report. Technical Report LBL-34086, Lawrence Berkeley Laboratory, 1993.
34. V. Paxson. Empiricall derived analytic models of wide-area TCP connections. *IEEE/ACM Trans. Networking*, **2**(4):316–336, 1994.
35. V. Paxson and S. Floyd. Wide-area traffic: the failure of Poisson modeling. *IEEE/ACM Trans. Networking*, **3**(1):226–244, 1995.
36. K. Park. G. Kim, and M. E. Crovella. On the relationship between file sizes, transport protocols, and self-similar network traffic. In *Proceedings IEEE International Conference on Network Protocols*, pp. 171–180, 1996.
37. V. N. Padmanabhan and J. C. Mogul. Improving HTTP latency. *Comput. Networks ISDN Syst.*, **28**(1/2):25–35, December 1995.
38. B. K. Ryu and A. Elwalid. The importance of long-range dependence of VBR video traffic in ATM traffic engineering: myths and realities. In *SIGCOMM Symposium on Communications Architectures and Protocols*, pp. 3–14, 1996.
39. D. W. Scott. On optimal and data-based histograms. *Biometrika*, **66**:605–610, 1979.
40. A. Shaikh, J. Rexford, and K. G. Shin. Evaluating the overheads of source-directed quality-of-service routing. Technical Report, University of Michigan, 1998.
41. W. Willinger, M. S. Taqqu, W. E. Leland, and D. V. Wilson. Self-similarity in high-speed packet traffic: analysis and modeling of Ethernet traffic measurements. *Statist. Sci.*, **10**(1):67–85, 1995.
42. W. Willinger, M. S. Taqqu, R. Sherman, and D. V. Wilson. Self-similarity through high-variability: statistical analysis of Ethernet LAN traffic at the source level. *IEEE/ACM Trans. Networking*, **5**:71–86, 1997.

16

ENGINEERING FOR QUALITY OF SERVICE

J. W. Roberts

France Télécom, CNET, 92794 Issy-Moulineaux, Cédex 9, France

16.1 INTRODUCTION

The traditional role of traffic engineering is to ensure that a telecommunications network has just enough capacity to meet expected demand with adequate quality of service. A critical requirement is to understand the three-way relationship between demand, capacity, and performance, each of these being quantified in appropriate units. The degree to which this is possible in a future multiservice network remains uncertain, due notably to the inherent self-similarity of traffic and the modeling difficulty that this implies. The purpose of the present chapter is to argue that sound traffic engineering remains *the* crucial element in providing quality of service and that the network must be designed to circumvent the self-similarity problem by applying traffic controls at an appropriate level.

Quality of service in a multiservice network depends essentially on two factors: the service model that identifies different service classes and specifies how network resources are shared, and the traffic engineering procedures used to determine the capacity of those resources. While the service model alone can provide differential levels of service ensuring that *some* users (generally those who pay most) have good quality, to provide that quality for a predefined *population* of users relies on previously providing sufficient capacity to handle their demand.

It is important in defining the service model to correctly identify the entity to which traffic controls apply. In a connectionless network where this entity is the datagram, there is little scope for offering more than "best effort" quality of service

Self-Similar Network Traffic and Performance Evaluation, Edited by Kihong Park and Walter Willinger
ISBN 0-471-31974-0

commitments to higher levels. At the other end of the scale, networks dealing mainly with self-similar traffic aggregates, such as all packets transmitting from one local-area network (LAN) to another, can hardly make performance guarantees, unless that traffic is previously shaped into some kind of rigidly defined envelope. The service model discussed in this chapter is based on an intermediate traffic entity, which we refer to as a "flow" defined for present purposes as the succession of packets pertaining to a single instance of some application, such as a videoconference or a document transfer.

By allocating resources at flow level, or more exactly, by rejecting newly arriving flows when available capacity is exhausted, quality of service provision is decomposed into two parts: service mechanisms and control protocols ensure that the quality of service of accepted flows is satisfactory; traffic engineering is applied to dimension network elements so that the probability of rejection remains tolerably small. The present chapter aims to demonstrate that this approach is feasible, sacrificing detail and depth somewhat in favor of a broad view of the range of issues that need to be addressed conjointly.

Other chapters in this book are particularly relevant to the present discussion. In Chapter 19, Adas and Mukherjee propose a framing scheme to ensure guaranteed quality for services like video transmission while Tuan and Park in Chapter 18 study congestion control algorithms for "elastic" data communications. Naturally, the schemes in both chapters take account of the self-similar nature of the considered traffic flows. They constitute alternatives to our own proposals. Chapter 15 by Feldmann gives a very precise description of Internet traffic characteristics at flow level, which to some extent invalidates our too optimistic Poisson arrivals assumption. The latter assumption remains useful, however, notably in showing how heavy-tailed distributions do not lead to severe performance problems if closed-loop control is used to dynamically share resources as in a processor sharing queue. The same Poisson approximation is exploited by Boxma and Cohen in Chapter 6, which contrasts the performance of FIFO (open-loop control) and processor sharing (PS) (closed-loop control) queues with heavy-tailed job sizes.

In the next section we discuss the nature of traffic in a multiservice network, identifying broad categories of flows with distinct quality of service requirements. Open-loop and closed-loop control options are discussed in Sections 16.3 and 16.4, where it is demonstrated notably that self-similar traffic does not necessarily lead to poor network performance if adapted flow level controls are implemented. A tentative service model drawing on the lessons of the preceding discussion is proposed in Section 16.5. Finally, in Section 16.6, we suggest how traditional approaches might be generalized to enable traffic engineering for a network based on this service model.

16.2 THE NATURE OF MULTISERVICE TRAFFIC

It is possible to identify an indefinite number of categories of telecommunications services, each having its own particular traffic characteristics and performance requirements. Often, however, these services are adaptable and there is no need for a network to offer multiple service classes each tailored to a specific application. In this section we seek a broad classification enabling the identification of distinct traffic handling requirements. We begin with a discussion on the nature of these requirements.

16.2.1 Quality of Service Requirements

It is useful to distinguish three kinds of quality of service measures, which we refer to here as *transparency, accessibility,* and *throughput.*

Transparency refers to the time and semantic integrity of transferred data. For real-time traffic delay should be negligible while a certain degree of data loss is tolerable. For data transfer, semantic integrity is generally required but (per packet) delay is not important.

Accessibility refers to the probability of admission refusal and the delay for setup in case of blocking. Blocking probability is the key parameter used in dimensioning the telephone network. In the Internet, there is currently no admission control and all new requests are accommodated by reducing the amount of bandwidth allocated to ongoing transfers. Accessibility becomes an issue, however, if it is considered necessary that transfers should be realized with a minimum acceptable throughput.

Realized throughput, for the transfer of documents such as files or Web pages, constitutes the main quality of service measure for data networks. A throughput of 100 kbit/s would ensure the transfer of most Web pages quasi-instantaneously (less than 1 second).

To meet transparency requirements the network must implement an appropriately designed service model. The accessibility requirements must then be satisfied by network sizing, taking into account the random nature of user demand. Realized throughput is determined both by how much capacity is provided and how the service model shares this capacity between different flows. With respect to the above requirements, it proves useful to distinguish two broad classes of traffic, which we term *stream* and *elastic.*

16.2.2 Stream Traffic

Stream traffic entities are flows having an intrinsic duration and rate (which is generally variable) whose time integrity must be (more or less) preserved by the network. Such traffic is generated by applications like the telephone and interactive video services, such as videoconferencing, where significant delay would constitute an unacceptable degradation. A network service providing time integrity for video signals would also be useful for the transfer of prerecorded video sequences and, although negligible network delay is not generally a requirement here, we consider this kind of application to be also a generator of stream traffic.

The way the rate of stream flows varies is important for the design of traffic controls. Speech signals are typically of on/off type with talkspurts interspersed by silences. Video signals generally exhibit more complex rate variations at multiple time scales. Importantly for traffic engineering, the bit rate of long video sequences exhibits long-range dependence [12], a plausible explanation for this phenomenon being that the duration of scenes in the sequence has a heavy-tailed probability distribution [10].

The number of stream flows in progress on some link, say, is a random process varying as communications begin and end. The arrival intensity generally varies according to the time of day. In a multiservice network it may be natural to extend current practice for the telephone network by identifying a busy period (e.g., the one hour period with the greatest traffic demand) and modeling arrivals in that period as a stationary stochastic process (e.g., a Poisson process). Traffic demand may then be expressed as the expected combined rate of all active flows: the product of the arrival rate, the mean duration, and the mean rate of one flow. The duration of telephone calls is known to have a heavy-tailed distribution [4] and this is likely to be true of other stream flows, suggesting that the number of flows in progress and their combined rate are self-similar processes.

16.2.3 Elastic Traffic

The second type of traffic we consider consists of digital objects or "documents," which must be transferred from one place to another. These documents might be data files, texts, pictures, or video sequences transferred for local storage before viewing. This traffic is elastic in that the flow rate can vary due to external causes (e.g., bandwidth availability) without detrimental effect on quality of service.

Users may or may not have quality of service requirements with respect to throughput. They do for real-time information retrieval sessions, where it is important for documents to appear rapidly on the user's screen. They do not for e-mail or file transfers where deferred delivery, within a loose time limit, is perfectly acceptable.

The essential characteristics of elastic traffic are the arrival process of transfer requests and the distribution of object sizes. Observations on Web traffic provide useful pointers to the nature of these characteristics [2, 5]. The average arrival intensity of transfer requests varies depending on underlying user activity patterns. As for stream traffic, it should be possible to identify representative busy periods, where the arrival process can be considered to be stationary.

Measurements on Web sites reported by Arlitt and Williamson [2] suggest the possibility of modeling the arrivals as a Poisson process. A Poisson process indeed results naturally when members of a very large population of users independently make relatively widely spaced demands. Note, however, that more recent and thorough measurements suggest that the Poisson assumption may be too optimistic (see Chapter 15). Statistics on the size of Web documents reveal that they are extremely variable, exhibiting a heavy-tailed probability distribution. Most objects are very small: measurements on Web document sizes reported by Arlitt and Williamson reveal that some 70% are less than 1 kbyte and only around 5% exceed 10 kbytes. The presence of a few extremely long documents has a significant impact on the overall traffic volume, however.

It is possible to define a notion of traffic demand for elastic flows, in analogy with the definition given above for stream traffic, as the product of an average arrival rate in a representative busy period and the average object size.

16.2.4 Traffic Aggregations

Another category of traffic arises when individual flows and transactions are grouped together in an aggregate traffic stream. This occurs currently, for example, when the flow between remotely located LANs must be treated as a traffic entity by a wide area network. Proposed evolutions to the Internet service model such as differentiated services and multiprotocol label switching (MPLS) also rely heavily on the notion of traffic aggregation.

Through aggregation, quality of service requirements are satisfied in a two-step process: the network guarantees that an aggregate has access to a given bandwidth between designated end points; this bandwidth is then shared by flows within the aggregate according to mechanisms like those described in the rest of this chapter. Typically, the network provider has the simple traffic management task of reserving the guaranteed bandwidth while the responsibility for sharing this bandwidth between individual stream and elastic flows devolves to the customer. This division of responsibilities alleviates the so-called scalability problem, where the capacity of network elements to maintain state on individual flows cannot keep up with the growth in traffic.

The situation would be clear if the guarantee provided by the network to the customer were for a fixed constant bandwidth throughout a given time interval. In practice, because traffic in an aggregation is generally extremely variable (and even self-similar), a constant rate is not usually a good match to user requirements. Some burstiness can be accounted for through a leaky bucket based traffic descriptor, although this is not a very satisfactory solution, especially for self-similar traffic (see Section 16.3.2).

In existing frame relay and ATM networks, current practice is to considerably overbook capacity (the sum of guaranteed rates may be several times greater than available capacity), counting on the fact that users do not all require their guaranteed bandwidth at the same time. This allows a proportionate decrease in the bandwidth charge but, of course, there is no longer any real guarantee. In addition, in these networks users are generally allowed to emit traffic at a rate over and above their guaranteed bandwidth. This excess traffic, "tagged" to designate it as expendable in case of congestion, is handled on a best effort basis using momentarily available capacity.

Undeniably, the combination of overbooking and tagging leads to a commercial offer that is attractive to many customers. It does, however, lead to an imprecision in the nature of the offered service and in the basis of charging, which may prove unacceptable as the multiservice networking market gains maturity. In the present chapter, we have sought to establish a more rigorous basis for network engineering where quality of service guarantees are real and verifiable.

This leads us to ignore the advantages of considering an aggregation as a single traffic entity and to require that individual stream and elastic flows be recognized for the purposes of admission control and routing. In other words, transparency, throughput, and accessibility are guaranteed on an individual flow basis, not for the aggregate. Of course, it remains useful to aggregate traffic within the network and flows of like characteristics can share buffers and links without the need to maintain detailed state information.

16.3 OPEN-LOOP CONTROL

In this and the next section we discuss traffic control options and their potential for realizing quality of service guarantees. Here we consider open-loop, or preventive, traffic control based on the notion of "traffic contract": a user requests a communication described in terms of a set of traffic parameters and the network performs admission control, accepting the communication only if quality of service requirements can be satisfied. Either ingress policing or service rate enforcement by scheduling in network nodes is then necessary to avoid performance degradation due to flows that do not conform to their declared traffic descriptor.

16.3.1 Multiplexing Performance

The effectiveness of open-loop control depends on how accurately it is possible to predict performance given the characteristics of variable rate flows. To discuss multiplexing options we make the simplifying assumption that flows have unambiguously defined rates like fluids, assimilating links to pipes and buffers to reservoirs. We also assume rate processes are stationary. It is useful to distinguish two forms of statistical multiplexing: bufferless multiplexing and buffered multiplexing.

In the fluid model, statistical multiplexing is possible without buffering if the combined input rate is maintained below link capacity. As all excess traffic is lost, the overall loss rate is simply $E[(\Lambda_t - c)^+]/E[\Lambda_t]$, where Λ_t is the input rate process and c is the link capacity. It is important to note that this loss rate only depends on the stationary distribution of Λ_t and not on its time-dependent properties, including self-similarity. The latter do have an impact on other aspects of performance, such as the duration of overloads, but this can often be neglected if the loss rate is small enough.

The level of link utilization compatible with a given loss rate can be increased by providing a buffer to absorb some of the input rate excess. However, the loss rate realized with a given buffer size and link capacity then depends in a complicated way on the nature of the offered traffic. In particular, loss and delay performance are very difficult to predict when the input process is long-range dependent. The models developed in this book are, for instance, generally only capable of predicting asymptotic queue behavior for particular classes of long-range dependent traffic.

An alternative to statistical multiplexing is to provide *deterministic* performance guarantees. Deterministic guarantees are possible, in particular, if the amount of data

$A(t)$ generated by a flow in an interval of length t satisfies a constraint of the form: $A(t) \leq \rho t + \sigma$. If the link serves this flow at a rate at least equal to ρ, then the maximum buffer content from this flow is σ. Loss can therefore be completely avoided and delay bounded by providing a buffer of size σ and implementing a scheduling discipline that ensures the service rate ρ [7]. The constraint on the input rate can be enforced by means of a leaky bucket, as discussed below.

16.3.2 The Leaky Bucket Traffic Descriptor

Open-loop control in both ATM and Internet service models relies on the leaky bucket to describe traffic flows. Despite this apparent convergence, there remain serious doubts about the efficacy of this choice.

For present purposes, we consider a leaky bucket as a reservoir of capacity σ emptying at rate ρ and filling due to the controlled input flow. Traffic conforms to the leaky bucket descriptor if the reservoir does not overflow and then satisfies the inequality $A(t) \leq \rho t + \sigma$ introduced above. The leaky bucket has been chosen mainly because it simplifies the problem of controlling input conformity. Its efficacy depends additionally on being able to choose appropriate parameter values for a given flow and then being able to efficiently guarantee quality of service by means of admission control.

The leaky bucket may be viewed either as a statistical descriptor approximating (or more exactly, providing usefully tight upper bounds on) the actual mean rate and burstiness of a given flow or as the definition of an envelope into which the traffic must be made to fit by shaping. Broadly speaking, the first viewpoint is appropriate for stream traffic, for which excessive shaping delay would be unacceptable, while the second would apply in the case of (aggregates of) elastic traffic.

Stream traffic should pass transparently through the policer without shaping by choosing large enough bucket rate and capacity parameters. Experience with video traces shows that it is very difficult to define a happy medium solution between a leak rate ρ close to the mean with an excessively large capacity σ, and a leak rate close to the peak with a moderate capacity [25]. In the former case, although the overall mean rate is accurately predicted, it is hardly a useful traffic characteristic since the rate averaged over periods of several seconds can be significantly different. In the latter, the rate information is insufficient to allow significant statistical multiplexing gains.

For elastic flows it is, by definition, possible to shape traffic to conform to the parameters of a leaky bucket. However, it remains difficult to choose appropriate leaky bucket parameters. If the traffic is long-range dependent, as in the case of an aggregation of flows, the performance models studied in this book indicate that queueing behavior is particularly severe. For any choice of leak rate ρ less than the peak rate and a bucket capacity σ that is not impractically large, the majority of traffic will be smoothed and admitted to the network at rate ρ. The added value of a nonzero bucket capacity is thus extremely limited for such traffic.

We conclude that, for both stream and elastic traffic, the leaky bucket constitutes an extremely inadequate descriptor of traffic variability.

16.3.3 Admission Control

To perform admission control based solely on the parameters of a leaky bucket implies unrealistic worst-case traffic assumptions and leads to considerable resource allocation inefficiency. For statistical multiplexing, flows are typically assumed to independently emit periodic maximally sized peak rate bursts separated by minimal silence intervals compatible with the leaky bucket parameters [8]. Deterministic delay bounds are attained only if flows emit the maximally sized peak rate bursts *simultaneously*. As discussed above, these worst-case assumptions bear little relation to real traffic characteristics and can lead to extremely inefficient use of network resources.

An alternative is to rely on historical data to predict the statistical characteristics of know flown types. This is possible for applications like the telephone, where an estimate of the average activity ratio is sufficient to predict performance when a set of conversations share a link using bufferless multiplexing. It is less obvious in the case of multiservice traffic, where there is generally no means to identify the nature of the application underlying a given flow.

The most promising admission control approach is to use measurements to estimate currently available capacity and to admit a new flow only if quality of service would remain satisfactory assuming that flow were to generate worst-case traffic compatible with its traffic descriptor. This is certainly feasible in the case of bufferless multiplexing. The only required flow traffic descriptor would be the peak rate with measurements performed in real-time to estimate the rate required by existing flows [11, 14]. Without entering into details, a sufficiently high level of utilization is compatible with negligible overload probability, on condition that the peak rate of individual flows is a small fraction of the link rate. The latter condition ensures that variations in the combined input rate are of relatively low amplitude, limiting the risk of estimation errors and requiring only a small safety margin to account for the most likely unfavorable coincidences in flow activities.

For buffered multiplexing, given the dependence of delay and loss performance on complex flow traffic characteristics, design of efficient admission control remains an open problem. It is probably preferable to avoid this type of multiplexing and to instead use reactive control for elastic traffic.

16.4 CLOSED-LOOP CONTROL FOR ELASTIC TRAFFIC

Closed-loop, or reactive, traffic control is suitable for elastic flows, which can adjust their rate according to current traffic levels. This is the principle of TCP in the Internet and ABR in the case of ATM. Both protocols aim to fully exploit available network bandwidth while achieving fair shares between contending flows. In the following sections we discuss the objectives of closed-loop control, first assuming a fixed set of flows routed over the network, and then taking account of the fact that this set of flows is a random process.

16.4.1 Bandwidth Sharing Objectives

It is customary to consider bandwidth sharing under the assumption that the number of contending flows remains fixed (or changes incrementally, when it is a question of studying convergence properties). The sharing objective is then essentially one of fairness: a single isolated link shared by n flows should allocate $(1/n)$th of its bandwidth to each. This fairness objective can be generalized to account for a weight φ_i attributed to each flow i, the bandwidth allocated to flow i then being proportional to $\varphi_i / \sum_{\text{all flows}} \varphi_j$. The φ_i might typically relate to different tariff options.

In a network the generalization of the simple notion of fairness is max–min fairness [3]: allocated rates are as equal as possible, subject only to constraints imposed by the capacity of network links and the flow's own peak rate limitation. The max-min fair allocation is unique and such that no flow rate λ, say, can be increased without having to decrease that of another flow whose allocation is already less than or equal to λ.

Max-min fairness can be achieved exactly by centralized or distributed algorithms, which calculate the explicit rate of each flow. However, most practical algorithms sacrifice the ideal objective in favor of simplicity of implementation [1]. The simplest rate sharing algorithms are based on individual flows reacting to binary congestion signals. Fair sharing of a single link can be achieved by allowing rates to increase linearly in the absence of congestion and decrease exponentially as soon as congestion occurs [6].

It has recently been pointed out that max-min fairness is not necessarily a desirable rate sharing objective and that one should rather aim to maximize overall utility, where the utility of each flow is a certain nondecreasing function of its allocated rate [15, 18]. General bandwidth sharing objectives and algorithms are further discussed in Massoulié and Roberts [21].

Distributed bandwidth sharing algorithms and associated mechanisms need to be robust to noncooperative user behavior. A particularly promising solution is to perform bandwidth sharing by implementing per flow, fair queueing. The feasibility of this approach is discussed by Suter et al. [29], where it is demonstrated that an appropriate choice of packets to be rejected in case of congestion (namely, packets at the front of the longest queues) considerably improves both fairness and efficiency.

16.4.2 Randomly Varying Traffic

Fairness is not a satisfactory substitute for quality of service, if only because users have no means of verifying that they do indeed receive a "fair share." Perceived throughput depends as much on the number of flows currently in progress as on the way bandwidth is shared between them. This number is not fixed but varies randomly as new transfers begin and current transfers end.

A reasonable starting point to evaluating the impact of random traffic is to consider an isolated link and to assume new flows arrive according to a Poisson process. On further assuming the closed-loop control achieves exact fair shares immediately as the number of flows changes, this system constitutes an $M/G/1$

processor sharing queue for which a number of interesting results are known [16]. A related traffic model where a finite number of users retrieve a succession of documents is discussed by Heyman et al. [13].

Let the link capacity be c and its load (arrival rate $\times$ mean size$/c$) be ρ. If $\rho < 1$, the number of transfers in progress N_t is geometrically distributed, $\Pr\{N_t = n\} = \rho^n(1-\rho)$, and the average throughput of any flow is equal to $c(1-\rho)$. These results are insensitive to the document size distribution. Note that the expected response time is finite for $\rho < 1$, even if the document size distribution is heavy tailed. This is in marked contrast with the case of a first-come-first-served $M/G/1$ queue, where a heavy-tailed service time distribution with infinite variance leads to infinite expected delay for any positive load. In other words, for the assumed self-similar traffic model, closed-loop control avoids the severe congestion problems associated with open-loop control. We conjecture that this observation also applies for a more realistic flow arrival process.

If flows have weights φ_i as discussed above, the corresponding generalization of the above model is discriminatory processor sharing as considered, for example, by Fayolle et al. [9]. The performance of this queueing model is not insensitive to the document size distribution and the results in Fayolle et al. [9] apply only to distributions having finite variance. Let $R(p)$ denote the expected time to transfer a document of size p. Figure 16.1 shows the normalized response time $R(p)/p$, as a function of p, for a two-class discriminatory processor sharing system with the following parameters: unit link capacity, $c = 1$; both classes have a unit mean,

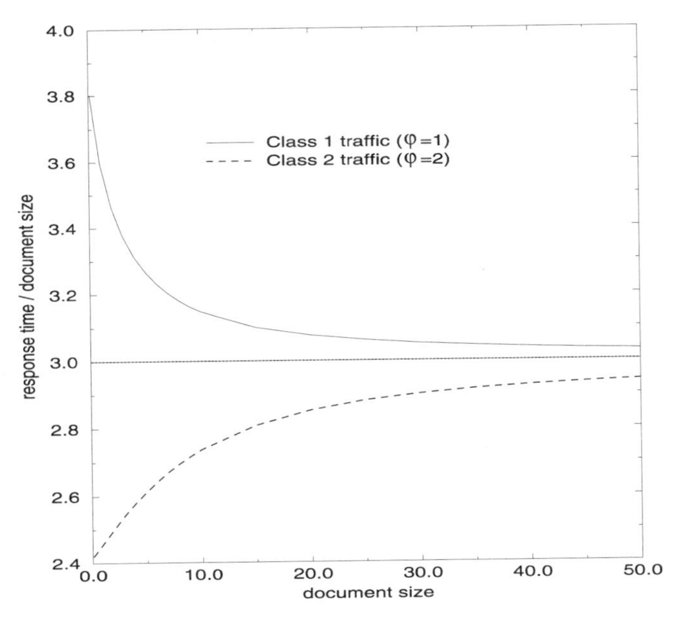

Fig. 16.1 Normalized response time $R(p)/p$ for discriminatory processor sharing.

exponential size distribution and an arrival rate of $\frac{1}{3}$; flows of class i have sharing parameter φ_i, where $\{\varphi_1, \varphi_2\} = \{1, 2\}$.

From the figure we note that the sharing parameters ensure effective discrimination for the transfer time of short documents but that throughput for both classes tends to the limit $c(1 - \rho)$ as document size increases. The limiting large object throughput is explained by the fact that, whatever its sharing parameter φ_i, a very long transfer utilizes all the bandwidth except that required by other users, equal on average to $c\rho$.

Results for hyperexponential distributions (not reported here) show that discrimination is more effective as the document size distribution variability increases. It is likely therefore that for a heavy-tailed distribution most document transfers will see an improvement in throughput with an increasing weight, although the improvement is less than proportional and still tends to disappear for exceptionally long documents.

Note that throughput of large objects is not affected by the rate assigned to the transfer of short objects, which start and finish within the transfer time of the former. Overall throughput can therefore be improved by giving priority to short objects. Indeed, it is known that the response time performance of a shared resource is optimized on using the shortest remaining processing time (SRPT) first scheduling discipline: a controller is assumed to know the remaining volume of data of all documents to be transferred and devotes link capacity exclusively to the smallest; if a new arrival concerns a document whose size is less than that of the document in service, the latter is preempted; any preempted transfer resumes service where it left off, as soon as its remaining volume is again smaller than that of any other pending request.

The performance of SRPT was studied by Schrage and Miller [30]. They derive expressions for the response time $R(p)$ of a document of size p under an assumption of Poisson arrivals and general service time distribution. Figure 16.2 shows a numerical evaluation of their formulas for exponential and infinite variance Pareto distributed document sizes, respectively. Link load is $\frac{2}{3}$, as in the example of Fig. 16.1. The p axis, in units of the mean document size, is on a log scale to capture the heavy tail particularly of the Pareto distribution. The normalized response time $R(p)/p$ is considerably less than that of perfectly fair sharing (i.e., the processor sharing model), equal here to 3 for all values of p. It is interesting to note that, in this system, the response time for medium to large documents improves on passing from short-range to long-range dependent processes.

Implementation of SRPT in the case of a single link would, of course, be very complex and the appropriate extension of this principle to a network remains unclear. However, it does provide a clear illustration that fairness, or weighted fairness, is not necessarily a useful objective in bandwidth sharing. In particular, both users and network provider stand to gain by employing a flow control protocol that discriminates in favor of short documents.

The processor sharing model illustrates how performance can deteriorate suddenly as offered load ρ increases through 1: if link bandwidth c is high, throughput performance is good even when ρ is close to 1. For heavier loads,

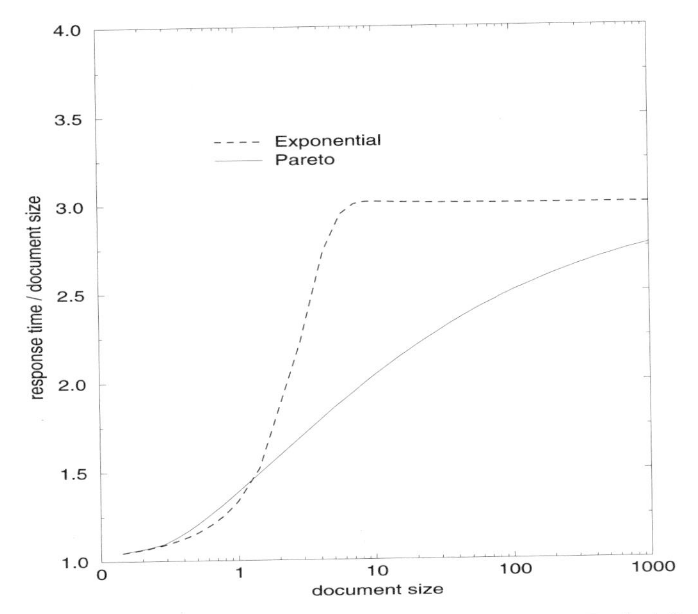

Fig. 16.2 Normalized response time $R(p)/p$ time for SRPT scheduling.

throughput is zero and the number of transfers in progress increases indefinitely. Of course, the model then ceases to be accurate, since many real users will abandon transfers as soon as they begin to notice the effects of such congestion. Since an abandoned or otherwise incomplete transfer serves no useful purpose and only adds to congestion, goodput can be improved by employing admission control.

16.4.3 Admission Control for Elastic Traffic

Admission control, by limiting the number of flows using any given link, ensures that throughput never decreases below some minimum acceptable level for flows that are admitted. Exactly what would constitute a minimum acceptable throughput is not clear. The choice depends on a trade-off between the extra utility of accepting a new flow and the risk that existing transfers would be prematurely interrupted if their rate were decreased. It does seem clear that such a minimum exists (through it may be different for different users) since otherwise a saturated network would be unstable [22].

Admission control does not necessarily imply a complex flow setup stage with explicit signaling exchanges between user and network nodes. This would be quite unacceptable for most elastic flows, which are of very short duration. We envisage a network rather similar to the present Internet where users simply send their data as and when they wish. However, nodes implementing admission control would keep a record of the identities of existing flows currently traversing each link in order to be able to recognize the arrival of a packet from a new flow. Such a packet would be accepted and its identifier added to the list of active flows if the number of flows currently in progress were less than a threshold, and would otherwise be rejected. A flow would be erased from the list if it sent no packets during a certain time-out interval.

Although many additional practical considerations would need to be addressed, such a control procedure does seem feasible technically given recent developments in router technology [17, 19]. Note, finally, that knowledge of the state of network links in terms of the number of flows currently in progress would also allow intelligent routing strategies where flows are not sent blindly to saturated links when other paths are available.

16.5 TOWARD A SIMPLE SERVICE MODEL

Given the above discussion on possible control options, it is tempting to speculate on the simplest service model capable of meeting identified requirements.

16.5.1 Service Classes

We envisage a service model with just two service classes, one based on open-loop control for stream traffic and the other using closed-loop control for elastic traffic. In this service model, flows destined for the first class declare just a peak rate that is actively policed by packet spacing at the network ingress. Measurement-based admission control would be used to ensure negligible data loss assuming bufferless multiplexing. Although, in practice, a small buffer is necessary to account for the nonfluid nature of traffic, delay and delay variation remain very small. Loss and delay performance are independent of any long-range dependence in the rate process of flows. A low loss rate (10^{-9}, say) is compatible with a reasonable average link utilization (50%, say) if the peak rate of flows is not more than a small fraction of the link bandwidth ($\frac{1}{100}$, say) [26, Chap. 16].

The necessary characteristics of the closed-loop control are less well understood. We can rely on users reacting intelligently to congestion signals, as in TCP, if the network additionally implements queue management mechanisms preventing uncooperative flows from adversely affecting the quality of service of other users. A promising solution is to perform per flow queueing with flow identification performed "on the fly," as suggested by Suter et al. [29]. The identification of the set of flows currently using a link allows the implementation of a simple admission control procedure whereby any packets from new flows are rejected when the number of flows in progress exceeds a link-capacity-dependent threshold.

Sharing link capacity dynamically between stream and elastic flows is advantageous for both types of traffic: a very low loss rate for stream traffic is not incompatible with reasonable utilization if elastic traffic constitutes a significant proportion of the total load; elastic flows gain greater throughput by being able to exploit the residual capacity necessarily left over by stream traffic to meet data loss rate and blocking probability targets. Admission control for both stream and elastic flows would take account of the measured stream load and the current count of the number of active elastic flows.

Simple head of line priority is sufficient to meet the delay requirements of stream traffic while per flow queueing is the perferred solution for elastic traffic. Fair queueing among elastic flows leads to fair bandwidth sharing. However, performance could be improved by implementing packet scheduling schemes giving priority to short documents. The performance of rate sharing schemes like fair queueing and SRPT does not appear to be adversely affected by the heavy-tailed nature of the document size distribution.

For any given application, a user might choose to set up a stream or an elastic flow. The choice depends on quality of service and cost. We have argued that open-loop control can meet the strict delay requirements of stream traffic while closed-loop control provides higher throughput for the transfer of elastic documents. The issue of providing price incentives to influence user choices is discussed in the next section (see also Odlyzko [24] and Roberts [27]).

16.5.2 The Impact of Charging

For largely historical reasons, most users of the Internet today are charged on a flat rate basis. They pay a fixed monthly charge that is independent of the volume of traffic they produce, although the charge does depend on the capacity of their network access line. The major advantage of *flat rate pricing* is its simplicity, leading to lower network operating costs. A weakness is its inherent unfairness, a light user having to pay as much as a heavy user. A more immediate problem is the absence of restraint inherent in this charging scheme, which may be said to contribute to the present state of congestion of the Internet.

Network usage can be controlled by the introduction of usage sensitive charging with rates determined by the level of congestion. This is the principle of *congestion pricing*. Congestion pricing ideally leads to an economic optimum, where available resources are used to produce maximum utility. While theoretically optimal schemes like the "smart market" [23] are unlikely to be implemented for reasons of practicality, it has been argued that the congestion control objective can be acheived simply by offering a number of differentially priced service classes with charges increasing with the expected level of quality of service [28]. Users determine the amount they are charged by their choice of service class. They have an incentive to choose more expensive classes in times of congestion. Such schemes suffer from a lack of transparency: How can users tell if the network provider isn't deliberately causing congestion? Why should they pay more to an inefficient provider? Are they currently paying more than they need to, given current traffic levels? Note that congestion pricing is not generally employed in other service industries subject to demand overloads such as electricity supply, public transportation, or the telephone network.

An alternative is to charge for use depending on the amount of resources used per transaction, accounting possibly for distance (number of hops) as well as volume.

We refer to such a charging scheme as *transaction pricing*. Transaction pricing is widely used in the telephone network (with the notable exception of local networks in North America), where switches and links are sized to ensure that congestion occurs only exceptionally. The price must be set at a value allowing the network operator to recover the cost of investment. Differential pricing according to the time of day is used to smooth out the demand profile to some extent but this is not generally viewed as a congestion control mechanism.

Choice between flat rate pricing, congestion pricing, and transaction pricing depends among other things on their ability to assure the economic viability of the network provider. Congestion pricing is intended to optimize the use of a network, not to recover the cost of installed infrastructure, which is regarded as a "sunk cost" in the economic optimization. If the network is well provisioned and always offers good quality of service, for example, costs must be entirely recovered by flat rate access charges. Transaction pricing has proved successful for telephone network operators, but then so has flat rate pricing in the case of North American local networks. Transaction pricing has the advantage of distributing the cost of shared network resources in relation to usage. In addition to being appealing from a fairness point of view, this is in line with the trend in telecommunications for "unbundling" and cost related pricing.

A second major issue is the complexity of implementing the different schemes. Any move from flat rate pricing appears as a major change for the Internet, requiring accounting and billing systems at least as complex as those of the telephone network. The cost of such systems must be weighed against any expected improvements in efficiency.

In proposing a simple two-class model, we have in mind a mixture of flat rate pricing and transaction pricing, where the role of the latter would be to allow users to be charged in relation to their use of shared resources. We argue [27] that, in a large network sized to offer good quality of service, resource provision is largely independent of whether the traffic is stream or elastic. This suggests a simple tariff based just on the number of bytes crossing an interface.

A likely evolutionary step is that cost-related charging be introduced for large users, including ISPs connected to a backbone, with individual small users continuing to pay only a flat rate charge.

The simple service model makes no distinction between elastic documents like Web pages intended for immediate display and documents like mail whose delivery is deferable. Users do not require minimal throughput for the latter and would arguably expect to pay less for their transport. A possible solution is that deferable documents transit via servers, operated by a "postal service," external to the transport network of routers and links. Users deliver a document directly to a local server, which then takes charge of forwarding it to its destination(s), generally via intermediate servers. The users pay the "postal service," which in turn pays the transport network. The service is cheaper for end users because the servers can send data in off-peak hours and negotiate special tariff arrangements with the network provider.

16.6 NETWORK SIZING

Traffic engineering for a multiservice network handling both stream and elastic traffic is still a largely unexplored field. In this section we suggest how it may be possible to generalize the methods and tools developed over the years for dimensioning the telephone network.

16.6.1 Provisioning for Stream Traffic

To determine the network capacity required to meet a target blocking probability for stream flows, it is necessary to make assumptions about the arrival process of new demands, their rate, and their duration. For illustration purposes, we consider a simple traffic model consisting of one link receiving traffic from a very large population of users. Details and more general models may be found in Roberts et al. [26] for example.

First assume that it is possible to identify m distinct homogeneous classes, flows of each class having a common rate distribution. Flows from class i arrive according to a Poisson process of intensity λ_i (requests per second) and have an expected duration of $1/\mu_i$ seconds. Their peak rate is p_i. For a fixed (fairly large) link capacity c, the impact of a flow of class i on the probability of data loss can be summarized in a single figure, the effective bandwidth: the effective bandwidth e_i is such that the probability of data loss is negligible (less than a target value) as long as $\sum n_i e_i \leq c$, where n_i is the number of class i flows in progress.

Although measurement-based admission control does not rely on the identification of the different classes (a new flow is denied access if its peak rate is greater than a real-time estimate of available bandwidth), for dimensioning purposes we can assume a flow of class j will be blocked if $\sum n_i e_i > c - e_j$. With this blocking condition and the assumption of Poisson arrivals, the distribution of the n_i has a well-known product form enabling computation of the blocking probability. Note that blocking probabilities and data loss rates are insensitive to the distribution of flow duration.

A reasonable approximation for the blocking probability of a flow with peak rate p_i when c is large with respect to the e_i is given by

$$B_i \approx \frac{p_i}{\delta} E(a/\delta, c/\delta), \tag{16.1}$$

where $a = \sum e_i(\lambda_i/\mu_i)$, $\delta = \sum e_i^2(\lambda_i/\mu_i)/a$ and

$$E(a, n) = \frac{a^n}{n!} \Big/ \sum_{i \leq n} \frac{a^i}{i!}$$

is Erlang's formula.

Formula (16.1) is a simplification of the formulas given by Lindberger [20]. It is less accurate but more clearly demonstrates the structural relationship between performance and traffic characteristics. Instead of identifying traffic classes with common traffic characteristics, it may prove more practical to estimate the essential parameters a and δ directly.

It is well known that application of Erlang's formula leads to scale economies: to achieve a low blocking probability and high utilization (a/c), it is necessary to have a large capacity c. For multirate traffic with blocking probabilities given by Eq. (16.1), the same requirement implies a high value of c/δ. The line labeled "stream" in Fig. 16.3 shows how achievable utilization a/c in a simple Erlang loss system varies with c for a target blocking probability of 0.01.

16.6.2 Provisioning for Elastic Traffic

Following the simple service model introduced in Section 16.5, we assume throughput quality of service is satisfied by limiting the number of elastic flows on a link and seek to dimension link capacity such that the blocking probability is less than some low target value ϵ.

Consider first an isolated link handling only elastic flows. Assuming Poisson arrivals, a minimum throughput requirement θ, exact fair shares (i.e., processor sharing service), and a link bandwidth of $c = n\theta$, the probability of blocking is equal to the saturation probability in an $M/G/1$ processor sharing queue of capacity n:

$$B_e = \rho^n(1 - \rho)/(1 - \rho^{n+1}) \tag{16.2}$$

where ρ is the link load.

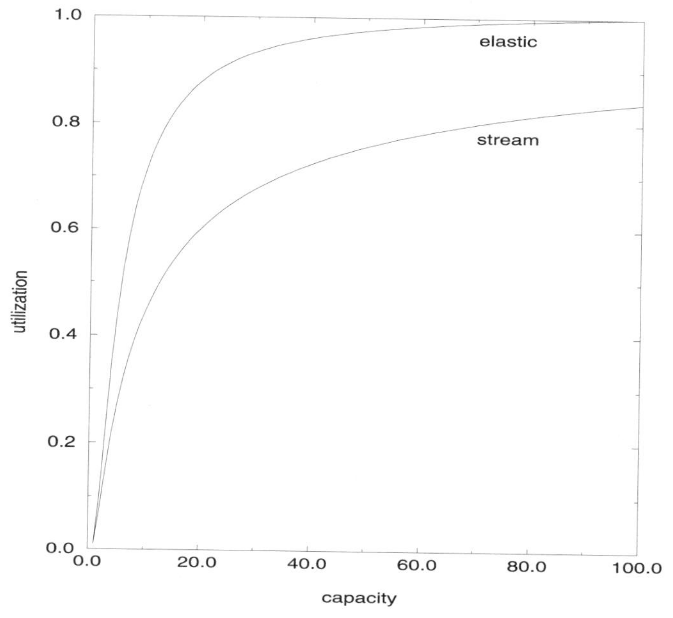

Fig. 16.3 Achievable utilization for stream and elastic traffic.

Since elastic flows use bandwidth more efficiently, blocking probability (16.2) can be considerably less than the corresponding probability for stream traffic requiring constant rate θ, as given by Erlang's formula $E(n\rho, n)$. The line labeled "elastic" in Fig. 16.3 shows achievable utilization ρ for elastic traffic such that B_e, given by Eq. (16.2), is equal to 0.01. These results clearly illustrate the scale economies effect and the greater efficiency of elastic sharing.

The advantage of elastic sharing with respect to rigid rate allocations is somewhat mitigated in a network where flows cannot always attain a full share of available link bandwidth because of congestion on other links of their path and their own limited peak rate. If, however, the flows can at least attain rate θ and this rate is guaranteed by admission control on every network link, the utilization predicted by the Erlang formula constitutes a lower bound. In other words, the Erlang formula can be used as a conservative dimensioning tool to determine the traffic capacity of a link dedicated to elastic traffic: a link of capacity c can handle a volume of elastic traffic A_e (flow arrival rate $\times$ average size) with minimum throughput θ and blocking probability less than ϵ if $E(A_e, c/\theta) \leq \epsilon$. Given the scale economies achieved with the Erlang formula, this simple dimensioning approach is efficient if c/θ is large (e.g., $A_e/c > 0.8$ if $c/\theta > 100$ for a 1% blocking probability).

An advantage of the above approach is that the integration of stream and elastic traffic is taken into account simply by including the latter as an additional traffic class in the multirate dimensioning methods alluded to in the previous section.

16.7 CONCLUSION

The realization of quality of service guarantees in a multiservice network depends more on sound traffic engineering than on the definition of a service model allowing priority access for an undefined number of privileged users.

We have argued that the service model should facilitate traffic engineering by distinguishing two broad categories of traffic: stream and elastic. For each category, the appropriate entity for traffic management is an individual flow (e.g., one videoconference, one file transfer) and not either an isolated packet or some aggregation of flows. A tentative simple service model is based on just two traffic classes.

One class destined for stream traffic is based on open-loop control and uses "bufferless multiplexing" with measurement-based admission control. This choice enables delay and loss rate performance guarantees, even for self-similar flows. The leaky bucket is not useful as a traffic descriptor and the only traffic parameter required here is the flow peak rate.

The second service class uses closed-loop control to share bandwidth between elastic flows. We advocate a lightweight form of admission control for elastic traffic, requiring that each link identify the flows it is currently transporting. Per flow queueing would be useful to enforce fairness, or to share bandwidth more efficiently by giving priority to short transfers, for example. In the simple bandwidth sharing models considered here, the heavy-tailed distribution of the size of transferred documents does not adversely affect the response time performance of closed-loop control.

We consider charging as a means to recover the network provider's costs rather than as a tool for congestion control. Prices would ideally be set to just ensure profitability when the network is dimensioned to handle all the offered traffic with good quality of service. There appears no essential reason to price stream and elastic traffic differently per byte transported. Users would naturally choose the service class best suited to their quality of service requirements: low delay for stream flows, high throughput for elastic flows.

We have given some indications of how traditional traffic engineering practice might be extended to a multiservice network based on the proposed simple service model. The basic principle is that transparency and throughput quality of service are assured by means of admission control acting at flow level, while the network is sized to produce a sufficiently low blocking probability.

REFERENCES

1. A. Arulambalam and X. Q. Chen. Allocating fair rates for available bit rate service in ATM networks. *IEEE Commun. Mag.*, **34**(11):92–100, 1996.
2. M. F. Arlitt and C. Williamson. Web server workload characterization: the search for invariants. In *Proc. ACM SIGMETRICS'96*, pp. 126–137, 1996.
3. D. Bertsekas and R. Gallager. *Data Networks*. Prentice Hall, Englewood Cliffs, NJ, 1987.
4. V. Bolotin. Telephone circuit holding time distribution. In J. Labetoulle and J. Roberts, eds., *The Fundamental Role of Teletraffic in the Evolution of Telecommunications Networks. Proceedings of ITC 14*. Elsevier, New York, 1994.
5. M. Crovella and A. Bestavros. Self-similarity in World Wide Web traffic: evidence and possible causes. In *Proc. ACM SIGMETRICS'96*, pp. 160–169, 1996.
6. D. M. Chiu and R. Jain. Analysis of the increase and decrease algorithms for congestion avoidance in computer networks. *Comput. Networks ISDN Syst.*, **17**:1–14, 1989.
7. R. L. Cruz. A calculus of network delay. Part I: network elements in isolation. *IEEE Trans. Inf. Theory*, **37**:114–131, 1991.
8. A. Elwalid, D. Mitra, and R. H. Wentworth. A new approach to allocating buffers and bandwidth to heterogeneous regulated traffic in an ATM node. *IEEE JSAC*, **13**(6):1115–1127, August 1995.
9. G. Fayolle, I. Mitrani, and R. Iasnogorodski. Sharing a processor among many jobs. *J. ACM*, **27**(3):519–532, 1980.
10. M. Frater. Origins of long range dependence in variable bit rate video traffic. In V. Ramaswami and P. E. Wirth, eds., *Teletraffic Contributions for the Information Age. Proceedings of ITC 15*. Elsevier, New York, 1997.
11. R. Gibbens, F. Kelly, and P. Key. A decision theoretic approach to call admission control in ATM networks. *IEEE JSAC*, **13**(6):1101–1114, August 1995.
12. M. Garrett and W. Willinger. Analysis, modeling and generation of self-similar VBR video traffic. In *Proc. SIGCOMM'94*, pp. 269–280, 1994.

13. D. Heyman, T. Lakshman, and A. Neidhart. A new method for analysing feedback-based protocols with application to engineering Web traffic over the Internet. In *Proc. ACM SIGMETRICS'97*, pp. 24–38, 1997.
14. S. Jamin, S. J. Shenker, and P. B. Danzig. Comparison of measurement-based admission control algorithms for controlled load service. In *Proc. INFOCOM'97*, April 1997.
15. F. Kelly. Charging and rate control for elastic traffic. *Eur. Trans. Telecommun.*, **8**:33–37, 1997.
16. L. Kleinrock. *Queueing Systems, Volume 2*. Wiley, New York, 1975.
17. V. P. Kumar, T. V. Lakshman, and D. Stiliadis. Beyond best effort: router architectures for differentiated services of tomorrow's Internet. *IEEE Commun. Mag.*, **36**(5):152–164, May 1998.
18. F. Kelly, A. Maulloo, and D. Tan. Rate control for communication networks: shadow prices, proportional fairness and stability. *J. Oper. Res. Soc.*, **49**:237–252, 1998.
19. S. Keshav and R. Sharma. Issues and trends in router design. *IEEE Commun. Mag.*, **36**(5):144–151, May 1998.
20. K. Lindberger. Dimensioning and design methods for integrated ATM networks. In J. Labetoulle and J. Roberts, eds., *The Fundamental Role of Teletraffic in the Evolution of Telecommunications Networks. Proceedings of ITC 14*. Elsevier, New York, 1994.
21. L. Massoulié and J. Roberts. Bandwidth sharing: objectives and algorithms. In *Proc. INFOCOM'99*, pp. 1395–1403, March 1999.
22. L. Massoulié and J. Roberts. Arguments in favor of admission control for TCP flows. In *Proceedings of ITC 16*, Elsevier, New York, 1999.
23. J. MacKie-Mason and H. Varian. Pricing the Internet. In B. Kahin and J. Keller, eds., *Public Access to the Internet*. Prentice Hall, Englewood Cliffs, NJ, 1995.
24. A. M. Odlyzko. The economics of the Internet: utility, utilization, pricing and quality of service. Preprint, 1998.
25. A. R. Reibman and A. W. Berger. Traffic descriptors for VBR videoconferencing over ATM networks. *IEEE/ACM Trans. Networking*, **3**(3):329–339, June 1995.
26. J. Roberts, U. Mocci, and J. Virtamo, eds. *Broadband Network Teletraffic (Final Report of COST 242)*. LNCS 1155. Springer-Verlag, New York, 1996.
27. J. Roberts. Quality of service guarantees and charging in multiservice networks. *IEICE Trans. Commun.* (Special issue on ATM traffic control and performance evaluation), **E81-B**(5):824–831, 1998.
28. S. Shenker, D. Clark, D. Estrin, and S. Herzog. Pricing in computer networks: reshaping the research agenda. *Telecommun. Policy*, **26**:183–201, 1996.
29. B. Suter, T. V. Lakshman, and D. Stiliadis. Design considerations for supporting TCP with per flow queueing. In *Proc. INFOCOM'98*, pp. 299–306, San Francisco, 1998.
30. L. Schrage and L. Miller. The $M/G/1$ queue with the shortest remaining processing time first discipline. *Oper. Res.*, **14**:670–684, 1966.

17

NETWORK DESIGN AND CONTROL USING ON/OFF AND MULTILEVEL SOURCE TRAFFIC MODELS WITH HEAVY-TAILED DISTRIBUTIONS

N. G. DUFFIELD AND W. WHITT
AT&T Labs–Research, Florham Park, NJ 07392

17.1 INTRODUCTION

In order to help design and control the emerging high-speed communication networks, we want source traffic models (also called offered load models or bandwidth demand models) that can be both realistically fit to data and successfully analyzed. Many recent traffic measurements have shown that network traffic is quite complex, exhibiting phenomena such as heavy-tailed probability distributions, long-range dependence, and self similarity; for example, see Cáceres et al. [7], Leland et al. [23], Paxson and Floyd [24], and Crovella and Bestavros [10].

In fact, the heavy-tailed distributions may be the cause of all these phenomena, because they tend to cause long-range dependence and (asymptotic) self-similarity. For example, the input and buffer content processes associated with an on/off source exhibit long-range dependence when the on and off times have heavy-tailed probability distributions; for example, see Section 17.9. Heavy-tailed distributions are known to cause self-similarity in models of (asymptotically) aggregated traffic; see Willinger et al. [27].

In this chapter we propose a way to analyze the performance of a network with multiple on/off sources and more general multilevel sources in which the on-time, off-time, and level-holding-time distributions are allowed to have heavy tails. To do

Self-Similar Network Traffic and Performance Evaluation, Edited by Kihong Park and Walter Willinger
ISBN 0-471-31974-0

so we must go be beyond the familiar Markovian analysis. To achieve the required analyzability with this added model complexity, we propose a simplified kind of analysis. In particular, we avoid the customary queueing detail (and its focus on buffer content and overflow) and instead concentrate on the instantaneous offered load. We describe the probability that aggregate demand (the input rate from a collection of sources) exceeds capacity (the maximum possible output rate) at any time. Focusing on the probability that aggregate demand exceeds capacity is tantamount to considering a bufferless model, which we believe is often justified. By also considering the probability that aggregate demand exceeds other levels, we provide a quite flexible performance characterization. This approach also can generate approximations describing loss and delay with finite capacity; for example, see Duffield and Whitt [14], Section 5. To a large extent, the present chapter is a review of our recent work [14, 15], to which we refer the reader for additional discussion. In Duffield et al. [16] the model is extended to include a nonhomogeneous Poisson connection arrival process. Then each active connection may generate traffic according to one of the source traffic models presented here. It is significant that we are able to obtain useful descriptions of the offered load in the nonstationary context.

17.2 A GENERAL SOURCE MODEL

Motivation for considering on/off and multilevel models as source models comes from traces of frame sizes generated by certain video encoders; for example, see Grasse et al. [19]. Shifts between levels in mean frame size appear to arise from scene changes in the video, with the distribution of scene durations heavy-tailed. Indeed, the expectation that scene durations will have heavy-tailed distributions is one of the motivations behind the renegotiated constant bit rate (RCBR) proposal of Grossglauser et al. [20].

Our approach is interesting for on/off and multilevel source models, but with little extra effort we can treat a wider class. The general model we consider has two components. The bandwidth demand for each source as a function of time, $\{B(t): t \geq 0\}$, is represented as the sum of two stochastic processes: (1) a macroscopic (longer-time-scale) *level process* $\{L(t): t \geq 0\}$ and (2) a microscopic (shorter-time-scale) *within-level variation process* $\{W(t): t \geq 0\}$, that is,

$$B(t) = L(t) + W(t), \quad t \geq 0. \tag{17.1}$$

We let the macroscopic level process $\{L(t): t \geq 0\}$ be a *semi-Markov process* (SMP) as in Çinlar [9, Chap. 10]; that is, the level process is constant except for jumps, with the jump transitions governed by a Markov process, while the level holding times (times between jumps) are allowed to have general distributions depending on the originating level and the next level. Given a transition from level j to level k, the holding time in level j has cumulative distribution function (cdf) F_{jk}. Conditional on the sequence of successive levels, the holding times are mutually independent. To obtain models compatible with traffic measurements cited earlier, we allow the holding-time cdf's F_{jk} to have heavy tails.

We assume that the within-level variation process $\{W(t): t \geq 0\}$ is a zero-mean piecewise-stationary process. During each holding-time interval in a level, the within-level variation process is an independent segment of a zero-mean stationary process, with the distribution of each segment being allowed to depend on the level. We allow the distribution of the stationary process segment to depend on the level, because it is natural for the variation about any level to vary from level to level.

We will require only a limited characterization of the within-level variation process; it turns out that the fine structure of the within-level variation process plays no role in our analysis. Indeed, that is one of our main conclusions. In several examples of processes that we envisage modeling by these methods, there will only be the level process. First, the level process may be some smoothed functional of a raw bandwidth process. This is the case with algorithms for smoothing stored video by converting into piecewise constant rate segments in some optimal manner subject to buffering and delay constraints; see Salehi et al. [25]. With such smoothing, the input rate will directly be a level process as we have defined it. Alternatively, the level process may stem from rate reservation over the period between level-shifts, rather than the bandwidth actually used. This would be the case for RCBR previously mentioned. In this situation we act as if the reservation level is the actual demand, and thus again have a level process.

A key to being able to analyze the system with such complex sources represented by our traffic model is exploiting asymptotics associated with multiplexing a large number of sources. The ever-increasing network bandwidth implies that more and more sources will be able to be multiplexed. This gain is generally possible, even in the presence of heavy-tailed distributions and more general long-range dependence; for example, see Duffield [12, 13] for demonstration of the multiplexing gains available for long-range dependent traffic in shared buffers. As the scale increases, describing the detailed behavior of all sources become prohibitively difficult, but fortunately it becomes easier to describe the aggregate, because the large numbers produce statistical regularity. As the size increases, the aggregate demand can be well described by laws of large numbers, central limit theorems, and large deviation principles.

We have in mind two problems: first, we want to do capacity planning and, second, we want to do real-time connection-admission control and congestion control. In both cases, we want to determine whether any candidate capacity is adequate to meet the aggregate demand associated with a set of sources. In both cases, we represent the aggregate demand simply as the sum of the bandwidth requirements of all sources. In forming this sum, we regard the bandwidth processes of the different sources as probabilistically independent.

The performance analysis for capacity planning is coarser, involving a longer time scale, so that it may be appropriate to do a steady-state analysis. However, when we consider connection-admission control and congestion control, we suggest focusing on a shorter time scale. We are still concerned with the relatively long time scale of connections, or scene times in video, instead of the shorter time scales of cells or bursts, but admission control and congestion control are sufficiently short-term that we propose focusing on the *transient* behavior of the aggregate demand process. In fact, even for capacity planning the transient analysis plays an important role. The transient analysis determines how long it takes to recover from rare congestion events. One application we have in mind is that of networks carrying rate-adaptive traffic. In this case the bandwidth process could represent the ideal demand of a source, even though it is able to function when allocated somewhat less bandwidth. So from the point of view of quality, excursion of aggregate bandwidth demand above available supply may be acceptable in the short-term, but one would want to dimension the link so that such excursions are sufficiently short-lived. In this or other contexts, if the recovery time from overload is relatively long, then we may elect to provide extra capacity (or reduce demand) so that overload becomes less likely. However, we do not focus specifically on actual design and control here; see Duffield and Whitt [14] for some specific examples. Our main contribution here is to show how the transient analysis for design and control can be done.

The remainder of this chapter is devoted to showing how to do transient analysis with the source traffic model. We suggest focusing on the future time-dependent mean conditional on the present state. The present state of each level process consists of the level and age (elapsed holding time in that level). Because of the anticipated large number of sources, the actual bandwidth process should be closely approximated by its mean, by the law of large numbers (LLN). As in Duffield and Whitt [14], the conditional mean can be thought of as a deterministic fluid approximation; for example, see Chen and Mandelbaum [8]. Since the within-level variation process has mean zero, the within-level variation process has no effect on this conditional mean. Hence, the conditional mean of the aggregate bandwidth process is just the sum of the conditional means of the component level processes. Unlike the more elementary $M/G/\infty$ model considered in Duffield and Whitt [14], however, the conditional mean here is not available in closed form.

In order to rapidly compute the time-dependent conditional mean aggregate demand, we exploit numerical inversion of Laplace transforms. It follows quite directly from the classical theory of semi-Markov process that explicit expressions can be given for the Laplace transform of the conditional mean. More recently, it has been shown that numerical inversion can be an effective algorithm; see Abate et al. [1].

For related discussions of transient analysis, design and control, see Chapters 13, 16, and 18 in this volume.

17.3 OUTLINE OF THE CHAPTER

The rest of this chapter is organized as follows. In Section 17.4, we show that the Laplace transform of the mean of the transient conditional aggregate demand can be expressed concisely. This is the main enabling result for the remainder of the chapter. The conditional mean itself can be very efficiently computed by numerically inverting its Laplace transform. To carry out the inversion, we use the Fourier-series method in Abate and Whitt [2] (the algorithm Euler exploiting Euler summation), although alternative methods could be used. The inversion algorithm is remarkably fast; computation for each time point corresponds simply to a sum of 50 terms. We provide numerical examples in Examples 17.6.2 and 17.8.1. Example 17.8.1 is of special interest, because the level-holding-time distribution there is Pareto.

In Section 17.5 we show that in some cases we can avoid the inversion entirely and treat much larger models. We can avoid the inversion if we can assume that the level holding times are relatively long compared to the times of interest for control. Then we can apply a single-transition approximation, which amounts to assuming that the Markov chain is absorbing after one transition. Then the conditional mean is directly expressible in terms of the level-holding-time distributions. Alternatively, we can perform a two-transition approximation, which only involves one-dimensional convolution integrals.

In Section 17.6 we describe the value of having more detailed state information, specifically the current ages of levels. With heavy-tailed distributions, a large elapsed holding time means that a large remaining holding time is very likely; for examples see Duffield and Whitt [14, Section 8] for background, and Harchol-Balter and Downey [21] for an application in another setting.

In Section 17.7 we turn to applications to capacity planning. The idea is to approximate the probability of an excursion in demand using Chernoff bounds and other large deviation approximations, then chart its recovery to a target acceptable level using the results on transience. Interestingly, the time to recover from excursions sufficiently close to the target level depends on the level durations essentially only through their mean. Correspondingly, the conditional mean demand relaxes linearly from its excursion, at least approximately so, for sufficiently small times. If the chance for a larger excursion is negligible (as determined by the large deviation approximation mentioned) then this simple description may suffice. An example is given in Section 17.8.

In Section 17.9 we show how long-range dependence in the level process arises through heavy-tailed level-holding-time distributions. Finally, we draw conclusions in Section 17.10.

17.4 TRANSIENT ANALYSIS

17.4.1 Approximation by the Conditional Mean Bandwidth

Throughout this chapter, the state information on which we condition will be either the current level of each source or the current level and age (current time) in that level of each source. No state from the within-level variation process is assumed. Conditional on that state information, we can compute the probability that each source will be in each possible level at any time in the future, from which we can calculate the conditional mean and variance of the aggregate required bandwidth by adding.

The Lindberg–Feller central limit theorem (CLT) for non-identically-distributed summands can be applied to generate a normal approximation characterized by the conditional mean and conditional variance; see Feller [18, p. 262]. For the normal approximation to be appropriate, we should check that the aggregate is not dominated by only a few sources.

Let $B(t)$ denote the (random) aggregate required bandwidth at time t, and let $I(0)$ denote the (known deterministic) state information at time 0. Let $(B(t)|I(0))$ represent a random variable with the conditional distribution of $B(t)$ given the information $I(0)$. By the CLT, the normalized random variable

$$\frac{(B(t)|I(0)) - E(B(t)|I(0))}{\sqrt{\mathrm{Var}(B(t)|I(0))}} \tag{17.2}$$

is approximately normally distributed with mean 0 and variance 1 when the number of sources is suitably large.

Since the conditional mean alone tends to be very descriptive, we use the approximation

$$(B(t)|I(0)) \approx E(B(t)|I(0)), \tag{17.3}$$

which can be justified by a (weaker) law of large numbers instead of the CLT. We will show that the conditional mean in Eq. (17.3) can be efficiently computed, so that it can be used for real-time control. From Eq. (17.2), we see that the error in the approximation (17.3) is approximately characterized by the conditional standard deviation $\sqrt{\mathrm{Var}(B(t)|I(0))}$. We also will show how to compute this conditional standard deviation, although the required computation is more difficult. If there are n sources that have roughly equal rates, then the conditional standard deviation will be $O(\sqrt{n})$, while the conditional mean is $O(n)$.

Given that our approximation is the conditional mean, and given that our state information does not include the state of the within-level variation process, the within-level variation process plays no role because it has zero mean. Let i index the source. Since the required bandwidths need not have integer values, we index the level by the integer j, $1 \leq j \leq J_i$, and indicate the associated required bandwidths in the level by b_j^i. Hence, instead of Eq. (17.1), the required bandwidth for source i can be expressed as

$$B^i(t) = b^i_{L^i(t)} + W_{L^i(t)}(t), \quad t \geq 0. \tag{17.4}$$

Let $P_{jk}^{(i)}(t|x)$ be the probability that the source-i level process is in level k at time t given that time 0 it was in level j and had been so for a period x (i.e., the age or elapsed level holding time at time 0 is x). If $\mathbf{j} \equiv (j_1, \ldots, j_n)$ and $\mathbf{x} \equiv (x_1, \ldots, x_n)$ are the vectors of levels and ages of the n source level processes at time 0, then the *state information* is $I(0) = (\mathbf{j}, \mathbf{x}) = (j_1, \ldots, j_n; x_1, \ldots, x_n)$ and the *conditional aggregate mean* is

$$E(B(t)|I(0)) \equiv M(t|\mathbf{j}, \mathbf{x}) = \sum_{i=1}^{n} \sum_{k_i=1}^{J_i} P_{j_i k_i}^{(i)}(t|x_i) b_{k_i}^i. \tag{17.5}$$

From Eq. (17.5), we see that we need to compute the conditional distribution of the level, that is, the probabilities $P_{jk}^{(i)}(t|x)$, for each source i. However, we can find relatively simple expressions for the Laplace transform of $P_{jk}^{(i)}(t|x)$ with respect to time because the level process of each source has been assumed to be a semi-Markov process.

We now consider a single source and assume that its required bandwidth process is a semi-Markov process (SMP). (We now have no within-level variation process.) We now omit the superscript i. Let $L(t)$ and $B(t)$ be the level and required bandwidth, respectively, at time t as in Eq. (17.4). The process $\{L(t): t \geq 0\}$ is assumed to be an SMP, while the process $\{B(t): t \geq 0\}$ is a function of an SMP, that is $B(t) = b_{L(t)}$, where b_j is the required bandwidth in level j. If $b_j \neq b_k$ for $j \neq k$, then $\{B(t): t \geq 0\}$ itself is an SMP, but if $b_j = b_k$ for some $j \neq k$, then in general $\{B(t): t \geq 0\}$ is not an SMP.

17.4.2 Laplace Transform Analysis

Let $A(t)$ be the age of the level holding time at time t. We are interested in calculating

$$P_{jk}(t|x) \equiv P(L(t) = k|L(0) = j, A(0) = x) \tag{17.6}$$

as a function of j, k, x, and t. The state information at time 0 is the pair (j, x). Let P be the transition matrix of the discrete-time Markov chain governing level transitions and let $F_{jk}(t)$ be the holding-time cdf given that there is a transition from level j to level k. For simplicity, we assume that $F_{jk}^c(t) = 1 - F_{jk}(t) > 0$ for all j, k, and t, so that all positive x can be level holding times. Let $P(t|x)$ be the matrix with elements $P_{jk}(t|x)$ and let $\hat{P}(s|x)$ be the Laplace transform (LT) of $P(t|x)$, that is, the matrix with elements that are the Laplace transforms of $P_{jk}(t|x)$ with respect to time:

$$\hat{P}_{jk}(s|x) = \int_0^\infty e^{-st} P_{jk}(t|x)\, dt. \tag{17.7}$$

We can obtain a convenient explicit expression for $\hat{P}(s|x)$. For this purpose, let G_j be the holding-time cdf in level j, unconditional on the next level, that is,

$$G_j(x) = \sum_k P_{jk} F_{jk}(x). \tag{17.8}$$

For any cdf G, let G^c be the complementary cdf, that is, $G^c(x) = 1 - G(x)$. Also, let

$$H_{jk}(t|x) = \frac{P_{jk}F_{jk}(t+x)}{G_j^c(x)} \quad \text{and} \quad G_j(t|x) = \sum_k H_{jk}(t|x) \tag{17.9}$$

for G_j in Eq. (17.8). Then let $\hat{h}_{jk}(s|x)$ and $\hat{g}_j(s|x)$ be the associated Laplace–Stieltjes transforms (LSTs):

$$\hat{h}_{jk}(s|x) = \int_0^\infty e^{-st}\, dH_{jk}(t|x) \quad \text{and} \quad \hat{g}_j(s|x) = \int_0^\infty e^{-st}\, dG_j(t|x). \tag{17.10}$$

Let $\hat{h}(s|x)$ be the matrix with elements $\hat{h}_{jk}(s|x)$. Let $\hat{q}(s)$ be the matrix with elements $\hat{q}_{jk}(s)$, where

$$Q_{jk}(t) = P_{jk}F_{jk}(t) \quad \text{and} \quad \hat{q}_{jk}(s) = \int_0^\infty e^{-st}\, dQ_{jk}(t). \tag{17.11}$$

Let $\hat{D}(s|x)$ and $\hat{D}(s)$ be the diagonal matrices with diagonal elements

$$\hat{D}_{jj}(s|x) \equiv [1 - \hat{g}_j(s|x)]/s, \quad \hat{D}_{jj}(s) \equiv [1 - \hat{g}_j(s)]/s, \tag{17.12}$$

where $\hat{g}_j(s)$ is the LST of the cdf G_j in Eq. (17.8).

Theorem 17.4.1. *The transient probabilities for a single SMP source have the matrix of Laplace transforms*

$$\hat{P}(s|x) = \hat{D}(s|x) + \hat{h}(s|x)\hat{P}(s|0), \tag{17.13}$$

where

$$\hat{P}(s|0) = (I - \hat{q}(s))^{-1}\hat{D}(s). \tag{17.14}$$

Proof. In the time domain, condition on the first transition. For $j \neq k$,

$$P_{jk}(t|x) = \sum_l \int_0^t dH_{jl}(u|x)P_{lk}(t-u|0),$$

so that

$$\hat{P}_{jk}(s|x) = \sum_l \hat{h}_{jl}(s|x)\hat{P}_{lk}(s|0),$$

while

$$P_{jj}(t|x) = G_j^c(t|x) + \sum_l \int_0^t dH_{jl}(u|x)P_{lj}(t-u|0),$$

so that

$$\hat{P}_{jj}(s|x) = \frac{1-\hat{g}_j(s|x)}{s} + \sum_l h_{jl}(s|x)\hat{P}_{lj}(s|0).$$

Hence, Eq. (17.13) holds. However, $P(t|0)$ satisfies a Markov renewal equation, as in Çinlar [9, Section 10.3]; that is, for $j \neq k$,

$$P_{jk}(t|0) = \sum_l \int_0^t dQ_{jl}(u)P_{lk}(t-u|0),$$

and

$$P_{jj}(t|0) = G_j^c(t) + \sum_l \int_0^\infty dQ_{jl}(u)P_{lj}(t-u|0),$$

so that

$$P(t|0) = D(t) + Q(t) * P(t|0)$$

where $*$ denotes convolution, and Eq. (17.14) holds. ■

To compute the LT $\hat{P}(s|0)$, we only need the LSTs $\hat{f}_{jk}(s)$ and $\hat{g}_j(s)$ associated with the basic holding-time cdf's F_{jk} and G_j. Abate and Whitt [3–5] give special attention to heavy-tail probability densities whose Laplace transforms can be computed and, thus, inverted. However, to compute $\hat{P}(s|x)$, we also need to compute $\hat{D}(s|x)$ and $\hat{h}(s|x)$, which require computing the LSTs of the *conditional* cdf's $H_{jk}(t|x)$ and $G_j(t|x)$ in Eq. (17.9). In general, even if we know the LST of a cdf, we do not necesssarily know the LST of an associated conditional cdf. However, in special cases, the LSTs of conditional cdf's are easy to obtain because the cdf's inherit their structure upon conditioning. For example, this is true for phase-type, hyperexponential and Pareto distributions; Duffield and Whitt [15, Section 4]. Moreover, other cdf's can be approximated by hyperexponential or phase-type cdf's; see Asmussen et al. [6] and Feldman and Whitt [17].

If the number of levels is not too large, then it will not be difficult to compute the required matrix inverse $(I - q(s))^{-1}$ for all required s. Note that, because of the probability structure, the inverse is well defined for all complex s with $\text{Re}(s) > 0$.

To illustrate with an important simple example, we next give the explicit formula for an on/off source. Suppose that there are two states with transition probabilities

$P_{12} = P_{21} = 1$ and holding time cdf's G_1 and G_2. From Eq. (17.9) or by direct calculation,

$$\hat{P}(s|0) \equiv \begin{pmatrix} \hat{P}_{11}(s|0) & \hat{P}_{12}(s|0) \\ \hat{P}_{21}(s|0) & \hat{P}_{22}(s|0) \end{pmatrix} = (I - \hat{q}(s))^{-1}\hat{D}(s)$$
$$= \frac{1}{s(1-\hat{g}_1(s)\hat{g}_2(s))} \begin{pmatrix} 1-\hat{g}_1(s) & \hat{g}_1(s)(1-\hat{g}_2(s)) \\ \hat{g}_2(s)(1-\hat{g}_1(s)) & 1-\hat{g}_2(s) \end{pmatrix}. \tag{17.15}$$

Suppose that the levels are labeled so that the initial level is 1. Note that all transitions from level 1 are to level 2. Hence, when considering the matrix $\hat{h}(s|x)$ in Eq. (17.10), it suffices to consider only the element $\hat{h}_{12}(s|x)$. Since

$$H_{12}^c(t|x) = G_1^c(t|x) = \frac{G_1^c(t+x)}{G_1^c(x)}, \tag{17.16}$$

then

$$\hat{h}_{12}(s|x) = \hat{g}_1(s|x) = \int_0^\infty e^{-st}\, dG_1(t|x). \tag{17.17}$$

Since $P_{11}(t|x) = 1 - P_{12}(t|x)$, it suffices to calculate only $P_{12}(t|x)$. Hence, in this context

$$\hat{P}_{12}(s|x) = \frac{\hat{g}_1(s|x)(1-\hat{g}_2(s))}{s(1-\hat{g}_1(s)\hat{g}_2(s))}. \tag{17.18}$$

We now determine the mean, second moment, and variance of the bandwidth process of a general multilevel source as a function of time, ignoring the within-level variation process. It is elementary that

$$m_j(t|x) = E(B(t)|L(0) = j, A(0) = x) = \sum_k P_{jk}(t|x)b_k, \tag{17.19}$$
$$s_j(t|x) = E(B(t)^2|L(0) = j, A(0) = x) = \sum_k P_{jk}(t|x)b_k^2, \tag{17.20}$$
$$v_j(t|x) = \mathrm{Var}(B(t)|L(0) = j, A(0) = x) = s_j(t|x) - m_j(t|x)^2. \tag{17.21}$$

We can calculate $m_j(t|x)$ and $s_j(t|x)$ by single inversions of their Laplace transforms, using

$$\hat{m}_j(s|x) \equiv \int_0^\infty e^{-st} m_j(t|x)\, dt = \sum_k P_{jk}(s|x)b_k$$

and

$$\hat{s}_j(s|x) = \sum_k \hat{P}_{jk}(s|x)b_k^2. \tag{17.22}$$

To properly account for the within-level variation process when it is present, we should add to its variance in level j, say, $w_j(t, x)$, to $v_j(t, x)$, but we need make no change to the mean $m_j(t, x)$. We anticipate that $w_j(t, x)$ will tend to be much less than $v_j(t, x)$ so that $w_j(t, x)$ can be omitted, but it could be included.

Finally, we consider the aggregate bandwidth associated with n sources. Again let a superscript i index the sources. The conditional aggregate mean and variance are

$$M(t|\mathbf{j}, \mathbf{x}) \equiv E(B(t)|I(0)) = \sum_{i=1}^{n} m_{j_i}^i(t|x_i) \tag{17.23}$$

and

$$V(t|\mathbf{j}, \mathbf{x}) \equiv \mathrm{Var}(B(t)|I(0)) = \sum_{i=1}^{n}[v_{j_i}^i(t|x_i) + w_{j_i}^i(t|x_i)], \tag{17.24}$$

where $\mathbf{j} = (j_1, \ldots, j_n)$ is the vector of levels and $\mathbf{x} = (x_1, \ldots, x_n)$ is the vector of elapsed holding times for the n sources with the single-source means and variances as in Eqs. (17.19) and (17.21).

It is significant that we can calculate the conditional aggregate mean at any time t by performing a single numerical inversion, for example, by using the Euler algorithm in Abate and Whitt [2]. We summarize this elementary but important consequence as a theorem.

Theorem 17.4.2. *The Laplace transform of the n-source conditional mean aggregate required bandwidth as a function of time is*

$$\hat{M}(s|\mathbf{j}, \mathbf{x}) \equiv \int_0^\infty e^{-st} M(t|\mathbf{j}, \mathbf{x})\, dt = \sum_{i=1}^{n} \sum_{k_i=1}^{J_i} \hat{P}_{J_i k_i}^{(i)}(s|x_i) b_{k_i}, \tag{17.25}$$

where the single-source transform $\hat{P}_{j_i k_i}^{(i)}(s|x_i)$ is given in Theorem 17.4.1.

Unlike for the aggregate mean, for the aggregate variance we evidently need to perform n separate inversions to calculate $v_{j_i}^i(t|x_i)$ for each i and then add to calculate $V(t|\mathbf{j}, \mathbf{x})$ in Eq. (17.24). (We assume that the within-level variances $w_{j_i}^i(t|x_i)$, if included, are specified directly.) Hence, we suggest calculating only the conditional mean in real time to perform control, and occasionally calculating the conditional variance to evaluate the accuracy of the conditional mean.

17.5 APPROXIMATIONS USING FEW TRANSITIONS

The most complicated part of the conditional aggregate mean transform $\hat{M}(s|\mathbf{j}, \mathbf{x})$ in Eq. (17.25) is the matrix inverse $(I - \hat{q}(s))^{-1}$ in the transform of the single-source transition probability in Eq. (17.14). Since the matrix inverse calculation can be a computational burden when the number of levels is large, it is natural to seek approximations that avoid this matrix inverse. We describe such approximations in this section.

The matrix inverse $(I - q(s))^{-1}$ is a compact representation for the series $\sum_{n=0}^{\infty} q(s)^n$. For $P(t|x)$, it captures the possibility of any number of transitions up to time t. However, if the holding times in the levels are relatively long in the time scales relevant for control, then the mean for times t of interest will only be significantly affected by a very few transitions. Indeed, often only a single transition need be considered.

The single-transition approximation is obtained by making the Markov chain absorbing after one transition. Hence, the single-transition approximation is simply

$$P_{jk}(t|x) \approx H_{jk}(t|x), \quad j \neq k, \quad \text{and} \quad P_{jj}(t|x) \approx G_j^c(t|x) + H_{jj}(t|x) \tag{17.26}$$

for $H_{jk}(t|x)$ in Eq. (17.9) and $G_j(t|x)$ in Eq. (17.9). From Eq. (17.26) we see that no inversion is needed.

Alternatively, we can develop a two-transition approximation. (Extensions to higher numbers are straightforward.) Modifying the proof of Theorem 17.4.1 in a straightforward manner, we obtain

$$P_{jk}(t|x) \approx \int_0^t G_k^c(T-u)\, dH_{jk}(u|x) + \sum_l \int_0^t P_{lk} F_{lk}(t-u)\, dH_{jl}(u|x) \tag{17.27}$$

for $j \neq k$ and

$$P_{jj}(t|x) \approx G_j^c(t|x) + \sum_l \int_0^t P_{lj} F_{lj}(t-u)\, dH_{jl}(u|x). \tag{17.28}$$

Expressed in the form of transforms, Eqs. (17.27) and (17.28) become

$$\hat{P}_{jk}(s|x) \approx \hat{h}_{jk}(s|x) \frac{(1 - \hat{g}_k(s))}{s} + \sum_l \hat{h}_{jl}(s|x) P_{lk} \frac{\hat{f}_{lk}(s)}{s} \tag{17.29}$$

for $j \neq k$ and

$$\hat{P}_{jj}(s|x) \approx \frac{1 - \hat{g}_j(s|x)}{s} + \sum_l \hat{h}_{jl}(s|x) P_{lj} \frac{\hat{f}_{lj}(s)}{s}. \tag{17.30}$$

Numerical inversion can easily be applied with Eqs. (17.29) and (17.30). However, since the time-domain formulas (17.27) and (17.28) involve single convolution integrals, numerical computation of Eqs. (17.27) and (17.28) in the time domain is also a feasible alternative. Moreover, if the underlying distributions have special structure, then the integrals in Eqs. (17.27) and (17.28) can be calculated analytically. For example, analytical integration can easily be done when all holding-time distributions are hyperexponential. In Duffield and Whitt [15, Section 3] we give a numerical example illustrating how the two approximations compare to the exact conditional mean for a single source with four levels.

17.6 THE VALUE OF INFORMATION

We can use the source model to investigate the value of information. We can consider how prediction is improved when we condition first, on, only the level and, second, on both level and age. The reference case is the steady-state mean

$$M = \sum_{i=1}^{n} m^i \quad \text{and} \quad m^i = \sum b_j^i p_j^i, \tag{17.31}$$

where p_j^i is the steady-state probability, that is,

$$p_j^i = \frac{\pi_j^i m(G_j^i)}{\sum_k \pi_k^i m(G_k^i)}, \tag{17.32}$$

with π^i the steady-state vector of the Markov chain P^i ($\pi^i = \pi^i P^i$) and $m(G_j^i)$ the mean of G_j^i for G_j in Eq. (17.8), for all sources i. With the steady-state mean, there is no conditioning. Section 17.2 gives the formula for conditioning on both level and age. Now we give the formulas conditioning only on the level; that is, we condition on the level, assuming that we are in the steady state. We omit the i superscript. Then the age in level j has the stationary-excess cdf

$$G_{je}(t) = \frac{1}{m(G_j)} \int_0^t G_j^c(u)\, du, \quad t \geq 0. \tag{17.33}$$

Let $P_{jk}(t)$ be the probability of being in level k at time t conditional on being in level j in the steady state at time 0. Let $\hat{P}_{jk}(s)$ be its Laplace transform. Let $m_j(t)$ be the conditional steady-state mean given level j at time 0 and let $\hat{m}_j(s)$ be its Laplace transform. Clearly,

$$m_j(t) = \sum_{k=1}^{J} P_{jk}(t) b_k \quad \text{and} \quad \hat{m}_j(s) = \sum_{k=1}^{J} \hat{P}_{jk}(s) b_k. \tag{17.34}$$

Hence, it suffices to calculate $\hat{P}_{jk}(s)$.

Theorem 17.6.1. *Assume that the level-holding-time cdf depends only on the originating level, that is,* $F_{jk}(t) = G_j(t)$. *The steady-state transition probabilities conditional on the level for a single SMP source have the matrix of Laplace transforms*

$$\hat{P}(s) = \hat{D}_e(s) + \hat{g}_e(s)\hat{P}(s|0), \tag{17.35}$$

where $\hat{P}(s|0)$ *is the matrix in Eq. (17.14),* $\hat{g}_e(s)$ *is the matrix with elements*

$$\hat{g}_{ejk}(s) = P_{jk}\hat{g}_{je}(s) = P_{jk}\frac{(1-\hat{g}_j(s))}{sm(G_j)}, \tag{17.36}$$

$\hat{D}_e(s)$ *is the diagonal matrix with diagonal elements*

$$\hat{D}_{ejj}(s) \equiv \frac{1-\hat{g}_{je}(s)}{s} = \frac{sm(G_j) - 1 + \hat{g}_j(s)}{s^2 m(G_j)}, \tag{17.37}$$

$\hat{g}_j(s)$ *is the level-j holding-time LST, and* $\hat{g}_{je}(s)$ *is the LST of its stationary-excess cdf in Eq. (17.33).*

Proof. Modify the proof of Theorem 17.4.1, inserting $P_{jl}G_{je}(t)$ for $H_{jl}(t|x)$ and $G_{je}^c(t)$ for $G_j^c(t|x)$. ■

Consider the on/off source in Section 17.4. Paralleling Eq. (17.18), it suffices to calculate only $P_{12}(t)$. Its Laplace transform is

$$\hat{P}_{12}(s) = \frac{\hat{g}_{1e}(s)(1-\hat{g}_2(s))}{s(1-\hat{g}_1(s)\hat{g}_2(s))}. \tag{17.38}$$

Example 17.6.2. To show the value of knowing the age, consider an on/off source with holding-time complementary cdf's

$$G_1^c(t) = 0.01e^{-0.01t} + 0.1e^{-0.1t} + 0.89e^{-t}, \quad G_2^c(t) = e^{-t}, \quad t \geq 0. \tag{17.39}$$

Let the bandwidths be $b_1 = 100$ and $b_2 = 0$. Since $m(G_1) = 2.89$ and $m(G_2) = 1.00$, the steady-state mean is

$$EB(\infty) = \frac{100m(G_1)}{m(G_1) + m(G_2)} = 74.29.$$

Let the initial level be 1. Since G_1 has an exponential component with mean 100, we anticipate the time to reach the steady state to be between 100 and 1000. In Fig. 17.1 we plot the conditional mean $m_1(t|x)$ for $x = 0.5$, 5.0, and 50.0, computed by

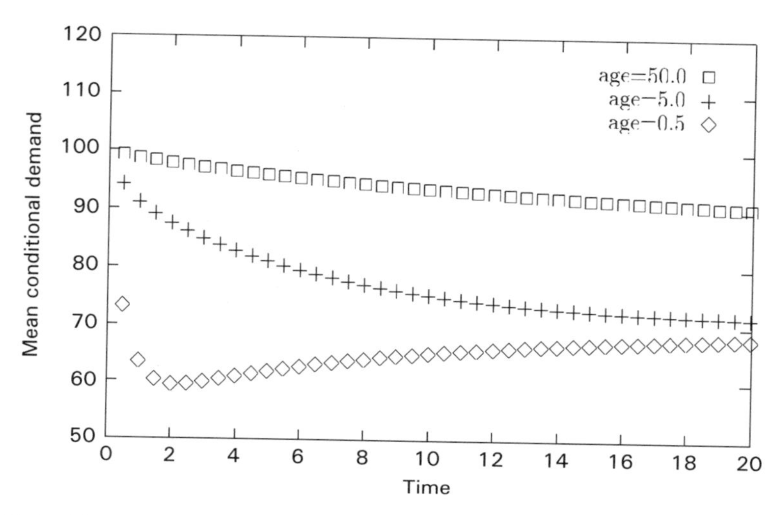

Fig. 17.1 The conditional mean aggregate demand as a function of the age of the holding time in level 1 for Example 17.6.2.

numerical transform inversion. Figure 17.1 shows that the age plays a very important role.

17.7 RECOVERY FROM CONGESTION IN THE STEADY STATE

For capacity planning, it is useful to consider the time required to recover from a high-congestion event, as well as the likelihood of the high-congestion event. The likelihood of a high-congestion event in the steady state can be estimated using a large deviation principle (LDP) approximation. The well-known Chernoff bound (e.g., see Dembo and Zeitouni [11]) gives an upper bound to the stationary tail probabilities of the aggregate level process, even for finitely many sources.

By Chebychev's inequality, for all $\theta > 0$,

$$\mathbf{P}[L(t) \geq x] \leq e^{-\theta x}\mathbf{E}[e^{\theta L(t)}] = e^{-\theta x}\prod_i \mathbf{E}[e^{\theta L_i(t)}] = e^{-\theta x}\prod_i \sum_j p^i_j e^{\theta b^i_j}, \qquad (17.40)$$

where b^i_j is the required bandwidth and p^i_j is the steady-state probability of level j in source i, as in Eq. (17.32). Thus,

$$\mathbf{P}[L(t) \geq x] \leq e^{-I(x)}, \quad \text{where} \quad I(x) = \sup_{\theta > 0}\left(\theta x - \sum_i \log \sum_j p^i_j e^{\theta b^i_j}\right). \qquad (17.41)$$

It can be shown [11] that such bounds are asymptotically tight (have a *large deviation* limit) as the number of sources increases, provided the spectrum of behavior of individual sources is sufficiently regular, yielding the exponential approximation

$$P(L(t) \geq x) \approx e^{-I(x)}. \tag{17.42}$$

Finding the rate function I will in general require numerical solution of the variational expression (17.41). It can be shown that the right-hand side (RHS) of Eq. (17.41) is a concave function of θ, and under mild conditions it is differentiable also. Hence, the supremum is achieved at the unique solution θ to the Euler–Lagrange equation

$$x = \sum_i \left(\frac{\sum_j b^i_j p^i_j e^{\theta b^i_j}}{\sum_{j'} p^i_{j'} e^{\theta b^i_{j'}}} \right). \tag{17.43}$$

Generally, it is not difficult to numerically determine the supremum in Eq. (17.41) by location of the solution to Eq. (17.43).

Example 17.7.1. In special cases the variational problem can be solved explicitly. This is possible in the case of n homogeneous two-level sources. Here we have $b^i_j = b_j$ with $j \in \{1, 2\}$, $0 \leq b_1 < b_2$, and $p_1 + p_2 = 1$. For this case,

$$I(x) = n \sup_{\theta > 0} (\theta x - \log(p_1 e^{\theta b_1} + p_2 e^{\theta b_2})) \tag{17.44}$$

$$= n \left(\frac{x}{b_2 - b_1} \log y(x) - \log(p_1 y(x)^{b_1/(b_2 - b_1)} + p_2 y(x)^{b_2/(b_2 - b_1)} \right) \tag{17.45}$$

with $y(x) = p_1(x - b_1)/(p_2(b_2 - x))$ for $b_1 \leq x \leq b_2$ and $I(x) = +\infty$ elsewhere.

We now show how to estimate the time to recover from the high-congestion event, where the high-congestion event is a large initial bandwidth x. We understand recovery to occur when the aggregate bandwidth is again less than or equal to the capacity c. In applications, we suggest examining the function of aggregate bandwidth giving *both* the probability of reaching that level and the recovery time from that level to assess whether or not capacity is adequate to meet demand.

We assume that recovery occurs when the aggregate bandwidth drops below a level c, where $x > c > m$, with x being the initial level and m being the steady-state mean. Given that we know the current level of each level process, we know that the remaining holding time (and also the age) is distributed according to the level-holding-time stationary-excess distribution in Eq. (17.33). We use the LDP to approximate the conditional distribution of the level process for each source (in the steady state). The idea is to perform the appropriate change of measure (tilting) corresponding to the rare event.

Given that $P(\sum_{i=1}^{n} B^i \geq x) \approx e^{-I(x)}$ for $I(x)$ in Eq. (17.41), the LDP approximation is

$$P\left(B^i = b_j^i \middle| \sum_{i=1}^{n} B^i \geq x\right) \approx \bar{p}_j^i \equiv \frac{p_j^i e^{b_j^i \theta^*}}{\sum_{k=1}^{J_i} p_k^i e^{b_k^i \theta^*}}, \tag{17.46}$$

where θ^* yields the supremum in Eq. (17.41). Put another way, comparing Eq. (17.46) with Eq. (17.43) we see that θ^* is chosen to make the expectation of $\sum_i B_i$ equal to x under the distribution $\bar{p}$. In the homogeneous case, equality in Eq. (17.46) in the limit as the number of sources increases is due to the conditional limit theorem of Van Campenhout and Cover [26]. The limit can be extended to cover suitably regular heterogeneity in the b_j^i, for example, finitely many types of source. We thus approximate the conditional bandwidth process by

$$\begin{aligned}(B(t)|B(0) = x) &\approx E(B(t)|B(0) = x) \\ &\approx = \sum_{i=1}^{n} \sum_{j=1}^{J_i} \bar{p}_j^i \int_0^\infty E(B_j^i(t)|B_j^i(0) = y)\, dG_{je}^i(y) \\ &= \sum_{i=1}^{n} \sum_{j=1}^{J_i} \bar{p}_j^i \sum_{k=1}^{J_i} b_k^i P_{jk}(t),\end{aligned} \tag{17.47}$$

for $\bar{p}_j^i$ in Eq. (17.46), which has Laplace transform

$$\sum_{i=1}^{n} \sum_{j=1}^{J_i} \bar{p}_j^i \sum_{k=1}^{J_i} b_k^i \hat{P}_{jk}(s). \tag{17.48}$$

The Laplace transform $\hat{P}_{jk}(s)$ in Eq. (17.48) was derived in Theorem 17.6.1. We can numerically invert it to calculate the conditional mean as a function of time. We then can determine when $E(B(t)|B(0) = x)$ first falls below c. In general, this conditional mean need not be a decreasing function, so that care is needed in the definition, but we expect it to be decreasing for suitably small t because the initial point $B(0)$ is unusually high.

17.8 A LINEAR APPROXIMATION

Assuming that the relevant time is not too large, we might approximate the conditional mean bandwidth using a Taylor-series approximation

$$E(B(t)|B(0) = x) = x + tr(x), \tag{17.49}$$

where

$$r(x) := E(B'(0)|B(0) = x) = \sum_{i=1}^{n} \sum_{j=1}^{J_i} \bar{p}_j^i \sum_{k=1}^{J_i} b_k^i P'_{jk}(0)$$
$$= \sum_{i=1}^{n} \sum_{j=1}^{J_i} \bar{p}_j^i \sum_{k=1}^{J_i} \frac{P_{jk}(b_k^i - b_j^i)}{m(G_j)}, \qquad (17.50)$$

which has the advantage that no numerical inversion is required.

Suppose the service capacity is $c > E(B(0))$ and we condition on $B(0) > c$. If $r(B(0)) < 0$, we can use Eq. (17.49) to approximate the first time to return to c, the *recovery time*, by

$$\tau = (x - c)/r(x). \qquad (17.51)$$

Suppose in addition that B is reversible; this will happen if the matrix P is reversible. Then since both the residual lifetime and the current age have distribution F_{jke}, $E(B(-t)|B(0)) = E(B(t)|B(0))$. Consequently, $B(0) = x$ is a local maximum, at $t = 0$, of $E(B(t)|B(0))$.

Now suppose that there are n independent sources. Then as in Duffield and Whitt [14], it follows by use of an appropriate functional law of large numbers that, as $n \to \infty$ under regularity conditions, the stochastic paths of the B process converge to this mean path. Thus we can identify, asymptotically as $n \to \infty$, $t = 0$ as a hitting time for the level x. Thus, we can use Eq. (17.42) to approximate the probability of this hitting time and τ in Eq. (17.51) to approximate the associated recovery time.

Example 17.8.1. Consider homogeneous two-level sources, that is, $j \in \{1, 2\}$, $0 \le b_1 < b_2$, $p_1 + p_2 = 1$ with mean lifetimes m_1, m_2, and $P_{11} = P_{22} = 1 - P_{12} = 1 - P_{21} = 0$. With n sources and $B(0) = x$ we can calculate the $\bar{p}_j$ in Eq. (17.46) directly from the relation $x = n(\bar{p}_1 b_1 + \bar{p}_2 b_2)$ for $x \in \{nb_1, (n - b)b_1 + b_2, \ldots, b_1 + (n - 1)b_2, nb_2\}$. Then

$$r(x) = n\bar{p}_1(b_2 - b_1)/m_1 + \bar{p}_2(b_1 - b_2)/m_2$$
$$= (nb_2 - x)/m_1 - (x - nb_1)/m_2. \qquad (17.52)$$

As a concrete example, we let $b_1 = 1$, $b_2 = 5$, $m_1 = 3$, $m_2 = 1$, giving a mean bandwidth per source of $(m_1 b_1 + m_2 b_2)/(m_1 + m_2) = 2$. We also let $n = 50$ and $c = 150$. The parameters of the example were chosen as a caricature of video traffic on an OC3 link: take b_i, x, c in Mb/s, m_i in seconds.

We present in Table 17.1 some values of n_1, n_2, the number of sources in each level for a given x, the approximate probability $e^{-I(x)}$ of demand exceeding x (using I from Eq. (17.44)), and τ. For comparison we give also the exact recovery time for the mean, calculated by using numerical transform inverstion methods [2], for particular

TABLE 17.1 Homogeneous Two-Level Sources[a]

# Sources Initially in Each Level		Initial Total Demand	Steady-state Probability of x	Recovery Time τ		
					Inversion	
n_1	n_2	x	$e^{-I(x)}$	Linear Approx.	Exponential Duration	Pareto Duration
25	25	150	7.5×10^{-4}	0	0	0
22	28	162	1.9×10^{-5}	0.15	0.16	0.19
19	31	174	2.4×10^{-7}	0.24	0.29	0.41
16	34	186	1.4×10^{-9}	0.31	0.41	0.64
13	37	198	3.5×10^{-12}	0.37	0.50	0.91

[a]Approximate hitting probabilities of aggregate demand x, together with recovery time τ of mean from x, by linear approximation, and exact for (1) exponential duration and (2) Pareto duration of higher level; see Example 17.8.1.

models of level durations with the same means: (1) both level durations exponentially distributed; (2) lower level exponential, upper level duration Pareto with exponent 1.5, and hence cdf $G^c(x) = (1 + 2x)^{-3/2}$ in order to give mean $m_2 = 1$. The Pareto density $g(x) = a(1 + x)^{-(1+a)}$ has a Laplace transform $ae^s s^a \Gamma(-a, s)$, where $\Gamma(a, z)$ is the incomplete gamma function

$$\Gamma(a, z) = \int_z^\infty t^{a-1} e^{-t} \, dt. \tag{17.53}$$

Hence, the required transform values for the Pareto distribution are readily computable. The algorithms for computing the incomplete gamma function typically involve continued fractions, as in Abate and Whitt [5].

In Fig. 17.2 we display the evolution of the conditioned mean in the linear approximation, and for the two distributions above with the same mean. As should be expected, the linear approximation is more accurate when the initial level x is closer to the capacity c. The linear approximation also behaves worse for the Pareto high-level durations than for the exponential high-level durations. The linear approximation tends to consistently provide a lower bound on the true recovery time for the mean. Even though the linear-approximation estimate of the recovery time diverges from the true mean computed by numerical inversion as the hitting level x increases, the probability of such high x can be very small. Even the largest errors in predicted recovery times in Table 17.1 are within one order of magnitude, and so might be regarded as suitable approximations. From our experiments, we conclude that the linear approximation is a convenient rough approximation, but that the numerical inversion yields greater accuracy.

In closing this section, we emphasize that a key point is the two-dimensional characterization of rare congestion events in terms of likelihood and recovery time. To further show how this perspective can be exploited, we plot in Fig. 17.3, for

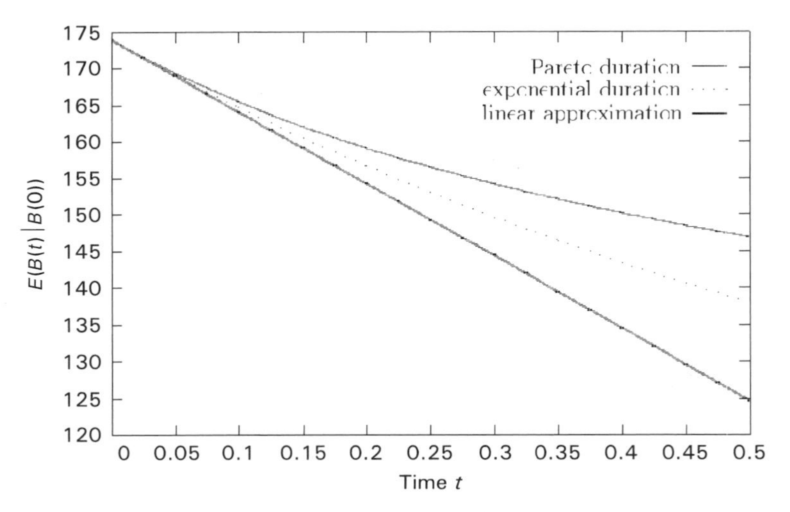

Fig. 17.2 Recovery curves for two-level sources: linear approximation and numerical transform inversion with exponential and Pareto durations in the higher level; see Example 17.8.1.

various offered loads, the approximate probability $e^{-I(x(\tau))}$ of a demand at least $x(\tau)$ as a function of τ, where $\tau = (c - x(\tau))/r(x(\tau))$; that is, $x(r)$ is the demand from which the recovery time to the level c is τ, using the linear approximation. Figure 17.3 shows that the two criteria together impose more constraints on what sets of

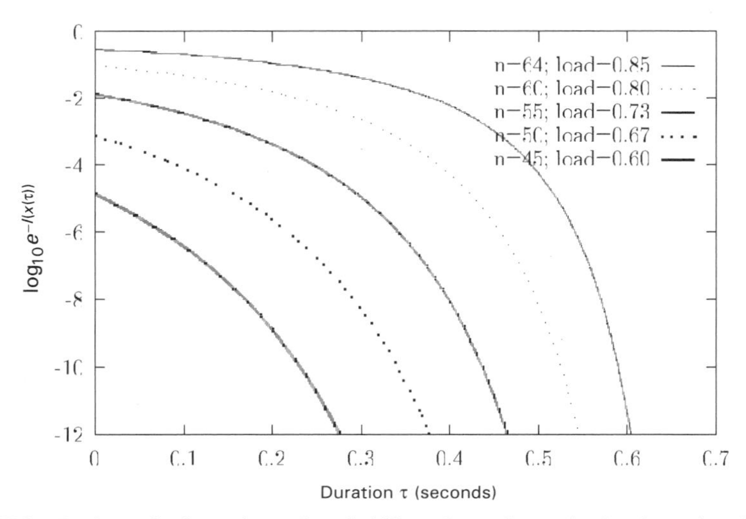

Fig. 17.3 Design criteria: estimated probability of overdemand of at least duration τ, for various offered loads; see Section 17.8.

sources are acceptable. Expressed differently, for the *same* probability of occurrence, rare congestion events can have very *different* recovery times.

17.9 COVARIANCE STRUCTURE

Useful characterizations of the aggregate and single-source bandwidth processes are their (auto)covariance functions. The covariance function may help in evaluating the fitting. We now show that we can effectively compute the covariance function for our traffic source model.

Let $\{B(t): t \geq 0\}$ and $\{B^i(t): t \geq 0\}$ be stationary versions of the aggregate and source-i bandwidth processes, respectively. Assuming that the single-source bandwidth processes are mutually independent, the covariance function of the aggregate bandwidth process is the sum of the single-source covariance functions; that is,

$$R(t) \equiv \mathrm{Cov}(B(0), B(t)) = \sum_{i=1}^{n} \mathrm{Cov}(B^i(0), B^i(t)). \tag{17.54}$$

Hence, it suffices to focus on a single source, and we do, henceforth dropping the superscript i.

In general,

$$R(t) = S(t) - m^2, \tag{17.55}$$

where the steady-state mean m is as in Eq. (17.31) and (17.32) and

$$\begin{aligned} S(t) \equiv EB(0)B(t) = \int_0^\infty dx \sum_{j=1}^{J} b_j p_j g_{je}(x) \sum_{k=1}^{J} b_k P_{jk}(t|x) \\ + \sum_{j=1}^{J} p_j G_{je}^c(t)\, \mathrm{Cov}(W_j(0), W_j(t)), \end{aligned} \tag{17.56}$$

where $g_{je}(x) = G_j^c(x)/m(G_j)$ is the density of G_{je}, and the second term captures the effect of the within-level variation process. In Eq. (17.56) b_j is the bandwidth in level j, p_j is the steady-state probability of level j, $g_{je}(x) = G_j^c(x)/m(G_j)$ with G_j the level-j holding-time cdf and $m(G_j)$ its mean, and $P_{jk}(t|x)$ is the transition probability, whose matrix of Laplace transforms is given in Theorem 17.4.1. We can thus calculate $S(t)$ by numerically inverting its Laplace transform

$$\begin{aligned} \hat{S}(s) \equiv \int_0^\infty e^{-st} S(t)\, dt = \sum_{j=1}^{J} \frac{b_j p_j}{m(G_j)} \int_0^\infty G_j^c(x) \sum_{k=1}^{J} b_k \hat{P}_{jk}(s|x)\, dx \\ + \sum_{j=1}^{J} p_j \int_0^\infty e^{-st} G_{je}^c(t)\, \mathrm{Cov}(W_j(0), W_j(t))\, dt. \end{aligned} \tag{17.57}$$

To treat the second term on the right in Eq. (17.57), we can assume an approximate functional form for the covariance of the within-level variation process $W(t)$. For example, if

$$\mathrm{Cov}(W_j(0), W_j(t)) = \sigma_j^2 e^{-\eta_j t}, \quad t \geq 0, \tag{17.58}$$

then

$$\sum_{j=1}^{J} p_j \int_0^\infty e^{-st} G_{je}^c(t)\, \mathrm{Cov}(W_j(0), W_j(t))\, dt = \sum_{j=1}^{J} p_j \sigma_j^2 \frac{(1 - \hat{g}_j(s + \eta_j))}{(s + \eta_j) m(G_j)}. \tag{17.59}$$

Thus, with approximation (17.58), we have a closed-form expression for the second term of the transform $\hat{S}(s)$ in Eq. (17.57). For each required s in $\hat{S}(s)$, we need to perform one numerical integration in the first term of Eq. (17.57), after calculating the integrand as a function of x.

A major role is played by the asymptotic variance $\int_0^\infty R(t)\, dt$. For example, the heavy-traffic approximation for the workload process in a queue with arrival process $\int_0^t B(u)\, du$, $t \geq 0$, depends on the process $\{B(t): t \geq 0\}$ only through its rate $EB(0)$ and its asymptotic variance; see Iglehart and Whitt [22]. The input process is said to exhibit long-range dependence when this integral is infinite. The source traffic model shows that long-range dependence stems from level-holding-time distributions with infinite variance.

Theorem 17.9.1. *If a level-holding-time cdf G_j has infinite variance, then the source bandwidth process exhibits long-range dependence, that is,*

$$\int_0^\infty R(t)\, dt = \infty.$$

Proof. In Eq. (17.56) we have the component

$$\int_0^\infty g_{je}(x) P_{jj}(t|x)\, dx,$$

which in turn has the component

$$\int_0^\infty g_{je}(x) G_j^c(t|x) = G_{je}^c(t),$$

but

$$\int_0^\infty G_{je}^c(t)\, dt = \infty$$

if G_j has infinite variance. (As can be seen using integration by parts, the integral is the mean of G_{je}; see Feller [18, p. 150]. In general, G_{je} has kth moment $m_{k+1}(G_j)/(k+1)m_1(G_j)$, where $m_k(G_j)$ is the kth moment of G_j.) ■

Note that if approximation (17.58) holds, then the level process contributes to long-range dependence, but the within-level variation process does not, because

$$\sum_{j=1}^{J} p_j \int_0^\infty G_{je}^c(t)\,\mathrm{Cov}(W_j(0), W_n(t))\,dt \approx \sum_{j=1}^{J} p_j \sigma_j^2 \frac{(1-\hat{g}_j(\eta_j))}{\eta_j m(G_j)} < \infty.$$

17.10 CONCLUSION

We have shown how transient analysis to support network design and control can be carried out for both on/off and multilevel source traffic models with general, possibly heavy-tailed, level-holding-time distributions. In Section 17.4 we analyzed the transient behavior of a general source traffic model composed of a semi-Markov level process and a zero-mean piecewise-stationary within-level variation process. We approximated the conditional aggregate demand from many sources given system state information by the conditional aggregate mean given level values and ages. The within-level variation process plays no role in this approximation. We showed that the conditional mean can be effectively computed using numerical transform inversion and developed several approximations to it (Sections 17.5 and 17.8). We showed how the model can be exploited to study the value of information (Section 17.6). We applied our techniques to examples in network design (Section 17.8).

Even though our approach is to focus on offered load, unaltered by loss and delay associated with finite capacity, we can apply the conditional mean approximation in Section 17.4 to develop an approximation to describe loss and delay from a finite-capacity system, just as described in Duffield and Whitt [14, Section 5] for the $M/G/\infty$ arrival process. Finally, our approach can be extended to a nonstationary setting in which connections arrive according to a nonhomogeneous Poisson process; see Duffield et al. [16]. Then each active connection may generate traffic according to the model considered here. It is significant that it is possible to obtain computationally tractable descriptions of the time-dependent aggregate demand.

REFERENCES

1. J. Abate, G. L. Choudhury, and W. Whitt. An introduction to numerical transform inversion and its application to probability models. In W. Grassman, ed., *Computational Probability*. Kluwer, Boston, pp. 257–323, 1999.
2. J. Abate and W. Whitt. Numerical inversion of Laplace transforms of probability distributions. *ORSA J. Comput.*, **7**:36–43, 1995.

3. J. Abate and W. Whitt. Modeling service-time distributions with non-exponential tails: beta mixtures of exponentials. *Stoch. Models*, **15**:517–546, 1999.
4. J. Abate and W. Whitt. Infinite series representations of Laplace transforms of probability density functions for numerical inversion. *J. Opns. Res. Soc. Japan* **42**, 1999, in press.
5. J. Abate and W. Whitt. Computing Laplace transforms for numerical inversion via continued fractions. *INFORMS J. Comput.*, **11**, 1999, in press.
6. S. Asmussen, O. Nerman, and M. Olsson. Fitting phase-type distributions via the EM algorithm. *Scand. J. Statist.* **23**:419–441, 1996.
7. R. Cáceres, P. G. Danzig, S. Jamin, and D. J. Mitzel. Characteristics of wide-area TCP/IP conversations. *Comput. Commun. Rev.*, **21**:101–112, 1991.
8. H. Chen and A. Mandelbaum. Discrete flow networks: bottleneck analysis and fluid approximations. *Math. Oper. Res.*, **16**:408–446, 1991.
9. E. Çinlar. *Introduction to Stochastic Processes*. Prentice-Hall, Englewood Cliffs, NJ, 1975.
10. M. E. Crovella and A. Bestavros. Self-similarity in World Wide Web traffic—evidence and possible causes. In *Proc. ACM SIGMETRICS'96*, pp. 160–169, 1996.
11. A. Dembo and O. Zeitouni. *Large Deviation Techniques and Applications*. Jones and Bartlett, Boston, 1993.
12. N. G. Duffield. Economies of scale in queues with sources having power-law large deviation scalings. *J. Appl. Probab.*, **33**:840–857, 1996.
13. N. G. Duffield. Queueing at large resources driven by heavy-tailed $M/G/\infty$-modulated processes. *Queueing Syst.*, **28**:245–266, 1998.
14. N. G. Duffield and W. Whitt. Control and recovery from rare congestion events in a large multi-sever system. *Queueing Syst.*, **26**:69–104, 1997. For shorter version see: Recovery from congestion in large multiserver systems. In *Proceedings ITC-15*, pp. 371–380, Washington, DC, 22–27 June 1997.
15. N. G. Duffield and W. Whitt. A source traffic model and its transient analysis for network control. *Stoch. Models*, **14**:51–78, 1998.
16. N. G. Duffield, W. A. Massey, and W. Whitt. A nonstationary offered-load model for packet networks. *Telecommun. Syst.*, in press, 1999.
17. A. Feldmann and W. Whitt. Fitting mixtures of exponentials to heavy-tail distributions to analyze network performance models. *Perf. Eval.*, **31**:245–279, 1998. Shorter version in *Proc. IEEE INFOCOM'97*, pp. 1098–1106.
18. W. Feller. *An Introduction to Probability Theory and Its Applications*, Vol. II, 2nd ed. Wiley, New York, 1971.
19. M. Grasse, M. R. Frater, and J. F. Arnold. Origins of long-range dependence in variable bit rate video traffic. In *Proceedings ITC-15*, pp. 1379–1388, Washington, DC, 22–27 June 1997.
20. M. Grossglauser, S. Keshav, and D. Tse. RCBR: A simple and efficient service for multiple time-scale traffic. In *Proc. ACM SIGCOMM'95*, pp. 219–230, 1995.
21. M. Harchol-Balter and A. Downey. Exploiting process lifetime distributions for dynamic load balancing. In *Proceedings of ACM SIGMETRICS'96 Conference on Measurement and Modeling of Computer Systems*, pp. 13–24, Philadelphia, 1996.
22. D. L. Iglehart and W. Whitt. Multiple channel queues in heavy traffic, II: sequences, networks and batches. *Adv. Appl. Probab.*, **2**:355–369, 1970.

23. W. E. Leland, M. S. Taqqu, W. Willinger, and D. V. Wilson. On the self-similar nature of Ethernet traffic. *IEEE/ACM Trans. Networking*, **2**:1–15, 1994.
24. V. Paxson and S. Floyd. Wide-area traffic: the failure of Poisson modeling. *IEEE/ACM Trans. Networking*, **3**:226–244, 1995.
25. J. Salehi, Z. Zhang, J. Kurose, and D. Towsley. Supporting stored video: reducing rate variability and end-to-end resource requirements through optimal smoothing. In *Proc. ACM SIGMETRICS*, pp. 222–231, 1996.
26. J. M. Van Campenhout and T. M. Cover. Maximal entropy and conditional probability. *IEEE Trans. Inf. Theory*, **4**:183–189, 1981.
27. W. Willinger, M. S. Taqqu, R. Sherman, and D. V. Wilson. Self-similarity through high-variability: statistical analysis of Ethernet LAN traffic at the source level. *IEEE/ACM Trans. Networking*, **5**:71–86, 1997.

18

CONGESTION CONTROL FOR SELF-SIMILAR NETWORK TRAFFIC

TSUNYI TUAN AND KIHONG PARK

Network Systems Lab, Department of Computer Sciences, Purdue University, West Lafayette, IN 47907

18.1 INTRODUCTION

Recent measurements of local-area and wide-area traffic [8, 28, 42] have shown that network traffic exhibits variability at a wide range of scales. What is striking is the ubiquitousness of the phenomenon, which has been observed in diverse networking contexts, from Ethernet to ATM, LAN and WAN, compressed video, and HTTP-based WWW traffic [8, 15, 23, 42]. Such scale-invariant variability is in strong contrast to traditional models of network traffic, which show burstiness at short time scales but are essentially smooth at large time scales; that is, they lack long-range dependence. Since scale-invariant burstiness can exert a significant impact on network performance, understanding the causes and effects of traffic self-similarity is an important problem.

In previous work [33, 34], we have investigated the causal and performance aspects of traffic self-similarity, and we have shown that self-similar traffic flow is an intrinsic property of networked client/server systems with heavy-tailed file size distributions, and conjoint provision of low delay and high throughput is adversely affected by scale-invariant burstiness. From a queueing theory perspective, the principal distinguishing characteristic of long-range-dependent (LRD) traffic is that the queue length distribution decays much more slowly—that is, polynomially—vis-à-vis short-range-dependent (SRD) traffic sources such as Poisson sources, which exhibit exponential decay. A number of performance studies [1, 2, 11, 29, 32, 34]

Self-Similar Network Traffic and Performance Evaluation, Edited by Kihong Park and Walter Willinger
ISBN 0-471-31974-0

have shown that self-similarity has a detrimental effect on network performance, leading to increased delay and packet loss rate. In Grossglauser and Bolot [18] and Ryu and Elwalid [37], the point is advanced that for small buffer sizes or short time scales, long-range dependence has only a marginal impact. This is, in part, due to a saturation effect that arises when resources are overextended, whereby the burstiness associated with short-range-dependent traffic is sufficient—and, in many cases, dominant—to cause significant buffer overflow.

What is still in its infancy, however, is the problem of *controlling* self-similar network traffic. By the control of self-similar traffic, we mean the problem of modulating traffic flow such that network performance including throughput is optimized. Scale-invariant burstiness introduces new complexities into the picture, which make the task of providing quality of service (QoS) while achieving high utilization significantly more difficult. First and foremost, scale-invariant burstiness implies the existence of concentrated periods of high activity at a wide range of time scales which adversely affects congestion control. Burstiness at fine time scales is commensurate with burstiness observed for traditional short-range dependent traffic models. The distinguishing feature is burstiness at coarser time scales, which induces extended periods of either overload or underutilization and degrades overall performance. However, on the flip side, long-range dependence, by definition, implies the existence of nontrivial correlation structure, which may be exploitable for congestion control purposes, information to which current algorithms are impervious.

In this chapter, we show the feasibility of "predicting the future" under self-similar traffic conditions with sufficient reliability such that the information can be effectively utilized for congestion control purposes. First, we show that long-range dependence can be on-line detected to predict future traffic levels and contention at time scales above and beyond the time scale of the feedback congestion control. Second, we present a traffic modulation mechanism based on *multiple time scale congestion control framework* (MTSC) [46] and show that it is able to effectively exploit this information to improve network performance, in particular, throughput. The congestion control mechanism works by selectively applying aggressiveness using the predicted future when it is warranted, throttling the data rate upward if the predicted future contention level is low, being more aggressive the lower the predicted contention level. We show that the *selective agressiveness* mechanism is of benefit even for short-range-dependent traffic; however, being significantly more effective for long-range dependent traffic, leading to comparatively large performance gains. We also show that as the number of connections engaging in selective aggressiveness control (SAC) increases, both fairness and efficiency are preserved. The latter refers to the total throughput achieved across all SAC-controlled connections.

The rest of the chapter is organized as follows. In Section 18.2, we give a brief overview of self-similar network traffic and the specific setup employed in this chapter. In Section 18.3, we describe the predictability mechanism and its efficacy at extracting the correlation structure present in long-range dependent traffic. This is followed by Section 18.4, where we describe the SAC protocol and a refinement of the predictability mechanism for on-line, per-connection estimation. In Section 18.5 we show performance results of SAC and show its efficacy under different long-range dependence conditions and when the number of SAC connections is varied. We conclude with a discussion of current results and future work.

18.2 PRELIMINARIES

18.2.1 Self-Similar Traffic: Basic Definitions

Let $(X_t)_{t \in \mathbb{Z}_+}$ be a time series, which, for example, represents the trace of data flow at a bottleneck link measured at some fixed time granularity. We define the aggregated series $X_i^{(m)}$ as

$$X_i^{(m)} = \frac{1}{m}(X_{im-m+1} + \cdots + X_{im}).$$

That is, X_t is partitioned into blocks of size m, their values are averaged, and i is used to index these blocks.

Let $r(k)$ and $r^{(m)}(k)$ denote the autocorrelation functions of X_t and $X_i^{(m)}$, respectively. X_t is *self-similar*—more precisely, *asymptotically second-order self-similar*—if the following conditions hold:

$$r(k) \sim \text{const} \cdot k^{-\beta}, \tag{18.1}$$

$$r^{(m)}(k) \sim r(k), \tag{18.2}$$

for k and m large, where $0 < \beta < 1$. That is, X_t is "self-similar" in the sense that the correlation structure is preserved with respect to time aggregation—relation (18.2)—and $r(k)$ behaves hyperbolically with $\sum_{k=0}^{\infty} r(k) = \infty$ as implied by Eq. (18.1). The latter property is referred to as *long-range dependence.*

Let $H = 1 - \beta/2$. H is called the *Hurst parameter*, and by the range of β, $\frac{1}{2} < H < 1$. It follows from Eq. (18.1) that the farther H is away from $\frac{1}{2}$ the more long-range dependent X_t is, and vice versa. Thus, the Hurst parameter acts as an indicator of the degree of self-similarity.

A test for long-range dependence can be obtained by checking whether H significantly deviates from $\frac{1}{2}$ or not. We use two methods for testing this condition. The first method, the *variance–time plot*, is based on the slowly decaying variance of a self-similar time series. The second method, the R/S plot, use the fact that for a self-similar time series, the *rescaled range* or R/S statistic grows according to a power law with exponent H as a function of the number of points included. Thus, the plot of R/S against this number on a log–log scale has a slope that is an estimate of H. A comprehensive discussion of the estimation methods can be found in Beran [4] and Taqqu et al. [39].

A random variable X has a *heavy-tailed* distribution if

$$\Pr\{X > x\} \sim x^{-\alpha}$$

as $x \to \infty$, where $0 < \alpha < 2$. That is, the asymptotic shape of the tail of the distribution obeys a power law. The *Pareto* distribution,

$$p(x) = \alpha k^{\alpha} x^{-\alpha-1},$$

with parameters $\alpha > 0$, $k > 0$, $x \geq k$, has the distribution function

$$\Pr\{X \leq x\} = 1 - (k/x)^{\alpha},$$

and hence is clearly heavy tailed.

It is not difficult to check that for $\alpha \leq 2$ heavy-tailed distributions have infinite variance, and for $\alpha \leq 1$, they also have infinite mean. Thus, as α decreases, a large portion of the probability mass is located in the tail of the distribution. In practical terms, a random variable that follows a heavy-tailed distribution can take on extremely large values with nonnegligible probability.

18.2.2 Structural Causality

In Park et al. [33], we show that aggregate traffic self-similarity is an intrinsic property of networked client/server systems where the size of the objects (e.g., files) being accessed is heavy-tailed. In particular, there exists a linear relationship between the heavy-tailedness measure of file size distributions as captured by α—the shape parameter of the Pareto distribution—and the Hurst parameter of the resultant multiplexed traffic streams. That is, the aggregate network traffic that is induced by hosts exchanging files with heavy-tailed sizes over a generic network environment running "regular" protocol stacks (e.g., TCP, flow-controlled UDP) is self-similar, being more bursty—in the scale-invariant sense—the more heavy-tailed the file size distribution are. This relationship is shown in Fig. 18.1. The relationship is robust with respect to changes in network resources (bandwidth, buffer capacity), topology, the influence of cross-traffic, and the distribution of interarrival times. We call this relationship between the traffic pattern observed at the network layer and the structural property of a distributed, networked system in terms of its high-level object sizes *structural causality* [33]. $H = (3 - \alpha)/2$ is the theoretical value predicted by the *on/off model* [42]—a $0/1$ renewal process with heavy-tailed on or off periods—assuming *independent* traffic sources with no interactions due to sharing of network resources.

Structural causality is of import to self-similar traffic control since (1) it provides an environment where self-similar traffic conditions are easily facilitated—just simulate a client/server network—(2) the degree of self-similar burstiness can be intimately controlled by the application layer parameter α, and (3) the self-similar network traffic induced already incorporates the actions and modulating influence of

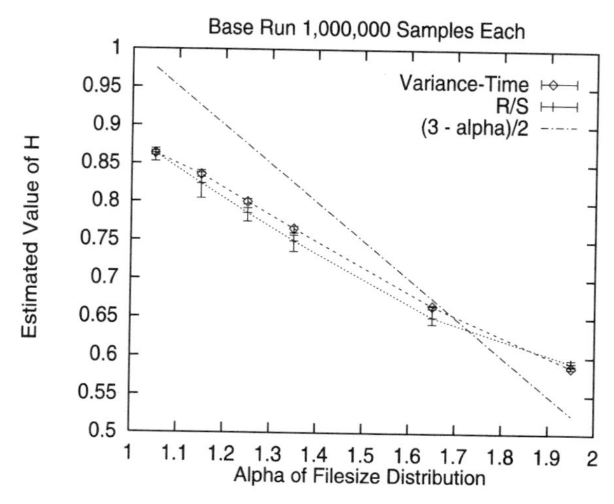

Fig. 18.1 Hurst parameter estimates (R/S and variance–time) for α varying from 1.05 to 1.95.

the protocol stack since the observed traffic pattern is a direct consequence of hosts exchanging files whose transport was mediated through protocols (e.g., TCP, flow-controlled UDP) in the protocol stack. This provides us with a natural environment where the impact of control actions by a congestion control protocol can be discerned and evaluated under self-similar traffic conditions.

18.3 PREDICTABILITY OF SELF-SIMILAR TRAFFIC

18.3.1 Predictability Setup

In this section, we show that the correlation structure present in long-range dependent (LRD) traffic can be detected and used to predict the future over time scales relevant to congestion control. Time series analysis and prediction theory have long histories with techniques spanning a number of domains from estimation theory to regression theory to neural network based techniques to mention a few [3, 17, 22, 40, 44, 45, 49]. In many senses, it is an "art form" with different methods giving variable performance depending on the context and modeling assumptions. Our goal is not to perform optimal time series prediction but rather to choose a simple, easy-to-implement scheme, and use it as a reference for studying congestion control techniques and their efficacy at exploiting the correlation structure present in LRD traffic for improving network performance. Our prediction method, which is described next, is a time domain technique and can be viewed as an instance of conditional expectation estimation.

Assume we are given a wide-sense stationary stochastic process $(\xi_t)_{t \in \mathbb{Z}}$ and two numbers $T_1, T_2 > 0$. At time t, we have at our disposal

$$a = \sum_{i \in [t-T_1, t)} q_i,$$

where q_i is a sample path of ξ_t over time interval $[t - T_1, t)$. For notational clarity, let

$$V_1 = \sum_{i \in [t-T_1, t)} \xi_i, \quad V_2 = \sum_{i \in [t, t+T_2)} \xi_i.$$

a may be thought of as the aggregate traffic observed over the "recent past" $[t - T_1, t)$ and V_1, V_2 are composite random variables denoting the recent past and near future. We are interested in computing the conditional probability

$$\Pr\{V_2 = b | V_1 = a\} \tag{18.3}$$

for b in the range of V_2. For example, if a represented a "high" traffic volume, then we may be interested in knowing what the probability of encountering yet another high traffic volume in the near future would be. Let

$$V^t_{\max} = \max \sum_{i \in [\tau-T-1, \tau)} q_i, \qquad V^t_{\min} = \min \sum_{i \in [\tau-T_1, \tau)} q_i$$

where $\tau = t\ kT_1$, $k = 0, 1, \ldots,$ $V^t_{\max}$ and $V^t_{\min}$ denote the highest and lowest traffic volume seen so far at time t, respectively.

To make sense of "high" and "low," we will partition the range between $V^t_{\max}$ and $V^t_{\min}$ into h *levels* with quantization step $\mu = (V^t_{\max} + V^t_{\min})/h$:

$$\begin{gathered}(0, V^t_{\min} + \mu), [V^t_{\min} + \mu, V^t_{\min} + 2\mu), [V^t_{\min} + 2\mu, V^t_{\min} + 3\mu), \ldots \\ [V^t_{\min} + (h-2)\mu, V^t_{\min} + (h-1)\mu), [V^t_{\min} + (h-1)\mu, \infty),\end{gathered}$$

We will define two new random variables L_1, L_2 where

$$\begin{aligned} L_k &= 1 \Leftrightarrow V_k \in (0, V^t_{\min} = \mu), \\ L_k &= 2 \Leftrightarrow V_k \in [V^t_{\min} + \mu, V^t_{\min} + 2\mu), \\ &\vdots \\ L_k &= h - 1 \Leftrightarrow V_k \in [V^t_{\min} + h - 2)\mu, V^t_{\min} + (h-1)\mu), \\ L_k &= h \Leftrightarrow V_k\ [V^t_{\min} + (h-1)\mu, \infty). \end{aligned}$$

In other words, L_k is a function of V_k, $L_k = L_k(V_k)$, and it performs a certain quantization. Thus if $L_k \approx 1$ then the traffic level is "low" relative to the mean, and if $L_k \approx h$, then it is "high."

In our case, eight levels ($h = 8$) were found to be sufficiently granular for prediction purposes. In practice, $V^t_{\max}$ and $V^t_{\min}$ are determined by applying a 3% threshold to the previously observed traffic volumes, i.e., the outliers corresponding to extraordinarily large or small data points are dropped to make the classification reasonable.

Returning to Eq. (18.3) and prediction, for certain values of T_1, T_2, we are interested in knowing the conditional probability densities

$$\Pr\{L_2 | L_1 = l\}$$

for $l \in [1, 8]$. If $\Pr\{L_2|L_1 = 8\}$ were concentrated toward $L_2 = 8$, and $\Pr\{L_2| L_1 = 1\}$ were concentrated toward $L_2 = 1$, then this information could be potentially exploited for congestion control purposes.

18.3.2 Estimation of Conditional Probability Density

To explore and quantify the potential predictability of self-similar network traffic, we use TCP traffic traces used in Park et al. [33] whose Hurst parameter estimates are shown in Fig. 18.1 as the main reference point. First, we use *off-line* estimation of *aggregate* throughput traffic, which is then refined to *on-line* estimation of aggregate traffic using *per-connection* traffic when performing predictive congestion control. Other traces including those collected from flow-controlled UDP runs yield similar results. The traces used are each 10,000 seconds long at 10 ms granularity. They represent the aggregate traffic of 32 concurrent TCP Reno connections recorded at a bottleneck router.

We observe that the aggregate throughput series exhibit correlation structure at several time scales from 250 ms to 20 s and higher. To estimate $\Pr\{L_2|L_1 = l\}$ from the aggregate throughput series X_t, we segment X_t into

$$N = \frac{10{,}000 \text{ (seconds)}}{T_1 + T_2 \text{ (seconds)}}$$

contiguous nonoverlapping blocks of length $T_1 + T_2$ (except possibly for the last block), and for each block $j \in [1, N]$ compute the aggregate traffic V_1, V_2 over the subintervals of length T_1, T_2.

For $l, l' \in [1, 8]$, let $h_l \in [0, N]$ denote the total number of blocks such that $L_1(V_1) = l$ and let $h_{l'} \in [0, h_l]$ denote the size of the subset of those blocks such that $L_2(V_2) = l'$. Then

$$\Pr\{L_2 = l' | L_1 = l\} = \frac{h_{l'}}{h_l}.$$

Figure 18.2 shows the estimated conditional probability densities for $\alpha = 1.05, 1.95$ traffic for time scales 500 ms, 1 s, and 5 s. In the following, $T_1 = T_2$.

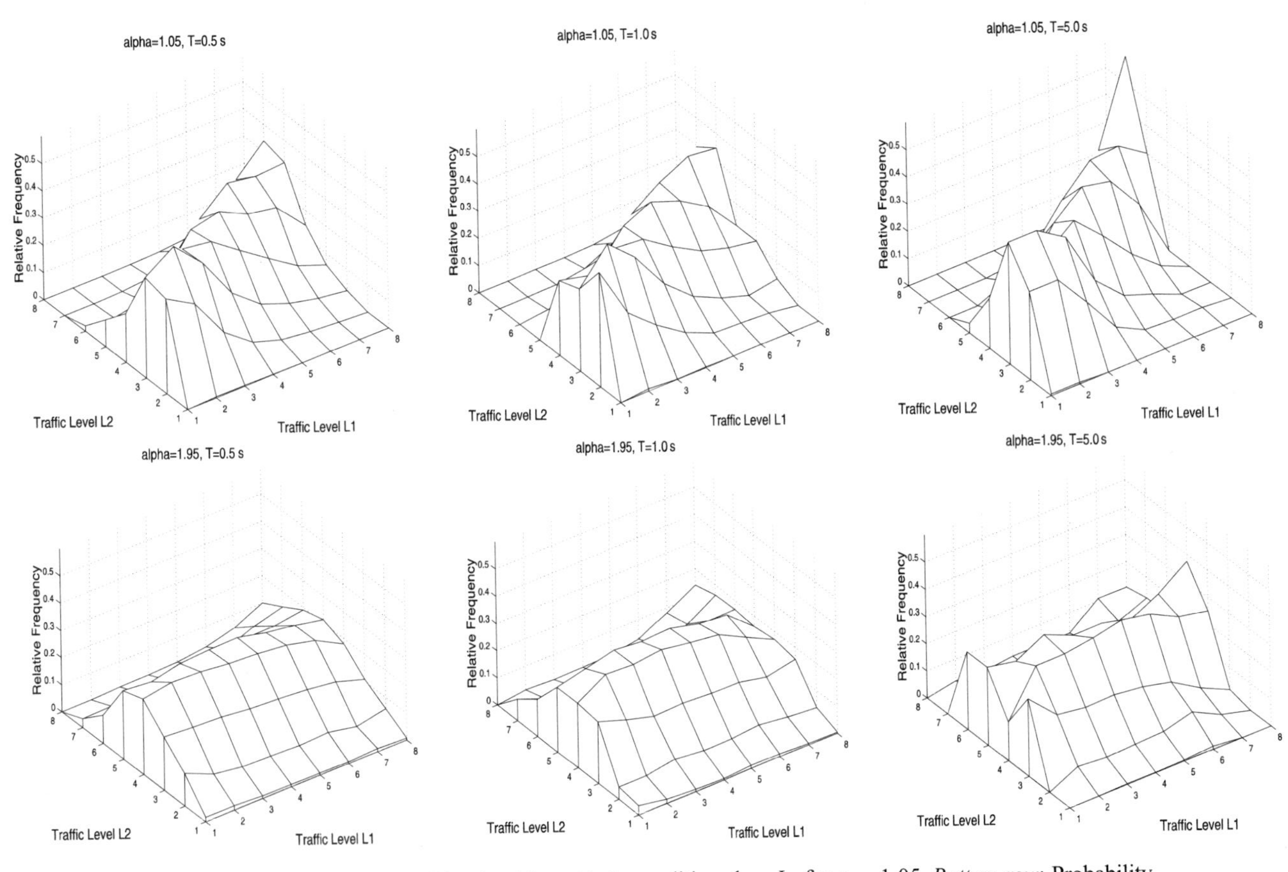

Fig. 18.2 *Top row*: Probability densities with L_2 conditioned on L_1 for $\alpha = 1.05$. *Bottom row*: Probability densities with L_2 conditioned on L_1 for $\alpha = 1.95$.

For the aggregate throughput traces with $\alpha = 1.05$—Figure 18.2 (top row)—the three-dimensional (3D) conditional probability densities can be seen to be skewed diagonally from the lower left side toward the upper right side. This indicates that if the current traffic level L_1 is low, say, $L_1 = 1$, chances are that L_2 will be low as well. That is, the probability mass of $\Pr\{L_2|L_1 = 1\}$ is concentrated *toward* 1. Conversely, the plots show that $\Pr\{L_2|L_1 = 8\}$ is concentrated *toward* 8. This is more clearly seen in Fig. 18.3(a), which shows two cross sections, that is, 2D projections, reflecting $\Pr\{L_2|L_1 = 1\}$ and $\Pr\{L_2|L_1 = 8\}$.

For the aggregate throughput traces with $\alpha = 1.95$ (Fig. 18.2 (bottom-row)), on the other hand, the shape of the distribution does not change as the conditioning variable L_1 is varied. This is more clearly seen in the projections of $\Pr\{L_2|L_1 = 1\}$ and $\Pr\{L_2|L_1 = 8\}$ shown in Fig. 18.3(b). This indicates that for $\alpha = 1.95$ traffic observing the past (over the time scales considered) does not help much in predicting the future beyond the information conveyed by the fixed *a priori distribution*. Given the definition of L_k, the Gaussian shape of the marginal densities is consistent with short-range correlations, making the central limit theorem approximately applicable over larger time scales. In both cases ($\alpha = 1.05, 1.95$), the shape of the distribution stays relatively constant across a wide range of time scales 500 ms to 20 s. For $\alpha = 1.35, 1.65$ the predictability structure lies "in-between" (not shown here).

18.3.3 Predictability and Time Scale

An important issue is how time scale affects predictability when traffic is long-range dependent. Going back to Fig. 18.2 (top row), one subtle effect that is not easily discernible is that as time scale is increased the conditional probability densities $\Pr\{L_2|L_1 = l\}$ become more concentrated. Given that $\Pr\{L_2|L_1 = l\}$ is a function of T_1, T_2, we would like to determine at what time scale predictability is maximized.

One way to measure the "information content"—that is, in the sense of randomness or unstructuredness—in a probability distribution is to compute its

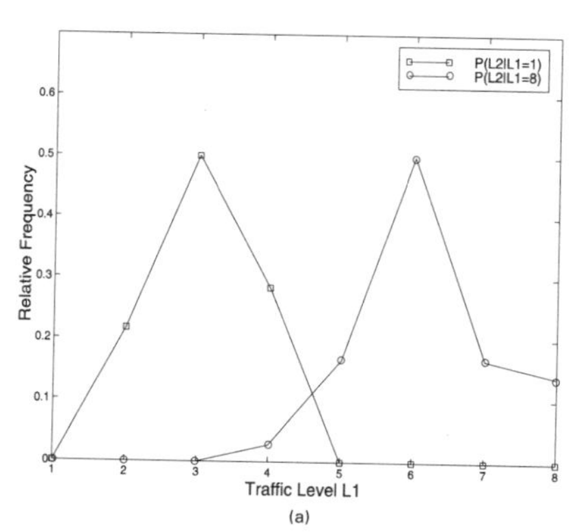

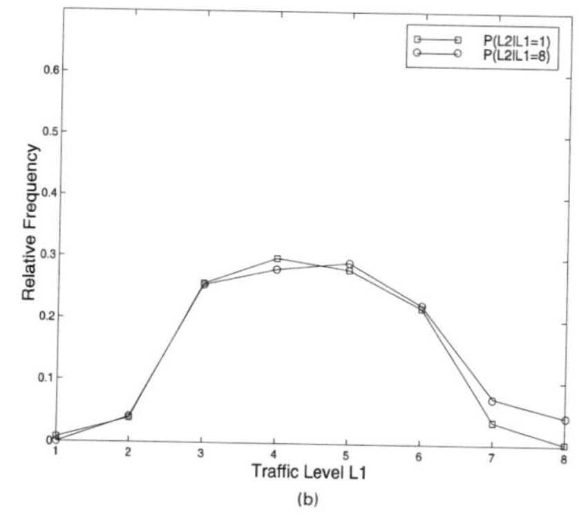

Fig. 18.3 (a) Shifting effect of conditional probability densities $P(L_2|L_1 = 1)$ and $P(L_2|L_1 = 8)$ for $\alpha = 1.05$. (b) For $\alpha = 1.95$, the corresponding probabilities remain invariant.

entropy. For a discrete probability density p_i, its *entropy* $S(p_i)$ is defined as $S(p_i) = \sum_i p_i \log(1/p_i)$. In the case of our conditional density $\Pr\{L_2|L_1 = l\}$,

$$S_l = -\sum_{l'=1}^{8} \Pr\{L_2 = l'|L_1 = l\} \log \Pr\{L_2 = l'|L_1 = l\}.$$

Thus, entropy is maximal when the distribution is uniform and it is minimal if the distribution is concentrated at a single point. Since we are given a set of eight conditional probability densities, one for each $L_1 = 1, 2, \ldots, 8$, we define the *average entropy* $\bar{S}$ as

$$\bar{S} = \sum_{l=1}^{8} S_l/8.$$

The average entropy remains a function of T_1, T_2: that is, $\bar{S} = \bar{S}(T_1, T_2)$.

Figure 18.4 plots $\bar{S}(T_1, T_2) = \bar{S}(T_1)$ (recall that $T_1 = T_2$) for the $\alpha = 1.05$ throughput series as a function of time scale or aggregation level T_1. Entropy is highest for small time scales in the range ~250 ms, and it drops monotonically as T_1 is increased. Eventually, $\bar{S}(T_1)$ begins to flatten out near the 3–5 second mark, reaching saturation, and stays so as time scale is further increased. From our analysis of various long-range dependent traffic traces, we find that the "knee" of the entropy curve is in the range of 1–5 seconds. Note that increasing T_1 further and further to gain small decreases in entropy brings forth with it an important problem, namely, if prediction is done over a "too long" time interval, then the information may not be effectively exploitable by various congestion control strategies. In the next section,

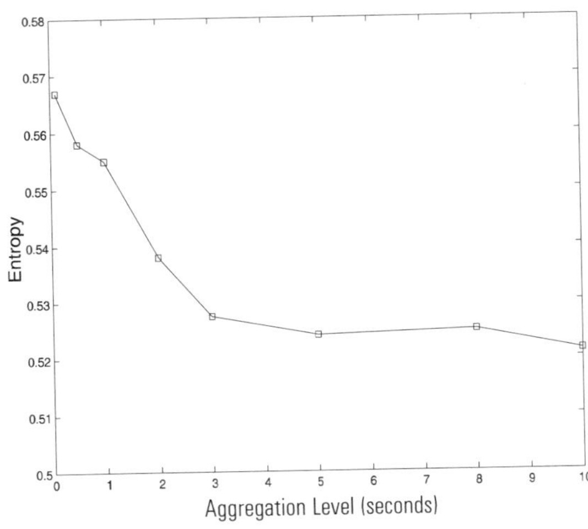

Fig. 18.4 Average entropy $\bar{S}(T_1)$ plot for $\alpha = 1.05$ traffic as a function of time scale T_1.

we show that one strategy—*selective aggressiveness*—is effective at exploiting the predictability structure found in the 1–5 second range.

18.4 SAC AND PREDICTIVE CONGESTION CONTROL

In this section we present a congestion control strategy called *selective aggressiveness control* (SAC) and show its efficacy at exploiting the predictability structure present in long-range dependent traffic for improving network performance. Our control scheme is a form of predictive congestion control based on the multiple time scale congestion control framework (MTSC) [46]. Explicit prediction of the long term network state l is performed at SAC's time scale (1–5 seconds). A certain control action $\varepsilon(l)$ is made by SAC based on this information about the future and is incorporated into the underlying congestion control to affect traffic control decisions. The overall structure is shown in Figure 18.5. SAC is aimed to be robust, efficient, and portable such that it can easily be incorporated into existing congestion control schemes.

SAC's modus operandi is to complement and help improve the performance of existing reactive congestion controls. Toward this end, we set up a simple, generic rate-based feedback congestion control as a reference and let our control module "run on top" of it. SAC always respects the decision made by the underlying congestion control with respect to the directional change of the traffic rate—up or down; however, it may choose to adjust the *magnitude* of change. That is, if, at any time, the underlying congestion control decides to *increase* its sending rate, SAC will never take the opposite action and decrease the sending rate. Instead, what SAC

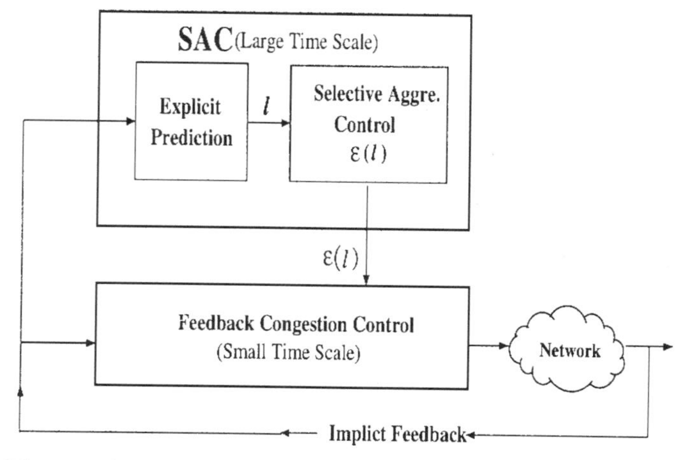

Fig. 18.5 The overall structure of predictive congestion control. SAC module is active at time scale (1–5 sec) exceeding the time scale of the underlying congestion control of its feedback loop.

will do is *amplify* or *diminish* the magnitude of the directional change based on its predicted future network state.

In a nutshell, SAC will try to aggresively soak up bandwidth if it predicts the future network state to be "idle," adjusting the level of aggressiveness as a function of the predicted idleness. We will show that the performance gain due to SAC is higher the more long-range dependent the network traffic is.

18.4.1 Underlying Congestion Control

18.4.1.1 Generic Feedback Congestion Control Congestion control has been an active area of networking research spanning over two decades with a flurry of concentrated work carried out in the late 1980s and early 1990s [5, 6, 16, 19, 24, 25, 27, 30, 31, 35, 36, 38]. Gerla and Kleinrock [16] laid down much of the early groundwork and Jacobson [24] has been instrumental in influencing the practical mechanisms that have survived until today. A central part of the investigation has been the study of stability and optimality issues [5, 13, 24, 25, 30, 31, 35, 38] associated with feedback congestion control. A taxonomy for classifying the various protocols can be found in Yang and Reddy [43].

More recently, the delay-bandwidth product problem arising out of high-bandwidth networks and quality of service issues stemming from support of real-time multimedia communication [7, 10, 12, 20, 21, 41] have added further complexities to the problem with QoS reigning as a unifying key theme. One of the lessons learned from congestion control research is that end-to-end rate-based feedback control using various forms of linear increase/exponential decrease can be effective, and asymmetry in the control law needs to be preserved to achieve stability.

We employ a simple, generic instance of rate-based feedback congestion control as a reference to help demonstrate the efficacy of selective aggressiveness control under self-similar traffic conditions. SAC is motivated, in part, by the simple yet important point put forth in Kim [26], which shows that the conservative nature of asymmetric controls can, in some situations, lead to throughput smaller than that achieved by a "nearly blind" aggressive control. By applying aggressiveness selectively—based on the prediction of future network contention—we seek to offset some of the cost incurred for stability.

Let λ denote packet arrival rate and let γ denote throughput. Our *generic linear increase/exponential decrease feedback congestion control* has a control law of the form[1]

$$\frac{d\lambda}{dt} = \begin{cases} \delta, & \text{if } d\gamma/d\lambda > 0, \\ -a\lambda, & \text{if } d\gamma/d\lambda < 0, \end{cases} \tag{18.4}$$

where $\delta, a > 0$ are positive constants. Thus, if increasing the data rate results in increased throughput (i.e., $d\gamma/d\lambda > 0$), then increase the data rate linearly. Conver-

[1] We use continuous notation for expositional clarity.

sely, if increasing the data rate results in a decrease in throughput (i.e., $d\gamma/d\lambda < 0$), then exponentially decrease the data rate. In general, condition $d\gamma/d\lambda < 0$ can be replaced by various measures of *congestion*.

Of course, difficulties arise because Eq. (18.4) is, in reality, a delay differential equation (the feedback loop incurs a time lag) and the sign of $d\gamma/d\lambda$ needs to be reliably estimated. The latter can be implemented using standard techniques.

18.4.1.2 Unimodal Load-Throughput Relation One item that needs further explanation is throughput γ. "Throughput" (in the sense of *goodput*) can be defined in a number of ways depending on the context, from *reliable throughput* (number of bits reliably transferred per unit time when taking into account reliability mechanism overhead), to *raw throughput* (number of bits transferred per unit time), to *power* (one of the throughput measures divided by delay). Raw throughput, denoted v, is both easy to measure (just monitor the number of packets, in bytes, arriving at the receiver per unit time) and to attain (in most contexts $v = v(\lambda)$ is a monotone increasing function of λ, e.g., $M/M/1/n$), but it does not adequately discriminate between congestion controls that achieve a certain raw throughput without incurring high packet loss or delay and those that do.

For example, achieving reliability using automatic repeat request (ARQ) with finite receiver and sender side buffers requires intricate control and coordination, and high packet loss can have a severe impact on the efficient functioning of such controls (e.g., TCP's window control). In particular, for a given raw throughput, if the packet loss rate is high, this may mean that a significant fraction of the raw throughput is taken up by duplicate packets (due to early retransmissions) or by packets that will be dropped at the receiver side due to "fragmentation" and buffer overflow. Thus, the reliable throughput associated with this raw throughput/packet loss rate combination would be low.

How severely packet loss impacts the throughput experienced by an application will depend on the characteristics of the application at hand. To better reflect such costs, we will use a throughput measure γ_k,

$$\gamma_k = (1 - c)^k v, \tag{18.5}$$

that (polynomially) penalizes raw throughput v by packet loss rate $0 \leq c \leq 1$, where the severity can be set by the parameter $k \geq 0$. Thus raw throughput v is a special instance of γ_k with $k = 0$. We will measure instantaneous throughput γ_k at the receiver and feedback to the sender for use in the control law (18.4). Figure 18.6 illustrates the relationship between γ_k and λ for an $M/M/1/n$ queueing system, which shows that for $c > 0$ the load-throughput curve $\gamma_k = \gamma_k(\lambda)$ is *unimodal*. Note that c is a monotone decreasing function of λ while v is monotone increasing. In the case of $M/M/1/n$ and most other network systems, raw bandwidth is upper bounded by the service rate or link speed—that is, $v \leq \mu$—and thus most load-throughput functions of interest (not just Eq. (18.5)) will be unimodal due to the above montonicity properties.

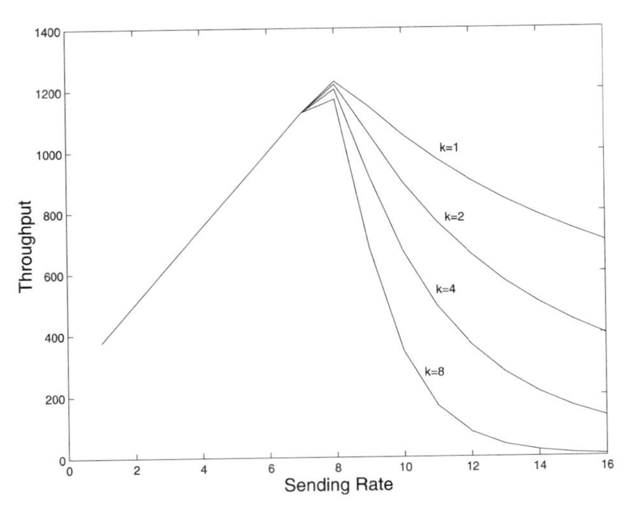

Fig. 18.6 Unimodal load-throughput curve $\gamma_k = \gamma_k(\lambda)$ for an $M/M/1/n$ system for $k = 1, 2,$ 4, 8.

18.4.2 Selective Aggressiveness Control (SAC)

Assuming that future network contention is predictable with a sufficient degree of accuracy, there remains the question of what to do with this information for performance enhancement purposes. The choice of actions, to a large measure, is constrained by the networking context and what degree of freedom it allows. In the traditional end-to-end congestion control setting, the network is a shared resource treated as a black box, and the only control variable available to a flow is its traffic rate λ.

In this chapter, the target mechanism to be improved using predictability is the performance loss stemming from conservative bandwidth usage during the *linear increase phase* of linear increase/exponential decrease congestion control algorithms [26]. Feedback congestion control protocols, including TCP, implement variants of this basic control law due to well-established stability reasons. In Kim [26], however, it was shown in the context of TCP Reno that the asymmetry stemming from *linear* increase after *exponential* back-off ends up significantly underutilizing bandwidth such that, in some situations, a simple nonfeedback control was shown to be more effective.[2]

Given that linear increase/exponential decrease is widely used in congestion control protocols including TCP, we seek to target the *linear increase* part of such protocols such that, when deemed beneficial, and only then, a more aggressive bandwidth consumption is facilitated. This *selective* application of aggressiveness, when coupled with predictive capability, will hopefully lead to a more effective use of

[2] This potential problem was also recognized in Jacobson's seminal paper [24] which, in part, motivated TCP Taboe's Slow Start feature.

bandwidth, resulting in improved performance. Without selective, controlled application of aggressiveness, however, the gain from aggressiveness may be canceled out (or even dominated) by its cost—aggressiveness, under high network contention conditions, can lead to deteriorated performance, even congestion collapse—thereby making predictability and its appropriate exploitation a nontrivial problem.

Our protocol—*selective aggressiveness control* (SAC)—is composed of two parts, prediction and application of aggression, and they are described next.

18.4.2.1 Per-Connection On-Line Estimation of Future Contention In the end-to-end feedback congestion control context, the two principal problems that a connection faces when estimating future network contention are:

1. The need to estimate "global" network contention using "local" *per-connection* information.
2. The need to perform *on-line* prediction.

First, with respect to requirement (1), since the network is a black box as far as an end-to-end connection or flow is concerned, we cannot rely on internal network support such as congestion notification via router support to reveal network state information. Instead, we need to gleam—in our case, predict—*future* network state using information obtained from a flow's input/output interaction with the network. For this to work, two assumptions need to hold in practice. First, due to the coupling induced by sharing of common resources, a connection's individual throughput when engaging in feedback congestion control (such as Eq. (18.4)) is correlated with the aggregate flow accessing the same resources. Second, aggregate traffic level, when partitioned according to the quantization scheme $L_k(V_k)$ of Section 18.3.1, is correlated to the contention level at the router that the aggregate traffic enters.

Second, with respect to requirement (2), it turns out that *on-line* estimation of the conditional probability density $\Pr\{L_2|L_1 = l\}$ is easily and efficiently accomplished using $O(1)$ cost update operations. On the sender side, SAC maintains a two-dimensional array or table

$$\texttt{CondProb}[\cdot][\cdot]$$

of size 8×9, one row for each $l \in [1, 8]$. The last column of `CondProb`, `CondProb[l][9]`, is used to keep track of h_l, the number of blocks observed thus far whose traffic level maps to l, that is, $L_1(V_1) = l$ (see Section 18.3.1). For each $l' \in [1, 8]$, `CondProb[l][l']` maintains the count $h_{l'}$. Since $\Pr\{L_2 = l'|L_1 = l\} = h_{l'}/h_l$, having the table `CondProb` means having the conditional probability densities.

All that is needed to maintain `CondProb` is a clock or alarm of period 2, which, starting at time $t = 0$, goes off at times

$$t = T_1, T_1 + T_2, T_1 + T_2 + T_1, T_1 + T_2 + T_1 + T_2, \ldots$$

If a feedback packet containing an instantaneous throughput γ measured at the receiver arrives during the period

$$[i(T_1 + T_2), i(T_1 + T_2) + T_1], \quad i \geq 0,$$

it is added to V_1. When the alarm goes off at $t = i(T_1 + T_2) + T_1$, V_1 is used to compute the updated $V^t_{\min}$, $V^t_{\max}$ and the quantization step μ which can be easily done incrementally by using $O(1)$ operations. Now $l = L_1(V_1)$ is computed using the updated $V^t_{\min}$ and $V^t_{\max}$, and `CondProb[l][9]` is incremented by 1. During interval

$$[i(T_1 + T_2) + T_1, (i + 1)(T_1 + T_2)], \quad i \geq 0,$$

a similar operation is performed, however, now, with respect to V_2. At the end of the interval, the updated $V^t_{\min}$, $V^t_{\max}$ are computed, and $l' = L_2(V_2)$ is computed. Finally, `CondProb[l][l']` is incremented by 1, and V_1, V_2 are reset to 0 to start the process anew. The number of operations within a time interval of length $T_1 + T_2$ is $O(1)$.

It should be noted that the conditional densities computed from `CondProb` at time t are *approximations* to the conditional probability densities computed off-line for the period $[0, t]$ since in the on-line algorithm running sums are used to compute and update $V^t_{\min}$ and $V^t_{\max}$. Results of on-line approximation of conditional probability densities are shown in Section 18.5.2.1.

18.4.2.2 Selective Application of Aggressiveness SAC aims to "expedite" the bandwidth consumption process during the linear increase phase of linear increase/exponential decrease feedback congestion control algorithms—in our case, represented by the *generic feedback congestion control algorithm* (18.4)—when such actions are warranted.

The actuation part of the interface between SAC and Eq. (18.4) is defined as follows. Let λ_t denote the newly updated rate value at time t—by Eq. (18.4)—and let $\lambda_{t'}$ be the most recently ($t' < t$) updated rate value previous to t.

SAC (Actuation Interface)

1. If $\lambda_t > \lambda_{t'}$ then update $\lambda_t \leftarrow \lambda_t + \epsilon_t$.
2. Else do nothing.

Here, $\epsilon_t \geq 0$ is an aggressiveness factor that is determined by SAC based on the current state of `CondProb`. Note that SAC kicks into action only during the linear increase phase of Eq. (18.4), that is, when $\lambda_t > \lambda_{t'}$. The magnitude of ϵ_t determines the degree of aggressiveness, and it is determined as a function of the predicted network state as captured by `CondProb` and its conditional probability densities.

At time t, the algorithm used to determine ϵ is as follows. Let S_t be the aggregate throughput reported by the receiver via feedback over time interval $[t - T_1, t]$.

SAC (ϵ Determination)

1. Let $l = L_1(S_t)$.
2. Compute $\bar{l}' = \mathbf{E}(L_2|L_1 = l) = \sum_{l'=1}^{8} l' \Pr\{L_2 = l'|L_1 = l\}$.
3. Set $\epsilon = \epsilon(\bar{l}')$.

Thus, the current traffic level S_t is normalized and mapped to the index $l = L_1(S_t)$, which is then used to calculate the expectation of L_2 conditioned on l', $\bar{l}'$. The latter is then finally used to index into a table $\epsilon(\bar{l}')$ is called the *aggressiveness schedule*. The intuition behind the aggressiveness schedule $\epsilon(\cdot)$ is that if the expected future contention level is low (i.e., $\bar{l}'$ close to 1) then it is likely that applying a high level of aggressiveness will pay off. Conversely, if the expected future contention level is high (i.e., $\bar{l}'$ near 8) then applying a low level of aggressiveness is called for. One schedule that we use is the *inverse schedule*,

$$\epsilon(\bar{l}') = 1/\bar{l}'.$$

Other schedules of interest include the *threshold schedule* with threshold $\theta \in [1, 8]$ and aggressiveness factor θ^*, where $\epsilon = \theta^*$ if $\bar{l}' \leq \theta$, and 0 otherwise.

Table 18.1 shows the `CondProb` table for two runs corresponding to $\alpha = 1.05$ (top) and $\alpha = 1.95$ (bottom) traffic conditions. The column containing h_l has been omitted and the entries show actual relative frequencies rather than $h_{l'}$ counts for illustrative purposes. Clearly, the conditional probability densities are skewed diagonally for $\alpha = 1.05$ traffic, whereas they are roughly invariant for $\alpha = 1.95$ traffic. The expected future contention level $\bar{l}' = \mathbf{E}(L_2|L_1 = l)$ and aggressiveness schedule (inverse) are shown as separate columns. For $\alpha = 1.05$ traffic, the expected future contention level $E[L_2|\cdot]$ varies over a wide range, which is a direct consequence of the predictability—that is, skewedness—present in the correlation structure. For $\alpha = 1.95$ traffic, however, $E[L_2|\cdot]$ is fairly "flat," indicating that conditioning on the present does not aid significantly in predicting the future.

18.5 SIMULATION RESULTS

18.5.1 Congestion Control Evaluation Setup

We use the LBNL Network Simulator, *ns* (version 2), as the basis of our simulation environment. *ns* is an event-driven simulator derived from Keshav's REAL network simulator supporting several flavors of TCP and router packet scheduling algorithms. We have modified *ns* in order to model a bottleneck network environment where several concurrent connections are multiplexed over a shared bottleneck link. A UDP-based unreliable transport protocol was added to the existing protocol suite, and our congestion control and predictive control were implemented on top of it.

An important feature of the setup is the mechanism whereby *self-similar traffic conditions* are induced in the network. One possibility is to have a host inject self-

TABLE 18.1 Snapshot of `CondProb` 10,000 s After Connection Has Been Established (Top, $\alpha = 1.05$; Bottom, $\alpha = 1.95$)

L_1 \ L_2	1	2	3	4	5	6	7	8	$E[L_2 \mid \cdot]$	$\epsilon(\cdot)$
1	0.667	0.333	0	0	0	0	0	0	1.3	0.769
2	0.003	0.568	0.306	0.093	0.027	0.003	0	0	2.6	0.384
3	0	0.126	0.468	0.262	0.116	0.023	0.003	0	3.4	0.294
4	0	0.035	0.205	0.368	0.305	0.077	0.201	0	4.2	0.238
5	0	0.003	0.078	0.296	0.356	0.205	0.060	0.002	4.9	0.204
6	0	0	0.012	0.099	0.285	0.418	0.182	0.003	5.7	0.175
7	0	0	0.018	0.079	0.245	0.443	0.213	0.003	5.8	0.172
8	0	0	0	0	0.333	0	0.500	0.167	6.5	0.153

L_1 \ L_2	1	2	3	4	5	6	7	8	$E[L_2 \mid \cdot]$	$\epsilon(\cdot)$
1	0.155	0.116	0.155	0.233	0.165	0.078	0.087	0.097	3.8	0.263
2	0.043	0.058	0.179	0.272	0.257	0.128	0.054	0.008	4.3	0.232
3	0.023	0.049	0.132	0.306	0.273	0.161	0.054	0.003	4.5	0.222
4	0.020	0.058	0.135	0.274	0.286	0.167	0.055	0.004	4.5	0.222
5	0.012	0.039	0.134	0.273	0.307	0.183	0.044	0.008	4.6	0.217
6	0.017	0.058	0.141	0.243	0.325	0.166	0.044	0.007	4.5	0.222
7	0.008	0.042	0.126	0.211	0.322	0.195	0.088	0.008	4.8	0.208
8	0	0	0.167	0.233	0.233	0.300	0.067	0	4.8	0.208

similar time series into the network. We follow a different approach based on the notion of *structural causality* (see Section 18.2.2) whereby we make use of the fact that a networked client/server environment, with clients interactively accessing files or objects with heavy-tailed sizes from servers across the network, leads to aggregate traffic that is self-similar [33]. Most importantly, this mechanism is robust and holds when the file transfers are mediated by transport layer protocols executing reliable flow-controlled transport (e.g., TCP) or unreliable flow-controlled transport. The separation and isolation of "self-similar causality" to the highest layer of the protocol stack allows us to interject different congestion control protocols in the transport layer, discern their influence, and study their impact on network performance. This is illustrated in Fig. 18.7.

Figure 18.8 shows a 2-server, n-client ($n \geq 33$) network configuration with a bottleneck link connecting gateways G_1 and G_2. The link bandwidths were set at 10 Mb/s and the latency of each link was set to 5 ms. The maximum segment size was fixed at 1 kB for all runs. Some of the clients engage in interactive transport of files with heavy-tailed sizes across the bottleneck link *to the servers* (i.e., the nomenclature of "client" and "server" are reversed here), sleeping for an exponential time between successive transfers. Others act as *infinite sources* (i.e., they have always data to send) executing the generic linear increase/exponential decrease feeback congestion control—with and without SAC—in the protocol stack trying to

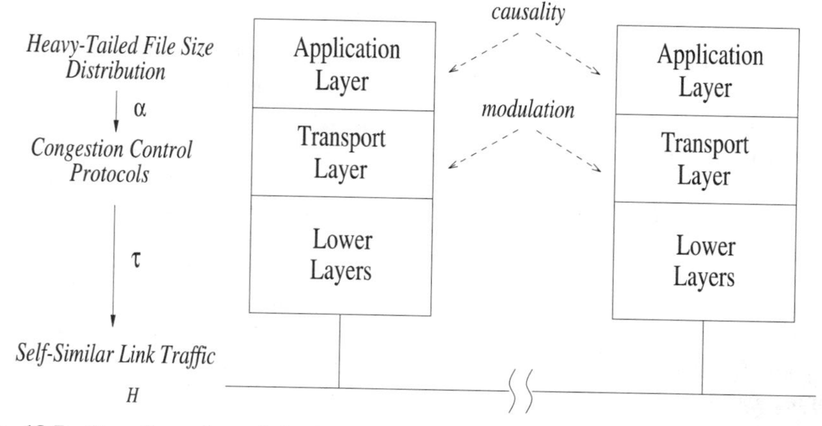

Fig. 18.7 Transformation of the heavy-tailedness of file size distribution property at the application layer via the action of the transport layer into its manifestation as self-similar aggregated traffic at the link layer.

maximize throughput. For any reasonable assignment of bandwidth, buffer size, mean file request size, and other system parameters, we found that by either adjusting the number of clients or the mean of the idle time distribution between successive file transfers appropriately, any target contention level could be achieved.

In a typical configuration, the first 32 connections served as "background traffic," transferring files from clients to servers (or sinks), where the file sizes were drawn from Pareto distributions with shape parameter $\alpha = 1.05, 1.35, 1.65, 1.95$. As in Park et al. [33], there was a linear relation between α and the long-range dependence of aggregate traffic observed at the bottleneck link (G_1, G_2) as captured by the Hurst parameter H. H was close to 1 when α was near 1, and H was close to $\frac{1}{2}$ when α was near 2. The 33rd connection acted as an infinite source trying to maximize throughput by running the generic feedback control, with or without SAC. In other settings, the number of congestion-controlled infinite sources was increased to observe their mutual interaction and the impact on fairness and efficiency. A typical run lasted for 10,000 or 20,000 seconds (simulated time) with traces collected at 10 ms granularity.

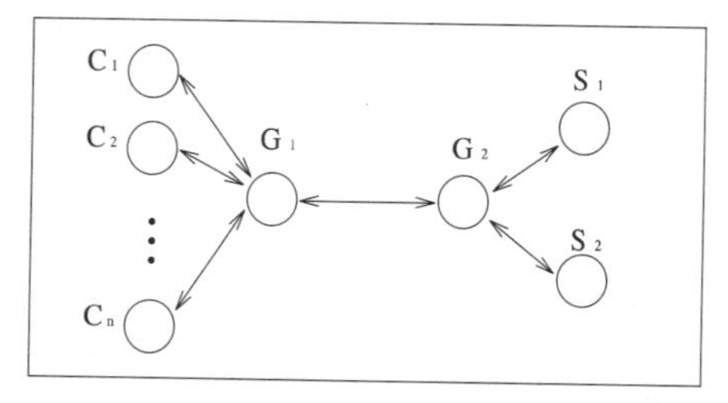

Fig. 18.8 Network configuration with bottleneck link (G_1, G_2).

18.5.2 Per-Connection On-Line Predictability

18.5.2.1 Predictability Structure One of the first items to test was estimation of the conditional probability densities $\Pr\{L_2|L_1\}$ using the per-connection, on-line method described in Section 18.4.2.1. We observe the same skewed diagonal shift characteristics as seen in the off-line case for $\alpha = 1.05$ traffic and the relatively invariant shape of the probability densities for $\alpha = 1.95$ traffic (we omit the 3D plots due to space constraints). Also, as in the off-line case, as we increase the time scale (i.e., T_1) from 500 ms to 1 s and higher, for $\alpha = 1.05$ traffic the probability densities becomes more concentrated, thus increasing the accuracy of prediction. Figure 18.9(a) shows the shifting effect of the conditional probabilities for $\alpha = 1.05$ traffic via a 2D projection that shows the marginal densities. Whereas the shifting effect is evident for $\alpha = 1.05$ traffic, for $\alpha = 1.95$ traffic (Fig. 18.9(b)) the probability densities stay largely invariant.

18.5.2.2 Time Scale In Section 18.3.2, in connection with the off-line estimation, we showed using entropy calculations that the conditional probability densities became more concentrated as time scale was increased, flattening out eventually near the 4–5 s mark. We observe a similar behavior with respect to the entropy curve in the on-line case. Locating the knee of the entropy curve is of import for on-line prediction and its use in congestion control since the size of the time scale will influence whether a certain procedure for exploiting predictability will be effective or not. As an extreme case in point, if the time scale of prediction were, say, 1000 s, it is difficult to imagine a mechanism that would be able to effectively exploit this information for congestion control purposes—too many changes may be occurring during such a time period that may be both favorable and detrimental to a constant control action yielding a zero net gain.

On the other hand, if control actions capable of spanning large time frames such as bandwidth reservation or pricing-based admission control were made part of the

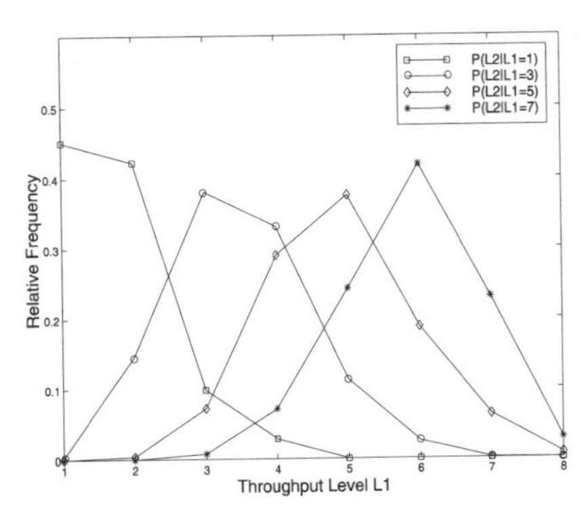

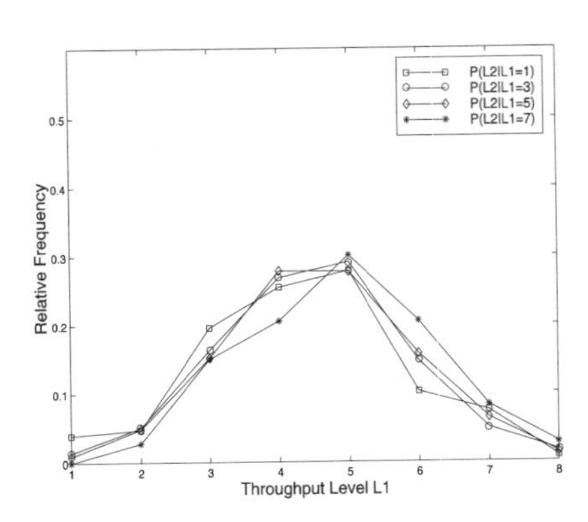

Fig. 18.9 (a) Shifting effect of conditional probability densities $P(L_2|L_1 = 1)$ and $P(L_2|L_1 = 8)$ under $\alpha = 1.05$ background traffic. (b) Shifting effect of conditional probability densities $P(L_2|L_1 = 1)$ and $P(L_2|L_1 = 8)$ under $\alpha = 1.95$ traffic.

model, then even large time scale predictability may be exploited, with some effectiveness, for performance enhancement purposes. In the present context, we set $T_1 = 2$ s when incorporating predictability into feedback congestion control via SAC. Our experience with different traffic traces suggests that the knee of the entropy curve, for many practical situations, may be located in the 1–6 s range.

18.5.2.3 Convergence Rate of On-Line Conditional Probabilities When using conditional probability densities computed from per-connection, on-line estimations, it becomes important to know when the estimations have converged or stabilized. If inaccurate information is used for selective aggressiveness control, it is possible that rather than helping improve performance, it may hurt performance.

SAC uses a distance measure to decide whether a particular conditional probability density is stable enough to be used for congestion control or not. Let $\texttt{CondProb}_{\texttt{t}}$, $\texttt{CondProb}_{\texttt{t}'}$ be two instances of the conditional probability density table measured at time instances $t > t'$ at least $T_1 + T_2$ apart. Then for each L_1 condition $l \in [1, 8]$, SAC computes

$$\|\texttt{CondProb}_{\texttt{t}}[\texttt{l}][\cdot] - \texttt{CondProb}_{\texttt{t}'}[\texttt{l}][\cdot]\|_2 < \Theta,$$

and if the check passes, allows this particular conditional probability density to be used in the actuation part of SAC when updating the data rate (see Section 18.4.2.2). Here, $\Theta > 0$ is an accuracy parameter and $\|\cdot\|_2$ is the "L_2" norm (not to be confused with the random variable L_2).

Figure 18.10 depicts the convergence rate of three conditional probability densities conditioned on $l = 1, 2$, and 8 by plotting the distance measure computed above. Figure 18.10(b) ($l = 2$) and Fig. 18.10(c) ($l = 8$) are the typical plots whose convergence rate is followed by the ones in the range $2 < l < 8$ as well. Figure 18.10(a) ($l = 1$), however, is atypical and only holds for condition $L_1 = 1$. This mainly stems from the fact that for condition $L_1 = 1$, very few sample points (<100) arise even during a 10,000 s run, and, as a result, the estimated probabilities are volatile. This, in turn, can be attributed to the fact that we are computing statistics for a long-range dependent process, and it is well known that for such processes a very large number of samples are needed to compute even its first-order statistics accurately. A related discussion can be found in Crovella and Lipsky [9].

The fact that this volatility stems from too few sample points also delimits the impact of this phenomenon. Namely, by Amdahl's law, the instances (over time) where the conditional probability $\Pr\{L_2 | L_1 = 1\}$ may have been invoked are so few to begin with that the loss from not having taken any actions at those instances is negligible with respect to overall performance.

18.5.3 Performance Measurement of SAC

18.5.3.1 Incremental Gain of Selective Aggressiveness

Unimodal Throughput Curve In this section, we evaluate the relative performance of SAC and its predictability gain. We measure the *incremental* benefit gained by

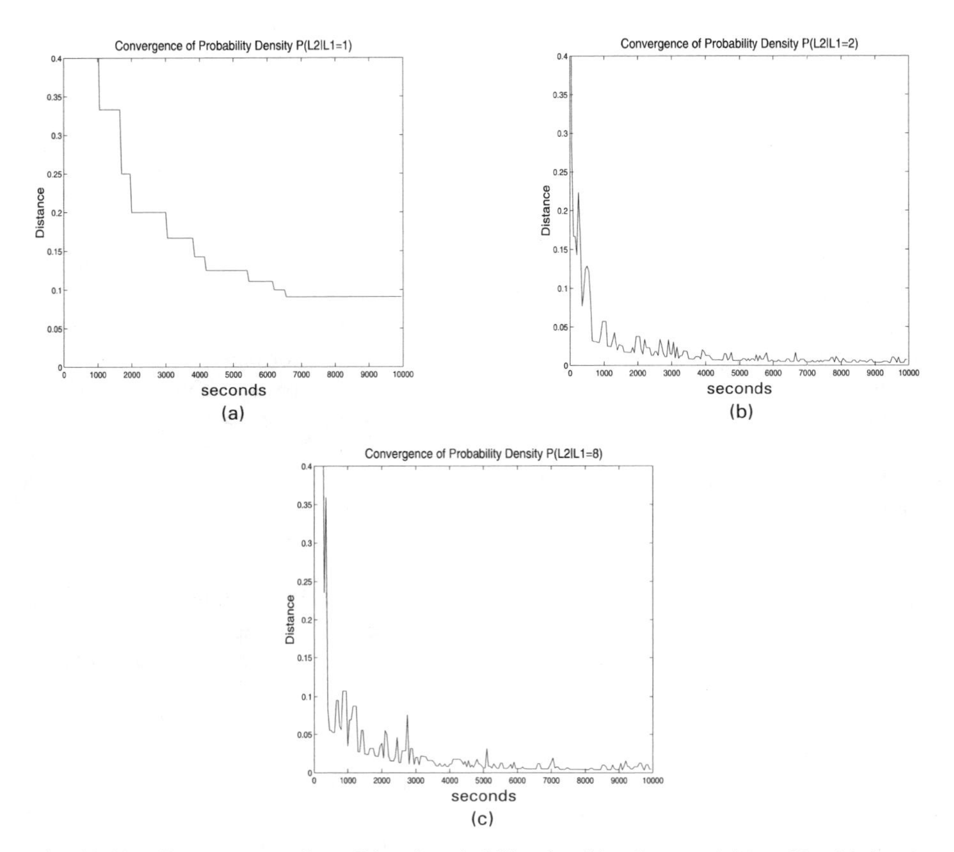

Fig. 18.10 Convergence of conditional probability densities for $\alpha = 1.05$ traffic: (a) $l = 1$, (b) $l = 2$, and (c) $l = 8$.

applying aggressiveness *selectively*, first by applying it only when the chances for benefit are highest (i.e., $\bar{l}' = \mathbf{E}(L_2|L_1 = l) = 1$ for some $l \in [1, 8]$), then second highest ($\bar{l}' = 2$), and so on. Eventually, we expect to hit a point when the cost of aggressiveness outweighs its gain, thus leading to a net decrease in throughput as the stringency of selectivity is further relaxed.

This phenomenon can be demonstrated using the *threshold aggressiveness schedule* of SAC (see Section 18.4.2.2) where aggressive action is taken if and only if $\bar{l}' \leq \theta$, where θ is the aggressiveness threshold. Figure 18.11 shows the throughput versus aggressiveness threshold curve for threshold values in the range $1 \leq \theta \leq 8$ for $\alpha = 1.05$ traffic. We observe that the gain is highest when going from $\theta = 1$ to 2, then successively diminishes until it turns to a net loss, thereby decreasing throughput. If $\theta = 8$, then this corresponds to the case where aggressiveness is applied at all times: that is, there is no selectivity.

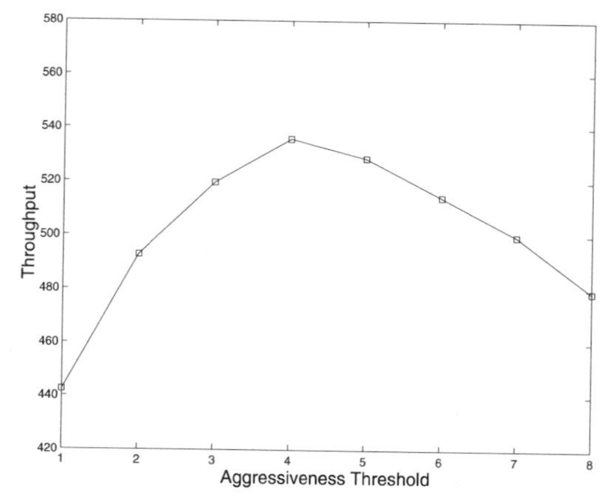

Fig. 18.11 Unimodal throughput curve as a function of aggressiveness threshold θ for $\alpha = 1.05$ traffic.

Monotone Throughput Curve Although the unimodal, dome-shaped throughput curve (as a function of the aggressiveness threshold) is a representative shape, two other shapes—monotonically increasing or decreasing—are possible depending on the network configuration. The shape of the curve is dependent on the relative magnitude of available resources (e.g., bandwidth) versus the magnitude of aggressiveness as determined by the aggressiveness schedule $\epsilon(\cdot)$. If resources are "plentiful" then aggressiveness is least penalized and it can lead to a monotonically increasing throughput curve. On the other hand, if resources are "scarce" then aggressiveness is penalized most heavily and this can result in a monotonically decreasing throughput curve. This phenomenon is shown in Fig. 18.12.

Figure 18.12 shows the throughput curves under the same network configuration except that the available bandwidth is decreased from the leftmost to the rightmost figure. This is affected by increasing the background traffic level from 2.5 Mb/s (Fig. 18.12(a)) to 5 Mb/s (Fig. 18.12(b)) to 7.5 Mb/s (Fig. 18.12(c)). We observe that the curve's shape transitions from monotone increasing to unimodal dome-shaped to monotone decreasing. In addition, due to the decrease in available bandwidth, overall throughput drops as the background traffic level is increased.

Figure 18.13 shows the change in the shape of the throughput curve as the aggressiveness schedule $\epsilon(\cdot)$ is shifted (or translated) upward—that is, made overall more aggressive—by 0.5, 2.0, 4.0, and 20.0 while keeping everything else fixed. We observe that an overall increase in the magnitude of aggressiveness can help improve throughput, transforming a monotone increasing throughput curve into a unimodal curve whose maximum throughput has increased. However, as the overall aggressiveness level is further increased, the cost of aggressiveness begins to outweigh its benefit and we observe a downward shift in the unimodal throughput curve.

SAC is designed to operate under all three network conditions, finding a near-optimum throughput in each case. The most challenging task arises when the

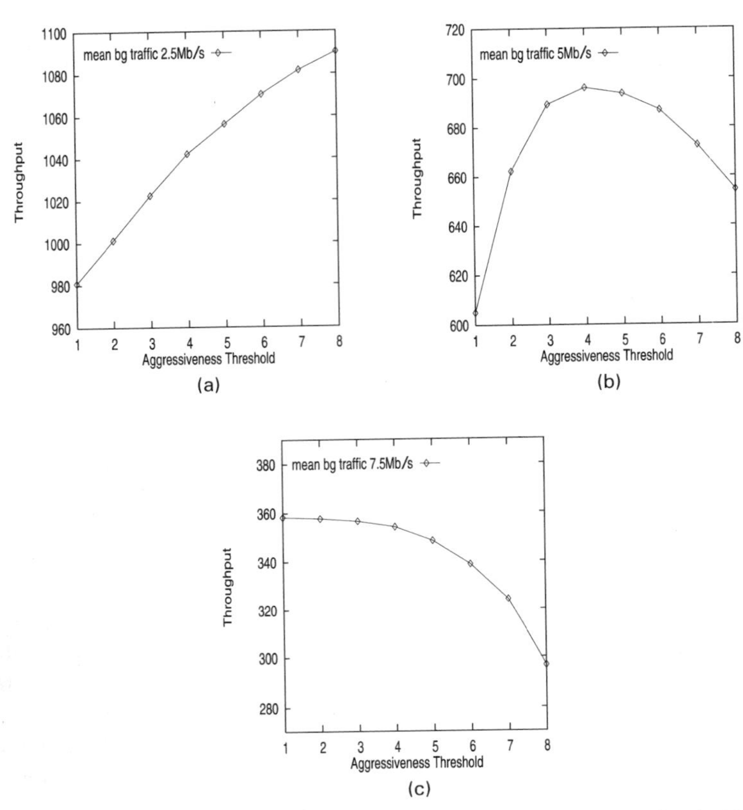

Fig. 18.12 Shape of throughput curve as a function of aggressiveness threshold for three levels of background traffic: (a) 2.5 Mb/s, (b) 5 Mb/s, and (c) 7.5 Mb/s.

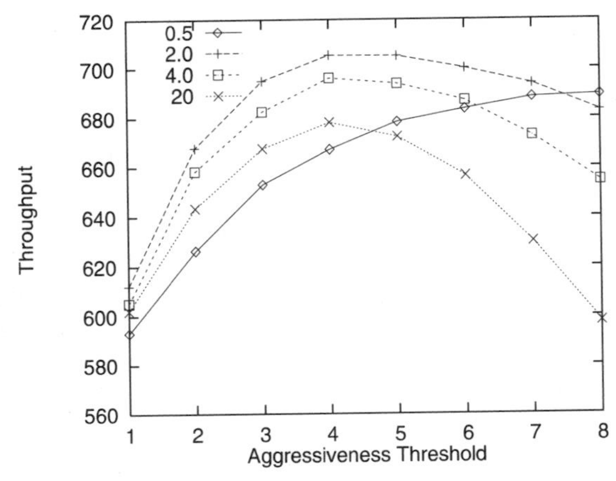

Fig. 18.13 Change in shape of throughput curve as the aggressiveness schedule $\epsilon(\cdot)$ is shifted (upward) by 0.5, 2.0, 4.0, 20.0.

network configuration leads to a unimodal throughput curve for which finding the maximum throughput is least trivial. That is, neither blindly applying aggressiveness nor absteining from it are optimal strategies. SAC's adaptivity is also useful in nonstationary situations where the network configuration can shift from one quasi-static throughput state to another.

18.5.3.2 Perfect Prediction, Uncertainty, and Aggressiveness Now that we have shown that selective aggressiveness can help but indiscriminate aggressiveness can hurt, we seek to understand three further aspects of SAC: (1) how much performance is gained by applying selective aggressiveness (vis-à-vis not applying at all), (2) how much performance is lost due to prediction inaccuracies, and (3) what is a practical aggressiveness schedule to use since we cannot assume to know the aggressiveness threshold for which maximum throughput is achieved (when using a threshold schedule).

The practical aggressiveness schedule that we found effective is the *inverse schedule* given by $\epsilon(x) = 1/x$. That is, the magnitude of aggressiveness is inverse-proportionally diminished as a function of the expected future traffic level. To measure the performance loss due to inaccuracies arising from the using per-connection on-line prediction of future traffic levels, we observe the performance of SAC when, instead of using the on-line `CondProb` table, a perfect knowledge of future aggregate traffic is assumed and employed in conjunction with the inverse schedule. Finally, to compare the net gain of having used a practical version of SAC—in this case, predicted future using per-connection on-line table and inverse aggressiveness schedule—we observe the generic linear increase/exponential decrease feedback congestion control without SAC active.

Figure 18.14(a) shows the original throughput versus threshold schedule curve superimposed with the throughput achieved by using SAC with perfect future knowledge and inverse aggressiveness schedule (topmost line), using SAC with predicted future and inverse schedule (middle line), and using the generic linear increase/exponential decrease feedback congestion control without SAC (bottom line). We observe that the generic feedback congestion control performs worst among the four—we are counting the family of SAC algorithms for the threshold schedule as one—which is mainly due to the costly nature of exponential backoff when coupled with conservative linear increase. For our purposes, the absolute magnitudes do not matter so much as the relative magnitudes, which demonstrate a qualitative performance relationship. SAC with perfect information and inverse epsilon schedule achieves the highest throughput (even higher than the peak threshold schedule throughput) and SAC with predicted future and inverse schedule achieves a performance level in between. Figure 18.13(b) shows the corresponding plots for $\alpha = 1.95$ traffic where a similar ordering relation is observed.

Figure 18.15 shows the packet loss rates corresponding to the throughput plots shown in Fig. 18.14. As expected, for the threshold schedule, packet loss rate increases monotonically as the aggressiveness threshold is increased. The generic (or regular) linear/exponential decrease congestion control incurs the least packet loss rate among the controls due to its conservativeness in the linear increase phase,

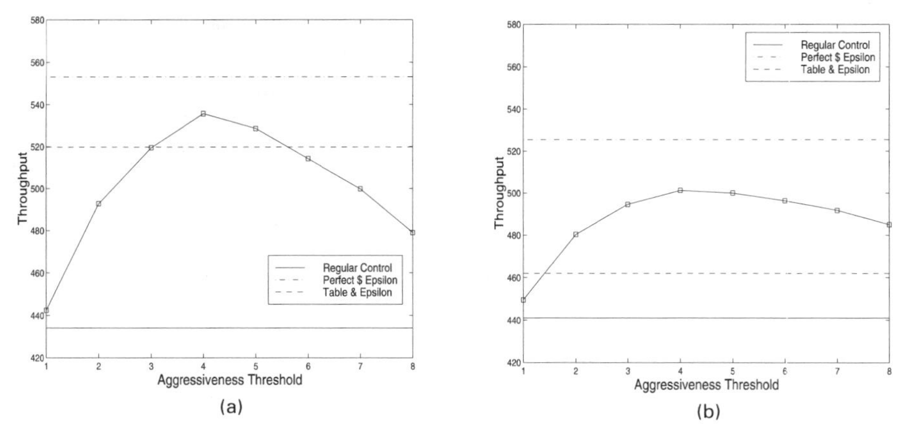

Fig. 18.14 (a) The horizontal lines show throughput when different control strategies are employed (*top line*: perfect prediction with inverse schedule; *middle line*: on-line table with inverse schedule; *bottom line*: generic linear increase/exponential decrease congestion control without SAC) for $\alpha = 1.05$ traffic. (b) Corresponding throughput plot for $\alpha = 1.95$ traffic.

albeit, at the cost of reduced throughput. Comparing Figs. 18.15(a) and 18.15(b), we observe that the overall packet loss rate for $\alpha = 1.05$ traffic is higher than that of $\alpha = 1.95$ traffic, which is expected due to the higher level of self-similar burstiness.

18.5.3.3 Impact of Long-Range Dependence The previous results, in addition to demonstrating a specific way to utilize correlation structure in self-similar traffic, showed that selective aggressiveness when coupled with predictability can lead to performance improvement above and beyond what a generic linear increase/

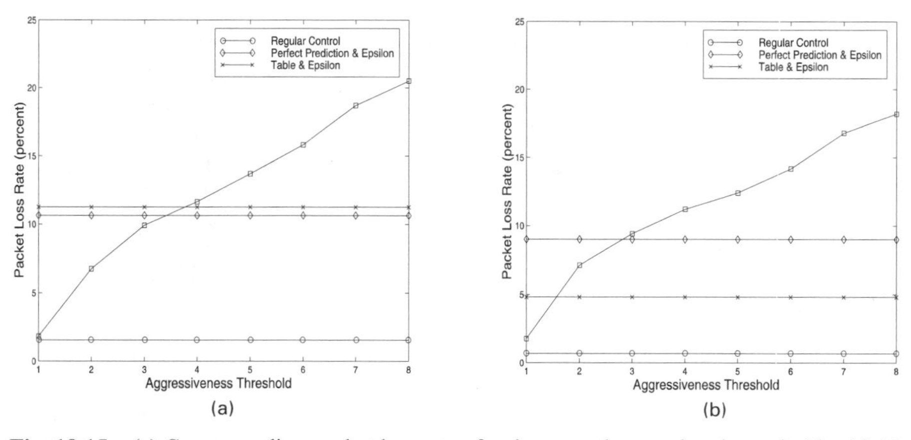

Fig. 18.15 (a) Corresponding packet loss rates for the control strategies shown in Fig. 18.13 for $\alpha = 1.05$ background traffic. (b) Corresponding packet loss rates for $\alpha = 1.95$.

exponential decrease feedback congestion control can achieve. The latter is of import since one of the practical applications of SAC is targeted at improving the performance of existing protocols.

In this section, we show that the relative performance gain due to selective aggressiveness control and predictability grows as long-range dependence increases. Figure 18.16 compares the relative performance gain stemming from employing predicted inverse schedule SAC for $\alpha = 1.05$ and $\alpha = 1.95$ background traffic. First, note that the throughput level for the generic feedback congestion control is higher for $\alpha = 1.95$ traffic than $\alpha = 1.05$ traffic. This is as expected since self-similar burstiness is known to lead to degraded performance unless resources are over-extended, at which point the burstiness associated with short-range dependent traffic is dominant in determining queueing behavior. More importantly, we observe that the throughput gain relative to the generic feedback congestion control is about 20% in the $\alpha = 1.05$ case versus about 4% for the $\alpha = 1.95$ case. This indicates that self-similar burstiness—although detrimental to network performance, in particular, QoS—possesses structure that can be exploited to dampen its negative impact. In fact, the more long-range dependent, the more structure there is to exploit effectively.

An important point to note is that we have held the mean of the background traffic levels for both $\alpha = 1.05$ and $\alpha = 1.95$ constant to achieve comparability. This is a nontrivial matter since for $\alpha = 1.05$, the mean traffic level estimated by using the Pareto distribution will overestimate the sample mean observed in practice, even if the system is run for 10,000 seconds. Figure 18.17 shows the predictability gain in terms of throughput achieved for four background traffic cases $\alpha = 1.05, 1.35, 1.65, 1.95$. Interestingly, the throughput gain shows a superlinear increase as α approaches 1 (i.e., becomes more long-range dependent).

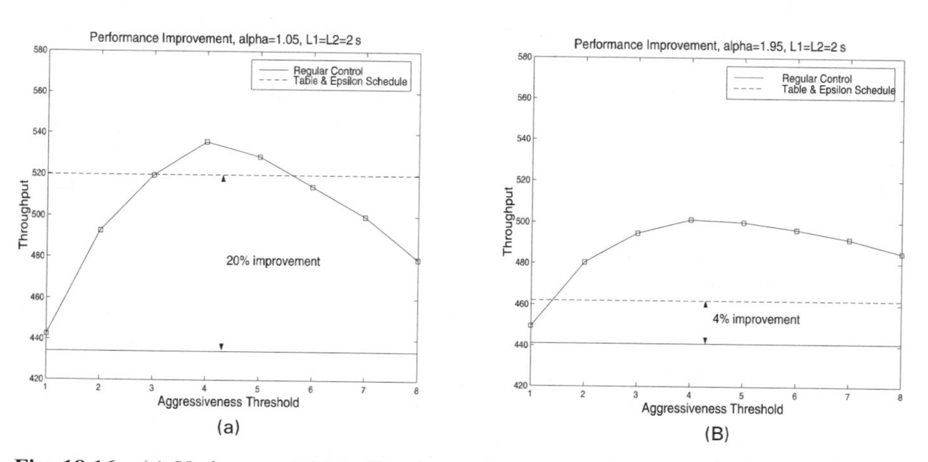

Fig. 18.16 (a) Under $\alpha = 1.05$ traffic, the performance improvement is about 20% when using SAC with on-line table and inverse schedule. (b) Under $\alpha = 1.95$ traffic, the performance improvement is only 4%.

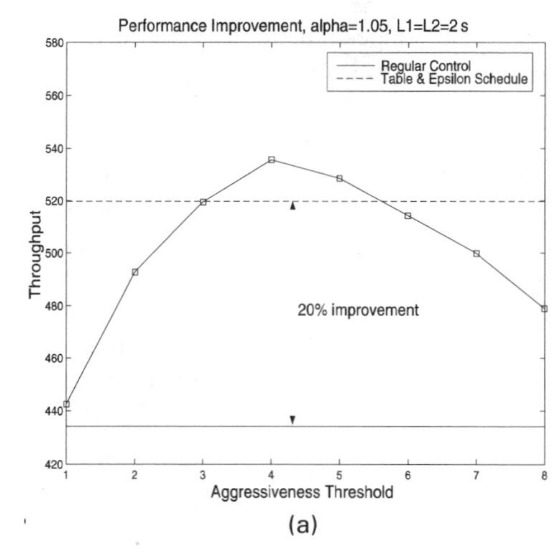

Fig. 18.17 Performance gain due to predictability for $\alpha = 1.05, 1.35, 1.65, 1.95$ background traffic.

18.5.3.4 Convergence Rate and Performance In Section 18.5.2.3 we have shown the rapid convergence rate of the on-line conditional probability table. The faster the convergence, the earlier the conditional probabilities can be employed for congestion control purposes, resulting in higher throughput gain. Other things being equal, early activation of SAC induces a trade-off relation between the benefit obtained by applying predictive information for congestion control and the cost of engaging SAC based on possibly inaccurate conditional probability estimates.

Figure 18.18(a) shows the impact of inaccuracies in the conditional probability density estimates on performance. The top graph plots throughput as a function of *training time*—that is, time spent in estimating the conditional probability densities—when the conditional probability table is subsequently fixed and used in a 10,000 second throughput measurement run. This allows us to assess the impact of inaccurate prediction estimates on throughput performance. As the top graph shows, convergence is rapid after which the incremental gain observed via further accuracies saturates. Due to rapid convergence, the net gain in throughput due to increased prediction accuracy is below 5%. Contrast this with the bottom two graphs of Fig. 18.18(a), which show 10,000 second throughput measurements when the conditional probability table used is that of $\alpha = 1.95$ traffic (trained over 10,000 seconds) and randomly, respectively. The gap shows that even inaccurate prediction estimates are significantly more useful than random or otherwise-structured information for performance enhancement purposes.

The fast convergence property can be further explained by Fig. 18.18(b), which plots the computed conditional expectation $\mathbf{E}(L_2|\cdot)$ as a function of training time. Recall that in both the threshold and inverse schedules $\mathbf{E}(L_2|\cdot)$ (quantized or not)—and not the conditional probability table proper—is used in computing the aggressiveness level. Figure 18.18(b) shows that the functional $\mathbf{E}(L_2|\cdot)$ quickly converges

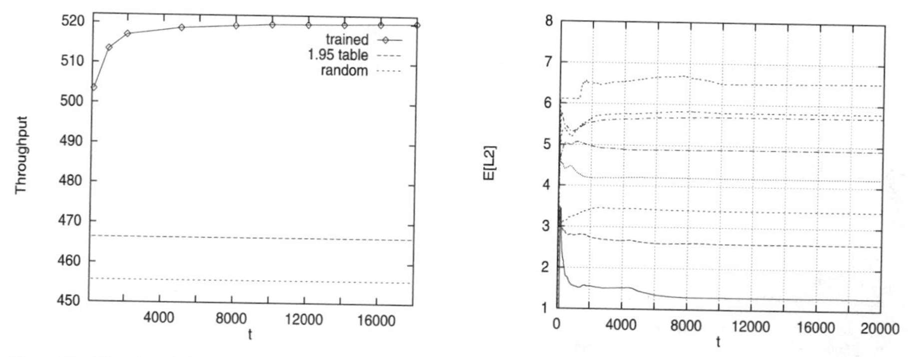

Fig. 18.18 (a) Throughput as a function of SAC conditional probability table training time. Also shown: SAC throughput with random and $\alpha = 1.95$ conditional probability table. (b) Convergence property of $\mathbf{E}(L_2|\cdot)$. Fast convergence to linear order $\mathbf{E}(L_2|L_1 = 1) < \mathbf{E}(L_2|L_1 = 2) < \cdots < \mathbf{E}(L_2|L_1 = 8)$.

to the linear ordering $\mathbf{E}(L_2|L_1 = 1) < \mathbf{E}(L_2|L_1 = 2) < \cdots < \mathbf{E}(L_2|L_1 = 8)$, as would be expected by the skewdness of the 3D conditional probability densities. The magnitudes of $\mathbf{E}(L_2|\cdot)$, after some undulation, stabilized to fixed values.

The quick establishment of the linear ordering property and the convergence of $\mathbf{E}(L_2|\cdot)$ to fixed values leaves open the possibility that *a priori* conditional probabilities may be used for predictive purposes, which is especially useful for short-lives connections for which per-connection conditional probability tables are impossible to establish.

18.5.3.5 Multiple Concurrent SAC Connections The SAC protocol is designed to run in shared network environments where different connections compete for available resources. In this section, we investigate the behavior of the SAC protocol with respect to fairness and efficiency when multiple connections engage in SAC. The results are based on the same setup as in Fig. 18.8 except that we increase the bottleneck link bandwidth to 20 Mb/s to accommodate up to 10 SAC connections. The main traffic rate of the first 32 connections—that is, non-SAC background traffic sources—is kept at 5 Mb/s. We increase the number of SAC connections from 1 to 10 (33rd connection and beyond) and observe whether bandwidth is shared fairly and efficiently. The latter refers to the question of whether the total throughput achieved across all SAC connections remains conserved—increased competition can create overhead and inefficiencies—as the number of SAC connections is increased.

Figure 18.19 depicts *average per-connection throughput* as a function of the aggressiveness threshold for one (top left), two (top right), four (bottom left), and ten (bottom right) SAC connections. Superimposed we also show the throughput achieved by the inverse schedule and generic feedback congestion control, respectively. First, we observe that the shape of the throughput curve changes from monotonically increasing to unimodal to monotonically decreasing as the number of SAC connections is increased. This is to be expected since, other things being equal,

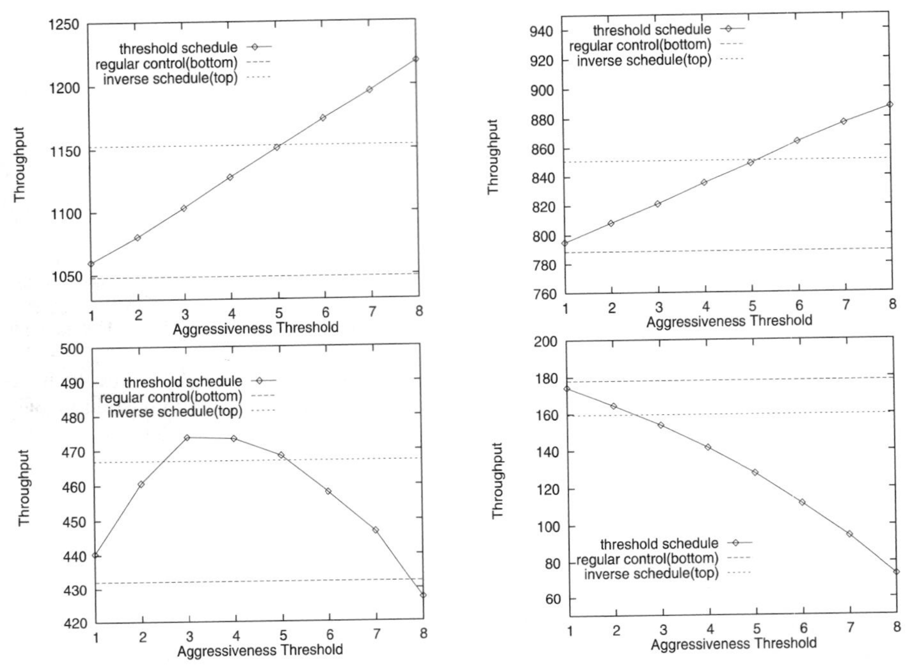

Fig. 18.19 Average (per-connection) throughput as a function of aggressiveness threshold for multiple SAC connections. The top horizontal line shows the throughput achieved with the inverse schedule. The bottom line represents the throughput of the generic linear increase/exponential decrease feedback congestion control. *Top left*: Single SAC connection; *top right*: two SAC connections; *bottom left*: four SAC connections; *bottom right*: ten SAC connections.

increasing the number of SAC connections amplifies the net aggressiveness level since there is no distributed control or cooperation among the SAC flows to maintain a constant overall aggressiveness level. Second, since we plot average per-connection throughput, we observe the per-connection throughput drop accordingly. Third, we observe that the performance of the threshold schedule eventually deteriorates below that of the generic feedback congestion control while the performance of the inverse schedule stays at a high level.

Table 18.2 shows the total throughput achieved across all SAC connections for the threshold schedule as the number of SAC connections is increased. For each threshold level, we observe a unimodal change in throughput as the number of connections is increased from 1 to 10 with the peak occurring earlier the higher the threshold level. This indicates a trade-off relation whereby, at first, the net increase in aggressiveness due to the increased number of SAC connections leads to a net increase in total SAC throughput. However, as the number of SAC connections is further increased, the amplification of the overall aggressiveness level asserts a negative impact on throughput eventually yielding a net decrease. A similar phenomenon is observed for the inverse schedule, which is shown in Table 18.3.

TABLE 18.2 Total Throughput Across All SAC Connections for the Threshold Schedule as the Number of SAC Connections Is Increased

Threshold Schedule	1	2	3	4	5	6	7	8
1 connection	1066.1	1080.3	1103.0	1127.4	1150.8	1173.6	1194.9	1218.4
2 connections	1591.7	1617.6	1644.2	1672.7	1698.5	1728.6	1755.4	1776.1
4 connections	1765.5	1845.7	1896.4	1893.7	1877.3	1832.2	1788.8	1712.6
8 connections	1800.3	1750.3	1653.1	1537.6	1376.2	1230.6	1051.7	869.4
10 connections	1748.9	1649.0	1540.2	1417.9	1276.4	1109.2	925.7	743.0

TABLE 18.3 Total Throughput Across All SAC Connections for the Inverse Schedule as the Number of SAC Connections Is Increased

Inverse Schedule	1 conn.	2 conn.	4 conn.	8 conn.	10 conn.
Throughput	1152.9	1702.0	1947.7	1676.4	1594.4

The onset of the peak is a function of available resources and it can be further delayed by decreasing the overall aggressiveness of each connection. Note that the multiple connection throughput behavior is achieved for the network configuration shown in Fig. 18.8, which, due to its uniform link latencies, can be prone to synchronization effects [14].

Figure 18.20 plots the individual throughput achieved by each SAC connection when a total of 10 are present. We observe that fairness, for the network configuration shown in Fig. 18.8, is well-preserved. As the configuration becomes less

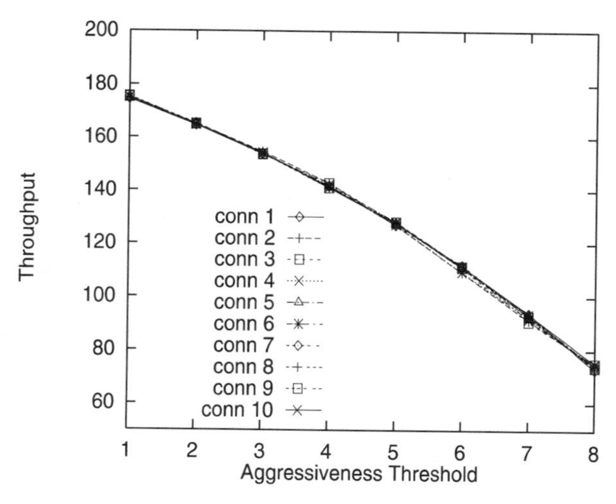

Fig. 18.20 Fair bandwidth access: individual throughput plot with 10 SAC connections present.

uniform, access discrepancies as with TCP and other feedback congestion control algorithms are bound to arise, which is a generic problem not specific to SAC.

18.6 CONCLUSION

In this chapter we have shown that a form of predictive congestion control called *selective aggressiveness control* (SAC) is able to exploit correlation structure present in long-range dependent traffic for performance improvement purposes. We have shown that predictability can be implemented on-line with sufficient accuracy such that when coupled with SAC mechanism, the cost associated with conservative bandwidth consumption during the linear increase phase of a generic linear increase/exponential decrease feedback congestion control algorithm can be offset to bring forth improved throughput. Furthermore, we show that the relative performance gain is higher the more long-range dependent the underlying network traffic. We have also shown that SAC preserves fairness to the degree expected of a feedback congestion control and its efficiency property is desirable within a range. An important consequence of multiple time scale congestion control is its ability to mitigate the dely-bandwidth product problem of broadband wide area networks by endowing proactivity. We refer the reader to [46, 48] for this result.

Current work is directed along four avenues: (1) in devising new mechanisms capable of exploiting predictability at longer time scales, (2) in further optimizing the various components of SAC, (3) in implementing the ideas of SAC in TCP, and (4) adopting the multiple time scale traffic control framework to the QoS-sensitive transport of real-time traffic using packet-level adaptive forward error correction. The TCP results are reported in [48], and the extension to quality of service is performed via multiple time scale redundancy control and described in [47].

REFERENCES

1. A. Adas and A. Mukherjee. On resource management and QoS guarantees for long range dependent traffic. In *Proc. IEEE INFOCOM'95*, pp. 779–787, 1995.
2. R. Addie, M. Zukerman, and T. Neame. Fractal traffic: measurements, modelling and performance evaluation. In *Proc. IEEE INFOCOM'95*, pp. 977–984, 1995.
3. T. Anderson. *The Statistical Analysis of Time Series*. Wiley, New York, 1994.
4. J. Beran. *Statistics for Long-Memory Processes*. Monographs on Statistics and Applied Probability. Chapman and Hall, New York, 1994.
5. J.-C. Bolot and A. U. Shankar. Analysis of a fluid approximation to flow control dynamics. In *Proc. IEEE INFOCOM'92*, pp. 2398–2407, 1992.
6. L. Brakmo and L. Peterson. TCP Vegas: end to end congestion avoidance on a global internet. *IEEE J. Select. Areas Commun.*, **13**(8):1465–1480, 1995.
7. I. Chlamtac and W. R. Franta. Rationale, directions, and issues surrounding high speed networks. *Proc. IEEE*, **78**(1):94–120, 1990.
8. M. Crovella and A. Bestavros. Self-similarity in World Wide Web traffic: evidence and possible causes. In *Proceedings of the 1996 ACM SIGMETRICS International*

Conference on Measurement and Modeling of Computer Systems, pp. 160–169, May 1996.

9. M. Crovella and L. Lipsky. Long-lasting transient conditions in simulations with heavy-tailed workloads. In *Proc. 1997 Winter Simulation Conference*, 1997.
10. R. Dighe, C. J. May, and G. Ramamurthy. Congestion avoidance strategies in broadband packet networks. In *Proc. IEEE INFOCOM'91*, pp. 295–303, 1991.
11. N. G. Duffield and N. O'Connel. Large deviations and overflow probabilities for the general single server queue, with applications. Technical Report DIAS-STP-93-30, DIAS Technical Report, 1993.
12. A. E. Eckberg. B-ISDN/ATM traffic and congestion control. *IEEE Network*, **6**:28–37, September 1992.
13. K. Fendick, M. Rodrigues, and A. Weiss. Analysis of a rate-based control strategy with delayed feedback. In *Proc. ACM SIGCOMM'92*, pp. 136–148, 1992.
14. S. Floyd and V. Jacobson. The synchronization of periodic routing messages. In *Proc. ACM SIGCOMM'93*, pp. 33–44, 1993.
15. M. Garret and W. Willinger. Analysis, modeling and generation of self-similar VBR video traffic. In *Proc. ACM SIGCOMM'94*, pp. 269–280, 1994.
16. M. Gerla and L. Kleinrock. Flow control: a comparative survey. *IEEE Trans. Commun.*, **20**(2):35–49, 1980.
17. G. C. Goodwin and K. S. Sin. *Adaptive Filtering, Prediction and Control*. Prentice Hall, Englewood Cliffs, NJ, 1984.
18. M. Grossglauser and J.-C. Bolot. On the relevance of long-range dependence in network traffic. In *Proc. ACM SIGCOMM'96*, pp. 15–24, 1996.
19. Z. Haas and J. Winters. Congestion control by adaptive admission. In *Proc. IEEE INFOCOM'91*, pp. 560–569, 1991.
20. Z. Haas. A communication architecture for high-speed networking. In *Proc. IEEE INFOCOM'90*, pp. 433–441, 1990.
21. D. Hong and T. Suda. Congestion control and prevention in ATM networks. *IEEE Network Mag.* **31**:10–16, July 1991.
22. K. Hornik, M. Stinchcombe, and H. White. Multilayer feedforward networks are universal approximators. *Neural Networks*, **2**(23):359–366, 1989.
23. C. Huang, M. Devetsikiotis, I. Lambadaris, and A. Kaye. Modeling and simulation of self-similar variable bit rate compressed video: a unified approach. In *Proc. ACM SIGCOMM'95*, pp. 114–125, 1995.
24. V. Jacobson. Congestion avoidance and control. In *Proc. ACM SIGCOMM '88*, pp. 314–329, 1988.
25. S. Keshav. A control-theoretic approach to flow control. In *Proc. ACM SIGCOMM'91*, pp. 3–15, 1991.
26. H. Kim. *A Non-Feedback Congestion Control Framework for High-Speed Data Networks*. Ph.D. thesis, University of Pennsylvania, 1995.
27. H. T. Kung, T. Blackwell, and A. Chapman. Credit-based flow control for ATM networks: credit update protocol, adaptive credit allocation, and statistical multiplexing. In *Proc. SIGCOMM'94*, pp. 101–114, 1994.
28. W. E. Leland, M. S. Taqqu, W. Willinger, and D. V. Wilson. On the self-similar nature of Ethernet traffic (extended version). *IEEE/ACM Trans. Networking*, **2**:1–15, 1994.

29. N. Likhanov and B. Tsybakov. Anlaysis of an ATM buffer with self-similar ("fractal") input traffic. In *Proc. IEEE INFOCOM'95*, pp. 985–992, 1995.
30. D. Mitra and J. Seery. Dynamic windows for high speed data networks: theory and simulations. In *Proc. ACM SIGCOMM'90*, pp. 30–37, 1990.
31. A. Mukherjee and J. Strikwerda. Analysis of dynamic congestion control protocols—a Fokker–Planck approximation. In *Proc. ACM SIGCOMM'91*, pp. 159–169, 1991.
32. I. Norros. A storage model with self-similar input. *Queueing Syst.*, **16**:387–396, 1994.
33. K. Park, G. Kim, and M. Crovella. On the relationship between file sizes, transport protocols, and self-similar network traffic. In *Proc. IEEE International Conference on Network Protocols*, pp. 171–180, 1996.
34. K. Park, G. Kim, and M. Crovella. On the effect of traffic self-similarity on network performance. In *Proc. SPIE International Conference on Performance and Control of Network Systems*, pp. 296–310, 1997.
35. Kihong Park. Warp control: a dynamically stable congestion protocol and its analysis. In *Proc. ACM SIGCOMM'93*, pp. 137–147, 1993.
36. K. K. Ramakrishnan and R. Jain. A binary feedback scheme for congestion avoidance in computer networks with a connectionless network layer. In *Proc. ACM SIGCOMM'88*, pp. 303–313, 1988.
37. B. Ryu and A. Elwalid. The importance of long-range dependence of VBR video traffic in ATM traffic engineering: myths and realities. In *Proc. ACM SIGCOMM'96*, pp. 3–14, 1996.
38. S. Shenker. A theoretical analysis of feedback flow control. In *Proc. ACM SIGCOMM'90*, pp. 156–165, 1990.
39. M. S. Taqqu, V. Tevrovsky, and W. Willinger. Estimators for long-range dependence: an empirical study. Preprint, 1995.
40. H. L. van Trees. *Detection, Estimation and Modulation Theory.* Wiley, New York, 1968.
41. Y. T. Wang and B. Sengupta. Performance analysis of a feedback congestion control policy under non-negligible propagation delay. In *Proc. ACM SIGCOMM'91*, pp. 149–157, 1991.
42. W. Willinger, M. Taqqu, R. Sherman, and D. Wilson. Self-similarity through high-variability: statistical analysis of Ethernet LAN traffic at the source level. In *Proc. ACM SIGCOMM'95*, pp. 100–113, 1995.
43. C. Yang and A. Reddy. A taxonomy for congestion control algorithms in packet switching networks. *IEEE Network*, **9**:34–45, July/August 1995.
44. G. Gripenberg and I. Norros. On the prediction of fractional brownian motion. *J. Applied Probability* **33**:400–410, 1996.
45. S.Ostring, H. Sirisena, and I. Hudson. Dual dimensional abr control scheme using prdictive filtering of self-similar traffic. In *Proc. ICC '99*, June 1999.
46. T. Tuan and K. Park. Multiple time scale congestion control for self-similar network traffic. *Performance Evaluation*, **36**:359–386, 1999.
47. T. Tuan and K. Park. Multiple time scale redundancy control for QoS-sensitive transport of real-time traffic. To appear in *Proc. IEEE INFOCOM '00*, 2000.
48. T. Tuan and K. Park. Performance evaluation of multiple time scale TCP under self-similar traffic conditions. Technical report, Dept. of Computer Sciences, Purdue University, 1999. CSD-TR-99-040.
49. D. Veitch and P. Abry. A wavelet based joint estimator of the parameters of long range dependence. *IEEE/ACM Transactions on Information Theory*, 45, Apr. 1999.

19

QUALITY OF SERVICE PROVISIONING FOR LONG-RANGE-DEPENDENT REAL-TIME TRAFFIC

ABDELNASER ADAS
Conexant, Inc., Newport Beach, CA 92660

AMARNATH MUKHERJEE
Knoltex Corporation, San Jose, CA 95157

19.1 INTRODUCTION

19.1.1 Overview

Network support for variable bit rate (VBR) video needs to consider (1) properties of workload induced (e.g., significant autocorrelations into far lags and heterogeneous marginal distributions), and (2) application-specific bounds on delay-jitter and statistical cell-loss probabilities. This chapter presents a quality-of-service (QoS) solution for such traffic at each multiplexing point in a network. Heterogeneity in both-offered workload and quality-of-service requirements are considered. The network is assumed to be cell-switched with virtual circuits (VCs) similar to that in ATM networks. Chapter 16 discusses an alternative approach for provisioning for long-range-dependent (LRD) traffic. See also the work of Heyman and Lakshman (Chapter 12) and Li and Li (Chapter 13).

19.1.2 Correlated Traffic and Its Implications

Studies on a range of video applications indicate that there exists a slowly decaying autocorrelation structure in the underlying stochastic processes [3, 7, 8, 12, 19, 21].

Self-Similar Network Traffic and Performance Evaluation, Edited by Kihong Park and Walter Willinger
ISBN 0-471-31974-0

Providing guarantees on maximum-delay, delay-jitter, and cell-loss probabilities (or some other measure of cell loss) in the presence of such traffic is nontrivial, especially if the coefficient of variation of the marginal distribution (or the distribution tail) is large. This is because such traffic significantly increases queue length statistics at a multiplexer [2, 5, 8, 16–18, 20].

As an illustration, consider the performance of a finite-buffer queue. Let the arrival process be a fractionally differenced autoregressive moving average process:

$$(1 - \phi_1 B)\Delta^d X_t = \epsilon_t,$$

where

- ϕ_1 $(0 < \phi_1 < 1)$ represents an exponentially decaying autocorrelation component;
- d $(0 < d < \frac{1}{2})$ represents a hyperbolically decaying autocorrelation component;
- $\{\epsilon_t\}$ is white noise, that is, uncorrelated; and
- B is the backward shift operator, that is, $BX_t = X_{t-1}$.

The correlation structure of X_t can be controlled by changing ϕ_1 and d. Figure 19.1 shows the mean cell loss versus number of frame buffers for three input correlation structures with approximately the same coefficient of variation (~0.24). A frame buffer is the maximum number of cells that can be transmitted by the output channel in a given time interval (frame time).

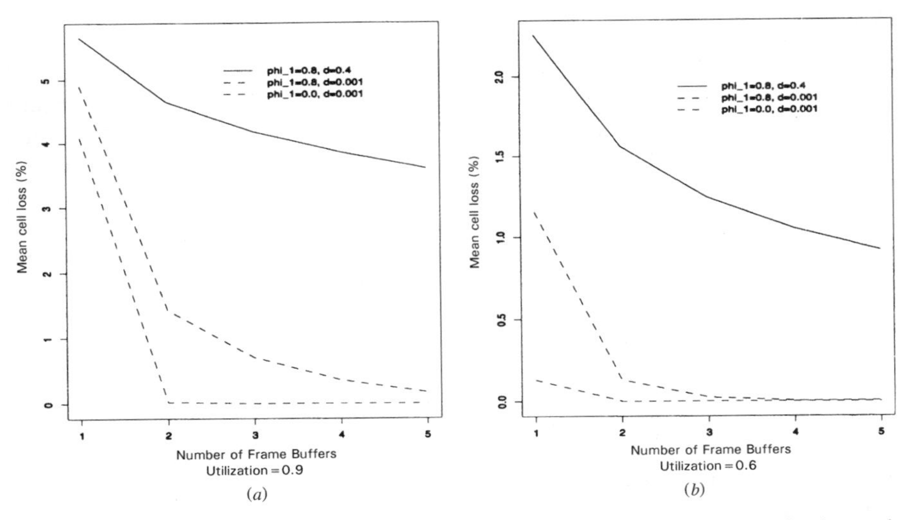

Fig. 19.1 Mean number of cells dropped for different dependency structures in the workload. (a) Mean utilization = 0.9; (b) mean utilization = 0.8. The model form used is a fractionally differenced ARIMA(1, d, 0) process [13]: $(1 - \phi_1 B)\Delta^d X_t = \epsilon_t$.

The solid line showing a slow decay is for a long-memory input traffic sequence. The dashed line in the middle is for traffic with short memory. The line with the smallest mean cell loss corresponds to a near white noise stream. Note that mean cell loss decays very slowly with increasing buffer size for traffic with a slowly decaying autocorrelation function. Qualitatively similar observations have been reported elsewhere [2, 5, 8, 16–18, 20].

19.1.3 Summary of the Proposed Architecture

This chapter introduces a per-virtual-circuit (per-VC) framing structure and a pseudo-earliest-due-date cell dispatcher to provide guaranteed delay-jitter bounds. Heterogeneous jitter bounds are supported through software-controlled frame sizes, which may be independently set of each VC. The framing structure is a generalization of per-link framing introduced by Golestani. The proposed framing structure eliminates correcting for phase mismatches between incoming frames and outgoing frames, necessary in per-link framing. This results in reduction in end-to-end delay bound and buffer requirements, and a simpler implementation.

Strong autocorrelations typically seen in video traffic make equivalent bandwidth computations for heterogeneous cell-loss bounds intractable. To address this, the framing strategy is combined with an active cell-discard mechanism with prioritized cell-dropping, the latter utilizing the history of dropped cells and target cell-loss bounds for each VC. Upper bounds on the equivalent bandwidth needed to support a given workload with a target quality of service are developed. These are validated through numerical and simulation results from variable bit rate MPEG-I video traces.

A high-level view of the proposed architecture is as follows.

1. *A Framing Structure on a Per-VC Basis.* To provide heterogeneous delay-jitter bounds, a framing structure is induced on VCs, similar to that in Golestani [9–11]. (Differences between the two approaches are described later.) Consider a virtual circuit (VC), i, with a desired delay-jitter bound M_i. The frame structure splits time for this VC into juxtaposed intervals of length M_i at each multiplexing point. Cells from VC i that arrive in a given frame at a multiplexer are buffered, and not transmitted until the beginning of the next frame time. If sufficient capacity is available to transmit these cells in the next interval, all cells arrive at the next hop within an interval of length M_i. This way, they are guaranteed to meet a delay-jitter bound of M_i. Also, if H_i is the number of hops for VC i, and D_i is a bound on the one-way propagation and processing delay, all cells that make it to the receiver are guaranteed an end-to-end delay bound of $H_iM_i + D_i$.

2. *Priority Scheduling.* Cells at an output queue that are ready to go contend for bandwidth and have competing delay-jitter bounds and cell-loss probability bounds. A priority scheduler addresses these concerns. For delay-jitter bounds, the scheduler follows an earliest due-date principle with modifications to enhance algorithmic efficiency. For cell-loss bounds, it uses a minimum guaranteed capacity,

C_i cells/frame for VC i, with the rest of the cells, if any, scheduled on a nonguaranteed basis. C_i is based on (i) marginal distribution of #cells/frame, (ii) maximum acceptable probability of cell loss in a frame, and (iii) the equivalent bandwidth of all VCs in this jitter class. The C_i are computed by the equivalent-bandwidth unit described in Item 4 below.

3. *An Active Cell-Discard Unit.* If there are excess cells left over from a frame at the multiplexer after the corresponding frame time is over, it is likely that this is due to persistence in the arrival process as suggested by the solid lines in Fig. 19.1. These cells are likely to cause increased delay for cells in successive frames. Since buffering does not reduce cell loss significantly for persistent traffic, we may elect to either toss them right away or mark them as low-priority cells and discard them on demand.

The active cell-discard unit reclaims (or marks as *old*) cells that do not get transmitted in their frame time. It is activated at the end of each frame.

An important side effect of using the active cell-discard unit is that it simplifies computation of equivalent bandwidth for correlated traffic (especially for heterogeneous cell-loss bounds), while achieving high utilization through statistical multiplexing.

4. *Equivalent Bandwidth Computations.* Algorithms for computing upper bounds on equivalent bandwidths are developed. They address heterogeneous cell-loss probabilities and heterogeneous jitter classes.

The computation decomposes traffic by their jitter requirements. All connections requiring a given delay-jitter are grouped into a class. For each jitter class, let ϵ_k ($k = 0, 1, 2, \ldots$) be the desired mean cell-loss ratio of a subset of connections k. An iterative algorithm approximates the total capacity needed to meet $\{\epsilon_k\}$. Also, virtual capacities C_k ($k = 0, 1, 2, \ldots$) are computed. All groups of connections specifying ϵ_k are guaranteed a bandwidth of C_k in every frame time if they need it. However, unused portions of virtual capacities are available to other connections.

5. *Miscellaneous*

- Frames may be implemented on a per-VC basis. Frames from different VCs need not be synchronized.
- Per-VC framing and active cell-discard may be implemented efficiently through associative matching of cell tags similar to that in processor pipelines (Section 19.2.2).
- The frame size for each VC is software setable, so delay-jitter bounds may be negotiated over a continuum (at the granularity of cell transmission time). Also, unlike per-link framing (see Section 19.1.4), the frame size of a given VC is not constrained by frame sizes of other active VCs.
- The minimum capacity guarantee per frame, C_i for each VC i provides protection from misbehaving or malfunctioning VCs.
- A call admission unit will use the equivalent bandwidth algorithms to determine if a specific cell can be admitted without violating quality-of-service guarantees of other calls (or if an important call must be admitted, which calls to disconnect). The call-admission unit is beyond the scope of this chapter.

19.1.4 Relationship with Stop-and-Go Queueing

Per-VC framing has been derived from Stop-and-Go Queueing described in Golestani [9–11]. The primary enhancements are as follows.

1. Framing is induced on a per-VC basis instead of a per-output-link basis; see Figs. 19.2 and 19.3. Per-VC framing eliminates the need for correcting for phase mismatches between incoming frames and outgoing frames at a multiplexer and significantly simplifies its implementation. As we shall see in Section 19.3, per-VC framing also reduces the maximum queueing delay by half and cuts buffer requirements by one-third at a switch, while retaining the same delay-jitter bound per-link framing.
2. Once cells from a frame become active (i.e., not dormant, waiting for their next frame time), they compete with active cells from other VCs for the output link. The algorithms that decide on which active cells to transmit and when, and which cells to drop, are necessitated by the need to meet heterogeneous

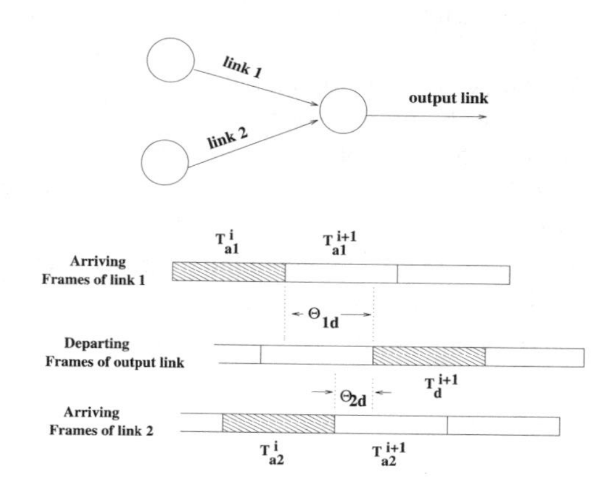

Fig. 19.2 Arriving frames and departing frames when framing is induced in a per-link basis. The phase mismatch between arriving and departing frames is corrected through delay circuits.

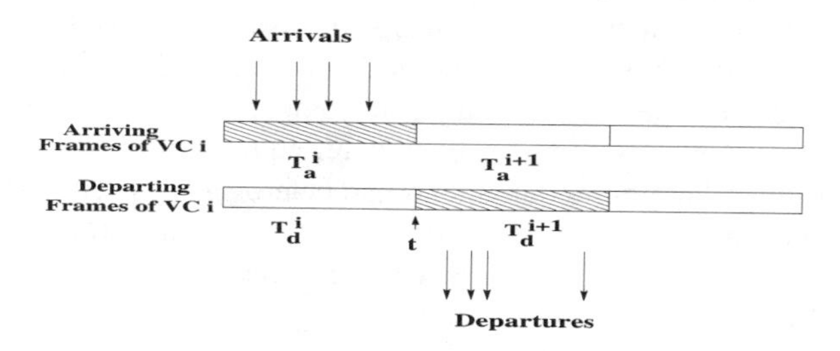

Fig. 19.3 Arriving and departing frames for VC i when framing is induced on a per-VC basis. Frames of different VCs need not be synchronized.

cell-loss bounds and heterogeneous delay-jitter bounds simultaneously. They also provide a firewall across connections (protection from misbehaving sources). These algorithms are new.

In Golestani [9], the objective was to support no-loss transmission with heterogeneous delay-jitter bounds. The latter were integral multiples of the smallest jitter bound supported. Golestani showed that a preemptive priority scheduler with highest priority to the smallest jitter class could meet all jitter bounds if sufficient capacity was available. Golestani [11] also presented a solution that allowed for cell losses for a single jitter class (fixed delay-jitter bound).

In the general case of meeting heterogeneous delay-jitter bounds with potential cell losses, however, the scheduler needs to follow (i) an earliest due-date principle and (ii) a cell-drop policy that takes into account current observations on dropped-cells per VC, and heterogeneous cell-loss bounds across VCs. See Section 19.2.3.

3. The original Stop-and-Go Queueing requirements that a traffic stream declare its (r, T)-smooth[1] parameter is dropped. This trades off higher utilization for a lossless network. For a long-memory input stream, the average rate over a small interval, T, can be significantly higher (or lower) that its overall average rate, so r would need to be the peak rate for lossless transmission and would result in significantly low utilizations. In the current proposal, cell losses, while allowed, will be reduced through statistical multiplexing across virtual circuits and controlled through equivalent bandwidth computations.
4. No-loss transmission can be guaranteed in the proposed architecture if desired; see Section 19.4.3. However, the emphasis is on efficient statistical multiplexing that can also guarantee specified cell-loss bounds.

19.1.5 Outline

The rest of this chapter is organized as follows. Section 19.2 presents the proposed architecture. It includes (1) per-VC framing with active cell-discard and (2) cell dispatching to meet the heterogeneous delay-jitter and cell-loss guarantees for heterogeneous VCs (with heterogeneous marginal distributions and autocorrelation structures). Section 19.3 presents maximum-delay bound, delay-jitter bound, and buffer requirements for per-VC framing and compares the results with per-link framing. Section 19.4 addresses upper bounds on equivalent bandwidth needed to meet heterogeneous delay-jitter requirements and heterogeneous cell-loss probability bounds, presents numerical and simulation examples, and shows that loss-free transmission may be achieved for desired VCs. Section 19.5 presents related work. Section 19.6 presents our conclusions.

[1]An (r, T)-smooth stream was defined as one where the average bit rate over a time interval T did not exceed r. Equivalently, the number of bits over $(nT, (n+1)T]$ did not exceed rT, for all integer n.

19.2 PROPOSED ARCHITECTURE

19.2.1 Framing on a Per-Virtual-Circuit Basis Versus Per-Link Basis

Enforcing framing on a per-link basis [9–11] results in a phase mismatch at a switch between arriving frames on input links and departing frames on output links. This phase mismatch is due to different propagation delays on different input links. As shown in Fig. 19.2, the arriving frames on input link 1 and departing frames on the output link have a phase mismatch of θ_{1d}, while the arriving frames on input link 2 have a phase mismatch with respect to the output link of θ_{2d}.

To correct for a phase mismatch, additional delay circuitry is necessary. Also, the admissible set of frame sizes is constrained. For example, all frame sizes are considered integer multiples of a base frame size in Golestani [9]. A simpler approach is to adopt a per-VC framing, without concern for what the frame sizes are, and whether or not the frames from different VCs are synchronized with respect to each other; see Fig. 19.3. As we will show in Section 19.3, per-VC framing, in conjunction with active cell-discard and an appropriate scheduler, retains the advantages of per-link framing, while improving on performance bounds and functional flexibility. For example, if VC i's frame size is M_i, and the number of hops is H_i, per VC framing provides the same delay-jitter bound, M_i, as per-link framing, a reduction in maximum-delay bound by an amount H_iM_i, and a maximum buffer requirement that is one-third lower. Also, in conjunction with the cell dispatcher described in Section 19.2.3, it guarantees heterogeneous cell-loss bounds for correlated traffic. Functional flexibility includes ability to set and modify admissible jitter classes at run time, and not be constrained to an integer multiple of a base frame size.

Hardware support for efficient and flexible implementation of per-VC framing with active cell-discard is discussed next.

19.2.2 Implementation of Per-Virtual-Circuit Framing with Active Cell Discard

The objective is to induce a framing structure on top of cells of a given VC, and for the multiplexer to actively discard (or mark as *old*) cells that are not served during their assigned frame time.

In order to allow for flexibility of application-specified jitter bounds, the frametime should be software setable (e.g., it may be negotiated during connection-open). It should then be set to the connection's delay-jitter tolerance. One may allow for adjusting the frame time during the lifetime of a VC, if desired.

Issues that need to be addressed for per-VC framing, and active cell-discard are as follows.

1. *Frame Identification Across Nodes (where Nodes Refer to Switches and End Points).* Cells transmitted during the tth frame ($t = 0, 1, 2, \ldots$) by a node must be recognized as belonging to frame t by the next downstream node. An alternating bit

sequence number distinguishing cells in adjacent frames is sufficient if the sequence number is generated at the transmitter. Old cells, if implemented, will be marked by the first multiplexer where a jitter deadline is missed.

2. *Frame-clock Generation.* For the ith VC, one needs a step-down counter, initialized under software control, to the maximum number of cells that constitutes its frame time. Let this number be M_i. The counter is to be fed with a clock that runs at the speed of cell transmission at the output link. On each clock cycle (at cell granularity), the counter must count down one tick until it hits zero. At this point, it will need to generate a *frame-clock* signal and reset itself to M_i.

3. *Cell Tagging.* A cell arriving during frame t for VC i will not be eligible for service until frame $t + 1$ for the same VC. It is, therefore, assigned a state, *dormant*, on arrival. See Fig. 19.4(a). When the next *frame-clock* signal arrives, the cell is ready to be transmitted, so its state needs to be changed to *active*. If it still remains in the queue when the following *frame-clock* signal arrives, it is old, and now there are two possibilities. One strategy is simply to discard the cell and reclaim its buffer. A second strategy is to change its state to *old* and keep it eligible for transmission on a best-effort basis. In ATM networks, cells need to be delivered in sequence, so it might be simplest to discard the old cells.

To simplify the discussion for what happens next, let us assume that *active* cells that are not transmitted in their frame time are discarded. Then, at any given time, cells belonging to frames t and $t + 2$ will never be simultaneously present at the multiplexer output queue, and all that is necessary is to distinguish between cells of frames t and $t + 1$. A single bit, therefore, suffices to distinguish between *active* and *dormant* cells.

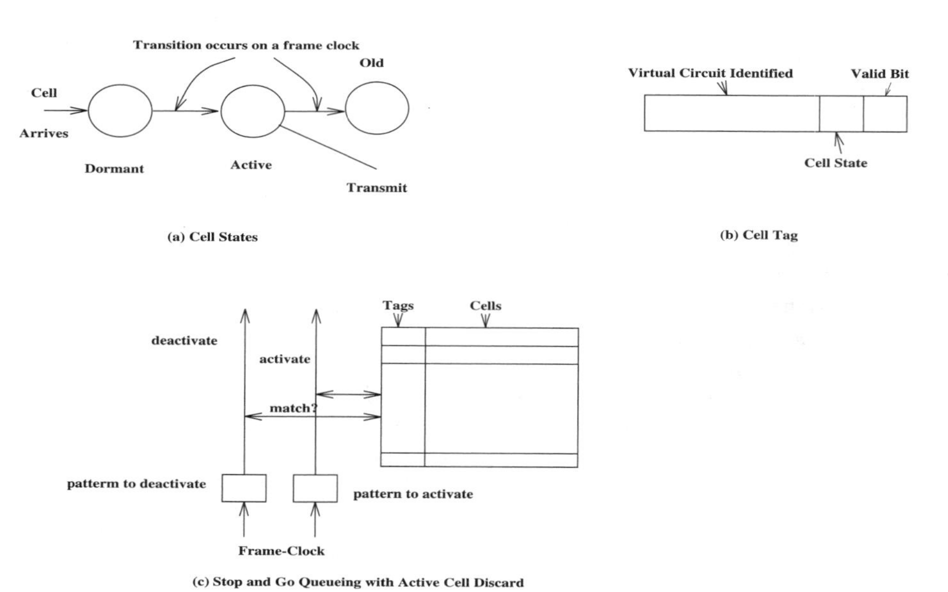

Fig. 19.4 Implementing per-VC framing with active cell-discard.

Assume that during frame t, *dormant* cells are represented by a 0 and *active* cells have been marked 1 in the previous cycle. On a new cell arrival, the multiplexer needs to attach to it a tag identifying its VC and its frame number (in this case 0), set its valid bit to 1, and forward it to the output queue. See Fig. 19.4(b). The valid bit's function is to help discard cells, similar to the action of flushing a cache memory on a context switch. In a fast cell-switched, VC network, a switch would implement a tagging scheme for VC identifiers anyway, so additional circuitry needed is small.

On the next *frame clock*, the entire output queue would be fed with two logical signals, one to deactivate the *active* cells that did not get transmitted during their allotted frame time (due to lack of available capacity), and one to activate the *dormant* cells. See Fig. 19.4(c). Both of these can be achieved by associatively matching cell tags with an identifier representing the appropriate VC and its state. The primary difference between this and off-the-shelf content-addressable memories is that more than one match is likely, especially for *dormant* cells. On a match, *activie* cells mark themselves invalid by setting their valid bits to 0; the *dormant* cells move to the *active* state and are ready to be transmitted. At this point, they move under the control of the cell dispatcher, which must decide on a strategy that is consistent with the overall goals of delay-jitter and statistical cell-loss bounds.

A convenient model for the buffer memory organization is to view it as a set of logical queues, one per VC, with a sequence number distinguishing *active* and *dormant* cells. All *old* cells may potentially be grouped into one logical queue, as discussed below.

19.2.3 Cell Dispatcher

The cell dispatcher is responsible for (1) *scheduling* and (2) *transmitting active* and (potentially) *old* cells. *Dormant* cells are not within its purview. From the dispatcher's perspective, the *active* cells for *each* VC are assumed to be logically organized as a queue (see Fig. 19.5(a)). The *old* cells (implemented optionally) are organized either as separate queues or as a single queue. In either case, they are served on a best-effort basis and may be reclaimed before they are served to accommodate new cell arrivals.

The dispatcher consists of two concurrent units, a scheduler and a transmitter. The scheduler allocates cell times to *active* cells of individual VCs and decides which cells are to be dropped if contentions for capacity arise. The transmitter transmits them (and *old* cells if all *active* queues are empty and *old* cells are waiting). The scheduler will guarantee transmission of at least C_i cells/frame for connection i ($i = 1, \ldots, K$), where K is the number of active VCs at the multiplexer. The computation of the C_i is based on cell-loss requirements for different VCs and their marginal distributions, and is presented in Section 19.4. The scheduler and the transmitter share a circular buffer that represents channel allocations in the future. This circular buffer is presented below as a linear array for convenience of exposition. Let this data-structure be called `channel_image`. `Channel_image` [n] records the ID of one VC. If `channel_image`[n] equals i, the transmitter will

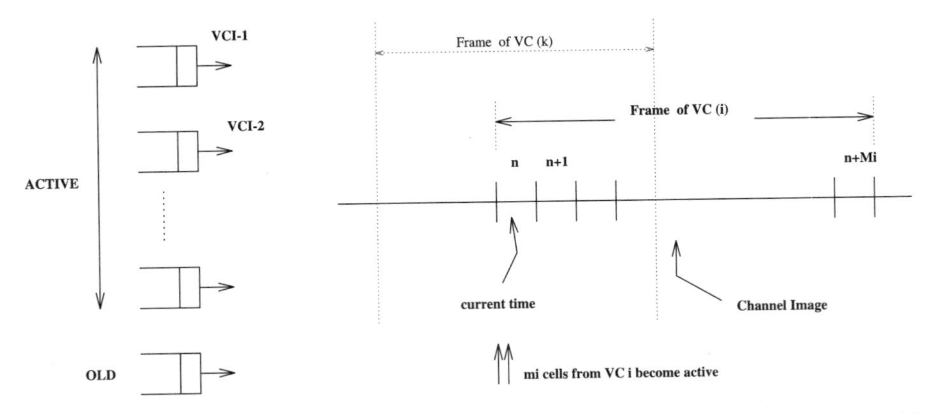

Fig. 19.5 Cell dispatcher's view. (a) Queue of *active* cells for each VC plus a queue of *old* cells. This is used by the transmitter unit. (b) `Channel_image`. This is shared by the scheduler and the transmitter units.

transmit from the head of the *active* queue corresponding to VC i at time n. This will be modified below after the basic algorithm is presented.

The scheduler is activated on every new frame activation, that is, on a *frame clock*. Let the new frame activation be at time n. (See Fig. 19.5(b).) Let the corresponding VC be i, the frame length (jitter bound) be M_i, the number of cells in the current *active* frame be m_i, and the minimum number of cells guaranteed to be transmitted from this VC in this frame be C_i. `Channel_image` records the action to be taken by the transmitter in future slots. The scheduler either marks the slots in `channel_image` with a VC identifier or leaves them empty. If it does mark a slot, it also records whether the transmission is to be guaranteed or not-guaranteed. If a slot is marked not-guaranteed, it may be reclaimed at some point in the future to serve a different VC (as described in Section 19.2.3.1).

The scheduler's task is as follows. Assume that it is activated on VC i's *frame clock*. The time window in the future over which the m_i *active* cells need to be transmitted is $[n+1,\ n+M_i]$. (The nth slot is kept aside for the transmission to begin transmitting.) The scheduler follows the following algorithm.

(a) Beginning with $n+M_i$, *going down* to $n+1$, the scheduler attempts to find the largest $k_i \leq m_i$ slots that are empty in `channel_image` $[n+M_i]$ through `channel_image` $[n+1]$, and marks each with the current VC, i.

(b)
```
If (ki ≤ Ci) {
  guaranteed_cells[i]=ki;
  not_guaranteed[i]=0;
}else{
  guaranteed_cells[i]=Ci;
  not_guaranteed[i]=ki-Ci;
}
```

(c) If $k_i = m_i$, the scheduler's allocation task for this frame is done. Else, it needs to pick at most $(m_i - k_i)$ slots and at least $(C_i - k_i)$ slots in `channel_image` $[n + 1]$ through `channel_image` $[n + M_i]$ that are marked, but not necessarily guaranteed, and overwrite them with i. Each slot overwritten represents a cell loss for the corresponding VC. Since this would give priority to some VCs over others when the number of *active* cells exceeds the capacity available (for meeting deadlines), the policy used must balance from fairness and cell-loss commitments. The dropping policy is described below.

19.2.3.1 Dropping Policy Let the negotiated mean cell-loss ratio of VC i be ϵ_i, and the estimated cell-loss ratio at time n be $\hat{\epsilon}_i[n]$. Let $S_i[n] = \epsilon_i / \hat{\epsilon}_i[n]$. Yang and Pan [24] have proposed a dropping policy where, if an incoming cell arrives to a full buffer (in our case, full `channel_image`), the scheduler will search the buffer for the VC j that has the largest $S_j[n]$ and discard one of its cells. If the arriving cell belongs to VC j itself, then that cell will be dropped. The authors show that using the largest $S_j[n]$ is optimal in bandwidth utilization among all stationary, space-conserving loss-scheduling schemes. It is also optimal among all stationary scheduling strategies when cell-loss requirements of all VCs are equal (see Yang and Pan [24]).

We modify this strategy to guarantee a minimum C_i cells/frame for individual VCs and replace negotiated mean cell-loss ratio with negotiated overflow probability. Continuing from the point just after (c) in the cell-dispatcher's algorithm:

(d) If $k_1 < C_i$, the scheduler needs to overwrite at least $C_i - k_i$ cells belonging to other VCs in `channel_image`. It selects VCs $\{j\}$ with (`not_guaranteed[j]` > 0), starting with the largest $S_j[n]$, and updating $S_j[n]$ as described in Yang and Pan [24]. In all cases, the number of erased cells is decremented from `not_guaranteed[j]`. `guaranteed_-cells[i]` is incremented by the corresponding amount. The latter should be equal to C_i at the end of this step. This is because the equivalent bandwidth algorithm described in Sectin 19.4.1.2 ensures that $\sum_i C_i$ is always less than or equal to available capacity. Therefore, if a cell is to be dropped,

$$\sum_{j \neq i} \texttt{not\_guaranteed[j]} \geq (C_i - k_i).$$

(e) If $m_i > C_i$, then from the previous step, C_i cells have been scheduled. An additional l_i cells $(0 \leq l_i \leq m_i - C_i)$ may be scheduled, in which case `not_guaranteed[i]` will be set to l_i. The scheduler schedules these cells only if there exists a j such that (`not_guaranteed[j]` > 0) and $(S_j[n] > S_i[n])$.

(f) $S_i[n]$ is updated with the number of dropped cells in this fame, if any, that is, with $(m_i - C_i - l_i)$.

The above cell dispatcher guarantees a minimum capacity C_i in each frame for VC i. Extra capacity, if available, will be used by other VCs without interference from the dispatcher's dropping policy—until the sum of arrivals fills up `channel_image`, that is, exceeds the total output link capacity. This way, if the number of arrivals in a frame for some VC is less than the allocated capacity for it, the extra capacity may be used by other VCs on a nonguaranteed basis.

If the output link capacity is exceeded, however, the dropping policy prioritizes the nonguaranteed cells of different VCs in accordance with their current $S_j[\cdot]$ values. With upper-bound computations of equivalent bandwidths (Section 19.4), and typical low cell-loss requirements expected, the dropping policy is not expected to be called upon too frequently. Simulation results comparing it with one that drops nonguaranteed cells on a last-in-first-out (LIFO) basis (i.e., newly active non-guaranteed cells that find a full channel-image) are given in Section 19.4.1.3.

19.2.3.2 Transmitter The transmitter works as follows. At time n,

1. If `channel_image`[n] is not empty, let $i =$ `channel_image` [n]. The transmitter transmits an *active* cell from the queue corresponding to VC i, and goes to Step 3.
2. If `channel_image`[n] is empty, however, the transmitter proceeds to the next nonempty slot $n' > n$. If no such n' exists, it proceeds to Step 4. Else it transmits an *active* cell from the queue corresponding to VC recorded in `channel_image`[n'] and marks this slot empty. Let this VC identifier be i.
3. ```
   If (guaranteed_cells[i] > 0)
     guaranteed_cells [i] = guaranteed_cells[i] − 1;
   else
     not_guaranteed[i] = not_guaranteed[i] − 1;
   ```
4. If all future slots are empty, it transmits an *old* cell if such an option is implemented. Note that this option cannot be implemented without additional considerations in a network that requires cells to be transmitted in order.

19.2.3.3 Rationale The algorithms for filling and emptying the `channel_image[]` data structure reduce the number of lost deadlines among current and future *active* cells. For the discussion below, refer to Fig. 19.5(b).

Assume that the current time is n. The scheduler reserves slots beginning with time $n + M_i$ *down to* $n + 1$. This ensures that slots near time n are available for VCs activated at some time in the future with deadlines earlier than $n + M_i$, that is, having smaller delay-jitter bounds. Since the objective is to ensure that *active* cells of VC i are transmitted at or before time $n + M_i$, this strategy would meet more deadlines than if its cells were allocated from time $n + 1$ upward.

The transmitter scans `channel_image[]` in ascending slot order (making necessary modulo arithmetic corrections for a circular buffer implementation). If slot n is nonempty, it transmits a cell from the corresponding VC recorded in
   ```

`channel_image[]` (see Section 19.2.3.2, transmitter Step 1). If, however, it finds this slot empty, it scans for the first nonempty slot in `channel_image[]`. If one is found, it transmits an *active* cell from the corresponding VC (see Section 19.2.3.2, transmitter step 2). This ensures that bandwidth gets utilized if there are *active* cells waiting in queue. The probability of contention for a future slot—should a new *frame clock* arrive from another VC—is lower as well. The scheduler's Steps (b)–(f) and the transmitter's Step 3 ensure the following.

- *Firewall Protection.* A minimum capacity of C_i is available to VC i in a frame if it needs it. This is true regardless of other VCs sharing the output link.
- *Efficient Capacity Utilization.* If VC i needs less than C_i slots in a frame, the unused capacity is available to other VCs sharing that output link.
- *Fairness.* If cells must be dropped, selecting VCs $\{j\}$ with larger $S_j[\,]$ balances cell losses among connections equitably over time—at least as measured by $S_j[\,] = \epsilon_j/\hat{\epsilon}_j$.

The scheduler–transmitter pair implements a pseudo-earliest-due-date schedule with a minor difference: if the scheduler finds a slot n' marked with VC j when it is allocating slots for VC i, it may allocate a cell $n'' < n$. for VC i, even though i's deadline is potentially greater than that of j. This does not violate the deadlines for either of two VCs, but does increase the likelihood of dropping a cell from VC j. The alternative is to shift j's slots to the left and allocate `channel_image[`n'`]` to i at a higher implementation cost.

The reason for keeping slot n out of reach of the scheduler's view is to enable the transmitter to schedule timely transmission during the current slot. Also, the transmitter may need to have priority over the scheduler in its access/writing to the `channel_image[]` data structure.

19.3 MAXIMUM DELAY, DELAY-JITTER, AND BUFFER REQUIREMENTS

19.3.1 Maximum Delay and Delay-Jitter Performance

Cells that arrive in the "arriving" frame T_a^i are transmitted on the output link in the next "departing" frame, T_d^{i+1}, if capacity is available, as shown in Fig. 19.3. If capacity is not available, they are either dropped or marked *old*. For per-VC framing, T_d^{i+1} starts at the same time that T_a^i ends.

Let the frame size for VC i be M_i, and define the end-to-end delay of a cell as the time difference between its arrival at the destination node and its arrival at the source node. The total end-to-end delay of a cell, W_i may be expressed as

$$W_i = D_i + X, \tag{19.1}$$

where D_i is a constant delay for all cells of VC i, and X is a random variable that varies from cell to cell. The value of X corresponds to the delay-jitter [10]. The range of X is the maximum delay-jitter [10] and is bounded by

$$-M_i < X < M_i. \tag{19.2}$$

The delay, D_i, consists of two terms:

$$D_i = \tau_i + H_i M_i, \tag{19.3}$$

where H_i is the total number of links between the source and the destination for VC i, and τ_i is the end-to-end propagation and processing delay for VC i. Compare Eq. (19.3) to the constant delay term of framing on a per-link basis given in Golestani [10]:

$$D_i = \tau_i + H_i * M_i + \sum_{h=1}^{H} \theta_{i_h, d_h}, \tag{19.4}$$

where θ_{i_h, d_h} is the phase mismatch between incoming and outgoing links for VC i at link h. This phase mismatch is due to different propagation delays among different incoming links. It is bounded by $0 \le \theta_{i_h, d_h} < M_i$. Therefore, D_i in per-link framing is bounded by

$$D_i \le \tau_i + 2H_i M_i. \tag{19.5}$$

Hence the maximum-delay bound in per-VC framing is reduced by $H_i M_i$. This is half the maximum queueing delay of per-link framing. Delay-jitter bounds for both are the same.

19.3.2 Buffer Requirements

Consider a VC i for which at most A_i cells may arrive in a frame length M_i. As depicted in Fig. 19.6, the worst-case buffering scenario will occur if:

1. During the arriving frame T_a^i, A_i cells arrive;
2. A new *frame clock* arrives soon after that, at time t; and
3. Another A_i cells arrive immediately following time t in frame T_a^{i+1}.

The framing structure guarantees that no more cells from VC i will arrive at the multiplexer until the next *frame clock* at $(t + M_i)$. Not counting *old* cells, a maximum of $2A_i$ cells, therefore, need to be buffered for VC i at each multiplexer. For framing on a per-link basis, the corresponding buffer requirement is $3A_i$ cells [10], that is, 1.5 times larger than that for per-VC framing.

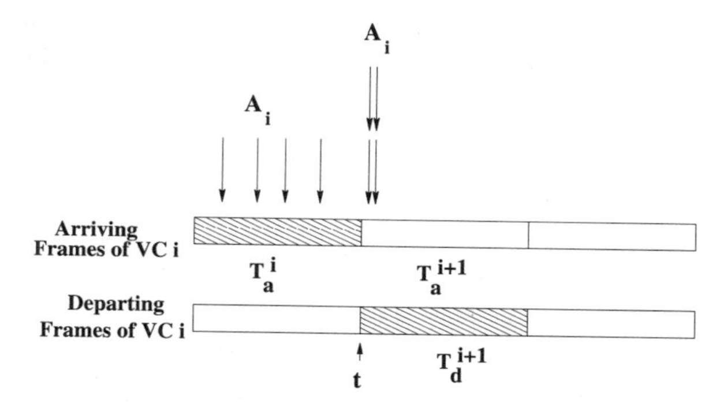

Fig. 19.6 Worst-case scenario for number of cells that need to be buffered.

19.4 EQUIVALENT BANDWIDTH

We next compute upper bounds on the capacity needed for guaranteeing desired overflow probabilities, $\{\epsilon_i\}$, in the presence of per-VC framing with active cell-discard. We also compute the minimum guaranteed capacity, $\{C_i\}$, used by the cell dispatcher in Section 19.2.3. The overflow probability, ϵ_i, is defined as

$$\epsilon_i = P\{\text{a cell from VC } i \text{ is dropped in a frame}\}.$$

19.4.1 Equivalent Bandwidth for Homogeneous Delay-Jitter Bounds

In the presence of active cell-discard, homogeneous delay-jitter bounds imply that all VCs have the same frame size. We consider the following two cases separately: (1) homogeneous cell-loss guarantees and (2) heterogeneous cell-loss guarantees. For the latter case, all virtual circuits with the same cell loss requirement are grouped into the same class.

19.4.1.1 Scenario 1: Equal Cell-Loss Probability Requirements This is the trivial case and is included for completeness. Let the number of connections be N, and let $\epsilon_i = \epsilon$ for all i. Let all frames be synchronized—which is a worst-case scenario for cell losses. Let $X_1, \ldots, X_N$ be the number of cell arrivals from connections 1 through N, in a given frame. Let

$$S_N = X_1 + \cdots + X_N.$$

Let $f_x(x)$ be the probability density function of X_k, $(k = 1, \ldots, N)$. Let $f_{S_N}(x)$ be the probability density function of S_N. Assuming that traffic across connections are mutually independent, $f_{S_N}(x)$ is the convolution of the $f_k(x)$ terms:

$$f_{S_N}(x) = f_1(x) \otimes f_2(x) \otimes \cdots \otimes f_N(x). \tag{19.6}$$

The minimum bandwidth required is the $(1-\epsilon)$th quantile of the distribution of S_N. For large N, using the central limit theorem, $f_{S_N}(x)$ is approximated by

$$f_{S_N}(x) \approx \frac{1}{\sqrt{2\pi}\sigma} e^{-(x-\mu)^2/2\sigma^2},$$

where $\mu = \sum_{k=1}^{N} \mu_k$ and $\sigma^2 = \sum_{k=1}^{N} \sigma_k^2$. μ_k and σ_k are the mean and standard deviation of X_k, respectively. In this case (large N), since C is the $(1-\epsilon)$th quantile of a normal distribution with mean μ and standard deviation σ, we may equivalently set $Z = (S_N - \mu)/\sigma$, and if $z_{1-\epsilon}$ is the $(1-\epsilon)$th quantile of a standard normal distribution, C is given by

$$C = \mu + z_{1-\epsilon}\sigma. \tag{19.7}$$

For small N, Eq. (19.6) needs to be used instead. Note that the value of C calculated in Eqs. (19.6) and (19.7) represents the capacity needed such that the cell-loss probability for the aggregate traffic $S_N \leq \epsilon$. A dropping policy is needed to distribute cell losses fairly among VCs so that the individual VCs meet their cell loss requirements.

The following set of experiments compare the proposal cell-drop policy of Section 19.2.3 with a simpler policy (described below). Rose's MPEG-I video traces [22] were used to drive the simulations. Three video connections were assumed to require an identical cell-loss probability bound ϵ. The necessary equivalent bandwidth for the aggregate was computed using Eq. (19.6), and scheduling was performed either by the dropping policy described in Section 19.2.3 or by the last-in-first-out (LIFO) policy. In LIFO, newly active nonguaranteed cells that found `channel_image[]` full were dropped. Table 19.1 shows the results for two different values of ϵ. Surprisingly, the dropping policy of Section

TABLE 19.1 Comparison of the Proposed Dispatcher and a Simple Last-In-First-Out Cell Dispatcher for Distributing the Loss Fairly Among Individual VCs[a]

Experiment Number	Video Trace	ϵ Desired	ϵ Delivered (Proposed Dispatcher)	ϵ Delivered (LIFO)
I	Terminator	0.01	0.0099	0.0026
	Goldfinger	0.01	0.0098	0.0121
	Soccer	0.01	0.0097	0.0104
II	Terminator	0.001	0.000923	0.00171
	Goldfinger	0.001	0.000981	0.000931
	Soccer	0.001	0.000926	0.000739

[a]The numbers indicate that additional controls over a simple LIFO dispatcher is necessary.

19.2.3 is able to guarantee ϵ for individual VCs while LIFO is not. Both dropping policies dropped exactly the same total number of cells.

19.4.1.2 Scenario 2: Heterogeneous Cell-Loss Proabability Requirements Consider two different classes of traffic with desired upper bounds on cell-loss probabilities, ϵ_1 and ϵ_2, respectively. Multiple VCs with same delay-jitter bound (frame size) and cell-loss probability requirement are considered part of the same class. Let $f_{X_i}(x_i)$ $(i = 1, 2)$ be the density function for the number of bits/frame transmitted by class i. The following two methods are upper-bound estimates, with progressively tighter bounds and increased computational complexity. The algorithm below generalizes to more than two classes.

Alternative 1 The simplest upper bound on equivalent bandwidth is the sum of equivalent bandwidths for each ϵ_i $(i = 1, 2)$, computed in isolation. The minimum bandwidth required for class i is the $(1 - \epsilon)$th quantile of X_i, that is, the smallest C_i that satisfies

$$\int_{C_i}^{\infty} f_{X_i}(x_i)\, dx_i \leq \epsilon, \tag{19.8}$$

and the equivalent bandwidth is $C = \sum_i C_i$. This approach does not consider the statistical multiplexing gains across different classes and overestimates the true equivalent bandwidth needed. The following algorithm gives a tighter upper bound.

Alternative 2 Let $f_{X_1,X_2}(x_1, x_2)$ be the joint density of $f_{X_1}(x_1)$ and $f_{X_2}(x_2)$. We assume that traffic classes are spatially independent, that is, $f_{X_1,X_2}(x_1, x_2) = f_{X_1}(x_1)f_{X_2}(x_2)$. An algorithm for computing a tighter equivalent bandwidth estimate as follows.

If there are K classes, the bandwidth needed for class k at iteration n is computed as follows (compare this to Eq. (19.9)):

$$P(X_k \leq C_k^n) + P\left(C_k^n < X_k \leq \sum_i C_i - \sum_{i \neq k} X_i\right) \geq 1 - \epsilon_k \tag{19.13}$$

The second term in Eq. (19.13) considers only the case when the total arrivals of all connections in a frame is less than the total capacity. It is possible to not drop any cell from connection k, even if the total number of arrivals is greater than the total capacity. Therefore, the equivalent bandwidth computation above is an upper bound.

ALGORITHM—ALTERNATIVE 2

1. Initialize C_i in accordance with Eq. (19.8). Let Δ be a (small) quantum of capacity that may be subtracted from C_i $(i = 1, 2)$. Let $C_i^0 = C_i (i = 1, 2)$, and let the iteration step $n = 1$.
2. Set $C_i^n = C_i^{n-1} - \Delta$ $(i = 1, 2)$. Set boolean variable `tryagain` to TRUE.
3. The probability that no cells will be dropped from class 1 is the probability that the arrivals from class 1, X_1, is less than C_1 OR X_1 is greater than C_1 but the total arrivals in the frame, $X_1 + X_2 \leq C_1 + C_2$. Therefore, we need to check if

$$P(X_1 < C_1^n) + P(C_1^n < X_1 \leq C_1^n + C_2^n - X_2) \geq 1 - \epsilon_1 \tag{19.9}$$

$$\Rightarrow F_{X_1}(C_1^n) + \int_0^{C_2^n} \int_{C_1^n}^{C_1^n C_2^n - x_2} f_{X_1}(x_1) f_{X_2}(x_2)\, dx_1\, dx_2 \geq 1 - \epsilon_1 \tag{19.10}$$

$$\Rightarrow \int_0^{C_2^n} [F_{X_1}(C_1^n + C_2^n - x_2) - F_{X_1}(C_1^n)] f_{X_2}(x_2)\, dx_2 \geq 1 - \epsilon_1 - F_{X_1}(C_1^n)$$

$$\Rightarrow \int_0^{C_2^n} F_{X_1}(C_1^n + C_2^n - x_2) f_{X_2}(x_2)\, dx_2 \geq 1 - \epsilon_1 - F_{X_1}(C_1^n)[1 - F_{X_2}(C_2^n)] \tag{19.11}$$

   If inequality (19.11) is not satisfied, mark `tryagain` FALSE.
4. Analogous to Eq. (19.11), check if the following is satisfied:

$$\int_0^{C_1^n} F_{X_2}(C_1^n + C_2^n - x_1) f_{X_1}(x_1)\, dx_1 \geq 1 - \epsilon_2 - F_{X_2}(C_2^n)[1 - F_{X_1}(C_1^n)] \tag{19.12}$$

   If Eq. (19.12) is not satisfied, mark `tryagain` FALSE.
5. If `tryagain` is TRUE go to Step 2. Else go to Step 6.
6. A tight upper bound on the equivalent bandwidth is $C_1^{n-1} + C_2^{n-1}$.

19.4.1.3 Numerical Results and Simulation

Equivalent Bandwidth Calculation Alternatives 1 and 2 were compared using the MPEG-I video traces mentioned earlier. Table 19.2 shows the equivalent bandwidths needed for three of the traces (Terminator, Goldfinger, and Soccer) with one VC per connection when all of them belonged to the same jitter class (same frame size) and needed different cell-loss probabilities. The marginal distribution of each trace appeared to be a Gamma distribution, so a fitted Gamma was used in the equivalent bandwidth computations. The numerical integration was evaluated using Mathematical Software package. Alternative 2 resulted in approximately 40% savings for

TABLE 19.2 Comparison of Alternatives 1 and 2[a]

Cell-Loss Probabilities[a] $(\epsilon_1, \epsilon_2, \epsilon_3)$	Alternative 1 Capacity (kbits/frame)	Alternative 2 Capacity (kbits/frame)	Bandwidth Reduction
$(1.5 \times 10^{-2}, 2 \times 10^{-2}, 10^{-2})$	256	150	0.41%
$(2 \times 10^{-3}, 3 \times 10^{-3}, 1 \times 10^{-3})$	388	220	0.43%
$(10^{-2}, 10^{-3}, 1 \times 10^{-4})$	460	280	0.39%

[a] ϵ_1, ϵ_2, and ϵ_3 represent the cell-loss probabilities for Terminator, Goldfinger, and Soccer, respectively.

these experiments. Note that without framing and active cell-discard, developing a solution for heterogeneous $(\epsilon_1, \epsilon_2, \epsilon_3)$ is nontrivial.

Effectiveness of the Firewall ($\{C_i\}$) *and Dropping Policy* The effectiveness of guaranteeing a minimum capacity C_i per frame for class i and the dropping policy is evaluated next to the context of heterogeneous ϵ values. This is achieved through trace-driven simulation. The scenario was as follows.

- MPEG-I video traces shown in Table 19.3 were used as input to the simulations.
- Their C_i were calculated using Alternative 2.
- The dispatcher was either the pseudo-earliest-due-date dispatcher described in Section 19.2.3 or LIFO for comparison purposes.

In the experiments, each class had one connection. Also, the traces were phase-shifted randomly to ensure that they were mutually unsynchronized—otherwise, one would have to consider correlated streams.

Recall from Section 19.2.3, that when `channel_image[]` was full, the cell dispatcher gave preference to classes $\{j\}$ that had not yet received at least C_j cell allocations in the current frame, and among those that did receive C_j, to those with smaller current estimates of the ratio $S_j[\cdot]$.

The performance of the algorithm is shown in Table 19.3. Results for a simple last-in-first-out (LIFO) dropping policy is also shown for comparison purposes. In

TABLE 19.3 Performance of the Proposed Cell Dispatcher for Heterogeneous Cell-Loss Probabilities, $\{\epsilon_i\}$

Experiment Number	Video Trace	ϵ_i Desired	ϵ_i Delivered (LIFO)	ϵ_i Delivered (Proposed Dispatcher)
I	Terminator	0.01	0.0008	0.0039
	Goldfinger	0.001	0.0012	0.00096
	Soccer	0.0001	0.0011	0.00009
II	Terminator	0.002	0.0014	0.0019
	Goldfinger	0.003	0.0020	0.0029
	Soccer	0.001	0.0018	0.0007

LIFO, newly *active* nonguaranteed cells that found the `channel_image` full were dropped. The results show a need for the priority scheduler.

19.4.2 Equivalent Bandwidth for Heterogeneous Frames

Let there be L delay-jitter classes, each with a different delay-jitter bound and corresponding frame size. Let n_l $(l = 1, \ldots, L)$ be the number of connections belonging to class l. For class l in isolation, the aggregate capacity needed to meet prescribed cell-loss probabilities may be computed in accordance with Section 19.4.1. Let C_l be the capacity needed for class l traffic. Then an upper bound on capacity needed for all classes is $\sum_l C_l$.

19.4.3 Loss-Free Transmission

Per-link framing was originally designed for loss-free transmission [9]. Per-VC framing can achieve the same if desired.

To see this, consider first a single jitter class, that is, let all VC have the same frame size, T. The total output link capacity allocated in #cells$/T$ will be $C_T \geq \sum_{i=1}^{K} C_i$, where K is the number of admitted VCs that have frame size T, and C_i is the number of cells guaranteed for VC i in each frame T.

Let j $(1 \leq j \leq K)$ be one of the admitted VCs. Let a new *frame clock* arrive for VC j at time t. The scheduling algorithm described in Section 19.2.3 will find at least C_j slots between $(t, t+T]$ in its `channel_image[]` that are either empty or marked as not guaranteed. This is because

- the total number of slots in `channel_image[t+1]` through `channel_image[t+T]` is at least C_T;
- $C_T \geq \sum_{i=1}^{K} C_i$;
- Since VC js new *frame clock* just arrived, at least C_j of the C_T slots in `channel_image[t+1]` through `channel_image[t+T]` are either empty or marked as not guaranteed.

Therefore, the scheduler will not need to overwrite any entry in `channel_image[]` that has already been scheduled as guaranteed for some other VC; that is, if VC j transmits at most C_j cells$/T$, all its cells will be transmitted.

This logic extends to multiple jitter classes—the capacity allocated to any VC is guaranteed to be available to it in every frame if it needs it.

19.5 RELATED WORK

A number of algorithms for quality-of-service provisioning inside a network have been proposed. Examples include Delay-Earliest-Due-Date (Delay-EDD) [6], Hierarchical Round Robin [14], Jitter-Earliest-Due-Date (Jitter-EDD) [23], Stop-and-Go Queueing [9, 10], Virtual Clock [25], and Weighted Fair Queueing [4].

Formulas for end-to-end delay bounds for Virtual Clock and Weighted Fair Queueing when input traffic is leaky-bucket constrained are given in Zhang [26]. Since video sources typically exhibit a slowly decaying autocorrelation function, the leaky-bucket rate parameters will need to be close to peak rates in order to achieve acceptable delay bounds. This will necessitate operating a network at low utilization levels.

In Delay-EDD, each source declares three parameters: X_{min}, X_{av}, and I, where X_{min} is its smallest packet interarrival time, X_{av} is its average packet interarrival time, and I is the averaging interval. Delay-EDD uses the worst case X_{min} to guaranteed a deterministic end-to-end delay bound. This algorithm requires operating a network at low utilization levels as well.

Virtual Clock, Weighted Fair Queueing, and Delay-EDD can bound end-to-end delay but not delay-jitter.

Jitter-EDD, Hierarchical Round Robin, and Stop-and-Go Queueing can guarantee both end-to-end delay bound and delay-jitter bound. Jitter-EDD uses the same traffic characterization (X_{min}, X_{av}, I) as Delay-EDD. Hierarchical Round Robin and Stop-and-Go Queing characterizes traffic using a rate r over a frame time T.

In this chapter, we have extended the framing strategy by:

1. Inducing frames on VCs instead of that on links. This reduces the end-to-end delay and the buffer requirement at intermediate multiplexers. It also enables the implementation of software setable heterogeneous jitter bounds.
2. Providing mechanisms for meeting heterogeneous cell-loss bounds through a coupling between equivalent bandwidth computations and a pseudo-earliest-due-date scheduler.

Implementing framing on a per-VC basis is simple and feasible and is described in Section 19.2.2. It requires a tag per cell to identify its state (*dormant*, *active*, or *old*), a *frame clock* to move cells among states, and mechanisms to prioritize and transmit *active* cells and discard *old* cells.

Framing on a per-link basis is discussed in Golestani [9, 10]. Synchronizing frames from multiple input links onto an outgoing frame requires some additional delay circuitry.

Framing on a per-VC basis is simpler and more flexible because of the following:

- There is no need to ensure that frames from different connections be synchronized in time.
- Frames do not need to be integer multiples of a base frame size.
- Frame sizes do not need to be etched in hardware. They can be set at connect-time and may admit adjustments in software. It is expected that the frame times would be determined by desired jitter bounds.

Is it feasible to provide the marginal distribution of every virtual circuit? This largely depends on the effort undertaken to characterize different classes of traffic.

Experience with current variable bit rate video traffic suggests that marginals follow a distribution that is approximately at Gamma/Pareto combination [8]. If this model form holds in the general case, or if specific model forms apply to specific classes of applications (e.g., all video-conferencing applications may belong to one class, all basketball games may belong to another class), the multiplexer needs to be informed of the parameter estimates and model forms for each VC. The accuracy of the model form and parameter estimates would determine the accuracy of the C_i computed.

19.6 CONCLUSION

Variable bit rate video traffic exhibits a slowly decaying autocorrelation function [3, 7, 8, 21]. The latter has been shown to significantly increase queue-length statistics at a multplexer, especially if the marginal distribution is heavy-tailed [2, 5, 8]. Increasing buffer size also makes guaranteeing bounds on maximum-delay, delay-jitter, and cell-loss probabilities difficult.

These observations motivate the need for per-VC framing with active cell-discard. These, in conjunction with the priority cell-dispatcher and the equivalent bandwidth computation unit proposed in this chapter, can provide the following guarantees.

- They can provide guaranteed bounds on maximum-delay, delay-jitter, and cell-loss probabilities at each multiplexer, even if the bound requirements are different for different VCs.
- They can meet these bounds for heterogeneous traffic types (different autocorrelation structures and marginals).

The proposed architecture does not resort to strict reservations, as, for example, in time-division-multiplexed systems, to achieve these goals. It exploits statistical multiplexing across VCs to increase multiplexer utilization, while simultaneously guaranteeing the quality-of-service bounds. In addition, it allows the delay-jitter bound to be set over a continuum for each VC.

Per-VC framing, in conjunction with the priority cell dispatcher and active cell-discard has the following properties:

- The framing structure guarantees a maximum end-to-end delay bound and a delay-jitter bound for any network topology [10].
- Compared to per-link framing, it reduces the end-to-end delay bound by H_iM_i, which is half the maximum queueing delay.
- It reduces the buffer requirement at each multiplexing point by one-third as compared to per-link framing.
- It supports hetereogeneous delay-jitter bounds.
- It simplifies computing upper bounds on equivalent bandwidth necessary for guaranteeing heterogeneous cell-loss probabilities.

- It is able to exploit statistical multiplexing across different connections and yet guard against large queue lengths caused by correlated traffic. This is because (1) the tail of the aggregate distribution of a sum of sources is smaller than the tail of an individual source (standard advantage of packet switching), (2) the framing structure protects packets in future frames from being delayed by packets in previous ones even if the arrival process is correlated, and (3) the dispatcher guarantees a minimum number of cell transmissions (C_i) for each VC i, per frame, and C_i takes into account the VCs required cell-loss ratio and its marginal distribution.
- The implementation of framing with associative tags adds only a moderate cost.
- Frames from different VCs need not be synchronized to meet desired quality-of-service objectives.

The cell dispatcher implements:

- A pseudo-earliest-due-date priority discipline for scheduling *active* cells with the help of the `channel_image[]` data-structure.
- A priority cell-drop mechanism that determines which *active* cells to drop in case `channel_image[]` is full. It does so based on (1) the ratio $S_j[\cdot] = \epsilon_j/\hat{\epsilon}_j$, and (2) the minimum capacities C_i to be provided to each VC per frame. The C_i are calculated by the equivalent bandwidth unit.

Cell-drop scheduling based on maximum $S_j[\cdot]$ has been shown to be optimal in bandwidth utilization among stationary, space-conserving loss-scheduling mechanisms by Yang and Pan [24]. In conjunction with the pseudo-earliest-due-date component, it attempts also to provide a minimum number of lost deadlines. The minimal capacity component, C_i, has two functions:

- It provides a firewall for VCs *from* connections j, with low $S_j[\cdot]$. The latter can result from long periods of small transmissions followed by a large surge of arrivals. Note that this is possible for well-behaved long-memory traffic that is not necessarily misbehaving or malicious.
- It enables computation of tighter equivalent bandwidths. In Alternative 2 in Section 19.4.1.2, Eq. (19.13), the tighter upper-bound computation assumes that an amount C_i will be available for VC i if it needs it.

ACKNOWLEDGMENTS

The authors wish to thank Jamal Golestani of Bell Laboratories and Biswanath Mukherjee of the University of California at Davis for their suggestions.

REFERENCES

1. A. Adas and A. Mukherjee. Providing heterogeneous quality of service bounds for correlated video at a multiplexor. *Perf. Eval.*, pp. 45–65, March 1999 (Preliminary version presented at *Proceedings of the SPIE, Performance and Control of Network Systems*, pp. 155–167, Dallas, November 1997.)
2. A. Adas and A. Mukherjee. On resource management and QoS guarantees for long-range dependent traffic. In *Proc. IEEE INFOCOM'95*, pp. 779–787, April 1995.
3. J. Beeran, R. Sherman, M. S. Taqqu, and W. Willinger. Long-range dependence in variable-bit-rate video traffic. *IEEE Trans. Commun.*, **43**:1566–1579, 1995.
4. A. Demers, S. Keshav, and S. Shenker. Analysis and simulation of a fair queuing algorithm. In *Proc. ACM SIGCOMM'89*, pp. 1–12, September 1989.
5. A. Erramilli, O. Narayan, and W. Willinger. Experimental queuing analysis with long-range dependent traffic. Preprint, September 28, 1994.
6. D. Ferrari and D. Verma. A scheme for real-time channel establishment in wide-area networks. *IEEE J. Select. Areas Commun.*, pp. 368–379, April 1990.
7. M. Garrett. Contributions toward real-time services on packet-switched networks. Ph.D. Thesis, Columbia University, 1993.
8. M. Garrett and W. Willinger. Analysis, modeling and generation of self-similar VBR video traffic. In *Proc. ACM SIGCOMM'94*, pp. 269–280, 1994.
9. S. J. Golestani. A Stop-and-Go Queuing framework for congestion management. In *Proc. ACM SIGCOMM'90*, pp. 8–18, September 1990.
10. S. J. Golestani. A framing strategy for congestion management. *IEEE J. Select. Areas Commun.*, pp. 1064–1077, September 1991.
11. S. J. Golestani. Duration-limited statistical multiplexing of delay-sensitive traffic in packet networks. In *Proc. IEEE INFOCOM'91*, pp. 323–332, 1991.
12. C. Huang, M. Devetsikiotis, I. Lambadaris, and A. Kayes. Self-similar modeling of variable bit-rate compressed video: a unified approach. In *Proc. ACM SIGCOMM'95*, Vol 25, Number 4, October 1995.
13. J. R. M. Hosking. Modeling persistent in hydrological time series using fractional differencing. *Water Resources Res.*, **20**(12):1898–1908, 1984.
14. C. R. Kalmanek, H. Kanakia, and S. Keshav. Rate controlled servers for very high speed networks. In *Proc. IEEE GLOBECOM'90*, 1990.
15. J. Kurose. On comuting per-session performance bounds in high-speed multihop computer networks. In *Proc. ACM SIGMETRICS'92*, pp. 128–139, June 1992.
16. S. Q. Li and C. L. Hwang. Queue response to input correlation functions: discrete spectral analysis. *IEEE/ACM Trans. Networking*, **1**(5):522–533, October 1993.
17. S. Q. Li and C. L. Hwang. Queue response to input correlation functions: continuous spectral analysis. *IEEE/ACM Trans. Networking*, **2**(6): 678–692, December 1994.
18. M. Livny, B. Melamed, and A. K. Tsiolis. The impact of auto-correlation on queueing systems. *Manage. Sci.*, pp. 322–339, March 1993.
19. A. Mukherjee and A. Adas. An MPEG2 traffic library and its properties. `http://www.knoltex.com/papers/mpeg2Lib.html`.
20. I. Norros. On the use of fractional Brownian motion in the theory of connectionless networks. *IEEE J. Select. Areas Communi.* **13**(6):August 1995.

21. P. Pancha and M. El Zarki. Variable bit rate video transmission. *IEEE Commun. Mag.*, **32**(5):54–66, May 1994.

22. O. Rose. Statistical properties of MPEG video traffic and their impact on traffic modeling in ATM systems. University of Wuerzburg, Institute of Computer Science Research Report Series, Report No. 101, February 1995.

23. D. C. Verma, H. Zhang, and D. Ferrari. Delay jitter control for real-time communication in a packet switching network. Technical Report 90–007, International Computer Science Institute, University of California, Berkeley, January 1991.

24. T. Yang and J. Pan. Measurement-based loss scheduling scheme. In *Proc. IEEE INFOCOM'96*, pp. 1062–1071, March 1996.

25. L. Zhang. A new architecture for packet switching network protocols. Ph.D. Thesis, MIT, Department of Electrical Engineering and Computer Science, August 1989.

26. H. Zhang. Service disciplines for guaranteed performance service in packet-switching networks. *Proc. of IEEE*, **83**(10):October 1995.

27. H. Zhang and E. W. Knightly. Providing end-to-end statistical performance guarantees with bounding interval dependent stochastic models. In *Proc. ACM SIGMETRICS'94*, pp. 211–220, May 1994.

20

TOWARD AN IMPROVED UNDERSTANDING OF NETWORK TRAFFIC DYNAMICS

R. H. Riedi

Department of Electrical and Computer Engineering, Rice University, Houston, TX 77251

Walter Willinger

Information Sciences Research Center, AT&T Labs–Research, Florham Park, NJ 07932

20.1 INTRODUCTION

Since the statistical analysis of Ethernet local-area network (LAN) traces in Leland et al. [20], there has been significant progress in developing appropriate mathematical and statistical techniques that provide a physical-based, networking-related understanding of the observed fractal-like or self-similar scaling behavior of measured data traffic over time scales ranging from hundreds of milliseconds to seconds and beyond. These techniques explain, describe, and validate the reported large-time scaling phenomenon in aggregate network traffic at the packet level in terms of more elementary properties of the traffic patterns generated by the individual users and/or applications. They have impacted our understanding of actual network traffic, to the point where we now know why aggregate data traffic exhibits fractal scaling behavior over time scales from a few hundreds of milliseconds onward. In fact, a measure of the success of this new understanding is that the corresponding mathematical arguments are at the same time rigorous and simple, are in full agreement with the networking researchers' intuition and with measured

Self-Similar Network Traffic and Performance Evaluation, Edited by Kihong Park and Walter Willinger
ISBN 0-471-31974-0

data, and can be explained readily to a non-networking expert. These developments have helped immensely in demystifying fractal-based traffic modeling and have given rise to new insights and physical understanding of the effects of large-time scaling properties in measured network traffic on the design, management, and performance of high-speed networks.

However, to provide a complete description of data network traffic, the same kind of understanding is necessary with respect to the dynamic nature of traffic over small time scales, from a few hundreds of milliseconds downward. Because of the predominant protocols and end-to-end congestion control mechanisms that play a central role in modern-day data networks and determine the flow of packets over those fine time scales and at the different layers in the TCP/IP protocol hierarchy, studying the fine-time scale behavior or local characteristics of data traffic is intimately related to understanding the complex interactions that exist in data networks such as the Internet between the different connections, across the different layers in the protocol hierarchy, over time as well as in space. In this chapter, we first summarize the results that provide a unifying and consistent picture of the large-time scaling behavior of data traffic and discuss the appropriateness of self-similar processes such as fractional Gaussian noise for modeling the fluctuations of the traffic rate process around its mean and for providing a complete description of the traffic on individual links within the network. Then we report on recent progress in studying the small-time scaling behavior in data network traffic and outline a number of challenging open problems that stand in the way of providing an understanding of the local traffic characteristics that is as plausible, intuitive, appealing, and relevant as the one that has been found for the global or large-time scaling properties of data traffic.

20.2 THE LARGE-TIME SCALING BEHAVIOR OF NETWORK TRAFFIC

In this section, we demonstrate why the empirically observed large-time scaling behavior or (asymptotic) self-similarity of aggregate network traffic is an additive property, with the additional requirement that the individual component processes that generate the total traffic exhibit certain high-variability or heavy-tailed characteristics.

20.2.1 Additive Structure and Gaussianity

When viewed over large enough time scales, the number of packets or bytes per time unit collected off a link in a network originates from all those connections that were active during the measurement period, utilized this link, and actively generated traffic during this time. In other words, if for "time scales" or "levels of resolution" $m \gg 1$, $X^{(m)} = (X^{(m)}(k): k \geq 0)$ denotes the overall traffic rate process, that is, the total number of packets or bytes per time unit (measured at time scale m) generated by all connections, then we can write

$$X^{(m)}(k) = \sum X_i^{(m)}(k), \quad k \geq 0, \tag{20.1}$$

where the sum is over all connections i that are active at time k and where $X_i^{(m)} = (X_i^{(m)}(k): k \geq 0)$ represents the total number of packets or bytes per time unit (again measured at time scale m) generated by connection i.[1] Thus, Eq. (20.1) captures the *additive* nature of aggregate network traffic by expressing the overall traffic rate process $X^{(m)}$ as a superposition of the traffic rate processes $X_i^{(m)}$ of the individual connections.

Assuming for simplicity that the individual traffic rate processes $X_i^{(m)}$ are independent from one another and identically distributed, then under weak regularity conditions on the marginal distribution of the $X_i^{(m)}$ (including, e.g., the existence of second moments), Eq. (20.1) guarantees that the overall traffic rate process (or its deviations from its mean) exhibits Gaussian marginals, as soon as the traffic is generated by a sufficiently large number of individual connections.

20.2.2 Self-Similarity Through Heavy-Tailed Connections

Focusing on the temporal dynamics of the individual traffic rate processes $X_i^{(m)}$, suppose for simplicity that connection i sends packets or bytes at a constant rate (say, rate 1) for some time (the "active" or "on" period) and does not send any packets or bytes during the "idle" or "off" period; we will return to the challenging problem of allowing for more realistic "within-connection" packet dynamics in Section 20.3. For example, in a LAN environment, a connection corresponds to an individual host-to-host or source–destination pair and the corresponding traffic patterns have been shown in Willinger et al. [38] to conform to an alternating renewal process where the successive pairs of on and off periods define the inter-renewal intervals. On the other hand, in the context of wide-area networks or WANs such as the Internet, we associate individual connections with "sessions," where a session starts at some random point in time, generates packets or bytes at a constant rate (say, rate 1) during the lifetime of the connection, and then stops transmitting packets or bytes. Here a session can be an FTP appplication, a TELNET connection, a Web session, sending e-mail, reading Network News, and so on, or any imaginable combination thereof. In fact, over $\frac{1}{2}$ to 1 hour periods, session arrivals on Internet links have been shown to be consistent with a homogeneous Poisson process; for example, see Paxson and Floyd [25] for FTP and TELNET sessions, and see Feldmann et al. [12] for Web sessions. Note that in the present setting, only global connection characteristics (e.g., session arrivals, lifetimes of sessions, durations of the on/off periods) play a role, while the details of how the packets arrive within a connection or within an on

[1]Note that the processes $X^{(m)}$ and $X_i^{(m)}$ are defined by averaging X and X_i over nonoverlapping blocks of size m.

period have been conveniently modeled away by assuming that the packets within a connection are generated at a constant rate.

To describe the stochastic nature of the overall traffic rate process $X^{(m)}$, the only stochastic elements that have not yet been specified are the distributions of the lengths of the on/off periods (in the case of the LAN example) or the distribution of the session durations (for the WAN case) associated with the individual traffic rate processes $X_i^{(m)}$. Based on measured on/off periods of individual host-to-host pairs in a LAN environment (e.g., see Willinger et al. [38]) and measured session durations from different WAN sites (e.g., see Feldman et al. [12], Paxson and Floyd [25] and Willinger et al. [37]), we choose these distributions to be heavy-tailed with infinite variance. Here, a positive random variable U (or the corresponding distribution function F) is called *heavy-tailed with tail index* $\alpha > 0$ if it satisfies

$$P[U > y] = 1 - F(y) \approx cy^{-\alpha}, \quad \text{as } y \to \infty, \tag{20.2}$$

where $c > 0$ is a finite constant that does not depend on y. Such distributions are also called *hyperbolic* or *power-law distributions* and include, among others, the well-known class of *Pareto distributions*. The case $1 < \alpha < 2$ is of special interest and concerns heavy-tailed distributions with finite mean but *infinite variance*. Intuitively, infinite variance distributions allow random variables to take values that vary over a wide range of scales and can be exceptionally large with nonnegligible probabilities. Hence, heavy-tailed distributions with infinite variance allow for compact descriptions of the empirically observed high-variability phenomena that dominate traffic-related measurements at all layers in the networking hierarchy; for example, see Feldman et al. [12].

Mathematically, the heavy-tailed property of, for example, the durations during which individual connections actively generate packets implies that the temporal correlations of the stationary versions of an individual traffic rate processes $X_i^{(m)}$ and, because of the additivity property (20.1), of the overall traffic rate process $X^{(m)}$ decay hyperbolically slowly; that is, they exhibit long-range dependence. More precisely, if $r^{(m)} = (r^{(m)}(k): k \geq 0)$ denotes the autocorrelation function of the stationary version of the overall traffic rate process $X^{(m)}$, then property (20.2) can be shown to imply *long-range dependence* (e.g., see Cox [4] and Willinger et al. [38]; for similar results obtained in the context of a fluid queueing system under heavy traffic, see Chapter 5 in this volume). That is, for all $m \geq 1$, $r^{(m)}$ satisfies

$$r^{(m)}(k) \approx ck^{2H-2}, \quad \text{as } k \to \infty, \quad 0.5 < H < 1, \tag{20.3}$$

where the parameter H is called the *Hurst parameter* and measures the degree of long-range dependence in $X^{(m)}$; in terms of the tail index $1 < \alpha < 2$ that measures the degree of "heavy-tailedness" in Eq. (20.2), H is given by $H = (3 - \alpha)/2$. Intuitively, long-range dependence results in periods of sustained greater-than-average or lower-than-average traffic rates, irrespective of the time scale over which the rate is measured. In fact, for a zero-mean covariance-stationary process, Eq. (20.3) implies (and is implied by) *asymptotic* (*second-order*) *self-similarity*; that is, after appropriate rescaling, the overall traffic rate processes $X^{(m)}$ have identical second-order statistical characteristics and "look similar" for all sufficiently large time scales m. In other words, Eq. (20.3) holds if and only if for all sufficiently large time scales m_1 and m_2, we have

$$m_1^{1-H} X^{(m_1)} \approx m_2^{1-H} X^{(m_2)}, \tag{20.4}$$

where the quality is in the sense of second-order statistical properties and where $\frac{1}{2} < H < 1$ denotes the self-similarity parameter and agrees with the Hurst parameter in Eq. (20.3).

The ability to explain the empirically observed self-similar nature of aggregate data traffic in terms of the statistical properties of the individual connections that make up the overall traffic rate process shows that (asymptotically) self-similar behavior (1) is an intrinsically additive property (i.e., aggregate over many connections), (2) is mainly caused by user/session/connection characteristics (i.e., Poisson arrivals of sessions, heavy-tailed distributions with infinite variance for the session sizes/durations), and (3) has little to do with the network (i.e., the predominant protocols and end-to-end congestion control mechanisms that determine the actual flow of packets in modern data networks). In fact, for the self-similarity property of data traffic over large time scales to hold, all that is needed is that the number of packets or bytes per connection is heavy tailed with infinite variance, and the precise nature of how the individual packets within a session or connection are sent over the network is largely irrelevant.

Note that this understanding of data traffic started with an extensive analysis of measured aggregate traffic traces, followed by the statistically well-grounded conclusion of their self-similar or fractal characteristics, and triggered the curiosity of networking researchers who wanted to know: "Why self-similar or fractal?" In turn, this question for a physical explanation of the large-time scaling behavior of measured data traffic resulted in findings about data traffic at the connection level that are, at the same time, mathematically rigorous, agree with the networking researchers' experience, are consistent with data, and are intuitive and simple to explain in the networking context. In this sense, the progression of results proceeded in an opposite way to how traffic modeling has traditionally been done in this area; that is, by first analyzing in great detail the dynamics of packet flows within individual connections and then appealing to some mathematical limiting result that allowed for a simple approximation of the complex and generally overparameterized aggregate traffic stream. In contrast, the self-similarity work has demonstrated that novel insights into and new and unprecedented understanding of the nature of actual data traffic can be gained by a careful statistical analysis of measured traffic at the aggregate level and by explaining aggregate traffic characteristics in terms of more elementary properties that are exhibited by measured data traffic at the connection level.

20.2.3 Self-Similar Gaussian Processes as Workload Models

Note that in the Gaussian setting discussed in Section 20.2.1, the self-similarity property (20.4) implies that for $\frac{1}{2} < H < 1$ and for all sufficiently large time scales

m, the traffic rate process $X^{(m)}$ (or, more precisely, the deviation from its mean) satisfies

$$m^{1-H}X^{(m)} \approx X, \qquad (20.5)$$

where in this case, the equality is understood in the sense of finite-dimensional distributions, and where $X = (X_k: k \geq 1)$ denotes *fractional Gaussian noise* (FGN), the only stationary (zero-mean) Gaussian process that is (*exactly*) *self-similar* in the sense that Eq. (20.5) holds for all $m \geq 1$. Equivalently, FGN is uniquely characterized as the stationary (zero-mean) Gaussian process with autocorrelation function $r(k) = \frac{1}{2}[(k+1)^{2H} - 2k^{2H} + (k-1)^{2H}]$, $k \geq 1$, $\frac{1}{2} < H < 1$.

For the purpose of modeling the dynamics of actual data traffic over a link within a network, FGN has the advantage of providing a complete description of the resulting traffic rate process; that is, specifying its mean, variance, and Hurst parameter H suffices to completely characterize the traffic. Given this advantage over other—typically incomplete—descriptions of network traffic dynamics, it is important to know under what conditions FGN is an adequate and accurate process for modelling the deviations around the mean of actual data traffic. To this end, Erramilli et al. [8] note that the FGN model can be expected to be an appropriate model for data traffic provided (1) the traffic is aggregated over a large number of independent and not too wildly fluctuating connections (i.e., ensuring Gaussianity of expression (20.1)), (2) the effects of flow control on any one connection are negligible (i.e., requiring, in fact, that we consider the traffic only over sufficiently large time scales where Eq. (20.4) holds), and (3) the time scales of interest for the performance problem at hand coincide with the scaling region (i.e., where Eq. (20.5) holds). In practice, these conditions are often satisfied in the backbone (i.e., high levels of aggregation) and for time scales that are larger than the typical round-trip time of a packet in the network.

20.2.4 Toward Self-Similar Non-Gaussian Workload Models?

One of the conditions mentioned above that justify the use of FGN as an adequate and accurate description of actual data traffic traversing individual links in a network states that the traffic over a specific link is made up of a large number of (more or less) independent connections, where each connection's own traffic rate cannot fluctuate too wildly; that is, $X_i^{(m)}$ is chosen from a distribution with finite variance. While this condition is generally applicable in many legacy LAN and WAN environments and can often be validated against measured traffic, due to changes in networking technologies, applications, and user behavior, it can no longer be taken for granted in today's networks. For example, advanced networking technologies such as 100 Mb/s Ethernets or gigabit Ethernets can be expected—despite the presence of TCP, for example—to allow the traffic rates of individual connections to vary over many orders of magnitude, from kilobits/second to megabits/second and beyond, depending on the networking conditions. Thus, for understanding modern-day network traffic, processes that combine heavy tails in time and space (i.e., the distributions of the durations as well as of the rates at which individual connections emit packets are heavy tailed with infinite variance) may become relevant in practice and may see genuine applications in the networking area in the near future.

To illustrate, let $X_i^{(m)}$ denote an on/off-type connection described earlier, where in addition to the duration of the on/off periods, the rate at which the connection emits packets during the on period is also heavy tailed with infinite variance (with tail index β, say). Focusing on this modification of the renewal model investigated by Mandelbrot [22] and Taqqu and Levy [34], Levy and Taqqu [21] recently showed that when studying the overall traffic rate process $X^{(m)}$ defined in Eq. (20.1)—that is, aggregating many such independent connections—one can obtain a dependent, stationary process that has a stable marginal distribution with infinite variance and that is self-similar as in Eq. (20.5) with self-similarity parameter H given by

$$H = \frac{\beta - \alpha + 1}{\beta}. \tag{20.6}$$

Here β denotes the index characterizing the heaviness of the tail of the traffic rate of the individual connections, and α denotes the tail index associated with the distributions of the durations of on and off periods, which we assume for simplicity to be identical. Observe that in the finite variance case ($\beta = 2$), relation (20.6) reduces to the familiar $H = (3 - \alpha)/2 \in (\frac{1}{2}, 1)$, which appears in connection with fractional Gaussian noise considered earlier. However, in contrast to FGN, the superposition process obtained under the assumption of heavy tails with infinite variance on the durations *and* rates is not Gaussian but has heavy-tailed marginals instead, implying that there is a much higher probability than in the Gaussian case that the overall traffic rate can differ greatly from the average value and that it can take extreme values (a phenomenon also known as *intermittency*). Being non-Gaussian, one of the obstacles at this stage for using these kinds of stable superposition processes in the context of modeling data traffic is that their statistical parameters α (which specifies the marginals) and H (Eq. (20.5)) do not define them completely; there exist a number of different dependent, stationary increment processes with stable marginals with the same α and same self-similarity parameter H—see, for example, Samorodnitsky and Taqqu [33]. This is in stark contrast to FGN, where knowing the second-order statistical characteristics (i.e., variance and Hurst parameter H) uniquely defines the process, due to Gaussianity.

20.3 THE SMALL-TIME SCALING BEHAVIOR OF NETWORK TRAFFIC

The analysis of measured network traffic and resulting understanding of some of its underlying structure outlined in Section 20.2 have led to the realization that while wide-area traffic is consistent with asymptotic self-similarity or large-time scaling behavior, its small-time scaling features are very different from those observed over large time scales. Thus, to provide an adequate and more complete description of actual network traffic, it is necessary to deal with these small-time scaling features and to ultimately understand their cause and effects. To this end, we summarize in this section our current understanding of this very recent development in network traffic analysis and modeling by introducing concepts that are novel to the networking area, for example, multifractals, conservative cascades, and multiplicative structure, and illustrate their relevance to networking.

20.3.1 Multifractals

From a networking perspective, it comes as no surprise that protocol-specific mechanisms and end-to-end congestion control algorithms operating on small time scales and at the different layers in the hierarchical structure of modern data networks give rise to structural properties that are drastically different from the large-time scaling behavior, which has been shown earlier to be mainly due to global user and/or session characteristics. Since these networking mechanisms determine largely the actual flow of packets across the networks, they are likely to cause the traffic to exhibit pronounced local variations and irregularities which, per se, cannot be expected to have any obvious connection to the self-similar behavior of the traffic over large time scales.

To quantify these local variations in measured traffic at a particular point in time t_0, let $Y = (Y(t): 0 \leq t \leq 1)$ denote the process representing the total number of packets or bytes sent over a link-up to time t, and for some $n > 0$, consider the traffic rate process $Y((k_n + 1)2^{-n}) - Y(k_n 2^{-n})$, $k_n = 0, 1, \ldots, 2^n - 1$; that is, the total number of packets or bytes seen on the link during nonoverlapping intervals of the form $[k_n 2^{-n}, (k_n + 1)2^{-n}]$. We say that the traffic has a local scaling exponent $\alpha(t_0)$ at time t_0 if the traffic rate process behaves like $(2^{-n})^{\alpha(t_0)}$, as $k_n 2^{-n} \to t_0$ $(n \to \infty)$. Note that $\alpha(t_0) > 1$ corresponds to instants with low intensity levels or small local variations (Y has derivative zero at t_0), while $\alpha(t_0) < 1$ is found in regions with high levels of burstiness or local irregularities. Informally, we call traffic with the same scaling exponent at all instants t_0 *monofractal* (this includes exactly self-similar traffic, for which $\alpha(t_0) = H$, for all t_0), while traffic with nonconstant scaling exponent $\alpha(t_0)$ is called *multifractal*.

More formally, the degree of local irregularity of a signal Y or its singularity structure at a given point in time t_0 can be characterized to a first approximation by comparison with an algebraic function, that is, $\alpha(t_0)$ is the best (i.e., largest) α such that $|Y(t') - Y(t_0)| \leq C|t' - t_0|^{\alpha}$, for all t' sufficiently close to t_0. Since our process Y has positive increments, this *singularity exponent* can be approximated through the somewhat simpler quantity

$$\alpha(t) = \lim_{n \to \infty} \alpha_n(t), \tag{20.7}$$

where—assuming the limit exists—for $t \in [k_n 2^{-n}, (k_n + 1)2^{-n})$,

$$\alpha_n(t) := \alpha^n_{k_n} := -\frac{1}{n} \log_2 |Y((k_n + 1)2^{-n}) - Y(k_n 2^{-n})|. \tag{20.8}$$

The aim of *multifractal analysis* (MFA) is to provide information about these singularity exponents in a given signal and to come up with a compact description of the overall singularity structure of signals in geometrical or in statistical terms. Before describing in more detail some of the commonly used MFA methods, we note that since wavelet decompositions contain information about the degree of local irregularity of a signal, it should come as no surprise that the singularity exponent $\alpha(t)$ is related to the decay of wavelet coefficients $w_{j,k} = \int Y(s)\psi_{j,k}(s)\,ds$ around the point t, where ψ is a bandpass wavelet function and where $\psi_{j,k}(s) := 2^{-j/2}\psi(2^{-j}s - k)$ (e.g., in the case of the well-known *Haar wavelet*, $\psi(s)$ equals 1 for $0 \le s \le 1$, -1 for $1 \le s \le 2$, and 0 for all other s; for a general overview of wavelets, we refer to Daubechies [5]). Indeed, assuming only that $\int \psi(s)\,ds = 0$ one can show as in Jaffard [18] that

$$2^{n/2} w_{-n,k_n} \le C \cdot 2^{-n\alpha(t)}, \quad \text{as } k_n 2^{-n} \to t. \tag{20.9}$$

Moreover, it is known that under some regularity conditions (for a precise statement see Jaffard [18] or Daubechies [5, Theorem 9.2]), relation (20.9) characterizes the degree of local irregularity of the signal at the point t. This suggests to define $\tilde{\alpha}(t)$ as in Eq. (20.8) but with $\alpha_n(t)$ replaced by $\tilde{\alpha}_n(t)$, where

$$\tilde{\alpha}_n(t) := \tilde{\alpha}^n_{k_n} := \frac{1}{-n \log 2} \log(2^{n/2} |w_{-n,k_n}|). \tag{20.10}$$

In general, this may give a different but nevertheless useful description of the singularity structure of Y, particularly for nonmonotonous processes (for an example, see Gilbert et al. [13]). Using wavelets may also have numerical advantages. The remainder of this section remains true if $\alpha(t)$ is replaced by $\tilde{\alpha}(t)$ and Eq. (20.8) by (20.10), that is increments by normalized wavelet coefficients.

Conceptually, the geometrical formulation of MFA in the time domain is the most obvious one. Its objective is to quantify what values of the limiting scaling exponent $\alpha(t)$ appear in a signal and how often one will encounter the different values. In other words, the focus here is on the "size" of the sets of the form

$$K_\alpha = \{t : \alpha(t) = \alpha\}. \tag{20.11}$$

To illustrate, since for FGN there exists only one scaling exponent (i.e., $\alpha(t) = H$), the set K_α is either the whole line (if $\alpha = H$) or empty, and FGN is therefore said to be "monofractal." Similarly, for the concatenation of several FGNs with Hurst parameters H^i in the interval $I^i = [i, i+1]$, we have $K_{H^i} = I^i$. In general, however, the sets K_α are highly interwoven and each of them lies dense on the line. Consequently, the right notion of "size" is that of the *fractal Hausdorff dimension* $\dim(K_\alpha)$, which is, unfortunately, impossible to estimate in practice and severely limits the usefulness of this geometrical approach to MFA. Therefore, we will focus below on different statistical descriptions of the multifractal structure of a given signal.

One such description involves the notion of the *coarse Hölder exponents* (20.8). To illustrate, *fix a path of* Y and consider a histogram of the α_k^n $(k = 0, \ldots 2^n - 1)$ taken at some finite level n. It will show a nontrivial distribution of values but is bound to concentrate more and more around the expected value as a result of the law of large numbers (LLN): values other than the expected value must occur less and less often. To quantify the frequency with which values other than the mean value occur, we make extensive use of the theory of large deviations. Generalizing the Chernoff–Cramer bound, the large deviation principle (LDP) states that probabilities of rare events (e.g., the occurrence of values that deviate from the mean) decay exponentially fast. To make this more precise consider a sequence of independent, identically distributed (i.i.d.) random variables $W, W_1, W_2, \ldots$ and set $V_n := W_1 + \cdots + W_n$. Using Chebyshev's inequality and the independence, we find, for any $q > 0$,

$$P[(1/n)V_n \geq a] = P[2^{qV_n} \geq 2^{nqa}] \leq \frac{\mathbb{E}2^{qV_n}}{2^{nqa}} = (\mathbb{E}[2^{qW}]2^{-qa})^n. \tag{20.12}$$

Since $q > 0$ is arbitrary, we can replace the right-hand side in Eq. (20.12) by its infimum over $q > 0$. A symmetry argument shows that $P[b \geq (1/n)V_n] \leq (\mathbb{E}[2^{qW}]2^{-qb})^n$, for all $q < 0$. Combining all this yields the following two upper bounds:

$$\frac{1}{n}\log_2 P[b \geq (1/n)V_n \geq a] \leq \begin{cases} \inf_{q>0}\{\log_2 \mathbb{E}[2^{qW}] - qa\}, \\ \inf_{q<0}\{\log_2 \mathbb{E}[2^{qW}] - qb\}. \end{cases} \tag{20.13}$$

For a discussion of this simple result, let $L(q) = \mathbb{E}[2^{q(W-a)}]$. Since $\log(\cdot)$ is a monotone function, finding the infimum of L is the same as finding the infimum of $\log(L)$. We note first that $L''(q) > 0$, for all $q \in \mathbb{R}$, hence L is a strictly convex function and must have a unique infimum for $q \in \mathbb{R}$. From $L(0) = 1$ we conclude that this infimum must be less than or equal to 1. Focusing now on $q > 0$, we infer from $L'(0) = \log(2)C\mathbb{E}[W] - a)$ that $\inf_{q>0}L(q)$ is assumed in $q = 0$ and equals 1 if and only if $\mathbb{E}[W] \geq a$. On the other hand, $\inf_{q>0}L(q) < 1$ if $\mathbb{E}[W] < a$. An analogous result holds for the second bound. In summary, if $b > \mathbb{E}W > a$ then the bounds on the right-hand side (RHS) in Eq. (20.13) are both zero and thus reflect the LLN, which says that $(1/n)V_n \to \mathbb{E}[W]$ almost surely. On the other hand, if $\mathbb{E}[W]$ is not contained in $[a, b]$ and when $P[b \geq (1/n)V_n \geq a]$ is the probability of $(1/n)V_n$ deviating far from its expected value, then exactly one of the bounds will be negative, proving (at least) exponential decay of this probability. LDP theorems extend this result to a more general class of random sequences V_n and establish conditions under which the bound in Eq. (20.13) is attained in the limit $n \to \infty$ [6, 7].

To apply the LDP approach to our situation, we fix a realization of Y and consider the location t, encoded by k_n via $t \in [k_n 2^{-n}, (k_n + 1)2^{-n})$, as the only randomness relevant for the LDP. Since k_n can take only 2^n different values, which we will assume to be all equally likely, the relevant probability measure for t is the counting measure P_t. The sequence of interest for our purpose is

$$V_n := -\log_2 |Y((k_n+1)2^{-n}) - Y(k_n 2^{-n})| = n\alpha^n_{k_n}.$$

Trying to obtain more precise information about the singularity behavior and aiming at simplifying Eq. (20.13), we not only let n tend to ∞ but also let $[a, b]$ shrink down to a single point $\alpha = (a+b)/2$, which unifies the two bounds in the limit. All this suggests that the following limiting "rate function" f will exist under mild conditions (see Riedi [27, Theorem 7]):

$$f(\alpha) := \lim_{\epsilon \to 0} \lim_{n \to \infty} \frac{1}{n} \log_2 f_n(\alpha, \epsilon), \tag{20.14}$$

with

$$f_n(\alpha, \epsilon) := 2^n P_t[\alpha + \epsilon > \alpha_n(t) > \alpha - \epsilon] = \#\{\alpha_n(t) \in (\alpha - \epsilon, \alpha + \epsilon)\}. \tag{20.15}$$

The counting in Eq. (20.15) relates to the notion of dimension: if $f(\alpha) = 1$ then all or at least a considerable part of the α^n_k are approximately equal to α, that is, $fn(\alpha, \epsilon) \simeq 2^n$. Such is the case for FGN with $\alpha = H$; but we also have $f(\alpha) = 1$ if only a certain constant fraction of the α_n values equals α, as is the case with the concatenation of FGNs described earlier [36]. Only if certain values of α_n are considerably more spurious than others will we observe $f(\alpha) < 1$. In fact, it can be shown [28, 29] that the rate function $f(\alpha)$ relates to $\dim(K_\alpha)$ and that we have

$$\dim(K_\alpha) \leq f(\alpha). \tag{20.16}$$

It is in this sense that f provides information on the occurrence of the various "fractal" exponents α and has been termed *multifractal spectrum*. Also, note that the rate function f is a random element because it is defined for every path of Y.

Although f can, in principle, be computed in practice, it is a very delicate and highly sensitive object, mainly because of its definition in terms of a double limit (see Eq. (20.14)). Fortunately, the LDP result suggests using the RHS of Eq. (20.13), with $\mathbb{E}[2^{qW}]$ replaced by $(\mathbb{E}[2^{qV_n}])^{1/n}$ as in Eq. (20.12), as an alternative method for estimating f that avoids double-limit operations and is generally more robust because it involves averages. In fact, consider the *partition function* $\tau(q)$ defined by

$$\tau(q) := \lim_{n \to \infty} -\frac{1}{n} \log_2(2^n \mathbb{E}_t[2^{qV_n}]) = \lim_{n \to \infty} -\frac{1}{n} \log_2 S_n(q), \tag{20.17}$$

where the *structure function* $S_n(q)$ is given by

$$S_n(q) := \sum_{k=0}^{2^n-1} |Y((k+1)2^{-n}) - Y(k2^{-n})|^q = \sum_{k=0}^{2^n-1} 2^{-qn\alpha^n_k}. \tag{20.18}$$

According to the theory of LDP we will have equality in Eq. (20.13) under mild conditions, at least in the limit as $n \to \infty$ and $b \to a$. Appealing to such results, it is possible to establish conditions under which $f(\alpha) = \inf(q\alpha - \tau(q))$. In fact, collecting the terms k in $S_n(q)$ with $\alpha_k^n(t)$ approximately equal to some given value, say, α, for varying α and noting that we have about $2^{nf(\alpha)}$ such terms yields

$$S_n(q) := \sum_{\alpha} \sum_{\alpha_n \simeq \alpha} 2^{-nq\alpha} \simeq \sum_{\alpha} 2^{-n(q\alpha - f(\alpha))} \simeq 2^{-n\ \inf_\alpha (q\alpha - f(\alpha))};$$

that is,

$$\tau(q) = f^*(\alpha) := \inf_{\alpha}(q\alpha - f(\alpha)), \tag{20.19}$$

where * denotes the Legendre transform of a function (for a mathematically rigorous argument, see Riedi [27, 28]).

While the partition function $\tau(q)$ is clearly easier to estimate than f, it has to be noted that f may contain more information than τ. In fact, the Legendre back-transform yields only

$$f(\alpha) \leq f^{**}(\alpha) = \tau^*(\alpha) = \inf_{q}(q\alpha - \tau(q)) \tag{20.20}$$

where f^{**} is the concave hull of f (compare Eq. (20.13)[2]). The questions are when and for which α the equality $f^{**}(\alpha) = f(\alpha)$ holds. A simple application of the LDP theorem of Gärtner–Ellis [7] provides an answer to these questions—under the assumption that $\tau(q)$ is differentiable everywhere (see Riedi [27]). In this particular case, we obtain the appealing formula

$$f(\alpha) = \tau^*(\alpha) = q\alpha - \tau(q) \quad \text{at } \alpha = \tau'(q). \tag{20.21}$$

Since $\tau(q)$ is the Legendre transform of f, it must always be *concave*. This follows also from the fact that $S_n(q)$ is a log-convex function of q. Consequently, $\tau(q)$ is differentiable in almost all q a priori. For FGN, however, we obtain the degenerate case of a concave function: with probability one, we have

$$\tau(q) = qH - 1, \quad q > -1. \tag{20.22}$$

This is consistent with the fact that $\alpha(t) = H$ for all t; that is, the set K_H has dimension 1. Formula (20.22) can be guessed directly from ergodicity and self-similarity:

$$S_n(q) \simeq \sum_{k=0}^{2^n - 1} \mathbb{E}|Y((k+1)2^{-n}) - Y(k2^{-n})|^q \approx 2^{n - nqH} \mathbb{E}|Y(1)|^q.$$

[2]The factors 2^n appearing in f and $\tau(q)$ are for convenience. The sign of $\tau(q)$ is chosen such as to render Eqs. (20.20) and (20.19) symmetrical. The signs of q in (20.13) and (20.20) are opposite to each other.

For the example considered earlier where we concatenated a number of FGNs, we find $\tau(q) = \min_k(qH_k - 1)$, which is again consistent with $\alpha(t)$ taking the values H_k on sets of dimension 1 (compare Eq. (20.33), see also Lévy Véhel and Riedi [36] for more details). This example shows also how noncavity in $\tau(q)$ can result in loss of information: $\tau(q)$ and its Legendre transform reflect only the minimal and the maximal of the H_k. In contrast, truly concave behavior of $\tau(q)$ indicates that there is a whole interval of α-values present in the signal and not just a few (hence the term *multifractal*).

20.3.2 Multiplicatively Generated Multifractals or Cascades

A construction that fragments a given set into smaller and smaller pieces according to some geometric rule and, at the same time, divides the measure of these pieces according to some other (deterministic or random) rule is called a *multiplicative process or cascade* (e.g., see Evertsz and Mandelbrot [9]). The limiting object generated by such a multiplicative process defines, in general, a singular measure or multifractal and describes the highly irregular way the mass of the initial set gets redistributed during this simple fragmentation procedure. The *generator* of the cascade specifies the mass fragmentation rule, and we consider in the following the class of *conservative cascades*, introduced by Mandelrot [23] characterized by a generator that preserves the total mass of the initial set at every stage of the construction (i.e., mass conservation). To illustrate, we will construct a binomial conservative cascade or measure μ on the interval $I := [0, 1]$. More precisely, we will construct its distribution function $Y(t) = \mu([0, t])$ and since the underlying generator will be random, Y will define a stochastic process. By construction it will have positive increments and $Y(0) = 0$ almost surely.

This iterative construction starts with a uniform distribution on the unit interval of total mass M^0 and then "redistributes" this mass by splitting it among the two subintervals of half-size in the ratio M_0^1 to M_1^1, where $M_0^1 + M_1^1 = 1$. Proceeding iteratively one obtains after n steps a distribution that is uniform on intervals $I_{k_n}^n := [k_n 2^{-n}, (k_n + 1)2^{-n}]$. The mass lying in $I_{k_n}^n$ is redistributed among its two dyadic subintervals $I_{2k_n}^{n+1}$ and $I_{2k_n+1}^{n+1}$ in the proportions $M_{2k_n}^{n+1}$ and $M_{2k_n+1}^{n+1}$, where $M_{2k_n}^{n+1} + M_{2k_n+1}^{n+1} = 1$ almost surely.

To summarize, for any n let us choose a sequence $k_1, k_2, \ldots, k_n$ such that the interval $I_{k_l}^l$ lies in $I_{k_i}^i$ whenever $i < l$. In other words, the k_i are the n first binary digits of any point $t \in I_{k_n}^n$. We call this a *nested sequence*, and it is uniquely defined by the value of k_n. Then we have

$$Y((k_n + 1)2^{-n}) - Y((k_n)2^{-n}) = \mu(I_{k_n}^n) = M_{k_n}^n \cdot M_{k_{n-1}}^{n-1} \cdot \ldots \cdot M_{k_1}^1 \cdot M_0^0. \qquad (20.23)$$

The various M_l^i, which collectively define the generator of the conservative cascade, may have distributions that depend on i and l and that are arbitrary, as long as they are positive and provided that for all i and all m,

$$M_{2m}^i + M_{2m+1}^i = 1, \qquad (20.24)$$

almost surely. Note that this mass conservation condition introduces a strong dependence between the two "children" of any present node. Furthermore, we will require that for all n and k_n ($n = 1, 2, \ldots$), all the multipliers appearing in Eq. (20.23) are mutually independent. We will call this property *nested independence*. As long as these two requirements on dependency are satisfied, one is completely free in how to introduce further correlation structure.

It is obvious from this iterative construction and from relation (20.23) that a multiplicatively generated "multifractal process" has approximately *lognormal* marginals. Indeed, as a sum of independent random variables, the logarithms of the increments of Y are approximately Gaussian, provided that the random variables $\log M_l^i$ have finite second moments.

Note that as we move from stage n to $n+1$ on our construction of a conservative cascade, the conservation property (20.24) ensures that the values of Y at dyadic points of order less than n are not changed. As we let n tend to infinity, we see from Eq. (20.23) that the increments of Y between dyadic points tend to zero, hence Y is continuous (μ has not atoms) and well defined. Moreover, Y has increments of all lags but no (meaningful) derivative in the usual sense. As we will see, $\alpha(t)$ equals the expected value $\bar{\alpha}$ almost everywhere with $\bar{\alpha} > 1$, hence in these points, the product in Eq. (20.23) behaves like $2^{-n\bar{\alpha}}$ and the conventional derivative Y' is zero. Thus, the essential growth of Y happens "in" the points where Y' does not exist. In other words, the true derivative of Y is a distribution or singular measure, that is, μ.

To study the singularity structure of Y using $\alpha(t)$, we calculate the partition function $\tau(q)$ of the binomial conservative measure "in expectation." To this end, we assume that the M_k^n ($k = 0, \ldots, 2^n - 1$) are identically distributed with $M^{(n)}$. Note that $M^{(n)}$ is necessarily symmetrically distributed around $\frac{1}{2}$ due to Eq. (20.24). Then, Eq. (20.23) is equally distributed as $M^{(n)} \cdot \ldots \cdot M^{(1)} \cdot M^0$ for each of the 2^n nested sequences $k_1, \ldots, k_n$ of length n. Using the "nested" independence we find

$$\mathbb{E}[S_n(q)] = 2^n \cdot \mathbb{E}(M^{(n)})^q \cdot \mathbb{E}(M^{(n-1)})^q \cdot \ldots \cdot \mathbb{E}(M^{(1)})^q \cdot \mathbb{E}(M^0)^q. \tag{20.25}$$

Assuming now further that the $M^{(n)}$ converge in distribution, say, to M, we have

$$T(q) := \lim_{n\to\infty} -\frac{1}{n} \log_2 \mathbb{E}S_n(q) = -1 - \log_2 \mathbb{E}[M^q]. \tag{20.26}$$

Using the relations (20.16), (20.20), and $\tau^* \leq T^*$ (see Riedi [28]), and combining them with results in Arbeiter and Patzschke [1], Barral [3], Falconer [10], and Riedi et al. [31], we get that, for every α,

$$\dim(K_\alpha) = f(\alpha) = \tau^*(\alpha) = T^*(\alpha) \quad \text{almost surely.} \tag{20.27}$$

To demonstrate how MFA applies to conservative cascades and what sort of numerical results it can yield in this case, we use the wavelet-based approach mentioned earlier. For convenience, we will also deal with the wavelet coefficients of the distribution μ rather than the ones of Y. The former are given by

$$w_{j,k} := \int \psi_{j,k}(t)\, d\mu(t). \tag{20.28}$$

Using the Haar wavelet, we get with Eq. (20.23) the explicit expression

$$2^{-n/2} w_{-n,k_n} = \mu(I_{2k_n}^{n+1}) - \mu(I_{2k_n+1}^{n+1}) = (M_{2k_n}^{n+1} - M_{2k_n+1}^{n+1}) \prod_{i=0}^{n} M_{k_i}^{i}. \tag{20.29}$$

Thus, we compare the increment-based MFA (in terms of α, S, and T) of Y to the wavelet-based MFA (in terms of $\tilde{\alpha}$, $\tilde{S}$, and $\tilde{T}$) of μ. Due to the fact that $M_{2k_n}^{n+1} - M_{2k_n+1}^{n+1} = 2M_{2k_n}^{n+1} - 1$, we have

$$\mathbb{E}\tilde{S}_n(q) = \mathbb{E} \sum_{k_n=0}^{2^n-1} |2^{n/2} w_{-n,k_n}|^q = 2^{nq} S_n(q) \cdot \mathbb{E}|2M^{(n+1)} - 1|^q.$$

This gives immediately

$$\tilde{T}(q) = -q + T(q). \tag{20.30}$$

More generally, this relation holds for any choice of mother wavelet, which is supported on [0, 1], provided the multipliers M_k^n are all identically distributed. This holds because the scaling properties (20.23) of μ allow us to write the wavelet coefficients in this case as $2^{n/2} \cdot M_{k_n}^n \cdot \ldots \cdot M_{k_1}^1$ times a random factor that is independent of $M_{k_i}^i$ and that is distributed as $w_{0,0}$ (compare also Bacry et al. [2]).

In order to be able to say more about $\tilde{\tau}(q)$ for the Haar wavelet, we make an assumption that guarantees that the Haar wavelent coefficients don't decay too fast (compare Eq. (20.9)), that is, the prefactor on the RHS in Eq. (20.29) doesn't become too small. Therefore, let us assume in addition that there is some $\epsilon > 0$ such that for all n, $|2M^{(n+1)} - 1| \geq \epsilon$ almost surely. Then for all t, $(1/n) \log(2M_{2k_n}^{n+1} - 1) \to 0$, and

$$\tilde{\alpha}(t) = -1 + \lim_{n \to \infty} \frac{1}{-n \log 2} \log(2^{-n/2} |\mu(I_{k_n}^n)|) = -1 + \alpha(t). \tag{20.31}$$

Observe that this is precisely the relation we expect between the scaling exponents of a process and its (distributional) derivative—at least in nice cases. Moreover, differentiating Eq. (20.30) and recalling Eq. (20.21), we get $\tilde{T}'(q) = -1 + T'(q)$, which is in agreement with Eq. (20.31). Thus, both the increment-based and wavelet-based MFA yield the same results for conservative binomial cascades with multipliers bounded away from $\frac{1}{2}$. For a more detailed wavelet-based analysis of conservative cascades, we refer to Gilbert et al. [14] and Riedi [28].

20.3.3 On the Multifractal Nature of Network Traffic

While multifractals are new to the networking area, they have been applied in the past—mainly for descriptive purposes—to such diverse fields as the statistical theory of turbulence, the study of strange attractors of certain dynamical sysems, and, more recently, physical-based rain and cloud modeling; see for example, Evertsz and Mandelbrot [9] and Holly and Waymire [17] and references therein. In the networking context, multifractals and their ability to account for time-dependent scaling laws offer great promise for describing irregular phenomena that are localized in time. The latter are typically associated with network-specific mechanisms that operate on small time scales and—depending on the state of the network—can be expected to have a more or less severe impact on how the packets within individual connections are sent across the network. Empirical evidence in support of complex within-connection or local traffic characteristics in measured wide-area traffic that can be traced to the dominant TCP/IP protocol hierarchy of IP networks has been reported in the original comprehensive analysis of WAN traces by Paxson and Floyd [25] and, more recently, in work by Feldman et al. [12]. The original findings of multifractal scaling behavior of measured aggregate WAN traffic are due to Riedi and Lévy-Véhel [30] (see also Lévy Véhel and Riedi [36]), followed by a similar study by Mannersalo and Norros [24] involving measured ATM WAN traces (for an earlier discussion on multifractal scaling and measured LAN traffic, see also Taqqu et al. [35]).

Motivated by the empirically observed multifractal scaling behavior in measured WAN traffic by Riedi and Lévy Véhel [30], Feldmann et al. [11] (see also Gilbert et al. [14]) present a more detailed investigation into the multifractal nature of network traffic and bring multifractals into the realm of networking by providing empirical evidence that WAN traffic is consistent with multifractal scaling because IP networks appear to act as conservative cascades. In particular, they demonstrate that (1) conservative cascades are inherent to wide-area network traffic, (2) multiplicative structure becomes apparent when studying data traffic at the TCP layer, and (3) the cascade paradigm appears to be a traffic invariant for WAN traffic that can coexist with self-similarity. By systematically investigating the causes for the observed multifractal nature of measured network traffic, they observe that the packet arrival patterns within individual TCP connections (where one or more TCP connections make up a session) appear to be consistent with a multiplicative structure. The latter, they argue, seems to be mainly caused by networking mechanisms operating on small time scales and results in aggregate network traffic that exhibits multifractal scaling behavior over a wide range of small time scales. Although it is tempting to invoke the TCP/IP protocol hierarchy of modern data networks for motivating the presence of an underlying conservative cascade construction (e.g., a Web session generates requests, each request gives rise to connections, each connection is made up of flows, flows consist of individual packets), Feldmann and co-workers demonstrate that the multiplicative structure associated with a conservative cascade construction is most apparent when studying network traffic at the TCP layer, where the network behavior (i.e., the way the packets within a TCP connection are sent across the network) is largely decoupled from the user behavior. Moreover, Feldmann and co-workers suggest that the transition from multifractal to self-similar scaling occurs around time scales on the order of the typical round-trip of a packet within the network under consideration.

While this work leaves open the "big" question—"Why are packets within individual TCP connections distributed in accordance with a conservative cascade construction?"—it clearly identifies the TCP layer as the most promising place in the networking hierarchy to search for the physical reasons behind the observed multifractal scaling behavior of measured network traffic and/or behind the conjecture that modern data networks act in a manner consistent with conservative cascades. Clearly, progress on these problems will require a close collaboration with networking experts. Realizing that it is difficult to think of any other area in the sciences where the available data provide such detailed information about so many different facets of behavior, there exists great potential for coming up with intuitively appealing, conceptually simple and mathematically rigorous statements as to the causes and effects of multifractals in data networking. Put differently, for multifractals to have a genuine impact on networking, their application has to move beyond the traditional descriptive stage and has to be able to answer questions as to why network traffic is multifractal (i.e., physical explanation in the network context) and how it may or may not impact network performance (i.e., engineering).

20.3.4 Multiplicative Structure and Lognormality

The observed multifractal nature of measured WAN traffic over small time scales and the empirical evidence discussed above in support of an underlying conservative cascade mechanism responsible for the multifractal scaling phenomenon imply that over those fine scales, network traffic is *multiplicatively* generated. In other words, at the microscopic level where the network (via the underlying protocols and end-to-end congestion control mechanisms) determines how the individual packets of a connection are sent across a given link in the network, the traffic rate process (i.e., total number of packets or bytes per small time unit) is the product of a large number of more or less independent "multipliers." In contrast, we have seen that at the macroscopic level or over large time scales, user and/or session characteristics are mainly responsible for the observed self-similar scaling behavior of network traffic and that over those time scales, the traffic rate process is *additive* in nature—that is, the sum of a large number of more or less independent "summands," where the individual summands or connections exhibit heavy-tailed distributions with infinite variance for their sizes or durations.

Intuitively, this distinction between the additive and multiplicative structure of measured network traffic over large and small time scales, respectively, can best be explained when considering an individual TCP connection. When viewed over large enough time scales, all we observe is the total workload M^0 (in bytes or packets) that is sent over the network during the connection's lifetime and, for simplicity, we assume in general that the connections traffic rate $X_c^{(m)}$ is constant and that the connection's duration is unity. However, when zooming in onto finer time scales, we observe that a certain fraction of the total workload was sent during the first half of the connection's lifetime and the rest in the second half. Continuing inductively, we find that the work-load emitted by the connection during a time interval of length 2^{-n} (which corresponds to a certain level of aggregation m) is of the form

$$X_c^{(m)} := M^n \cdot M^{n-1} \cdot \ldots \cdot M^1 \cdot M^0, \tag{20.32}$$

where the multipliers M^k reflect the "state of the network" and determine the amount of workload that the connection can send across the link at any given point in time. Small multipliers suggest heavy competition for the link, while large multipliers indicate that the connection can temporarily transmit at close to full speed.

As we have seen earlier, the idea of successively fragmenting the total work-load into parts leads naturally to a *multiplicative* process or cascade. While the networking application justifies our choice of considering conservative cascades, our focus on an underlying binomial structure for the cascades is for simplicity. On mild independence assumptions on the multipliers (they should form a certain martingale) we are assured that we can talk about the limit of infinitely fine scales ($n \to \infty$) and that this limit has interesting statistical properties. In fact, by experimenting with turning a constant bit-rate connection into a highly bursty one via an appropriately chosen conservative binomial cascade construction [14], we find that the latter can closely match the way networking mechanisms operating on small time scales determine the actual flow of packets/bytes over the duration of a TCP connection. Moreover, when the traffic rate over a small time interval is described in terms of a conservative binomial cascade, it is explicitly multiplicative in nature (Eq. (20.32)); and as a result, the marginals of the traffic rate process over small time scales will automatically be approximately *lognormal* (e.g., apply CLT to the random variables $\log M^k$).

20.4 TOWARD COMPLETE DESCRIPTIONS OF NETWORK TRAFFIC

The empirical finding that measured WAN traffic contains an additive component as well as a multiplicative component provides new motivation for and insights into developing a more complex description of the dynamic nature of actual network traffic. In the following, we discuss a simple workload model that exhibits self-similar as well as multifractal scaling but is not consistent with measured network traffic. Then we illustrate the changes that are required to turn this simple model into one that is consistent with actual traffic, not only with respect to the large-time and small-time scaling behavior of measured aggregate traffic rate processes, but also at the different layers in the IP protocol hierarchy.

20.4.1 A Simple Multifractal Workload Model

To start, we consider the workload model discussed in Section 20.2.3, where user-initiated sessions (1) arrive in accordance with a Poisson process, (2) bring with them a workload (e.g., number of bytes, packets, flows, or TCP connections, session duration) that is heavy tailed with infinite variance, and (3) distribute the workload over the lifetime of the session at a constant rate. A result by Kurtz [19] states that, over large enough time scales, the fluctuations of the aggregate traffic rate around its mean value are well described by FGN, for a very general class of within-session traffic rate processes that includes the special case of constant bit-rate sessions. Recall that the self-similar scaling property over large time scales (or equivalently, long-range dependence) is essentially due to the fact that the session sizes exhibit infinite variance, and that approximate Gaussianity follows from an application of the CLT, that is, from aggregating over a large number of independent sessions whose individual traffic rates are sufficiently "tame."

To incorporate multiplicative structure into this simple traffic description, we simply modify property (3) above and require that the constant within-session traffic rate processes are replaced by multiplicative processes or, more precisely, by independent and identically distributed multifractals generated by appropriately chosen conservative binomial cascades with associated partition function $\tau(q)$ (or the more informative multifractal spectrum f). This modified workload process is a generalization of Kurtz's model by allowing within-session traffic rates to be multifractals. Since Kurtz's model is known to be insensitive to the particular within-session traffic dynamics, the self-similar scaling property over large time scales remains intact, even for multifractal within-session structure, and represents the additive component of network traffic, which is mainly due to the global characteristics of user-initiated sessions. However, when viewed over small time scales, this modified workload process will also exhibit multifractal scaling, not only at the session level, where it does so by definition, but also at the aggregate level. In fact, it can be shown that the superposition of i.i.d. conservative binomial cascades also exhibits multifractal structure, with a multifractal spectrum that is identical to the one of a "typical" session-related conservative binomial cascade. To illustrate, let μ and ν be two multifractals generated by two (possibly different) conservative binomial cascades. It is easy to see that independent of their supports, for the multifractal $\mu + \nu$ obtained by superposing μ and ν, we have

$$\tau^{\mu+\nu}(q) = \min(\tau^{\mu}(q), \tau^{\nu}(q)), \tag{20.33}$$

for all $q \geq 0$. For a proof, simply use that for all positive a, b, and q, we have $(a^q + b^q)/2 \leq (\max\{a, b\})^q \leq (a + b)^q \leq (2 \max\{a, b\})^q \leq 2^q(a^q + b^q)$, and hence

$$S_n^{\mu}(q) + S_n^{\nu}(q) \leq 2 \cdot S_n^{\mu+\nu}(q) \leq 2^{q+1}(S_n^{\mu}(q) + S_n^{\nu}(q)).$$

If the supports of μ and ν are disjoint, we have $S_n^{\mu+\nu}(q) = S_n^{\mu}(q) + S_n^{\nu}(q)$ and Eq. (20.33) holds for all q. However, for more general cascades with overlapping support, we will typically see $\tau(q) > \min(\tau^{\mu}(q), \tau^{\nu}(q))$ for negative q.

Assuming now that μ and ν have the same $\tau(q)$ and taking the Legendre transform, we see that the superposition $\mu + \nu$ has the same spectrum $f(\alpha)$ in the increasing part, that is, for small α ($\alpha < \bar{\alpha}$), which corresponds to the bursty part of the multifractal. For α larger than the expected value $\bar{\alpha}$, corresponding to the smoother parts, we may observe a smaller $f(\alpha)$. In other words, the superposition has a tendency toward more bursts and fewer smooth parts. This is natural since bursts of one multifractal may overwhelm some smooth parts of the other.

Thus, the built-in multifractal within-session structure causes the overall traffic rate process to be multiplicative over small time scales, thereby accounting in a parsimonious manner for the effect that the network has on the small-time scale dynamics of traffic rates on individual links within the network.

20.4.2 The Additive and Multiplicative Nature of Network Traffic

Note that the above generation of Kurtz's workload model that allows for multifractal within-session traffic rates is not consistent with measured data. In fact, Feldmann et al. [1] present empirical evidence that the observed within-session structure is itself a complicated mixture of additive and multiplicative components, and only by investigating network traffic at the TCP level (e.g., in terms of port-to-port flows) is it possible to clearly isolate the multiplicative structure in measured network traffic. Using the findings from yet another empirical traffic study (see Feldmann et al. [12]), we also know that the overall number of TCP connections per time unit exhibits self-similar scaling behavior for time scales on the order of seconds and beyond. Thus, to get a workload model for wide-area traffic that combines additive and multipicative structure and is consistent with measured data, we simply modify the multifractal version of Kurtz's process and require that (1) TCP connections arrive in accordance to a self-similar process, that is, the fluctuations around the mean of the total number of TCP connection arrivals per time unit follow a FGN; (2) The TCP connections' workload is heavy tailed with infinite variance; and (3) the workload of a TCP connection is distributed over the connection's lifetime in a multifractal fashion, that is, according to a conservative binomial cascade.

To see that the latter model has the desired large-time and small-time scaling properties and hence is in agreement with the observed additive and multiplicative properties of actual network traffic, we keep (2) and (3) as is, but note that the self-similar scaling property for the aggregate TCP connection traffic rate can be accomplished by relying on the underlying session structure of the original Kurtz's model. That is, user-initiated sessions continue to arrive in a Poisson fashion, but the session workload is now expressed in terms of the number of TCP connections that make up a particular session and remains heavy tailed with infinite variance; for consistency, we assume that the TCP connections within a session arrive in such a way that they don't overlap with one another. It is then easy to see that this two-tier approach to describing aggregate WAN traffic yields the additive traffic component via the TCP-connection-within-structure and the multiplicative component via the dynamics prescribed for the packets within individual TCP connections. Moreover, this two-tier approach is also fully consistent with measured Internet traffic at the different layers in the TCP/IP protocol heirarchy (e.g., see Feldmann et al. [12]).

20.4.3 Toward a Comprehensive Study of Network Performance

The attractive feature of the above structural model for wide-area traffic is that it is consistent with measured traffic at all levels of interest and that it accounts in a parsimonious manner for both the global or large-time scale as well as local or small-time scale characteristics observed in measured WAN traffic. While the global scaling behavior is already part of Kurtz's original model (via the relationship between heavy-tailed sizes or durations of the individual sessions and the asymptotic self-similarity of the aggregate packet stream) and is captured by the Hurst parameter H, the original model does not incorporate local scaling behavior. However, we have seen earlier that by choosing an appropriate generator for the generic underlying conservative binomial cascade for the within-connection traffic rate process, we are able to obtain the same overall multifractal scaling as captured by the multifractal spectrum associated with the generic cascade model for the individual TCP connections.

The practical relevance for such a structural workload model is that it allows for a more complete description of network traffic than exists to date in cases where higher-order stastistics or multiplicative aspects of the traffic play an important role but cannot be adequately accounted for by traditional, strictly second-order descriptions of network traffic. By aiming for a complete description of traffic, a more comprehensive analysis of network performance-related problem becomes feasible and desirable. In the past, thorough analytical studies of which aspects of network traffic are imortant for which aspects of network performance have often been prevented due to a lack of models that provide provably complete descriptions of the traffic processes under study. This situation can lead to misconceptions and misunderstandings of the relevance of certain aspects of traffic for certain aspects of performance (e.g., see Grossglauser and Bolot [15], Heyman and Lakshman [16], and Ryu and Elwalid [32].)

In a first attempt to allow for a more complete description of network traffic, Riedi at al. [31] (see also Ribeiro et al. [26]) emphasize performance aspects of description traffic models with additive and multiplicative structures. Working in the wavelet domain, they discuss [31] a multiplicative model based on binomial cascades, which exhibits the multifractal properties observed in measured network traffic at small scales and, in addition, matches the self-similar behavior of traffic over large time scales. Their model becomes approximately additive at large scales, as the variance of the cascade generator decreases with increasing scale, explaining why a purely multiplicative model can be consistent with an additive property in the limit of large scales. Riedi et al. [31] also provide initial evidence that models allowing for a more complete description of network traffic, in particular its multifractal behavior, typically outperform additive Gaussian models in the context of specific performance problems [26].

20.5 CONCLUSION

One of the implications of the discovery of self-similar or multifractal scaling behavior in measured network traffic has been the realization that network traffic modeling and performance analysis can and should no longer be viewed as exercises in data fitting and queueing theory or simulations. Instead, relevant traffic modeling has become a natural by-product of a renewed effort that aims at gaining a physical (i.e., network-related) understanding of the empirically observed scaling phenomena. Moreover, the novel insights gained from such a physical-based understanding of actual network traffic dynamics often allows for a qualitative assessment of their potential impact on network performance, when more quantitative methods appear to be mathematically intractable or are not yet available. While traditional performance modeling has mainly lived in the confines of mathematically tractable queueing models, the observed scaling properties of measured network traffic and the constantly changing nature of today's networks strongly suggest a shift away from focusing exclusively on quantitative methods for assessing the wide range of network performance-related problems toward achieving instead a more qualitative understanding of the implications of the dominant features of measured network traffic on relevant networking issues. While supporting such a qualitative knowledge—where possible—through a quantitative analysis is clearly desirable, we believe that the development of an ubiquitous, stable, robust, and high-performance networking infrastructure of the future will depend crucially on a qualitative rather than quantitative understanding of networks and network traffic dynamics.

Finally, in terms of practical relevance, we also argue that by incorporating—via multifractals—local scaling characteristics of the traffic into a workload model, it may become in fact feasible to adequately describe traffic in a closed system (like the Internet) with an open model. The vast majority of currently used models for network traffic completely ignore the fact that the dynamic nature of packet traffic over a given link is the result of a combination of source/user behavior and highly nonlinear interactions between the individual users and the network. The search for a physical explanation of the observed multifractal nature of measured traffic at the packet level is intimately related to trying to sort out these complicated interactions and to abstract them to a level that is intuitively appealing, conforms to networking reality, and captures and explains in a mathematically rigorous manner empirically observed phenomena. Clearly, a prerequisite for succeeding in this endeavor is a close collaboration with networking experts who are familiar with the details of the various protocols and control mechanisms that operate at the different layers within the hierarchical structure of modern-day data networks and who are aware of the problems that are associated with the highly dynamic, constantly changing, and extremely heterogeneous nature of today's communication networks.

REFERENCES

1. M. Arbeiter and N. Patzschke. Self-similar random multifractals. *Math. Nachr.*, **181**:5–42, 1996.
2. E. Bacry, J. Muzy, and A. Arneodo. Singularity spectrum of fractal signals from wavelet analysis: exact results. *J. Statist. Phys.*, **70**:635–674, 1993.
3. J. Barral. Continuity, moments of negative order, and multifractal analysis of Mandelbrot's multiplicative cascades. Université Paris Sud, thèse No. 4704, 1997.

4. D. R. Cox. Long-range dependence: a review. In H. A. David and H. T. David, ed., *Statistics: An Appraisal*, pp. 55–74. Iowa State University Press, Ames, 1984.
5. I. Daubechies. *Ten Lectures on Wavelets*. SIAM Philadelphia, 1992.
6. J.-D. Deuschel and D. W. Stroock. *Large Deviations*. Academic Press, New York, 1994.
7. R. Ellis. Large deviations for a general class of random vectors. *Ann. Prob.*, **12**:1–12, 1984.
8. A. Erramilli, O. Narayan, and W. Willinger. Experimental queueing analysis with long-range dependent packet traffic. *IEEE/ACM Trans. Networking*, **4**:209–223, 1996.
9. C. J. G. Evertsz and B. B. Mandelbrot. Multifractal measures. In H.-O. Peitgen, H. Jurgens, and D. Saupe, eds., *Chaos and Fractals: New Frontiers in Science*. Springer-Verlag, New York, 1992.
10. K. J. Falconer. The multifractal spectrum of statistically self-similar measures. *J. Theor. Probab.*, **7**:681–702, 1994.
11. A. Feldman, A. C. Gilbert, and W. Willinger. Data networks as cascades: investigating the multifractal nature of Internet WAN traffic. *Comput. Commun. Rev.*, **28**(4):42–55, 1998. (*Proc. ACM/SIGCOMM'98*, Vancouver, Canada, September 1998).
12. A. Feldman, A. C. Gilbert, W. Willinger, and T. G. Kurtz. The changing nature of network traffic: scaling phenomena. *Comput. Commun. Rev.*, **28**(2):5–29, April 1998.
13. A. C. Gilbert, W. Willinger, and A. Feldmann. Visualizing multifractal scaling behavior: a simple coloring heuristic. In *Proceedings of the 32nd Asilomar Conference on Signals, Systems and Computers*, Pacific Grove, CA, November 1998.
14. A. C. Gilbert, W. Willinger, and A. Feldmann. Scaling analysis of conservative cascades, with applications to network traffic. *IEEE Trans. Inf. Theory, Special Issue on "Multiscale Statistical Signal Analysis and Its Applications*," **45**(3):971–991, April 1999.
15. M. Grossglauser and J.-C. Bolot. On the relevance of long-range dependence in network traffic. *Comput. Commun. Rev.*, **26**(4):15–24, 1996. (*Proc. ACM/SIGCOMM'96*, Stanford, CA, 1996.)
16. D. P. Heyman and T. V. Lakshman. What are the implications of long-range dependence for VBR-video traffic engineering? *IEEE/ACM Trans. Networking*, **4**:301–317, 1996.
17. R. Holley and E. C. Waymire. Multifractal dimensions and scaling exponents for strongly bounded random cascades. *Ann. Appl. Probab.*, **2**:819–845, 1992.
18. S. Jaffard. Local behavior of Riemann's function. *Contemp. Math.*, **189**:287–307, 1995.
19. T. G. Kurtz. Limit theorems for workload input models. In F. P. Kelly, S. Zachary, and I. Ziedins, eds., *Stochastic Networks: Theory and Applications*. Clarendon Press, Oxford, 1996.
20. W. E. Leland, M. S. Taqqu, W. Willinger, and D. V. Wilson. On the self-similar nature of Ethernet traffic (extended version). *IEEE/ACM Trans. Networking*, **2**:1–15, 1994.
21. J. B. Levy and M. S. Taqqu. Renewal reward processes with heavy-tailed interarrival times and heavy-tailed rewards. Preprint, 1997.
22. B. B. Mandelbrot. Long-run linearity, locally Gaussian processes, *H*-spectra and infinite variances. *Int. Econ. Rev.* **10**:82–113, 1969.
23. B. B. Mandelbrot. Intermittent turbulence in self-similar cascades: divergence of high moments and dimension of the carrier. *J. Fluid. Mech.*, **62**:331–358, 1974.
24. P. Mannersalo and I. Norros. Multifractal analysis of real ATM traffic: a first look. *COST257TD*, 1997.

25. V. Paxson and S. Floyd. Wide area traffic: the failure of Poisson modeling. *IEEE/ACM Trans. Networking*, **3**:226–244, 1995.
26. V. Ribeiro, M. S. Crouse, R. Riedi, and R. G. Baraniuk. Simulation of non-Gaussian long-range-dependent traffic using wavelets. In *Proc. SIGMETRICS'99*, **27**(1):1–12, June 1999.
27. R. H. Riedi. An improved multifractal formalism and self-similar measures. *J. Math. Anal. Appl.*, **189**:462–490, 1995.
28. R. H. Riedi. Multifractal processes. TR 99-06, ECE Dept., Rice University, 1999.
29. R. H. Riedi and B. B. Mandelbrot. Exceptions to the multifractal formalism for discontinuous measures. *Math. Proc. Cambridge Philos. Soc.*, **123**:133–157, 1998.
30. R. Riedi and J. Lévy Véhel. Multifractal properties of TCP traffic: a numerical study. Technical Report No. 3129, INRIA Rocquencourt, France, Feb. 1997. Also available at `www.dsp.rice.edu`.
31. R. Riedi, M. S. Crouse, V. Ribeiro, and R. G. Baraniuk. A multifractal wavelet model with application to TCP network traffic. *IEEE Trans. Inf. Theory, Special Issue on "Multiscale Statistical Signal Analysis and Its Applications,"* **45**(3):992–1018, April 1999.
32. B. K. Ryu and A. Elwalid. The importance of long-range dependence of VBR video traffic in ATM traffic engineering: myths and realities. *Comput. Comm. Rev.*, **26**(4):3–14, 1996, *Proc. ACM/SIGCOMM'96*, Stanford, CA, 1996.
33. G. Samorodnitsky and M. S. Taqqu. *Stable non-Gaussian Processes: Stochastic Models with Infinite Variance*. Chapman and Hall, New York, 1994.
34. M. S. Taqqu and J. Levy. Using renewal processes to generate long-range dependence and high variability. In E. Eberlein and M. S. Taqqu, eds., *Dependence in Probability and Statistics*, pp. 73–89, Birkhäuser, Boston, 1986.
35. M. S. Taqqu, V. Teverovsky, and W. Willinger. Is network traffic self-similar or multifractal? *Fractals*, **5**:63–73, 1997.
36. J. Lévy Véhel and R. Riedi. Fractional Brownian motion and data traffic modeling: the other end of the spectrum. In *Fractals in Engineering*, pp. 185–202, Springer, Berlin, 1997.
37. W. Willinger, V. Paxson, and M. S. Taqqu. Self-similarity and heavy tails: structural modeling of network traffic. In R. Adler, R. Feldmann, and M. S. Taqqu, eds., *A Practical Guide to Heavy Tails: Statistical Techniques for Analyzing Heavy Tailed Distributions*. Birkhäuser Verlag, Boston, 1998.
38. W. Willinger, M. S. Taqqu, R. Sherman, and D. V. Wilson. Self-similarity through high-variability: statistical analysis of Ethernet LAN traffic at the source level. *IEEE/ACM Trans. Networking*, **5**:71–86, 1997.

21

FUTURE DIRECTIONS AND OPEN PROBLEMS IN PERFORMANCE EVALUATION AND CONTROL OF SELF-SIMILAR NETWORK TRAFFIC

KIHONG PARK

Network Systems Lab, Department of Computer Sciences, Purdue University, West Lafayette, IN 47907

21.1 INTRODUCTION

Since the seminal study of Leland et al. [41] on the self-similar nature of network traffic, significant advances have been made in understanding the statistical properties of measured network traffic—in particular, Internet workloads—why self-similar burstiness is an ubiquitous phenomenon present in diverse networking contexts, mathematical models for their description and performance analysis based on queueing, and traffic control and resource management under self-similar traffic conditions. Chapter 1 gives a comprehensive overview including a summary of previous works, and the individual chapters give a detailed account of a cross section of relevant works in the area. Chapter 20 provides a discussion of traffic and workload modeling, with focus on long versus short time scales and nonuniform scaling observed in wide area IP traffic [23,24].

This chapter presents a broad outlook into the future in terms of possible research avenues and open problems in self-similar network traffic research. The specific items described in the chapter are but a subset of interesting research issues and are meant to highlight topics that can benefit from concerted efforts by researchers in the community due to their scope and depth. The research problems are organized

Self-Similar Network Traffic and Performance Evaluation, Edited by Kihong Park and Walter Willinger
ISBN 0-471-31974-0

around recent developments and the landscape of previous accomplishments, grouped into three areas—workload characterization, performance analysis, and traffic control. Physical modeling, which can be viewed as a fourth category, is grouped with workload characterization.

Workload Characterization The original focus of self-similar burstiness in local area and wide area network traffic has expanded into the generalized framework of *workload modeling*, which captures source behavior and structural properties of network systems, not necessarily restricted to network and link layers. This stems, in part, from the realization that network performance—as measured by packet drop, queueing delay, and jitter at multiplexing points in the network—is affected by a multitude of factors including variability of streamed real-time VBR video, connection arrival patterns and their durations, the make-up of files being transported, control actions in the protocol stack, and user behavior that drives network applications. Increasingly, these activities transpire under the umbrella of the World Wide Web (WWW), and characterizing the structural properties—static and dynamic—of the global wired/wireless Internet that impact network performance has become an important goal. The research challenge lies in identifying, quantifying, and modeling *invariant* —or "slowly changing"—system traits, in the midst of a rapidly growing network infrastructure, that are relevant to network performance.

Performance Analysis Performance analysis of queueing systems with self-similar input has yielded the fundamental insight that queue length distribution decays polynomially vis-à-vis the more accustomed case of exponential decay with Markovian input. In the resource provisioning context, this is interpreted to mean that resource dimensioning using buffer sizing is an ineffective policy relative to bandwidth allocation. There remain a number of challenges. First, the queueing results are asymptotic in nature where buffer capacity—in some form—is taken to infinity to achieve tractability. Little is known about finite buffer systems except for observations on the dependence of packet loss rate on the"effective" time scale induced by buffer size, and its delimiting impact on correlation structure at larger time scales with respect to its influence on queueing [28, 60]. Second, performance evaluation with self-similar traffic has concentrated on first-order performance measures—that is, packet loss rate and queueing delay—which is but one, albeit important, yardstick. In the modern network environment with multimedia and other QoS-sensitive traffic streams comprising a growing fraction of network traffic, second-order performance measures in the form of "jitter" such as delay variation and packet loss variation are of import to provisioning user-specified QoS. Self-similar burstiness is expected to exert a negative influence on second-order performance measures and multimedia traffic controls—for example, packet-level FEC—that are susceptible to concentrated packet loss. Third, performance analysis is carried out in equilibrium, which may be problematic for self-similar workloads given their slow convergence properties. As a related point, the bulk of TCP connections is known to be short-lived, and there is a disconnect between steady-state techniques and performance evaluation of short- and medium-duration flows.

The same problem exists when using simulation as the principal performance evaluation tool.

Traffic Control Traffic control for self-similar traffic has been explored on two fronts: (1) as an extension of performance analysis in the resource provisioning context, and (2) from the multiple time scale traffic control perspective where correlation structure at large time scales is actively exploited to improve network performance. The resource provisioning aspect advocates a small buffer/large bandwidth resource dimensioning policy, which—when coupled with the central limit theorem—yields predictable multiplexing gains when a large number of independent flows are aggregated. Whereas resource provisioning is open-loop in nature, multiple time scale traffic control seeks to achieve performance gains by exploiting correlation structure in self-similar traffic at time scales exceeding the time horizon of the feedback loop to impart proactivity to reactive controls (e.g., TCP). This is relevant in broadband wide area networks where the delay–bandwidth product problem is especially severe, and mitigating the performance degradation due to outdated feedback is critical to facilitating scalable, adaptive traffic control. The initial success of this approach [62, 67–69] (see Chapter 18 for an application to rate-based congestion control) leads to a generalization to *workload-sensitive traffic control*, where facilitation of workload sensitivity is expanded along several traffic control dimensions including the two core features for harnessing predictability at large time scales: long-range correlation in network traffic and heavy-tailedness of connection durations. Workload-sensitive traffic control is a broad area that can benefit from concerted efforts at several fronts, spanning novel mechanisms for detecting and exploiting large time scale predictability structure, short-duration connection management, packet scheduling, end system support, and dynamic admission control with self-similar call arrivals and/or heavy-tailed connection durations.

21.2 OPEN PROBLEMS IN WORKLOAD CHARACTERIZATION

21.2.1 Physical Modeling

Unlike many systems of study including economic, social, and certain physical sciences (e.g., astronomy, earth and atmospheric science), network systems admit to design, implementation, and controlled experimentation of the underlying physical system at nontrivial scales—for example, protocol deployment in autonomous systems belonging to a single service provider—which facilitates an intimate, mechanistic understanding of the system at hand. Model selection is not bound by "black box" evaluations, and physical models that can explicate traffic characteristics in terms of elementary, verifiable system properties and network mechanics, in addition to data fitting, provide an opportunity to be exploited. The challenge lies in combining relevant features from workload modeling, network architecture—protocols and transmission technology—user behavior, and analytical modeling into a consistent, effective description of network systems, in particular, the Internet. As such, physical modeling is a research program that transcends workload modeling, encompassing both performance analysis and traffic control.

21.2.2 Multifractal Traffic Characterization

Since the collection and analysis of the Bellcore Ethernet LAN data [41], follow-up works [1, 15, 27, 57] have shown the robustness of self-similar burstiness in network traffic. This has led to the heuristic description: *Poisson connection arrivals with heavy-tailed connection duration times lead to self-similar burstiness in multiplexed network traffic*. This is a rough, "first-order" description of the empirical facts—for example, TCP connection arrivals exhibit self-similarity (see Chapter 15 on TCP workload modeling)—which serves to point toward the principal causal attribute of self-similarity: heavy-tailed activity durations. Recent analysis of WAN IP traffic [23, 24] has revealed multifractal structure in the form of nonuniform scaling across short and long time scales (see Chapter 20 for a comprehensive discussion). That is, on top of the monofractal picture captured by the heuristic statement above, there exists further variability within each connection—in particular, heavy-tailed TCP connection lifetimes—that fall outside the scope of monofractal self-similarity, which principally concerns large time scale structure in network traffic. The refined, short time scale structure can be described by cascade constructions—also used in the generation of deterministic fractals such as two-dimensional grey-scale fractal images [5]—where variability (within a connection) is obtained by recursive application of "measure redistribution" according to some fixed rule (cf. Chapter 1, Fig. 1.2 (middle)). Several problems remain unsolved.

Multiplicative Scaling and Causality What causes multiplicative scaling observed for short-range correlation structure? Is it related to fragmentation in the TCP/IP protocol stack (including the MAC layer)? TCP's feedback control (ARQ and window-based congestion control)? ACK compression? Topological considerations? If a combination, are there dominant factors? Cascades—although suggestive of certain physical causes—are ultimately a data modeling construct and fall short of establishing a mechanistic description of the underlying workload. From a workload generation or synthesis perspective, given the possible dependence of multiplicative scaling in short time scale traffic structure on feedback control, an open-loop generation of traffic may be unsatisfactory for closed-loop traffic control and its performance evaluation purposes.

Impact of Refined Short Time Scale Modeling Is multiplicative scaling a robust, invariant phenomenon as self-similarity is for large time scale structure? Can modeling of short time scale structure lead to a better understanding of dynamic properties of network protocols? Does a refined model of short-range structure lead to a more accurate prediction of network performance? In other words, is refined modeling of short time scale structure in network traffic a "relevant" research activity? It is clear that in some contexts (see, e.g., Chapter 12 for a discussion of short-range versus long-range dependence issues) short-range structure can dominate performance. Refined traffic modeling, in general, if not checked with respect to its potential to advance fundamental understanding, can become a "data fitting" activity—the subject of time series analysis—yielding limited new networking insights. The standards required of refined traffic modeling work must therefore be evermore stringent.

21.2.3 Spatial Workload Characterization

Physical modeling [15, 51], which reduces the root cause of self-similarity in network traffic to heavy-tailed file size distributions on file systems and Web servers is a form of *spatial* workload modeling. That is, the *temporal* property of network traffic—which is a primary factor determining performance—is related to the spatial or structural property of networked distributed systems. Following are a number of extensions to the spatial workload modeling theme that may exhibit features related to "correlation at a distance," a characteristic of self-similarity in network traffic.

Mobility Model In an integrated wired/wireless network environment, understanding the movement pattern of mobiles is relevant for effective resource management and performance evaluation. Current models are derived from transportation studies [19, 34, 40], which possess a coarse measurement resolution or, more commonly, make a range of user mobility assumptions including random walk, Poisson number of base stations/cells visited as a function of time, and exponential stay durations whose validity is insufficiently justified. It would not be too surprising to find correlation structure at large time and/or space scales—a user, after the morning commute, may stay at her office for the remainder of the day except for brief excursions, students on a campus move from class to class at regular intervals and in clusters, users congregate in small regions (e.g., to take in a baseball game at a stadium) in numbers significantly exceeding the average density, traffic obeys predictable flow patterns—which, in turn, can impact performance due to sustained load on base stations connected to wireline networks. A measurement-based mobility model (and tools for effective tracing [48]) that accurately characterizes user mobility is an important component of future workload modeling.

Logical Information Access Pattern With the Internet and the World Wide Web becoming interwoven in the socioeconomic fabric of everyday life, it becomes relevant to characterize the information access pattern by information content (in addition to geographical location) so as to facilitate efficient access and dissemination. Popular Web sites—that is, URLs—may be accessed more frequently than less popular URLs in a statistically regular fashion, for example, with access frequency obeying power laws as a function of some popularity index (e.g., ranking). Hypertext documents and hyperlinks can be viewed as forming a directed graph, and the resulting graph structure of the World Wide Web can be analyzed with respect to its connectivity in an analogous manner as has been carried out recently for Internet network topology [22]. An *information topology* project that parallels efforts in

Internet topology and distance map discovery (e.g., IDMaps [26, 37]), and identifies how logical information is organized on the World Wide Web—including possible invariant scaling features in its connectivity structure and access pattern—may have bearing on network load/temporal traffic properties and, consequently, network performance.

User Behavior Most network applications are driven by users—for example, via interaction with a Web browser GUI—and thus the connection, session, or call arrival process is intimately tied with user behavior, in particular, as it relates to network state. Starting with the time-of-day, user behavior may be a function of network congestion leading to self-regulation (a user may choose to continue his Web surfing activities at a later time if overall response time is exceedingly high, a form of backoff), congestion pricing may assign costs above and beyond those exacted by performance degradation, users may switch between different service classes in a multiservice network [10, 20], users may perform network access and control decisions cooperatively or selfishly leading to a noncooperative network environment characteristic of the Internet, users may observe behavioral patterns when navigating the Web, and so forth. The challenge lies in identifying robust, invariant behavioral traits—possibly exhibiting scaling phenomena—and quantifying their influence on network performance.

Scaling Phenomena in Network Architecture The recent discovery of power law scaling in network topology [22] points toward the fact that scaling may not be limited to network traffic and system workloads. On the other hand, power law scaling in the connectivity structure of the Internet stretches the meaning of "workload characterization" if it is to be included under the same umbrella. More importantly, it is unclear whether the diffusive connectivity structure implied by power laws affects temporal traffic properties and network performance in unexpected, nontrivial ways. For example, routing in graphs with exponential scaling in their connectivity structure is different from routing in graphs with power law scaling, but that is not to say that this has implications for traffic characterization and performance above and beyond its immediate scope of influence—number of paths between a pair of nodes, their make-up, and generation of "realistic" network topologies for benchmarking. If the distribution of link capacities were to obey a power law, then it is conceivable that this may exert a *traffic shaping* effect in the form of variable stretching-in-time of a transmission, which can inject heavy tailedness in transmission or connection duration that is not present in the original workload. The challenge in architectural characterization lies in identifying robust, invariant properties exhibiting scaling behavior and relating these properties to network traffic, load, and performance where a novel and robust relationship is established.

21.2.4 Synthetic Workload Generation

An integral component of workload modeling is synthetic workload generation. In many instances, in particular, those where the workload model is constructive in nature, the process of generating network traffic is suggested by the model under consideration. There are two issues of special interest to self-similar traffic and workload generation that can benefit from further investigation.

Closed-loop Workload Generation Many traffic generation models are *time series* models that output a sequence of values—interpreted as packets or bytes per time unit—which are then fed to a network system (simulation or physical). These "open-loop" synthetic traffic generation models can be used to evaluate queueing behavior of finite/infinite buffer systems with self-similar input, they can be used to generate background or cross traffic impinging on bottleneck routers, and they can serve as traffic flows that are controlled in an open-loop fashion including traffic shaping. In network systems governed by feedback traffic controls where source output behavior is a function of network state, open-loop traffic generation is, in general, ill-suited due to its a priori fixed nature, which does not incorporate dependence on network state. Traffic emitted from a source is influenced by specific control actions in the protocol stack—for example, TCP Tahoe, Reno, Vegas, and flow-controlled UDP—and capturing network state and protocol dependence falls outside the scope of open-loop traffic generation models. A closed-loop traffic generation model for self-similar traffic that captures both network and protocol dependence—based on physical modeling—works by generating file transmission events with heavy-tailed file size distribution at the application layer and lets each file transmission event pass through the protocol stack (e.g., TCP in the transport layer), which then results in packet transmission events at the network/link layer. The consequent traffic flow reflects both the control actions such as reliability, congestion control, fragmentation, and buffering undertaken in the protocol stack, as well as feedback from the network. The closed-loop workload generation framework allows the effect of different control actions on network traffic and performance to be discerned and evaluated. Several issues remain: Which connection arrival model should be used at the application layer (e.g., exponential versus heavy-tailed interconnection arrival times) and for what purpose? Should the arrival time of the next connection be counted from the time of completion of the previous connection or independently/concurrently? How sensitive are the induced traffic properties and network performance to details in the application layer workload model (cf. Chapter 14 for related results)? Are there conditions under which traffic generated from closed-loop workload models can be approximated by open-loop traffic synthesis models? For example, the use of independent loss process models for tractable analysis of TCP dynamics [50] is an instance of open-loop approximation. It is important to delineate the conditions under which open-loop approximation is valid as it is possible to "throw out the baby with the bath water."

Sampling from Heavy-tailed Distributions The essential role played by heavy-tailedness in self-similar traffic models renders sampling from heavy-tailed distributions a key component of synthetic workload generation models (e.g., on/off, $M/G/\infty$, physical models). As discussed in Chapters 3 and 1, sampling from heavy-tailed distributions suffers from convergence and underestimation problems where the sample mean $\bar{X}_n$ of a heavy-tailed random variable X converges *very slowly* to the population mean (cf. Fig. 3.2 of Chapter 3). For researchers accustomed to light-tailed distributions—for example, exponential (Markovian models) or Gaussian (white noise generation)—where convergence is exponentially fast, it is possible to use heavy-tailed distributions in performance evaluation studies without explicitly considering their idiosyncracies and potentially detrimental consequences on the conclusions advanced. For example, a common mistake arises when comparing short-range- and long-range dependent traffic models with respect to their impact on queueing, where self-similar input is generated using heavy-tailed random variables. The traffic rate is assumed equal by virtue of the configured distribution parameters. As a case in point, short-range and long-range dependent traffic may be generated from an on/off source where the off periods are exponential and i.i.d., but the on period is exponential with parameter $\lambda > 0$ for short-range dependent input and Pareto with shape parameter $1 < \alpha < 2$, location parameter $k > 0$ for long-range dependent input. For α close to 1, k and λ may be chosen such that the population mean values of on periods in the two cases are equal; that is, choose k and λ such that

$$\frac{1}{\lambda} = \frac{k\alpha}{\alpha - 1}.$$

Unless the number of samples is "extremely" large—and the traffic series correspondingly long—the actual traffic rate of sample paths in long-range dependent traffic will be nonnegligibly smaller than the corresponding traffic rate of short-range dependent traffic. Thus observations on packet loss and other performance measures may stem from sampling errors—in particular, smaller traffic intensity due to insufficient samples in the self-similar case—than differences in correlation structure of the inputs. How to remedy the problem? Cognizance of the problem is necessary, but not sufficient, to address the potential pitfalls of the problem. We can consider three approaches: (1) Perform *sufficient* sampling such that statistics from sample paths approach that of population statistics. This is the most straightforward approach. The main drawback is that events (e.g., lifetime of connections) and performance measurements of interest may occur at time scales significantly smaller than that required to reach steady state. Also, the sheer sample size and corresponding time requirement put a heavy computational burden on simulation and experimental studies; (2) perform various forms of sample path normalization. For example, the traffic intensity λ_{on} (bps) during on periods can be varied such that the actual traffic rate matches that of a prespecified target. This is most suited for open-loop workload generation (e.g., CBR or VBR traffic over UDP). The main justification of this approach is that λ_{on} does not affect the correlation structure of the generated traffic series. For closed-loop workload generation, one may vary k such that sample path normalization with respect to first-order properties is achieved. Again, correlation structure or second-order properties are not affected by k. This is a heuristic approach and the values of λ_{on} and k depend on the sample size and must be empirically calibrated. Since first-order performance measures such as packet loss rate and average queueing delay are heavily impacted by offered load—in some instances dominating the influence of second-order structure—it is pertinent to perform sample path normalization if the effect of correlation structure on performance is to be discerned. The fundamental soundness of this approach, however, requires further investigation. When sample sizes are insufficient to yield matching sample and population statistics, second-order properties of the generated traffic may be impacted as well. How severe is the sampling problem with respect to second-order structure? Are "corrections" viable? If a certain number of samples is needed to achieve sample paths with statistics approaching that of the population distribution, what fundamental justification is there to allow short-cutting the required sampling process? Perhaps long stretches of time where the sample mean of long-range dependent traffic is significantly smaller than that of short-range dependent traffic is the natural state of affairs (i.e., with respect to network traffic), during which first-order properties dominate second-order properties in impacting performance. In the long run, there are bound to be stretches of time where the opposite is true. This is an intrinsic problem with no simple answers; (3) as a continuation of the second approach, the investigation of speed-up methodologies is the subject of rare event simulation [4, 61], where various techniques including extreme value theory, large deviations theory, and importance sampling are employed to establish conditions under which simulation speed-up is possible. In the case of light-tailed distributions, simulation speed-up using importance sampling is well understood; however, the heavy-tailed case is in its infancy and remains a challenge [4].

21.2.5 Workload Monitoring and Measurement

Systematic, careful monitoring of Internet workloads is a practically important problem. It would be desirable to have a measurement infrastructure that filters, records, and processes workload features at sufficient accuracy, which, in turn, is essential to reliably identifying invariant features and trends in Internet workloads. It is unclear whether there are open research problems related to workload monitoring and measurement instrumentation above and beyond a range of expected engineering issues—for example, placement of instrumentation, what to log, efficient probing (resource overhead, minimally disturb Schrödinger's cat), efficient storage, synchronization, and so forth. It is possible that there are hidden subtleties but, if so, they await to be uncovered. Given the recent interest in Internet topology, distance map, and "weather map" discovery (see, e.g., Francis et al. [26]), integration and coordination of various measurement and estimation related activities may deserve serious consideration. A laissez-faire approach without coordinated efforts may be impeded by protective walls set up by service providers with respect to autonomous systems under private administrative control, which can render certain measurement efforts difficult or infeasible.

21.3 OPEN PROBLEMS IN PERFORMANCE ANALYSIS

21.3.1 Finite Buffer Systems and Effective Analysis

Queueing Analysis of Finite Buffer Systems Most queueing results with self-similar input are asymptotic in nature where either buffer capacity is assumed infinite and the tail probability of queue length distribution in steady state $\Pr\{Q_\infty > x\}$ is estimated as $x \to \infty$, or buffer capacity b is assumed finite but buffer overflow probability is computed as b becomes unbounded. Little is known about the finitary case, and Grossglauser and Bolot [28] and Ryu and Elwalid [60] provide approximate, heuristic arguments regarding the impact of finite time scale implied by bounded x and b. Large deviation techniques [64] are too coarse to be effectively applied to finite x and b, and not surprisingly, the unbounded case or bufferless queueing case (i.e., $b = 0$) is more easily amenable to tractable analysis. The bufferless case can provide indirect insight on performance with "small" buffer capacities and complements the conclusions advanced in the asymptotic case (cf. Chapter 17 for a discussion of bufferless queueing with self-similar input). The divide between our understanding of unbounded and zero memory systems, on the one hand, and finitary systems of interest, on the other, limits the applicability of these techniques both quantitatively and qualitatively—above and beyond polynomial decay of queue length distribution and its broad interpretation as amplified buffering cost—to resource provisioning and control. The difficulty underlying analysis of finite buffer systems with non-Markovian, in particular, self-similar input is a fundamental problem at the heart of probability theory and, perhaps, beyond the scope of applied probability. Fundamental advancement in understanding, proof techniques, and tools is needed to overcome the challenges—a longer term venture. For networking applications, this points toward the need for experimental queueing analysis to fill the void in the interim. As discussed in Section 21.2.4, there are a number of problems and issues associated with performance evaluation under workloads involving sampling from heavy-tailed distributions due to slow convergence of sample statistics to population statistics. When empirical performance evaluation is carried out with synthetic traffic—in addition to measurement traces—which are then used to support generalizations and comparative evaluations, extreme care needs to be exercised to check the influence of sampling. This is a highly nontrivial problem on its own and provides an opportunity for theoretical advances in rare event simulation with heavy-tailed workloads [4] to facilitate experimental queueing analysis and performance evaluation.

Tight Buffer/Packet Loss Asymptotics Significant effort has been directed at deriving tight upper and lower bounds for the tail of the queue length distribution of various queueing systems (e.g., on/off, $M/G/\infty$ or FBM input and constant service rate server) with long-range-dependent input [11, 17, 18, 38, 42, 43, 49, 56, 66]. Most of the approaches can be viewed in the framework of large deviation analysis, where the queue length process is shown to obey a large deviation principle (LDP) with specific rate function and time scale, assuming the arrival process satisfies LDP.

Irrespective of the limited applicability to, and impact on, network design and control, refined characterization of large buffer asymptotics is of independent interest and relevant for advancing the foundations of queueing theory. We refer the reader to the queueing analysis chapters in this book (see, e.g., Chapters 8, 9, and 10) for a detailed discussion of related research issues. Chapter 9, in addition, provides an excellent overview of recent results.

21.3.2 Second-order Performance Measures

Impact of Self-similarity on Jitter Performance evaluation under self-similar traffic conditions has focused on first-order performance measures such as packet loss rate and mean queueing delay. Second-order performance measures that relate to jitter—for example, delay variance and packet loss variance—are of import to multimedia data transport and QoS provisioning. Even under conditions where self-similarity has limited impact on performance with respect to first-order measures, persistent periods of high and low contention implied by self-similar burstiness can exert a negative effect on second-order performance measures. Figure 21.1 shows packet drop traces at a bottleneck router where 32 TCP Reno connections are multiplexed on a common output port. Each connection transports heavy-tailed files, and the four traces stem from the same set-up except that the shape parameter (or tail index) of the heavy-tailed distribution (Pareto) is varied in the range 1.05, 1.35, 1.65, 1.95. The plots depict packet drop traces at 100 second time aggregation. We observe significant variation in packet drops across different 100 second time segments for the $\alpha = 1.05$ tail index case, which diminishes as α approaches 2. This is even more pronounced when the flows are open-loop controlled over UDP. Given the difficulties underlying queueing analysis of finite buffer systems with self-similar input with respect to first-order performance measures, the challenges facing queueing analysis for second-order measures are even greater. There is, however, a special case—bufferless queueing—that is more amenable to tractable analysis (see Chapter 17 for a discussion of bufferless queueing in the context of predictive control). Bufferless queueing, which can be viewed as an extreme form of the small buffer/large bandwidth resource provisioning strategy, derives its justification from (1) the high delay penalty associated with long-range dependent traffic and (2) the observation that a large number of independent flows—correlated in time or not—when aggregated over space (i.e., flows) is approximately Gaussian by the central limit theorem. Thus, for second-order stationary processes (most traffic models fall under this class) this yields a simple handle on the marginal distribution, which can be used to compute deviation probabilities—the aggregate traffic exceeds a given link capacity—using the tail of the Gaussian. Real-time multimedia traffic such as video and audio can tolerate some packet loss (described by first-order measures), but for the same packet loss rate, concentrated packet drops exert a more detrimental impact on QoS than dispersed losses (similarly for delay variations in buffered systems). A comprehensive understanding of the packet loss process and, in general, delay variation in buffered systems is needed to complement the hereto one-sided

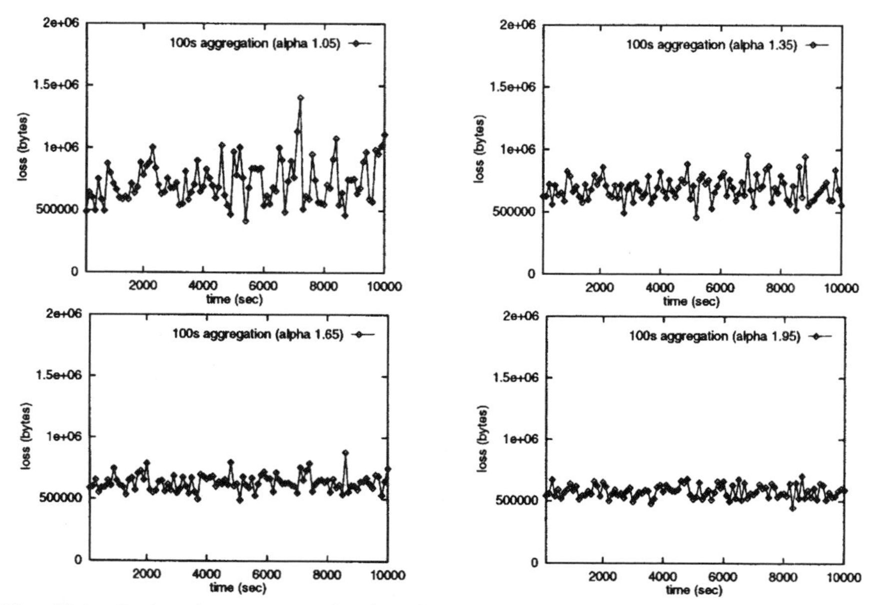

Fig. 21.1 Packet drop trace at bottleneck router multiplexing 32 TCP connections that transport heavy-tailed files, at 100 second time aggregation. *Top row*: File size distribution with shape parameter $\alpha = 1.05$ (left) and $\alpha = 1.35$ (right). *Bottom row*: Corresponding traces for shape parameter $\alpha = 1.65$ (left) and $\alpha = 1.95$ (right).

focus on first-order performance measures when evaluating the effect of self-similar burstiness on network performance.

Impact of self-similar Burstiness on Packet-Level FEC As a continuation of the second-order performance measure issue, we draw attention to the impact of self-similar burstiness on packet-level forward error correction (FEC). Packet-level FEC is a form of error control, which injects redundancy into a packet stream such that reliable transmission is achieved in the presence of packet drops or erasures. That is, given $k \geq 1$ data packets (for simplicity assume fixed size), encoding results in $n = k + h$ packets ($h \geq 0$) that satisfy the property that from any k subset—that is, not more than h packets are lost—the original k packets can be recovered through decoding. When transporting multimedia traffic with real-time constraints over wide area networks where end-to-end latency renders ARQ infeasible, FEC facilitates QoS control where reliability is proactively affected. The effectiveness of FEC hinges on the "k-out-of-n" property being realized over an end-to-end path, which is impeded by concentrated packet drops. As pointed out by Biersack [7] and McAuley [46], burstiness resident in VBR video can significantly impact error recovery. Packet-level FEC injects further complexity into the dependency structure of loss processes due to queueing. In real-time data transport (e.g., MPEG video at a specified frame rate), packets belonging to a frame must, in addition, arrive in a timely manner [53, 54]. Otherwise, they are considered to be as useless as if they had been dropped by the network. It has been shown that correlated erasures stemming from queueing can significantly impede the efficacy of packet-level FEC when compared to independent packet drops [2, 12–14]. Little is known, however, about performance analysis of packet-level FEC under self-similar burstiness—either from the source traffic itself or interference from cross traffic—and this looms as an important challenge. As in the general second-order performance measure case, a starting point is bufferless queueing under self-similar input where the packet loss process with respect to a block of n consecutive packets—the packets belonging to a self-similar traffic stream are viewed as totally ordered—is analyzed. For simplicity, n can be considered fixed although, in general, n is variable.

21.3.3 Short-range Versus Long-range Correlation

Related to refined traffic modeling (see Section 21.2) is the issue of short-range versus long-range correlation in determining queueing behavior and performance in network systems. This book is meant, in part, to clarify some of the surrounding issues given the mixed—and sometimes conflicting—messages and conclusions advanced in various works [16, 21, 28, 32, 52, 60]. It has been shown that buffer capacity and time scale, traffic intensity and bandwidth, marginals, payload type (e.g., aggregated data traffic, VBR video), and the performance measure of interest—first-order versus second-order statistics—collectively determine what the relative import of short time scale and long time scale structure is on performance. Recent works along the line "here is a traffic model with controllable short-range and long-range structure, and short-range or, alternatively, long-range structure is dominant in impacting performance" oftentimes provide insufficient comparative evaluation of related works yielding one-sided and, on the surface, contradictory conclusions. The reader may take away the message that there are few unconditional truths in performance evaluation with self-similar traffic, modeling of Internet traffic admits a large degree of freedom in choosing models (i.e., assumptions) and parameters, and queueing with self-similar input is but one—albeit important—facet of self-similar network traffic research. On the other hand, as a science with engineering applications to network design, resource provisioning and control, further clarification efforts that focus on carefully qualified comparative evaluations are needed to distill the facts, assumptions, and derived conclusions into a mutually consistent and coherent description—unless the works contain technical errors, by definition, this is possible—where assumptions and opinions are delineated from scientific facts. Without these efforts, ambiguities and resulting confusion may put forth avoidable barriers to effectively applying the lessons and knowledge learned from self-similar traffic research to networking practice.

21.3.4 Queueing Analysis of Feedback Control Systems

Analysis of feedback-controlled queueing systems is a difficult problem. Tractability is achieved by considering queueing systems with state-dependent arrival rates and admission control systems where arrivals are admitted/rejected according to a decision rule [33, 65]. In both cases, injection of control is carefully administered such that the Markov property is preserved. Since the bulk of current Internet traffic is governed by TCP—a complicated feedback congestion control—and its state-dependent actions may influence the very traffic being measured and analyzed, it is important that the influence of feedback control on traffic characteristics and performance be ascertained. For example, in the multifractal characterization of network traffic [23], it is conjectured that multiplicative scaling observed in short-range structure is influenced by TCP's control actions. If so, why is this the case and what are the underlying mechanisms? Park et al. [51] (see also Chapter 14) show that multiplexing of concurrent TCP connections at a bottleneck router transporting heavy-tailed files leads to self-similar burstiness, but the empirical slope of the Hurst parameter curve as a function of tail index is less than $-\frac{1}{2}$, the slope value implied by the relation $H = (3 - \alpha)/2$ stemming from the on/off model where connections are assumed independent [70]. Does coupling among feedback-controlled connections sharing common resources lead to changes in traffic properties? Does it matter whether the feedback congestion control is TCP, rate-based control, or adaptive FEC? TCP, because of its idiosyncracies and historical evolution, is not an easy protocol to analyze. Tractable analysis of its dynamics, to date, is only achieved by making an independence assumption on the loss process [45, 50], which, in general, is a heavy price to pay for tractability. A more fundamental avenue of exploration is the analysis of feedback congestion controls including linear increase/exponential decrease controls—a tractable exercise without injecting decoupling by assumption—which have been investigated, principally, for infinite source models [9, 25, 47, 55, 63, 71]. For heavy-tailed workloads where most file transfers are small and a few very large, little is known about the consequent system behavior. For large file transfers, steady-state may be reached due to its approximation of an infinite source. For small transfers—the bulk of connections—analysis remains a challenge. In a similar vein, the impact of feedback on traffic properties when multiplexing a number of heavy-tailed sources has not been investigated sufficiently. The dynamics and effect of feedback congestion control have been studied with respect to fairness, stability, and synchronization issues; traffic properties represent another dimension to their multifaceted influence on network performance.

21.3.5 Impact of Packet Scheduling

In modern routers, flow protection and multiservice support are provided in the form of configurable service classes—per-flow or aggregate-flow—over various scheduling disciplines including GPS, priority queues, and RED. From a performance analysis perspective, the question arises as to what impact self-similar burstiness has on packet scheduling, from subtle effects such as the influence on resource sharing and performance across service classes due to work conservation, to more active design questions such as optimal scheduling with respect to target objective functions. A comparative evaluation of FCFS, PS, and LCFS-PR under heavy traffic conditions (see Chapter 6) provides insight into the role of scheduling with self-similar input. Performance analysis of EDF, fixed priority schemes, and related scheduling algorithms under self-similar input remain to be investigated. Refined analysis of work conservation and its impact on GPS with respect to providing inter-flow (in general, inter-service class) protection under self-similar workloads looms as a challenging problem. On a broader level, the influence of self-similarity and heavy-tailedness on scheduling need not be restricted to routers. Empirical evidence of heavy-tailedness across UNIX process life time distribution [35, 51], UNIX file size distribution [35, 51], and Web document size distribution [3, 15] points toward CPU scheduling policies that make active use of the heavy-tailed property. For example, given the empirical observation that most tasks require short service times whereas a few require very long service times, a shortest job first (SJF) scheduling policy—known to be optimal with respect to average waiting time—is expected to yield amplified performance gain vis-à-vis FCFS and other workload-insensitive schedulers. The technical challenge lies in achieving tractable analysis of relevant performance measures such as waiting time under heavy-tailed workloads when using SJF or other workload-sensitive schedulers. The service time of a task may be known a priori—for example, if related to the size of documents at Web servers—or it may be estimated on-line. Heavy tailedness implies predictability—if a task has been active for some time, then it is likely to persist into the future (see Chapter 1, Section 1.4)—which can be used to perform on-line identification of long-running tasks.

21.4 OPEN RESEARCH PROBLEMS IN TRAFFIC CONTROL

21.4.1 Closed-loop Traffic Control

Multiple Time Scale Traffic Control Self-similarity and long-range dependence imply predictability structure at large time scales that may be exploitable for traffic control purposes. Most feedback traffic controls are impervious to this information, principally, due to the fact that the time scale of the feedback loop—that is, round-trip time—is an order of magnitude (or more) smaller than the time scale at which "long-range" correlation structure, in practice, manifests itself: millisecond versus second range. The multiple time scale traffic control framework was introduced by Tuan and Park [67] and shown to be effective at yielding significant performance improvement when large time scale correlation is exploited for traffic control. In Tuan and Park [67] (see also Chapter 18), large time scale correlation structure was on-line estimated and utilized to modulate the bandwidth consumption behavior of linear increase/exponential decrease rate-based congestion control for throughput maximization. In Tuan and Park [68], the approach was adapted to window-based congestion control and reliable transport using TCP (e.g., Reno and Vegas) with similar performance gains. In Tuan and Park [69], multiple time scale traffic control was extended to adaptive redundancy control where adaptive packet-level FEC is used for end-to-end QoS control of real-time traffic [8, 54]. An important benefit of multiple time scale traffic control is the mitigation of performance cost of reactive controls in broadband wide area networks with a large delay–bandwidth product due to outdated feedback. The mechanism for exploiting large time scale structure employed by Tuan and Park [67–69] couples the feedback traffic control (i.e., short time scale module) with the large time scale module via a well-defined modular interface, and is called selective aggressiveness control (SAC). Two methods are distinguished: selective slope control and selective level control. In selective slope control—used in rate-based and window-based TCP congestion control—the slope of the linear increase phase is varied as a function of expected contention level during large time intervals (e.g., 2 seconds), increasing the slope if the contention level is predicted to be low (thereby amplifying aggressiveness), and vice versa if the oppositive is true. In selective level control—used in adaptive redundancy control—a "DC" level, which is held constant during a large time scale interval, is shifted from high to low (and vice versa) across successive time intervals as a function of predicted contention level. This is depicted in Fig. 21.2. Selective aggressiveness control is but one approach to engaging large time scale predictability structure for traffic control. There may be other approaches and mechanisms, equally or perhaps even more effective, which can be used to harness long-term predictability for traffic control. Their identification and evaluation is a subject for continued exploration. We note that there are four challenges to be overcome in this endeavor: (1) correlation structure is dispersed over large distances in time, (2) information is probabilistic, (3) the mechanism should be implementable over existing traffic controls, and (4) it should not "harm" the underlying traffic control (if not help it) with respect to performance.

Multilayered Feedback Control Multiple time scale traffic control, by coupling short time scale and large time scale control modules, leads to a multilayered feedback control. The large time scale module dynamically estimates the optimal slope and level values to use for particular network contention levels, which are then used to modulate the small time scale feedback control module. Even if the underlying small time scale feedback control (e.g., linear increase/exponential decrease rate-based control, TCP, or AFEC) is stable, this does not imply that the

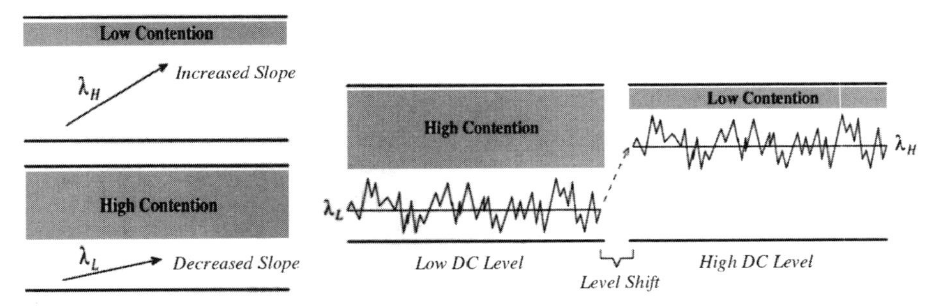

Fig. 21.2 *Left*: Selective slope control—that is, slope shift—during linear increase phase for high- and low-contention periods. *Right*: Selective level control—that is, "DC" level shift—between high- and low-contention periods.

coupled system is stable. It is expected, however, that, by exploiting timescale separation between the two control modules, stability of the overall system can be achieved under fairly weak conditions. In particular, if T_L, T_S denote the time scales of the large and small time scale modules, respectively, $T_S \ll T_L$ (in a suitable sense), and the small time scale feedback control is assumed to converge fast (e.g., within a small factor of T_S by exponential convergence), then the overall system may be shown to be stable (but not asymptotically stable) by the quasi-stationarity argument that the small time scale control converges "well within" T_L—a locally stationary regime from its perspective—leading to a concatenation of locally well-behaved trajectories (short transient followed by convergence), assuming the large time scale control itself is stable. The most challenging problem arises when sufficient separation between small and large time scales does not hold. Complicated dynamics can ensue and identification of sufficiency conditions for stability with relevant counterparts in networking looms as a challenging problem.

Workload-sensitive Traffic Control Multiple time scale traffic control can be extended to workload-sensitive traffic control where traffic controls are made, to varying degrees, cognizant of workload properties in the broadest sense when this is deemed beneficial to do so. For example, when TCP is invoked by HTTP in the context of Web client/server interactions, the size of the file being transported—known at the server—may be conveyed or made accessible to protocols in the transport layer, including the selection of alternative protocols, for more effective data transport (see, e.g., Heddaya and Park [31] for a discussion of a specific mechanism in the context of congestion control). For short files, which constitute the bulk of connection requests in heavy-tailed file size distributions of Web servers, elaborate feedback control geared toward steady-state efficiency may be by-passed in favor of lightweight mechanisms in the spirit of optimistic control, which can, in some instances [39], result in improved "effective bandwidth." Recently, the heavy-tailed characteristics of IP flow durations [23, 24] have been used to selectively perform routing table updates based on connection lifetime classification, where it is shown that desensitizing route updates triggered by short-lived flows can enhance routing stability [62]. Endowing workload sensitivity on traffic control need not be restricted to congestion control, error control, and routing (for that matter, closed-loop control) and represents a broad area for future exploration facilitated by the recent discoveries and advancement in traffic characterization.

Optimal Prediction of Long-range Correlation Structure Predicting the future traffic level from past observations is an important component to affecting traffic control under self-similar traffic conditions. Chapters 17 and 18 provide heuristic approaches to estimating the future traffic level based on conditional expectation—optimal with respect to mean square error—but due to its nonlinearity, effective optimal prediction remains a technical challenge [6]. With long-range dependent time series and their slow convergence properties, it becomes difficult to devise effective predictors that can rigorously be shown to have desirable properties. Traditional estimation theory achieves tractability through assumption of Markovian input (Kalman filters) or restriction on observation models (Wiener filters). On the positive side, the conditional expectation predictor $E[X^{(m)}(i)|X^{(m)}(i-1)]$ has proved effective in empirical evaluations with respect to yielding traffic level estimates that are close to the true traffic level. Furthermore, the performance gain obtained by engaging approximate information has been shown to approach the gain achievable when perfect information is available [67]. Thus there is room for further performance gain due to improved prediction, but its magnitude is expected to be incremental. On a related front, inference of network state through minimally intrusive actions is relevant for effectively incorporating the prediction mechanisms in network protocols. For example, in Tuan and Park [68] a TCP connection's interaction with other traffic flows at bottleneck routers and the consequent impact on its output behavior—observable at the sender—is used to infer the contention level at bottleneck routers. That is, no separate probing mechanism is engaged to estimate network state. The effectiveness of this minimally intrusive scheme is shown to be dependent on the tracking ability of the underlying feedback control [68]; state estimation suffers most heavily during the linear increase phase after backoff, which results in better tracking performance for TCP Vegas over TCP Reno. Inference can be performed by other means including arrival behavior of ACK packets, and further exploration of effective measurement schemes is of interest. When explicit probing is employed, the question arises as to how to perform accurate sampling and estimation of network state without significantly affecting Schrödinger's cat in the process. Accurate sampling is made difficult by the slow convergence properties of self-similar traffic, and trade-offs between accuracy and probing duration can benefit from further investigation (also relevant to measurement-based admission control [36]).

21.4.2 Open-loop Control and Resource Provisioning

Resource Reservation and Admission Control The performance analysis techniques and issues discussed in Section 21.3 have direct bearing on the ability to compute performance bounds for routers fed with self-similar input, estimation of effective bandwidth and statistical multiplexing gain, bandwidth and buffer capacity dimensioning/provisioning, and admission control. Chapter 19 presents a specific open-loop architecture based on per-VC framing, and Chapter 16 discusses design and performance evaluation considerations of open-loop architectures under self-similar traffic conditions. The same observations regarding the need for finitary analysis, incorporation of second-order performance measures, relative impact of short-range and long-range correlation structure, and influence of scheduling advanced in Section 21.3 hold for the resource provisioning context. Similarly, the small buffer/large bandwidth resource provisioning strategy is expected to play a dominant role in facilitating guaranteed services—deterministic or statistical—in the context of open-loop control.

Dynamic Admission Control In resource provisioning, admission control is exercised in a static manner, where the set of connections requesting service is known beforehand. In dynamic admission control, a connection i has an arrival time s_i and duration τ_i associated with it. The server has finite capacity C. Assuming Poisson call arrivals and exponential holding times, this defines a Markov Decision Process [59], which can be solved using standard techniques. In general, there may be multiple service classes and a reward (or utility) function r_i for servicing connection i with a particular service level. We can distinguish two ways by which self-similarity is introduced: (1) The arrival process is allowed to be self-similar (see Chapter 15 for a discussion of TCP session arrivals), and (2) the connection duration is allowed to be heavy-tailed. The latter is consistent with a canonical source model introduced by Likhanov et al. [42] (see also Chapters 8 and 1), where connection arrivals are Poisson but connection durations are heavy-tailed. The optimal decision rule, under heavy-tailed call durations, is a function of r_i, which, in turn, may depend on τ_i. By exploiting the memoryless property of connection arrivals, it is expected that optimal decision procedures can be derived for a range of specific admission control models in the framework of Markov renewal processes. The most challenging problem arises when connection arrivals are self-similar. The correlated nature of arrivals renders computation of expected reward conditioned on current system state difficult. For non-Markovian arrival processes, it is possible to achieve tractability by adopting a worst-case approach—that is, stochastic structure in the arrival process is ignored—based on competitive analysis [44, 58]. The goodness of an on-line algorithm is evaluated with respect to its performance vis-à-vis an optimal off-line algorithm where the ratio of their respective solutions—the competitive ratio—is evaluated over all input instances, that is, sample paths. As far as evaluating the impact of different arrival processes (in particular, self-similar processes) is concerned, competitive analysis yields limited insight due to its worst-case property.

Optimistic Traffic Control for Short-lived Connections Long-range dependence and its exploitation for traffic control—in particular, feedback traffic control—is, by definition, best suited for flows or connections whose lifetime or connection duration is long-lasting. Trivially, if the connection is too short, then reliable network state estimation is bound to fail or yield highly variable results. Empirical workload measurements have shown that connection durations and file sizes are heavy-tailed, which implies that the majority of connections are short-lived although a few connections contribute the bulk of traffic in terms of volume. Thus, in spite of the import of effectively managing long-lived connections due to their disproportionate contribution, if performance measures such as transmission completion time for short-lived connections are considered, then performance improvement may be achievable by utilizing workload information that is presently ignored. For example, if a server knows the size of the file to be transmitted, then conditioned on the file size, either an optimistic control in the spirit of open-loop control may be invoked for small files to reduce overhead, with reactive actions undertaken only in the fraction of instances here needed (e.g., retransmissions for reliable transport), or full-fledged feedback congestion control with multiple timescale extension engaged when the requested file is large. If the file size distribution is known to be heavy-tailed but the size of each transfer instance is not known, then optimistic control may be exercised by default until such time when the connection in question—by connection duration prediction—is concluded to be long-lived (see, e.g., Kim [39] for a comparison of aggressive open-loop congestion control versus linear increase/ exponential decrease feedback congestion control). From a multiple time scale traffic control perspective, the fact that a connection is short-lived need not deter one from utilizing a priori information about the long-term network state if available. One possible extension is the use of network state information that is not tied to any particular connection but shared across many. The effectiveness of this scheme hinges on maintaining persistent, shared information at a host (e.g., Web server or client) where multiple connections are multiplexed in space or time. Since network state may depend on the location of each individual destination host and the route taken, a challenge lies in maintaining such information in a compact, easily accessible, and updated form. The conditional expectation predictor, for long-range dependent traffic, can be taken to say that the past observation should be used as the future prediction—that is, $E[X^{(m)}(i)|X^{(m)}(i-1)] = X^{(m)}(i-1)$—which facilitates a slowly changing, invariant prediction table that can be indexed if the past observation $X^{(m)}(i-1)$ is known.

REFERENCES

1. R. Addie, M. Zukerman, and T. Neame. Fractal traffic: measurements, modelling and performance evaluation. In *Proc. IEEE INFOCOM '95*, pp. 977–984, 1995.
2. O. Ait-Hellal, E. Altman, A. Jean-Marie, and I. Kurkova. On loss probabilities in presence of redundant packets and several traffic sources. *Perf. Eval.*, **36**:485–518, 1999.
3. M. F. Arlitt and C. L. Williamson. Web server workload characterization: the search for invariants. In *Proc. SIGMETRICS '96*, pp. 126–137, May 1996.
4. Soren Asmussen. Rare events simulation for heavy-tailed distributions. In *Proc. RESIM '99 Workshop*, pp. 139–157. University of Twente, Enschede, The Netherlands, March 11–12 1999.
5. M. Barnsley. *Fractals Everywhere*. Academic Press, San Diego, 1988.
6. J. Beran. *Statistics for Long-Memory Processes*. Monographs on Statistics and Applied Probability. Chapman and Hall, New York, 1994.
7. E. Biersack. Performance evaluation of forward error correction in ATM networks. In *Proc. ACM SIGCOMM '92*, pp. 248–257, 1992.
8. J. Bolot, S. Fosse-Parisis, and D. Towsley. Adaptive FEC-based error control for Internet telephony. In *Proc. IEEE INFOCOM '99*, pp. 1453–1460, 1999.
9. J. Bolot and A. Shankar. Analysis of a fluid approximation to flow control dynamics. In *Proc. IEEE INFOCOM '92*, pp. 2398–2407, 1992.
10. S. Chen and K. Park. An architecture for noncooperative QoS provision in many-switch systems. In *Proc. IEEE INFOCOM '99*, pp. 864–872, 1999.
11. J. Choe and N. Shroff. A central-limit-theorem based approach for analyzing queue behavior in ATM networks. *IEEE/ACM Trans. Networking*, **6**(5):659–671, 1998.
12. I. Cidon, A. Khamisy, and M. Sidi. Analysis of packet loss processes in high-speed networks. *IEEE Trans. Inf. Theory*, **39**(1):98–108, 1993.

13. I. Cidon, A. Khamisy, and M. Sidi. Dispersed messages in discrete-time queues: delay, jitter, and threshold crossing. In *Proc. IEEE INFOCOM '94*, pp. 218–223, 1994.
14. I. Cidon, A. Khamisy, and M. Sidi. Delay, jitter and threshold crossing in ATM systems with dispersed messages. *Perf. Eval.*, **29**(2):85–104, 1997.
15. M. Crovella and A. Bestavros. Self-similarity in World Wide Web traffic: evidence and possible causes. *IEEE/ACM Trans. Networking*, **5**:835–846, 1997.
16. N. G. Duffield, J. T. Lewis, N. O'Connel, R. Russell, and F. Toomey. Statistical issues raised by the Bellcore data. In *Proc. 11th IEE Teletraffic Symposium*, 1994.
17. N. G. Duffield and N. O'Connel. Large deviations and overflow probabilities for the general single server queue, with applications. Technical Report DIAS-STP-93-30, DIAS Technical Report, 1993.
18. N. Duffield. On the relevance of long-tailed durations for the statistical multiplexing of large aggregations. In *Proc. 25th Allerton Conference on Communication, Control and Computing*, pp. 741–750, 1997.
19. ECMT. *European Transport Trends and Infrastructural Needs*. OECD Publications, 1995.
20. R. Edell and P. Varaiya. Providing internet access: what we learn from the INDEX trial. *IEEE Network*, 1999.
21. A. Erramilli, O. Narayan, and W. Willinger. Experimental queueing analysis with long-range dependent packet traffic. *IEEE/ACM Trans. Networking*, **4**:209–223, 1996.
22. M. Faloutsos, P. Faloutsos, and C. Faloutsos. On power-law relationships of the Internet topology. In *Proc. ACM SIGCOMM '99*, 1999.
23. A. Feldmann, A. C. Gilbert, P. Huang, and W. Willinger. Dynamics of IP traffic: a study of the role of variability and the impact of control. In *Proc. ACM SIGCOMM '99*, 1999.
24. A. Feldmann, A. C. Gilbert, and W. Willinger. Data networks as cascades: investigating the multifractal nature of Internet WAN traffic. In *Proc. ACM SIGCOMM '98*, pp. 42–55, 1998.
25. K. Fendick, M. Rodrigues, and A. Weiss. Analysis of a rate-based control strategy with delayed feedback. In *Proc. ACM SIGCOMM '92*, pp. 136–148, 1992.
26. P. Francis, S. Jamin, V. Paxson, L. Zhang, D. Gryniewicz, and Y. Jin. An architecture for a global Internet host distance estimation service. In *Proc. IEEE INFOCOM '99*, pp. 210–217, 1999.
27. M. Garret and W. Willinger. Analysis, modeling and generation of self-similar VBR video traffic. In *Proc. ACM SIGCOMM '94*, pp. 269–280, 1994.
28. M. Grossglauser and J. -C. Bolot. On the relevance of long-range dependence in network traffic. In *Proc. ACM SIGCOMM '96*, pp. 15–24, 1996.
29. M. Harchol-Balter. Process lifetimes are not exponential, more like $1/t$: implications on dynamic load balancing. Technical report, EECS, University of California, Berkeley, 1996. CSD-94-826.
30. M. Harchol-Balter and A. Downey. Exploiting process lifetime distributions for dynamic load balancing. In *Proc. SIGMETRICS '96*, pp. 13–24, 1996.
31. A. Heddaya and K. Park. Congestion control for asynchronous parallel computing on workstation networks. *Parallel Comput.*, **23**:1855–1875, 1997.
32. D. Heyman and T. Lakshman. What are the implications of long-range dependence for VBR-video traffic engineering? *IEEE/ACM Trans. Networking*, **4**(3):301–317, June 1996.

33. A. Hordijk and F. Spieksma. Constrained admission control to a queueing system. *Adv. Appl. Prob.*, **21**:409–431, 1989.
34. P. Hu and J. Young. *1990 NPTS Databook: National Personal Transportation Survey.* Federal Highway Administration, 1993.
35. G. Irlam. Unix file size survey—1993. Available at http://www.base. com/gordoni/ufs93. html, September 1994.
36. S. Jamin, P. Danzig, S. Shenker, and L. Zhang. A measurement-based admission control algorithm for integrated services packet networks. In *Proc. ACM SIGCOMM '95*, pp. 2–13, 1995.
37. S. Jamin, C. Jin, Y. Jin, D. Raz, Y. Shavitt, and L. Zhang. On the placement of Internet instrumentation. To appear in *Proc. IEEE INFOCOM '00*, 2000.
38. P. R. Jelenlovic and A. A. Lazar. The effect of multiple time scales and subexponentiality in MPEG video streams on queueing behavior. *IEEE J. Select. Areas Commun.*, **15**(6):1052–1071, 1997.
39. H. Kim. *A Non-Feedback Congestion Control Framework for High-Speed Data Networks.* Ph. D. thesis, University of Pennsylvania, 1995.
40. R. Kitamura. Time-of-day characteristics of travel: an analysis of 1990 NPTS data. Federal Highway Administration, Chapter 4, 1995.
41. W. Leland, M. Taqqu, W. Willinger, and D. Wilson. On the self-similar nature of ethernet traffic. In *Proc. ACM SIGCOMM '93*, pp. 183–193, 1993.
42. N. Likhanov, B. Tsybakov, and N. Georganas. Analysis of an ATM buffer with self-similar ("fractal") input traffic. In *Proc. IEEE INFOCOM '95*, pp. 985–992, 1995.
43. Z. Liu, P. Nain, D. Towsley, and Z. Zhang. Asymptotic behavior of a multiplexer fed by a long-range dependent process. *J. Appl. Probab.*, 1999.
44. M. Manasse, L. McGeoch, and D. Sleator. Competitive algorithms for online problems. In *Proc. ACM STOC '88*, pp. 322–332, 1988.
45. M. Mathis, J. Semke, J. Mahdavi, and T. Ott. The macroscopic behavior of the TCP congestion avoidance algorithms. *Comput. Commun. Rev.*, 27, 1997.
46. A. McAuley. Reliable broadband communication using a burst erasure correcting code. In *Proc. ACM SIGCOMM '90*, pp. 197–306, 1990.
47. A. Mukherjee and J. Strikwerda. Analysis of dynamic congestion control protocols—Fokker–Planck approximation. In *Proc. ACM SIGCOMM '91*, pp. 159–169, 1991.
48. B. Noble, G. Nguyen, M. Satyanarayanan, and R. Katz. Mobile network tracing. RFC 2041, 1996.
49. I. Norros. A storage model with self-similar input. *Queueing Syst.*, **16**:387–396, 1994.
50. J. Padhye, V. Firoiu, D. Towsley, and J. Kurose. Modeling TCP throughput: a simple model and its empirical validation. In *Proc. ACM SIGCOMM '98*, 1998.
51. K. Park, G. Kim, and M. Crovella. On the relationship between file sizes, transport protocols, and self-similar network traffic. In *Proc. IEEE International Conference on Network Protocols*, pp. 171–180, 1996.
52. K. Park, G. Kim, and M. Crovella. On the effect of traffic self-similarity on network performance. In *Proc. SPIE International Conference on Performance and Control of Network Systems*, pp. 296–310, 1997.
53. K. Park and W. Wang. AFEC: an adaptive forward error correction protocol for end-to-end transport of real-time traffic. In *Proc. IEEE IC3N*, pp. 196–205, 1998.

54. K. Park and W. Wang. QoS-sensitive transport of real-time MPEG video using adaptive forward error correction. In *Proc. IEEE Multimedia Systems '99*, pp. 426–432, 1999.
55. K. Park. Warp control: a dynamically stable congestion protocol and its analysis. In *Proc. ACM SIGCOMM '93*, pp. 137–147, 1993.
56. M. Parulekar and A. Makowski. Tail probabilities for a multiplexer with self-similar traffic. In *Proc. IEEE INFOCOM '96*, pp. 1452–1459, 1996.
57. V. Paxson and S. Floyd. Wide-area traffic: the failure of Poisson modeling. In *Proc. ACM SIGCOMM '94*, pp. 257–268, 1994.
58. S. Plotkin. Competitive routing of virtual circuits in ATM networks. *IEEE J. Select. Areas Commun.*, **13**(6):1128–1136, 1995.
59. M. L. Puterman. *Markov Decision Processes: Discrete Stochastic Dynamic Programming*. Wiley, New York, 1994.
60. B. Ryu and A. Elwalid. The importance of long-range dependence of VBR video traffic in ATM traffic engineering: myths and realities. In *Proc. ACM SIGCOMM '96*, pp. 3–14, 1996.
61. P. Shahabuddin. Rare event simulation in stochastic models. In *Proc. 1995 Winter Simulation Conference*, pp. 178–185, 1995.
62. A. Shaikh, J. Rexford, and K. Shin. Load-sensitive routing of long-lived IP flows. In *Proc. ACM SIGCOMM '99*, 1999.
63. S. Shenker. A theoretical analysis of feedback flow control. In *Proc. ACM SIGCOMM '90*, pp. 156–165, 1990.
64. A. Shwartz and A. Weiss. *Large Deviations for Performance Analysis*. Chapman and Hall, London, 1995.
65. S. Stidham. Optimal control of admission to a queueing system. *IEEE Trans. Automat. Contr.*, **AC-30**(8):705–713, 1985.
66. B. Tsybakov and N. D. Georganas. Self-similar traffic and upper bounds to buffer overflow in an ATM queue. *Perf. Eval.*, **36**(1):57–80, 1998.
67. T. Tuan and K. Park. Multiple time scale congestion control for self-similar network traffic. *Perf. Eval.*, **36**:359–386, 1999.
68. T. Tuan and K. Park. Performance evaluation of multiple time scale TCP under self-similar traffic conditions. Technical Report CSD-TR-99-040, Department of Computer Sciences, Purdue University, 1999.
69. T. Tuan and K. Park. Multiple time scale redundancy control for QoS-sensitive transport of real-time traffic. To appear in *Proc. IEEE INFOCOM '00*, 2000.
70. W. Willinger, M. Taqqu, R. Sherman, and D. Wilson. Self-similarity through high-variability: statistical analysis of Ethernet LAN traffic at the source level. In *Proc. ACM SIGCOMM '95*, pp. 100–113, 1995.
71. C. Yang and A. Reddy. A taxonomy for congestion control algorithms in packet switching networks. *IEEE Network*, **9**:34–45, 1995.

INDEX